Innovative Materials and Methods for the Removal of Pollutants from the Environment

Innovative Materials and Methods for the Removal of Pollutants from the Environment

Editors

Chiara Bisio
Monica Pica

MDPI • Basel • Beijing • Wuhan • Barcelona • Belgrade • Manchester • Tokyo • Cluj • Tianjin

Editors

Chiara Bisio
Department of Science and Technological Innovation
Università del Piemonte Orientale "A. Avogadro"
Alessandria
Italy

Monica Pica
Department of Pharmaceutical Sciences
Università di Perugia
Perugia
Italy

Editorial Office
MDPI
St. Alban-Anlage 66
4052 Basel, Switzerland

This is a reprint of articles from the Special Issue published online in the open access journal *Molecules* (ISSN 1420-3049) (available at: www.mdpi.com/journal/molecules/special_issues/removal_pollutants_environment).

For citation purposes, cite each article independently as indicated on the article page online and as indicated below:

LastName, A.A.; LastName, B.B.; LastName, C.C. Article Title. *Journal Name* **Year**, *Volume Number*, Page Range.

ISBN 978-3-0365-1696-7 (Hbk)
ISBN 978-3-0365-1695-0 (PDF)

Contents

About the Editors vii

Preface to "Innovative Materials and Methods for the Removal of Pollutants from the Environment" ix

Geetha Palani, A. Arputhalatha, Karthik Kannan, Sivarama Krishna Lakkaboyana, Marlia M. Hanafiah, Vinay Kumar and Ravi Kumar Marella
Current Trends in the Application of Nanomaterials for the Removal of Pollutants from Industrial Wastewater Treatment—A Review
Reprinted from: *Molecules* **2021**, *26*, 2799, doi:10.3390/molecules26092799 1

Monica Pica
Treatment of Wastewaters with Zirconium Phosphate Based Materials: A Review on Efficient Systems for the Removal of Heavy Metal and Dye Water Pollutants
Reprinted from: *Molecules* **2021**, *26*, 2392, doi:10.3390/molecules26082392 17

Sheng Cheng, Tao Chen, Wenbin Xu, Jian Huang, Shaojun Jiang and Bo Yan
Application Research of Biochar for the Remediation of Soil Heavy Metals Contamination: A Review
Reprinted from: *Molecules* **2020**, *25*, 3167, doi:10.3390/molecules25143167 39

Daniele C. da Silva Alves, Bronach Healy, Luiz A. de Almeida Pinto, Tito R. Sant'Anna Cadaval and Carmel B. Breslin
Recent Developments in Chitosan-Based Adsorbents for the Removal of Pollutants from Aqueous Environments
Reprinted from: *Molecules* **2021**, *26*, 594, doi:10.3390/molecules26030594 61

Sun-Wook Jeong and Yong Jun Choi
Extremophilic Microorganisms for the Treatment of Toxic Pollutants in the Environment
Reprinted from: *Molecules* **2020**, *25*, 4916, doi:10.3390/molecules25214916 107

Souad Rakass, Hicham Oudghiri Hassani, Ahmed Mohmoud, Fethi Kooli, Mostafa Abboudi, Eman Assirey and Fahd Al Wadaani
Highly Efficient Methylene Blue Dye Removal by Nickel Molybdate Nanosorbent
Reprinted from: *Molecules* **2021**, *26*, 1378, doi:10.3390/molecules26051378 123

Vanessa Miglio, Chiara Zaccone, Chiara Vittoni, Ilaria Braschi, Enrico Buscaroli, Giovanni Golemme, Leonardo Marchese and Chiara Bisio
Silica Monolith for the Removal of Pollutants from Gas and Aqueous Phases
Reprinted from: *Molecules* **2021**, *26*, 1316, doi:10.3390/molecules26051316 143

Wondwosen Sime Geleta, Esayas Alemayehu and Bernd Lennartz
Volcanic Rock Materials for Defluoridation of Water in Fixed-Bed Column Systems
Reprinted from: *Molecules* **2021**, *26*, 977, doi:10.3390/molecules26040977 157

Stephan Wasielewski, Eduard Rott, Ralf Minke and Heidrun Steinmetz
Application of Natural Clinoptilolite for Ammonium Removal from Sludge Water
Reprinted from: *Molecules* **2020**, *26*, 114, doi:10.3390/molecules26010114 177

Shaojun Jiang, Jiachen Wu, Lianxin Duan, Sheng Cheng, Jian Huang and Tao Chen
Investigating the Aging Effects of Biochar on Soil C and Si Dissolution and the Interactive Impact on Copper Immobilization
Reprinted from: *Molecules* **2020**, *25*, 4319, doi:10.3390/molecules25184319 . . . . **199**

Hanna Siwek and Krzysztof Pawelec
Competitive Interaction of Phosphate with Selected Toxic Metals Ions in the Adsorption from Effluent of Sewage Sludge by Iron/Alginate Beads
Reprinted from: *Molecules* **2020**, *25*, 3962, doi:10.3390/molecules25173962 . . . . **215**

Mohammed Abdullah Issa and Zurina Z. Abidin
Sustainable Development of Enhanced Luminescence Polymer-Carbon Dots Composite Film for Rapid Cd^{2+} Removal from Wastewater
Reprinted from: *Molecules* **2020**, *25*, 3541, doi:10.3390/molecules25153541 . . . . **229**

Saad S. M. Hassan, Ayman H. Kamel, Amr A. Hassan, Abd El-Galil E. Amr, Heba Abd El-Naby, Mohamed A. Al-Omar and Ahmed Y. A. Sayed
$CuFe_2O_4$/Polyaniline (PANI) Nanocomposite for the Hazard Mercuric Ion Removal: Synthesis, Characterization, and Adsorption Properties Study
Reprinted from: *Molecules* **2020**, *25*, 2721, doi:10.3390/molecules25122721 . . . . **245**

Somaia G. Mohammad, Sahar M. Ahmed, Abd El-Galil E. Amr and Ayman H. Kamel
Porous Activated Carbon from Lignocellulosic Agricultural Waste for the Removal of Acetampirid Pesticide from Aqueous Solutions
Reprinted from: *Molecules* **2020**, *25*, 2339, doi:10.3390/molecules25102339 . . . . **259**

Siham S. Hassan, Ahmed S. El-Shafie, Nourhan Zaher and Marwa El-Azazy
Application of Pineapple Leaves as Adsorbents for Removal of Rose Bengal from Wastewater: Process Optimization Operating Face-Centered Central Composite Design (FCCCD)
Reprinted from: *Molecules* **2020**, *25*, 3752, doi:10.3390/molecules25163752 . . . . **275**

Abhishek RoyChowdhury, Pallabi Mukherjee, Saumik Panja, Rupali Datta, Christos Christodoulatos and Dibyendu Sarkar
Evidence for Phytoremediation and Phytoexcretion of NTO from Industrial Wastewater by Vetiver Grass
Reprinted from: *Molecules* **2020**, *26*, 74, doi:10.3390/molecules26010074 . . . . **295**

Wenjie Sui, Tairan Pang, Guanhua Wang, Cuiyun Liu, Ashak Mahmud Parvez, Chuanling Si and Chao Li
Stepwise Ethanol-Water Fractionation of Enzymatic Hydrolysis Lignin to Improve Its Performance as a Cationic Dye Adsorbent
Reprinted from: *Molecules* **2020**, *25*, 2603, doi:10.3390/molecules25112603 . . . . **311**

Ming-Hsien Hsueh, Chia-Nan Wang, Meng-Chang Hsieh, Chao-Jung Lai, Shi-Hao Wang, Chia-Hsin Hsieh, Tsung-Liang Wu and Jo-Hung Yu
An Analysis of Exhaust Emission of the Internal Combustion Engine Treated by the Non-Thermal Plasma
Reprinted from: *Molecules* **2020**, *25*, 6041, doi:10.3390/molecules25246041 . . . . **327**

About the Editors

Chiara Bisio

Chiara Bisio is associate professor of physical chemistry at the Università del Piemonte Orientale (Italy) and researcher associated to the Institute of Chemical Technologies "Giulio Natta"CNR-SCITEC. She is member of the scientific committee of CRIMEDIM- Center for Research and Training in Disaster Medicine, Humanitarian Aid, and Global Health of the Università del Piemonte Orientale. Her research interests are related to the optimization of solids with controlled morphology, chemical composition and surface properties with applications as heterogeneous catalysts for the decontamination and detection of chemical and biological hazardous agents, and for the removal of environmental pollutants. She authored 87 research papers (1560 citations, h-index 24), 2 book chapters, and 2 patents. C. Bisio received, in 2019, the "Chiusoli"Gold Medal Award from the Industrial Chemistry Division of the Italian Society of Chemistry.

Monica Pica

Monica Pica, PhD, is Assistant Professor at the Department of Pharmaceutical Sciences at University of Perugia, Italy. Her current research topics include synthesis and characterization of inorganic layered compounds, Zr(IV) phosphates and phosphonates, heterogeneous catalysis, polymer nanocomposites. She authored 58 articles (1332 citations, h-index = 21), among them 4 reviews. She is member of the Italian Chemical Society.

Preface to "Innovative Materials and Methods for the Removal of Pollutants from the Environment"

It is with great pleasure we welcome you to this *Molecules* Special Issue devoted to "Innovative Materials and Methods for the Removal of Pollutants from the Environment".

In the last decade, environmental pollution has reached worrying proportions, and intensive efforts have been devoted to seeking solutions that can be effectively employed for the removal of pollutants from different environmental compartments. This Special Issue aims to gather the most recent and significant contributions related to innovative materials and processes for environmental purposes, with special emphasis on the depollution of air, water, and soil. Biological approaches for pollutants removal are also considered as a valid alternative to contrast environmental threats.

The 18 articles (5 reviews and 13 research papers) composing this Special Issue can be thus considered as a representative selection of the current research and novel applications of inorganic, organic, or hybrid materials with porous or layered structures for the removal of pollutants through adsorption processes or (catalytic) transformation of noxious compounds into species with reduced environmental impact. Bioremediation approaches have been also proposed to treat contaminated media, for instance by altering environmental conditions to stimulate growth of microorganisms aiming to the degradation of toxic species.

Three of the five reviews articles report an overview of the available research strategies for the removal of pollutants from aqueous environments and industrial wastewaters. These contributions, together with research articles dedicated to water safety and depollution, allow to acquire knowledge about the research efforts currently carried out to preserve and improve quality of one of the most important natural resources in the world.

Most of the articles composing the Special Issue are related to the use of adsorption processes for the removal of toxic species from polluted water. In this respect, the use of nanomaterials for the removal of different classes of pollutants normally found in industrial wastewaters (i.e. dyes, heavy metals and pesticides) has been reviewed, also taking into account their possible regeneration and reuse strategies, thus designing future possible applications (S.K. Lakkaboyana et al., DOI: 10.3390/molecules26092799).

Among the different materials that can be employed for wastewater treatments, layered zirconium phosphate (ZrP) has proven to be really versatile materials and their possible use as sorbents for heavy metal cations or dye molecules in wastewater treatments has been reviewed. Special emphasis has been given to the the possibility of combining ZrP with polymers or other inorganic materials (i.e. photocatalytic species), aiming at having suitable systems for real and large scale applications (M. Pica, DOI: 10.3390/molecules26082392).

Chitosan and derived composites are attracting considerable interest as environmentally adsorbents and have the potential to remove water contaminants, such as heavy metal ions, cationic and anionic dyes, phenol and other organic molecules, such as antibiotics. The development of chitosan-based adsorbents is reviewed starting from the extraction of chitin from seafood wastes, followed by its conversion to chitosan, and then describing the emerging chitosan/carbon-based composite materials and their application for water pollutants removal (C. B. Breslin et al., DOI: 10.3390/molecules26030594).

Several research papers are related to materials for the removal of toxic molecules from aqueous media. Removing methylene blue (MB) dye from aqueous solutions was examined by the use of nickel molybdate (-NiMoO4) as adsorbent. The effect of different parameters, such as the pH, the adsorbent dose, the temperature, the contact time, and the initial dye concentration, have been studied, aiming at improving the removal conditions, thus indicating that -NiMoO4 nanosorbent is an outstanding candidate that can be exploited for removing MB dye in wastewater (S. Rakass et al., DOI:10.3390/molecules26051378).

The application of mesoporous silica monoliths for the removal of organic pollutants from water and gas phase have also been proposed. Besides the adsorption properties of the monoliths for gaseous toluene, as a model of volatile organic compounds (VOCs), the same monoliths have been considered for their potential application for the removal of Rhodamine B in water, showing good performances and stability of the siliceous sorbents in aqueous media (C. Bisio et al., DOI:10.3390/molecules26051316).

An eco-friendly approach for acetampirid pesticide removal based on use of micro- and mesoporous activated carbon (TPAC), synthesized via chemical treatment of tangerine peels, is also proposed. The effects of various parameters on the adsorption of acetampirid including adsorbent dose (0.02–0.2 g), pH 2–8, initial adsorbate concentration (10–100 mg/L), contact time (10–300 min), and temperature (25°C–50°C) were studied. The results of the presented study show that TPAC may be effectively used for the removal of acetamiprid pesticides from aqueous solutions (A. E.-G. E. Amr et al., DOI:10.3390/molecules25102339).

Another approach is related to the use of pineapple leaves (PAL) for the adsorption of rose bengal (RB) from contaminated water samples. Raw pineapple leaves (RPAL) and the thermally activated bio-waste leaves have been studied and the adsorption processes have been followed by means of a multitechnique approach (M. El-Azazy et al., DOI:10.3390/molecules25163752).

Another research paper reports the use of Carpathian clinoptilolite powder to remove ammonium from sludge water arising from the dewatering of anaerobic digested sludge with high loads of ammonium in wastewater treatment plants (WWTP). This paper aims also to recover ammonium as a valuable resource to substitute ammonia from the energy intensive Haber–Bosch process for fertilizer production (S. Wasielewski et al., DOI:10.3390/molecules26010114).

Wastewaters are characterized by a high content of phosphate and toxic metals. Many studies have confirmed the sorption affinity of alginate adsorbents for these ions. The adsorption of phosphate from effluent of sewage sludge on biodegradable alginate matrices cross-linked with Fe3+ ions (Fe_Alg) was also investigated (H. Siwek et al., DOI:10.3390/molecules25173962).

A versatile synthetic approach has been developed to prepare polyvinyl alcohol (PVA)/nitrogen-doped carbon dots (CDs) composite film (PVA-CDs) for removal of toxic cadmium ions from waters. The PVA-CDs film possess synergistic functionalities through increasing the content of hydrogen bonds for chemisorption compared to the pure CDs. The film was successfully applied to real water samples with an outstanding removal efficiency for tap and drinking water, respectively (M.H. Issa, DOI:10.3390/molecules25153541).

Copper ferrite nano-particles (CuFe2O4) have been also modified with polyaniline to form CuFe2O4/PANI nano-composite and used as new adsorbents for the removal of the hazardous mercuric ions from aqueous solutions. Adsorption conditions have been modified and the method at the optimum conditions exhibited high performance with high mercury removal performances (S.S.M. Hassan et al., DOI:/10.3390/molecules25122721).

Continuous fixed-bed column adsorption system was also proposed for defluoridation of water using volcanic rocks, virgin pumice (VPum) and virgin scoria (VSco), as adsorbents. The effects of particle size of adsorbents, solution pH, and flow rate on the adsorption performance of the column were assessed at room temperature, constant initial concentration, and bed depth. The results revealed that the developed column is effective in handling water containing excess fluoride and the possiblity to regenerate confirmed that defluoridation of groundwater employing volcanic rocks is a safe and sustainable method (B. Lennartz et al., DOI:10.3390/molecules26040977).

Some of the articles that appear in this collection are related to the removal of toxic contaminants from soils. It is well-known that soil contamination by heavy metals threatens the quality of agricultural products and human health, so it is necessary to choose certain economic and effective remediation techniques to control the continuous deterioration of land quality.

An overview on the application of biochar (produced by the thermochemical transformation of plant- and animal-based biomass) for the remediation of heavy-metal-contaminated soil is proposed. Different conditions, such as biomass type, pyrolysis temperature, heating rate and residence time are the pivotal factors governing the performance characteristics of biochar. The potential risks of in-situ remediation strategy of biochar are expounded upon, which provides the directions for future research to ensure the safe production and sustainable utilization of biochar (T. Chen, DOI:10.3390/molecules25143167).

A research paper also deals with the use of biochars produced from Litchi branch for the removal of Cu from contaminated soils. The study demonstrates that BCs, especially those produced at a higher temperature, are promising for Cu immobilization and to release available silicon in soils. However, the remediation efficiencies were restricted by the soil type contamination status and remediation time (T. Chen et al., DOI:10.3390/molecules25184319).

The remediation of toxic components, such as radioactive waste, toxic heavy metals, and other harmful substances, under extreme conditions is quite difficult due to their restricted accessibility. Thus, novel treatment methods for the removal of toxic pollutants using extremophilic microorganisms that can thrive under extreme conditions have been investigated during the past several decades. Recent trends in bioremediation using extremophilic microorganisms and related approaches to develop them are reviewed, with relevant examples and perspective (Y.J. Choi et al., DOI:10.3390/molecules25214916).

Another contribution regards lignin fractionation by enzymatic hydrolysis is proposed as an effective approach to reduce the heterogeneity of lignin and improve the adsorption and recycle performances of lignin as a cationic dye adsorbent. The proposed fractionation method is proved to be a novel and efficient non-chemical modification approach that significantly improves adsorption capacity and recyclability of lignin (W. Sui et al., DOI:10.3390/molecules25112603).

A study to explore the potential phytoremediation of industrial NTO (3-nitro-1,2,4-triazol-5-one)-wastewater using vetiver grass (Chrysopogon zizanioides L.) is also proposed. Evidences that vetiver can effectively remove NTO from wastewater, and also translocated NTO from root to shoot are also presented (D. Sarkar et al., DOI:10.3390/molecules26010074).

Finally, some contribution is also related to air pollution causing serious challenges to modern society. A study proposes a non-thermal plasma (NTP) system for placement in the exhaust system of internal combustion engines to reduce the toxic contaminants (HC, CO, and NOx) of exhaust gases. The proposed NTP system, installed in the exhaust system, can significantly reduce air pollutants (M-H. Hsueh et al., DOI:10.3390/molecules25246041).

Of course, we wish to thank all the authors for their valuable contribution, as well as the reviewers for their work and relevant input. Finally, we would like to thank the editorial staff at

Molecules for making this themed Special Issue possible.

We created, for this book, an original cover image, dedicated to the efforts of chemistry to defend the beauty of environment, represented by flowers, against every prejudice that considers chemistry an enemy of life. We hope this Special Issue makes enjoyable reading and that it can be a source of inspiration and future fruitful collaboration.

Chiara Bisio, Monica Pica
Editors

molecules

MDPI

Review

Current Trends in the Application of Nanomaterials for the Removal of Pollutants from Industrial Wastewater Treatment—A Review

Geetha Palani [1], A. Arputhalatha [2], Karthik Kannan [3], Sivarama Krishna Lakkaboyana [4,*], Marlia M. Hanafiah [5,6], Vinay Kumar [7] and Ravi Kumar Marella [8]

1 Physics Research Centre, Dhanalakshmi College of Engineering, Tambaram 601301, Chennai, India; kesangee@gmail.com
2 Department of Physics, Arizona State University, Tempe, AZ 85287, USA; llathu@yahoo.com
3 School of Advanced Materials and Engineering, Kumoh National Institute of Technology, 61 Daehak-ro, Gum-si, Gyeongbuk 39177, Korea; karthikkannanphotochem@gmail.com
4 School of Ocean Engineering, Universiti Malaysia Terengganu, Kuala Nerus 21030, Terengganu DarulIman, Malaysia
5 Department of Earth Sciences and Environment, Faculty of Science and Technology, Universiti Kebangsaan Malaysia, Bangi 43600, Selangor, Malaysia; mhmarlia@ukm.edu.my
6 Centre for Tropical Climate Change System, Institute of Climate Change, Universiti Kebangsaan Malaysia, Bangi 43600, Selangor, Malaysia
7 Department of Biotechnology, Indian Institute of Technology Roorkee, Roorkee 247667, Uttarakhand, India; vkmjnu@gmail.com
8 Department of Chemistry (H & S), PACE Institute of Technology & Sciences, Ongole 523001, Prakasam, India; ravikumarmarella@gmail.com
* Correspondence: svurams@gmail.com

Citation: Palani, G.; Arputhalatha, A.; Kannan, K.; Lakkaboyana, S.K.; Hanafiah, M.M.; Kumar, V.; Marella, R.K. Current Trends in the Application of Nanomaterials for the Removal of Pollutants from Industrial Wastewater Treatment—A Review. *Molecules* **2021**, *26*, 2799. https://doi.org/10.3390/molecules26092799

Academic Editors: Chiara Bisio and Monica Pica

Received: 17 March 2021
Accepted: 27 April 2021
Published: 10 May 2021

Abstract: In the recent decades, development of new and innovative technology resulted in a very high amount of effluents. Industrial wastewaters originating from various industries contribute as a major source of water pollution. The pollutants in the wastewater include organic and inorganic pollutants, heavy metals, and non-disintegrating materials. This pollutant poses a severe threat to the environment. Therefore, novel and innovative methods and technologies need to adapt for their removal. Recent years saw nanomaterials as a potential candidate for pollutants removal. Nowadays, a range of cost-effective nanomaterials are available with unique properties. In this context, nano-absorbents are excellent materials. Heavy metal contamination is widespread in underground and surface waters. Recently, various studies focused on the removal of heavy metals. The presented review article here focused on removal of contaminants originated from industrial wastewater utilizing nanomaterials.

Keywords: nanomaterials; rare earth metals; wastewater treatment; pollutants

1. Introduction

One of the common natural resources in the world is water, which is indispensable for the endurance of every human and human development. With the rapid increase of urbanization and industrialization, water usage is expanding quickly, and the problem of water shortage has become imperative to solve for developing economies. A huge quantity of polluted wastewater is released from various industries, which includes the manufacturing of batteries, mining, toxins, and electroplating. Wastewater toxins cause numerous antagonistic impacts on living creatures and to the surrounding environment [1,2]. It ended up being a more productive and more affordable tool for treating industrial wastewater. Contaminant types present in the industrial wastewater rely upon the production process [3]. Toxins of industrial wastewater usually include large constituents of organic compounds, increased pH level, harmful substantial metals, huge saltiness, and increased turbidity from the presence of impurities of inorganic compounds. Adsorption, flotation, chemical precipitation, membrane filtration, flocculation, and coagulation are included in the industrial wastewater treatment [4,5].These wastewater treatments are sometimes insufficient in eliminating explicit impurities, for example, harmful heavy metals, oil, and microorganisms.

Wastewater is generated from various sources such as residential areas, commercial areas, industrial properties, and agriculture lands. Composition of wastewater differs extensively, and it is majorly dependent on the source from which it is generated. Common constituents of wastewater are inorganic substances such as solutes, heavy metals, metal ions, ammonia, and gases, complex organic compounds such as excreta, plants material, food, protein, natural organic matter, and nitrate, and other pollutants present in surface water, ground water, and/or industrial water. Typically, industrial waste can be divided into two categories, hazardous and nonhazardous. Nonhazardous industrial wastes do not cause environmental and health hazards and are produced from cardboard, plastic, iron, glass, stone, and organic waste. In contrast, hazardous wastes are industrial waste that can be harmful to public health or the environment, such as flammable, biodegradable, and hazardous materials [6]. Industrial waste is classified as wastewater, solid waste, or air leaks. There is some overlap in the physical properties of the substances present in these three categories, as wastewater can contain suspended solids and suspended liquids, and precipitation of solid waste can include gas, liquid, and some liquids. Particles and air exposures may consist of a fluid that emits air fluid and a substance known as particle emission [7]. Industrial waste, which has a significant concentration of non-recyclable or recyclable metals, is usually a good candidate for landfill, which is the dumping of waste into the ground area. Figures 1 and 2 shows wastewater and industrial wastewater general classifications.

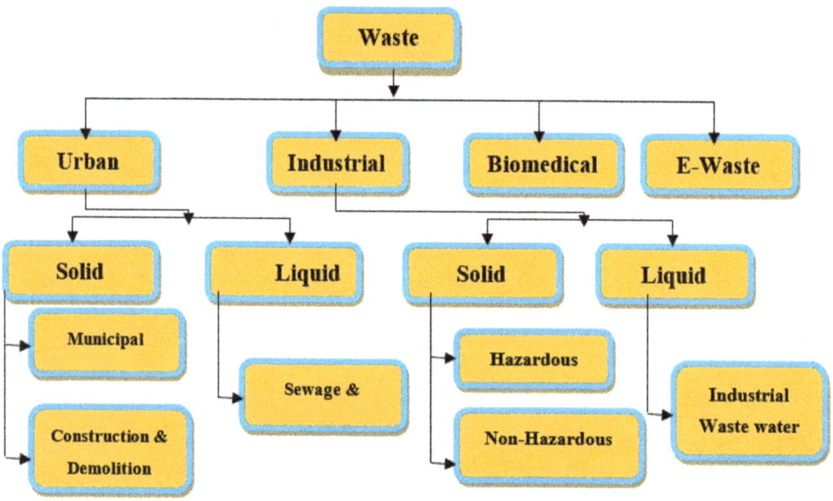

Figure 1. Anthropogenic wastes classification, (Reprinted with permission from [8]).

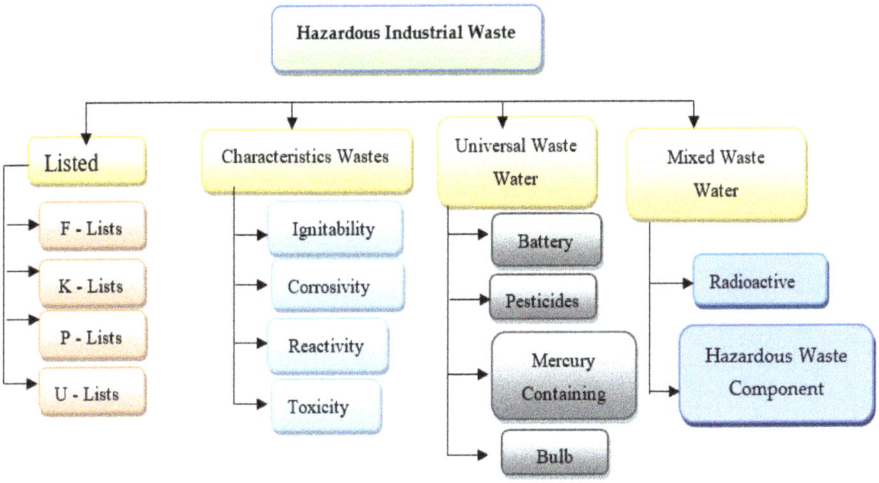

Figure 2. Industrial hazardous wastes general classification (reprinted with permission from [8]). F-list (wastes from common manufacturing and industrial processes), K-list (wastes from specific industries), P-list, and U-list (wastes from commercial chemical products).

When left untreated, these constituents may pose a threat to living beings and the environment, which makes it crucial to treat wastewater before disposal. Various physical, chemical, and biological treatment processes are used for wastewater treatment. Until now, a variety of research on nanomaterials was done to research heavy metal water treatments to find their applications, and they show incredible potential as an irreplaceable option to adsorb heavy metals from wastewater [8]. For the removal of heavy metals from polluted wastewater, these properties are very useful. According to the type of nanomaterial, wastewater treatment is classified into three fundamental groups [9]: nano-adsorbents, nanomembranes, and nano-catalysts. Some of the common sources of wastewater types are below [10]:

- Wastewater from municipal/domestic: discharged wastewater from habitations, foundations such as schools and medical clinics, and business offices, for example, shopping centers, restaurants, and so forth.
- Wastewater from industries: industrial processes removing wastewater, for example, drug, textile, and poultry industries.
- Infiltration/inflow: water that, in the long run, enters the sewer from establishment channels, pipes leaking, submerged manholes, groundwater invasion, etc.
- Storm water: rainfall runoff and snowmelt

Industrial wastewater treatment needs different successive complicated studies to fulfill the standard of water reusability. In looking at nanomaterials in waste streams and possibilities for recovery, one of the first items that need to be understood is the amount of industrial nanomaterial being produced and how much is then discharged, as shown in Table 1 [11]. The amount discharged to waste was based on presumptive material flow and then back calculated by knowing the reported amount produced rather than the actual measurement of the true nanomaterials concentration in a waste stream. For example, iron, zinc, and copper oxides would likely not last very long in the nanoparticle form, as they are more soluble. On the other hand, TiO_2 is one of the least soluble among nanomaterials and would likely stay in the nanoparticle form for much longer. SiO_2 would also be very similar to TiO_2 with a low solubility.

Table 1. Nanomaterials and discharges (reprinted with permission from [11]).

Si. No	Nano Material	Discharge
1	Silica (SiO_2)	95,000
2	Titania (TiO_2)	88,000
3	Alumina (Al_2O_3)	34,900
4	Zinc oxides	34,100
5	Nano-clays	10,400
6	Cu and Cu oxides	497
7	Ag	424

One of the most significant advancements in the 21st century was nanotechnology. Nanomaterials criteria include well-organized structure, filtration competence, small in size, and high surface to volume ratio. Some special properties of nanomaterials under the nanoscale are effects on the surface region, large quantum tunnel effects, small size effects, and quantum effects. These properties add to their adsorption capacity and reactivity, which are unprecedented and are great for heavy metal ions removal [12,13]. Industrial pollution continues to be a major factor in worsening the environment around us, the water we use, the air we breathe, and the land we live in. The growing power of industrialization not only consumes large agricultural land but at the same time causes environmental degradation as well as land degradation.

For raw materials management and manufacturing related to human activities, materials are driven from the liquids of industrial wastewaters (IWW) [14,15]. IWW acts as one of the pollutants of environmental water. From a recent survey, a huge amount of industrial wastewater is mixed into lakes, beaches, and streams. In the end, this produces contamination complications and entanglements in the surrounding water and leads to the eco-system returning as a negative output that negatively impacts human life. Industrial wastewater results in human population spill, and the climate ends up being awful in many scenarios. Huge quantities of these kinds of wastewaters are naturally incredibly solid, highly inorganic, effectively biodegradable, or inhibitory potential. In regard to these qualities, total dissolved solids (TSS), biological oxygen demand (BOD), and chemical oxygen demand (COD) might be high [16]. Industrial wastewater comprises wastewater from each sector of industry that produces its own exact blend of impurities.

As with the industrial wastewater's shifting character, industrial wastewater processing must be arranged explicitly for the specific sort of produced liquid. The metal industries discharge heavy metals and some of their compounds; also, the electroplating industry is a critical cause of contamination [17]. Higher amounts of Ag compounds are produced by photograph handling workshops, and, at the same time, printing plants release inks and dyes. The chlorine substances generally rely upon mash and paper industries; generally, they contain chloride compounds and dioxins. A very large quantity of phenols and oils are released by the petrochemical industry. Food handling effluents of the plant are loaded with organic and solid issues. Commonly, industrial wastewater is categorized into two classes: (1) organic industrial wastewater and (2) inorganic industrial wastewater. The main compound, inorganic industrial wastewater, exists in steel and coal industries, non-metallic minerals industry, metals manufacturing surface processing, and in commercial adventures [18]. Generally, wastewater, solid substances, and oils are delivered, and they contain incredibly poisonous solutes. This type may impact gas washing wastewater or cyanide blast-furnace industrial wastewater of metal processing with alkaline or acids solutions in which wastewater exudes from the gas refinement of Al work, which contains fluoride.

According to local regulations, non-metallic minerals exist in tiny and normal sizes alongside metal handling plants and are situated so they may discharge their wastewater into municipal wastewater systems; their effluents should be treated before liberation. The contaminants that come from the chemical compounds/industries which act on a large-scale by organic industrial wastewater mostly exploit substances for chemical responses [19]. The fluids incorporate substances of organics with variable properties and beginnings. The below-listed plants and industries primarily produce organic industrial wastewaters (OIW) [20]:factories manufacturing pharmaceuticals, beauty products, synthetic detergents, herbicides, and pesticides, leather and tanneries factories, textile factories, paper and cellulose manufacturing plants, factories related to oil-refining, and metal processing industry.

Now, the steel industry is viewed as a fundamental and crucial industry. Steel industries produce wastewater in huge quantities that contain many disintegrated chemicals in the sludge and undisclosed substances [20]. Producing iron out of its metals involves intensely decreased responses in impact heaters. Necessary cooling waters such as cyanide and ammonium hydroxides are soiled. Acidified rinse waters are present in wastewaters and waste corrosive blends. Regardless of the large number of plants working in acid plants recovery, where the mineral acid is reduced from the salts of iron, a huge amount of acid ferrous sulfate/chloride remains undisposed [21]. The paper production and wood-pulping products distribute polluters when natural fluids are delivered into emanating waters [22]. These liquids subsequently have extraordinary mutagenic effects that cause physiological weakness and damage aquatic organisms. The textile industry altogether additionally adds to water contamination in changing produced and natural threads into fabrics and different products. While fabricating many of the materials, wet chemical techniques are compulsory to legitimately sanitize, color, plan, or finalize the item [23]. This represents wastewater development, which usually contains loads to eliminate pollution from the crude materials themselves, yet excess reagent compounds are utilized in preparing such processes. Essential polluters in textile wastewaters are exceptionally chemical oxygen, heat, suspended solids, corrosiveness, and some dissolvable substances [24].

Industrial Wastewater Treatment (IWT) Processes

IWT is commonly arranged as physical, chemical, and organic processes. The usually embraced advances might be divided into [25]: (i) pre-treatment, (ii) primary; (iii) secondary and tertiary; (iv) refinement; and (v) purification. Usually, a basic level of treatment is based on size divisions utilizing physical methods, for example, filtration/sedimentation, for fundamental cleaning. More than 99% of removal can be achieved by tertiary treatment, which involves the final polishing of the effluent by toxic removal of pollutants to certain levels [26–28]. Wastewater produced by the primary process is not reasonable for release or reuse. The primary goal is to deliver the water suitable for optional and tertiary separations. A noticeable illustration of this is pH clarification/modification before membrane adsorption or separation or ion exchange. Separation processes, which are ordinarily viewed as basic treatments, include separation based on size, including the actual driving force that affects separation. Screening, cyclone separations, sedimentation, precipitations, thickening, centrifugation, and filtration are included in primary methods. Among the main homogenization/leveling, pre-treatments are meant to balance, and wastewater homogenization from inlets, particularly where industrial production creation occurs, is irregular and variable for consumption toxins [28].

Further developed processes of separation are utilized with immense changes in equipment and process nature in secondary and tertiary treatment stages. The separation process generally incorporates evaporation, absorption, distillation, extraction, ion exchange, biological processes, adsorption, crystallization, cavitations, and separation of membranes [29]. The process of separations involved here can be arranged depending on driving forces, such as thermal driving force, and pressure-driven processes, such as membrane separation-microfiltration (MF), ultra filtration (UF), nano-filtration (NF), and electrical forces, e.g., electro-dialysis or RO. Physicochemical methods are the main processes of separation that assume an imperative function in the field of wastewater treatment. This main class incorporates a huge assortment of processes, such as flocculation/coagulation, cavitations, oxidations, separation reactions, and extractions [30]. The process of separation includes ion exchange, and adsorption also comes under physicochemical methods of treatment for utilizing either electrostatic attraction or surface forces. Ion exchange, coagulation, and adsorption are methods of membranes separation that belong to charge-based separations, where the removal process is generally contaminated under charge neutralization and is applicable explicitly for the removal of charged bodies/ions from the solution. Based on the idea of the profluent, at least one process of separation is involved in meeting the end goals of water reuse/discharge [31]. Figure 3 shows the steps in wastewater treatment processes as a flow chart.

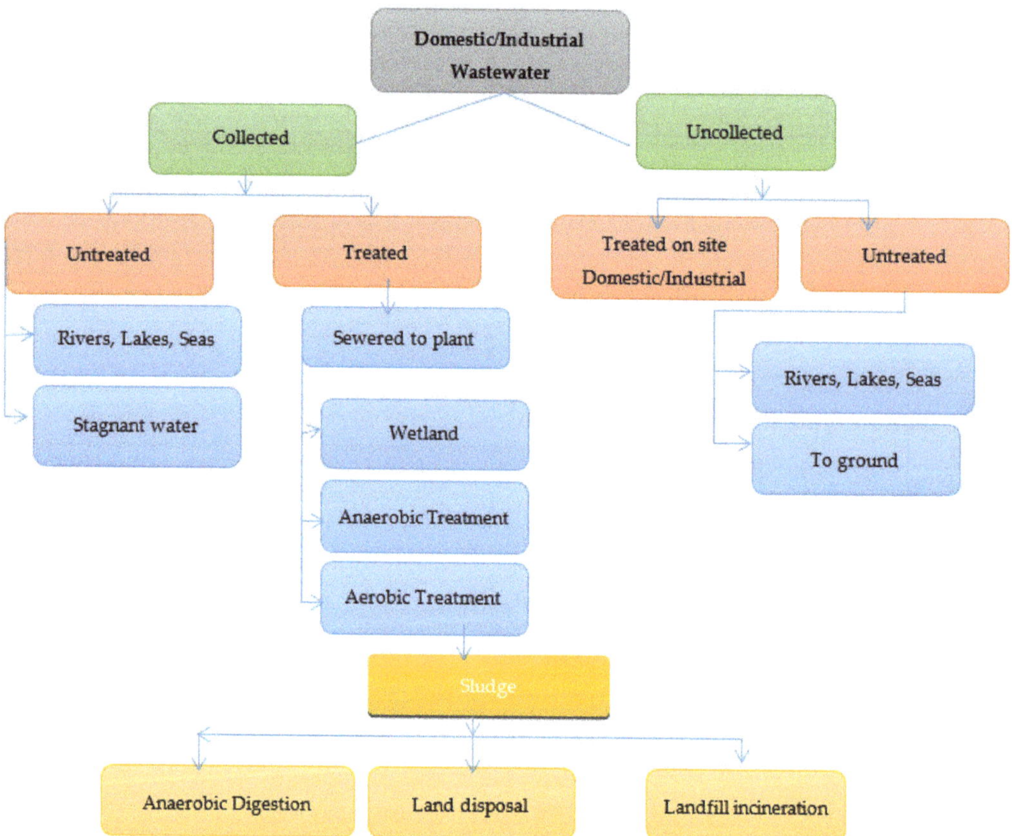

Figure 3. Flow chart showing steps in wastewater treatment processes.

2. Removal of Dyes

From industries, waters dyes and pesticides are being released frequently, and their findings, particularly at low concentrations, need the development of complex advances, for example, separation or filtration of compound combinations combined with detection utilizing multi-technique methods [32,33]. These estimations are subsequently tedious due to the many middle-level handling steps associated with preparing the sample. Utilization of silicon-graphene (sg) nanoporous composites takes into consideration an extreme cut of all these methods, as the compounds can pre-concentrate the analyte into the porous structure and widen the analyte signal if a proper method is utilized [34]. Table 2 shows the application and the harmfulness of various dyes [35,36].

Table 2. Application and harmfulness of various dyes, reprinted with permission from [35,36].

Dyes	Example	Advantages	Toxicity
Acid	Methyl orange, Sunset yellow	Wool, paper, leather, silk	Carcinogenic
Cationic	Rhodamine 6g, Methylene blue	Paper modified polyesters	Carcinogenic
Direct	Congo red, Direct red 23	Cotton, paper, leather	Bladder cancer
Disperse	Disperse red, Disperse orange 3	Nylon, acrylic fibers	Skin allergenic
Reactive	Reactive red 198 Reactive red 120	Nylon, wool, cotton	Dermatitis, allergic conjunctivitis
Vat	Vat orange 28, Vat orange 50	Cellulosic fibers	

By using Raman spectroscopy [28], for example, steps in sampling the water can be removed. The specific hierarchal porous silicon graphene composites design and the thick coating formed shape permit a "lab-on-a-chip" device to be combined with the Raman scattering method. Samples of water can be straightforwardly stored over the silicon graphene coatings, and toxins can be identified with the utilization of Raman spectroscopy. Raman signal enhancement identified with the analysis can be achieved through pure graphene-enhanced Raman scattering effect (GERS) or a combination of the surface-enhanced Raman scattering (SERS), e.g., graphene-mediated, surface-enhanced Raman scattering (G-SERS) given by plasmonic nanoparticles [37,38]. The affectability of such devices can be obtained by molecular imprinting methods. This substance approach permits framing exceptionally tiny cavities in the porous silica structures with sub-atomic recognition capabilities. Using molecular templating techniques, silicon graphene is able to form thick films that can be utilized for detecting dyes that mostly show the largest analytical enhancement factor of 14.64 for Rhodamine 6G dyes for a concentration of 10^{-3} M [39,40]. Additionally, the same team prepared porous SG templated films for the detection of paraoxon, an organophosphate pesticide, with the concentration of 10^{-5} M. Aside from identification, composites of silicon graphene additionally react as sorbents for the equivalent molecules of the analyte. Adsorption of organ phosphorus pesticides (OPPs) on silicon graphene composites was researched [41].

The adsorption was helped because of the substance co-operations between the composite and the functional groups; moreover, the strong π bonding between composites and the phenyl ring preferred the adsorption. The performance of adsorption of silicon graphene compounds was explored for various pesticides, and the capacity of removal varied. As referred to recently, for this situation, the expansion of magnetic nanoparticles to silicon graphene composites assumes a significant part in facilitating the recovery of the composites. Additionally, with an extra modification in the hydrophobic surface, the adsorption sites of pesticides expand [42]. Dye adsorption onto silicon graphene composites is likewise preferred by a mixture of physicochemical adsorption routes that rely upon the sorbent physicochemical properties.

Silicon graphene nanocomposites of the multifunctional compound were prepared by Kubo et al., inducing a super paramagnetic nanoparticles material to the mesoporous silica embedding graphene oxide. The nanoparticle functionalization permitted a simple recovery of the composites by an outer magnet. Fe_3O_4 addition, nonetheless, caused sharp diminishing of the area of the surface by 72% and ~15% of pore sizes, which decreased from 305 mg/g to 125 mg/g in regard to methylene blue (MB) removal capacity [43,44]. Considering this, amino group surface functionalization of the multifunctional silicon graphene nanocomposites might help increase the interactions for the pollutant removal.

The sorbent physical properties, e.g., pore size, surface area, and shape, give more sites for the dyes to diffuse and secure through the pores. The performance of adsorption is reliant additionally on the physicochemical properties of the dyes and the environmental media [28,45], such as different structure charges and dyes, which show unmistakable sorption conduct in various pH conditions.

3. Heavy Metals Removal

Heavy metals are those metals which have very high density and are highly poisonous, even at low concentrations. These heavy metals include mercury (Hg), cadmium (Cd), arsenic(As), chromium (Cr), thallium (Tl), zinc (Zn), nickel (Ni), copper (Cu), and lead (Pb). Municipal wastewater, industrial wastewaters, landfill leaches, mining wastes, and urban runoff are the main sources of contamination [46,47]. Industries are the major source of heavy metals in wastewater. The electroplating industry's wastewater contains a great number of metals such as Cd, Zn, Pb, Cr, Ni, C, Ag, Pd, and titanium. The wood processing industry is also a major source of metal in wastewater. Paint and enameling industries also release their effluents containing nickel into water bodies [48]. PCB manufacturing industries are also a very significant source of producing metal waste [49]. Nanomaterials of metal compounds displayed preference for heavy metals removal over activated carbon, e.g., titanium dioxide nanoparticles in arsenic adsorption and nanosized magnetite. The photo catalyst usage of, for example, nanoparticles titanium dioxide, was explored in detail to decrease the toxic metal ions in water. In a survey, titanium dioxide nanocrystalline showed adequacy in eliminating various types of As, and it was demonstrated to be the most viable photo catalyst aside from industrially accessible nanoparticles of titanium dioxide, which showed almost extreme efficiency of arsenic removal at a relatively neutral pH value [50]. A titanium dioxide nanocomposite and nanoparticles of titanium dioxide added on a graphene sheet was additionally utilized to decrease chromium VI to chromium III in daylight. Comparatively, chromium treatment was completed by utilizing nanoparticles of palladium in another survey; conventional technologies for heavy metal removal are shown in Figure 4. The removing capacity of arsenic (heavy metal) was likewise tested by utilizing Fe_2O_3 and Fe_3O_4 as adsorbents by most of the analysts. Removal of As was also additionally explored by utilizing a high, particular surface area of iron oxide nanocrystals [51].

Recently, this process was recognized widely for removal of heavy metals from wastewater. Many cheap adsorbents were developed recently. These adsorbents are widely used for treatment of wastewater containing heavy metals [52]. These adsorbents are derived from the waste products generated from industrial activities, waste generated from agriculture, and natural materials [53]. Adsorption can be defined as a mass transfer process which transfers the substance from the liquid phase to the surface of a solid and becomes bound by physical and chemical interactions. Figure 5, illustrates some conventional methods for removal of metal. It is a three-step treatment process: (1) the pollutant is transferred to the sorbent surface from bulk solution, (2) adsorption occurs on the particle surface, (3) transportation occurs within the sorbent particle. These techniques are very cost effective: adsorption on modified natural materials, adsorption on industrial by-products, bio-sorption [54].

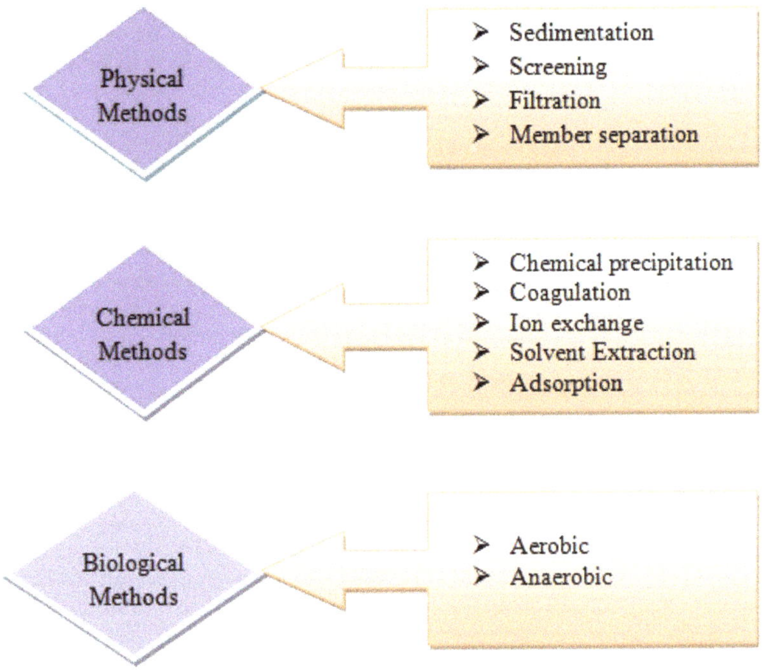

Figure 4. Conventional technologies for heavy metal removal, reprinted with permission from [52].

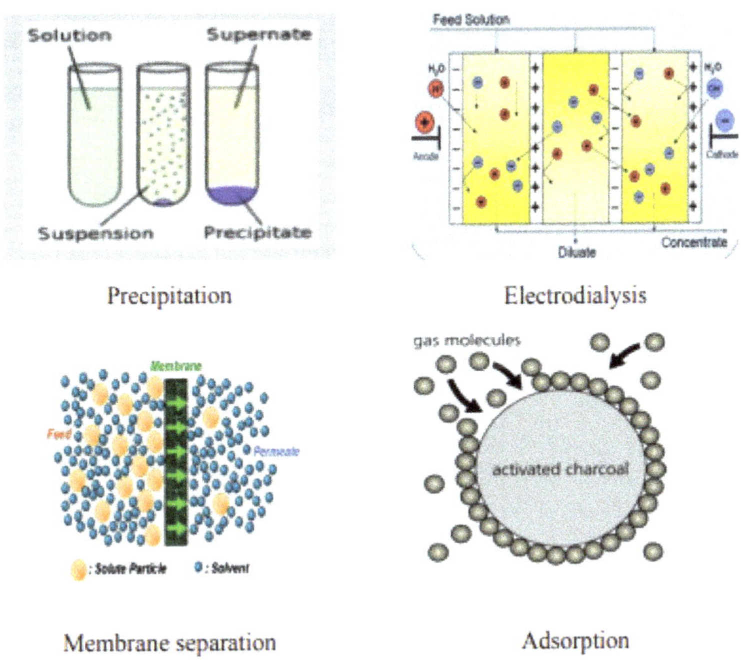

Figure 5. Conventional methods for removal of metal, reprinted with permission from [52].

4. Removal of Pesticides

Low-cost adsorbents development for pesticide maintenance is a significant part of environmental sciences research. Wastes from industries such as carbon slurry, fly ash, and sludge are delegated as easy materials due to their minimal price, and local accessibility pesticides removal can be utilized as adsorbents. Fly ash, lignite, coal-fired, thermal, power station solid waste, is an easy adsorbent that demonstrated huge adsorption limit with regard to organic contaminations [55]. Some researchers explained the sorption of pesticides and fly ash capability and suggested its utilization for pesticides removal from wastewater samples and water. Coal fly ash altogether has a high maintenance limit with regard to metolachlor, atrazine, and metribuzin.

When compared with metribuzin and metolachlor, atrazine was the most sorbed. The fly ash sorption herbicide efficiency relied upon the herbicide concentration in the mixes, and the highest removal of herbicide was seen at lower concentrations. These are common experiences in samples of water. At some point, when the water enters, pollution of biodegradables goes through concentrated processes of biological disintegration. Pollution of bio-resistant items fundamentally presents, to a lesser extent, an issue on the off chance that they are inactive biologically or are inorganic, inert, material framing sludges that are environmentally appropriate for benthos production and biological processes improvement. Inorganic poisonous contamination [56] is caused by heavy metals, organic pesticides, and so on, where nature speaks to the most unsafe kind of bio-resistant pollution.

The health risk from unsafe and hazardous chemical substances present in drinking water is regularly characterized as the likelihood that an unfriendly impact on health will be caused by substance exposure. The microelements interact and transport particularly heavy metals; streams courses indicate the most complex aquatic systems. In the agricultural region, water pollution is mainly caused by fertilizers, pesticides, and waste from poultry processing plants, drainage from livestock farms, and so on. Particularly, pollution made by pesticides on the row by heavy metals from fertilizers and nitrates contamination is very dangerous. Water mineralization indicates groundwater contamination of soil improvement on territories. Large mineral exploitation stores frequently prompt water quality disintegration on the more extensive area around strip mines and mines [57]. A great deal of water is commonly cleared during the depositing of drainage, and it upsets the groundwater regime and disturbs the hydro-chemical equilibrium.

5. Key Parameters for the Pollutant Removal

Commonly, contaminations can exist in viscous crude oil forms that are mostly either miscible or immiscible in water or as heavy metals, dyes, and pesticides of dispersible molecules. From the accessible information, there is more to viscous oil capturing than the actual process, and it intensely relies upon the huge composites' porosity and density for dissemination of fluid density via pores. Fewer viscosity solvents stream into the pores easily, but highly viscous liquids are used to diffuse slowly. The multifunctional composites design can be heated up through various mechanisms by using Joule effect, which permits one to reduce the oil viscosity, allowing faster diffusion of oil into the pores and promoting quicker removal. On the account of water-dispersible contaminations, the elimination depends on electrostatic interactions and chemicals with the sorbent. These may be improved by expanding the number of dynamic sites such as surface area and charge by chemical functionalization.

Silicon graphene multifunctional composites can be surface-functionalized to enhance the modifications with the analyte. The technique of sub-atomic is a wiser approach to silicon graphene composites with improved performance [58,59]. The molecular cavities focus on molecules and improve the detecting capabilities of dyes. The molecular imprinting is likewise useful for designing nanocomposites of silicon graphene for removing heavy metals. Moreover, conditions of the environment play a critical function in the pollutant's adsorption.

In general, the process of sorption relies on the physical state and the physicochemical properties of the composite and the pollutant chemical composition in the water. The properties evaluation versus the performance of the material is significant due to the absence of broad availability of characterization data and associated sorption capacities, such as specific surface areas [60,61]. Regardless of these obstructions, we distinguished the patterns and were effective in building up connections that helped increase the performance of the composites.

6. Nanomaterials Reuse and Retaining

Nano-adsorbent reuse and materials recovery from an aqueous solution are profoundly hard and may cause problems to the environment; as the compounds are adsorbed, pre-treatment samples must be created, and the process of separation technologies must ensue [62]. As another option, ballistic electrons discharged from the nanostructured adsorbent material and the nonporous carbonaceous utilizing microwave irradiation may likewise annihilate the adsorbed compound. Nanomaterial holding and reutilization by membrane filtration are enabled device designs of nanotechnology, key features that allow persistent chemical usage because of the expense and general health concerns. Besides, membranes of ceramics are beneficial compared to polymeric membranes in photo catalytic applications, as they exceptionally oppose chemical oxidants and ultraviolet.

Nanomaterials can be controlled on different membranes and resins, thus separation can be removed further. Vast studies are needed to propel easy, budget free techniques for nanomaterials immobilization without influencing their performances. In any case, the liberation potential is required to be generally reliant on the separation process and the technique of immobilization employed. For nanomaterials, which liberate metal ions, their disintegration must be restrained carefully [63,64]. The nanomaterial liberation detection is a significant, specialized obstacle for hazardous assessment and remains challenging, retaining particular nanomaterials reutilization to correct the cost process. For a nanomaterial applicable to treat wastewater, there are two investigative requirements. Future research with good, consistent conditions is needed to utilize various nanomaterials efficiencies.

7. Environmental Significance and Future Application

As of now, there is no doubt of the nanomaterials usage efficiency in industrial wastewater treatment, however, nanomaterials impose a huge number of genuine cons because, during treatment processes, they may discharge into the surrounding environment, and they can withstand serious risks for a long time. In such a manner, there is a requirement for more studies and surveys to decrease the toxicity in the environment [40,65]. Among different metal oxide nanoparticles, titanium dioxide, for instance, is generally utilized and have some significant constraints and some toxic effects towards human health and the environment, which causes difficulty in creating a sustainable environmental pollutants removal system. In such a manner, initiatives for new findings and research should be investigated to overcome these difficulties, and several scientists and analysts are continuously attempting to defeat these obstructions [66]. A few researchers just determined a better approach to diminish the band gap to utilize solar energy proficiently [67]. More research is needed to build up the practical techniques for the synthesis of nanomaterials and to determine effective areas of application in discovering the material proficiency. Table 3 summarizes the advantages and the disadvantages of the different physicochemical technologies for treatment of rare-earth elements in water and wastewater.

The market requirement is to restrict the costs of the procedure concerning the environmentally friendly nature of the referenced nanomaterials. Critical accentuation for this utilization should be given to green technology by material by-products of agricultural wastes. Numerous works in this area are needed in order to improve. The main concentration of the existing research works is that their utilization is yet to occur in the research [68,69]. Only a restricted amount of studies for brief market analyses and economic aspects is accessible. The principle market objective for the future is to improve the process of treatment on an industrial scale, which requires a significant monetary and technological method. In such a way, colleges and research labs can assume an effective part through a more official way to deal with the transfer of technology and copyright protection [70–72].

Table 3. Summary of the advantages and the disadvantages of the different physicochemical technologies for treatment of rare earth elements in water and wastewater [72].

Si.No.	Removal Technologies	Advantages	Disadvantages
1.	Chemical precipitation	Simple and safe operation Low capital cost Most metals can be removed	The increased operational ease disposal of sludge Slow metal precipitation kinetics
2.	Electrocoagulation	The high particulate removal efficiency Relatively low cost Compact treatment facility	Sacrificial anodes need to be replaced periodically Sludge production High operating cost
3.	Flotation	Low sludge generation Low cost and low energy requirements	To develop the removal efficiency treatments are required
4.	Ion exchange	Selection of metals High regeneration of materials Less time consuming High removal capacity	for metal removal not all resin ion exchange is suitable Regeneration creates a sludge disposal problem High Capital cost
5.	Biosorption	Use of inexpensive biosorbents Regeneration of biosorbent and low operating cost Selectivity and efficiency High metal binding capacity	The potential for biological process improvement is limited. Separation of biosorbents is difficult after adsorption Early saturation
6.	Ionic imprinted polymer	Stable and easy to prepare Inexpensive Metal selective	Polydispersity nature of the recognition sites Difficult to characterize

8. Conclusions

Water safety is of critical importance among major areas in the universe because of population increase, droughts drawn out, environmental changes, etc. From the literature review, wastewater or water treatment utilizing nanomaterials is becoming a prominent field in research. When compared with other planets, water makes our planet better. However, the overall pure water accessibility is part of what is causing current and unsurprising demands for water. The resources of drinking water are not satisfying domestic developmental, fundamental, or basic needs in many regions of the world. In particular regions, there is a lack of pure water to satisfy the fundamental requirement for sanitation, and human water needs are positively a breaking point in regard to the well-being of humans and for other creatures in the world. Academicians, research institutes, research fellows, and young scientists must discover a new way to eliminate these constraints. This universe is confronting many difficulties in doing this, particularly given a fluctuating and environmental future; a rise in population is driving community enlargement, globalization, and urbanization. How to preeminently defeat these difficulties requires investigation in every aspect of water management.

Nanomaterial treatment for water pollutants is becoming a trend, and it is drastically improving in this advanced time because of entirely awful states of water and freshwater demand in the entire universe. A significant requirement for progressive innovation for water treatment draws near, explicitly to affirm good quality drinking water and to eliminate micro and macro contaminations and toxins. We must improve developments of industrial production through deftly replaceable approaches for water treatment. Nanotechnology has manifested incredible accomplishments for water decontamination, controlling difficulties and making some progress for the future. Approaches of nanomaterials such as nanostructured catalytic membranes, nanosorbents, and so on are extremely productive, require less time, are eco-friendly techniques, and require less energy, however, every one of these techniques is inexpensive, and they are not utilized at this point for the industrial purpose of wastewater purification at an enormous scope.

Due to the high reaction rate, nanomaterials show high efficiency. In any case, there are a few shortcomings that should be avoided. There is no digital, computerized monitoring methods that offer predictable measurement on real-life facts regarding nanoparticles prevalence, which are available in limited quantities in water. Besides, to decrease hazards to health, international research universities and research institutes ought to plan legitimate terms and conditions to solve these circumstances. Moreover, the nanoengineered mechanical restriction for approaching water is that it is inconsistently adaptable to mass growths, and in today's scenario, few cases exist that are not obtrusive with moderate treatment approaches. Moreover, there are incredible requirements to incorporate some nanomaterials modifications, which include having large productivity and being cost effective, simple to deal with, and environmentally friendly. It is likewise important for wastewater treatment to grasp the economic difficulties and the commercialization of these innovations. Various uses of nanomaterial can give a gigantic proposal to flexible drinking water to the entire universe.

Author Contributions: Conceptualization, G.P., K.K.; methodology, G.P.; software, S.K.L.; validation, S.K.L., A.A. and M.M.H.; formal analysis, K.K.; investigation, G.P., K.K., S.K.L., V.K.; resources, M.M.H., A.A., K.K., R.K.M.; data curation, R.K.M.; writing—original draft preparation, G.P.; writing—review and editing, G.P., K.K., S.K.L.; visualization, S.K.L.; supervision, A.A, K.K., S.K.L., M.M.H.; project administration, M.M.H. All authors have read and agreed to the published version of the manuscript.

Funding: This research received no external funding.

Institutional Review Board Statement: Not applicable.

Informed Consent Statement: Not applicable.

Data Availability Statement: Not applicable.

Acknowledgments: Marlia M. Hanafiah was funded by the Universiti Kebangsaan Malaysia (DIP-2019-001; GUP-2020-034).

Conflicts of Interest: The authors declare no conflict of interest.

References

1. Ashraf, M.A.; Hanafiah, M.M. Sustaining life on earth system through clean air, pure water, and fertile soil. *Environ. Sci. Pollut. Res.* **2019**, *26*, 13679–13680. [CrossRef] [PubMed]
2. Reddy, M.S.; Sivaramakrishna, L.; Reddy, A.V. The use of an agricultural waste material, Jujuba seeds for the removal of anionic dye (Congo red) from aqueous medium. *J. Hazard. Mater.* **2012**, *203–204*, 118–127. [CrossRef] [PubMed]
3. Lakkaboyana, S.K.; Soontarapa, K.; Marella, R.K.; Kannan, K. Preparation of novel chitosan polymeric nanocomposite as an efficient material for the removal of Acid Blue 25 from aqueous environment. *Int. J. Biol. Macromol.* **2021**, *168*, 760–768. [CrossRef] [PubMed]
4. Banch, T.J.H.; Hanafiah, M.M.; Alkarkhi, A.F.M.; Abu Amr, S.S. Factorial Design and Optimization of Landfill Leachate Treatment Using Tannin-Based Natural Coagulant. *Polymers* **2019**, *11*, 1349. [CrossRef]
5. Lakkaboyana, S.K.; Khantong, S.; Asmel, N.K.; Yuzir, A.; Yaacob, W.Z.W. Synthesis of copper oxide nanowires-activated carbon (AC@ CuO-NWs) and applied for removal methylene blue from aqueous solution: Kinetics, isotherms, and thermodynamics. *J. Inorg. Organomet. Polym. Mater.* **2019**, *29*, 1658–1668. [CrossRef]
6. Hanafiah, M.M.; Mohamad, N.H.S.M.; Abd Aziz, N.I.H. Salvinia molesta and Pistia stratiotes as phytoremediation agents in sewage wastewater treatment. *Sains Malays.* **2018**, *47*, 1625–1634.
7. Carboni, D.; Jiang, Y.; Faustini, M.; Malfatti, L.; Innocenzi, P. Improving the Selective Efficiency of Graphene-Mediated Enhanced Raman Scattering through Molecular Imprinting. *ACS Appl. Mater. Interfaces.* **2016**, *8*, 34098–34107. [CrossRef]
8. Hanafiah, M.M.; Hashim, N.A.; Ahmed, S.; Ashraf, M.A. Removal of chromium from aqueous solutions using a palm kernel shell adsorbent. *Desalin. Water Treat.* **2018**, *118*, 172–180. [CrossRef]

9. Saleh, T.A.; Gupta, V.K. Processing methods, characteristics and adsorption behavior of tire derived carbons: A review. *Adv. Colloid Interface Sci.* **2014**, *211*, 93–101. [CrossRef]
10. Hanafiah, M.M.; Yussof, M.K.M.; Hasan, M.; Abdulhasan, M.J.; Toriman, M.E. Water Quality Assessment of Tekala River, Selangor, Malaysia. *Appl. Ecol. Environ. Res.* **2018**, *16*, 5157–5174. [CrossRef]
11. Bond, R.G.; Straub, C.P. *Handbook of Environmental Control: Waste Water: Treatment and Disposal*; CRC Press: Boca Raton, FL, USA, 1974.
12. Rivero-Huguet, M.; Marshall, W.D. Reduction of hexavalent chromium mediated by micron- and nano-scale zero-valent metallic particles. *J. Environ. Monit.* **2009**, *11*, 1072–1079. [CrossRef]
13. Basu, T.; Ghosh, U.C. Nano-structured iron(III)–cerium(IV) mixed oxide: Synthesis, characterization and arsenic sorption kinetics in the presence of co-existing ions aiming to apply for high arsenic groundwater treatment. *Appl. Surf. Sci.* **2013**, *283*, 471–481. [CrossRef]
14. Lakkaboyana, S.K.; Soontarapa, K.; Asmel, N.K.; Kumar, V.; Marella, R.K.; Yuzir, A.; Zuhairi WZ, W.Y. Synthesis and characterization of $Cu(OH)_2$-NWs-PVA-AC Nano-composite and its use as an efficient adsorbent for removal of methylene blue. *Sci. Rep.* **2021**, *11*, 5686. [CrossRef]
15. Gavrilescu, M. Fate of Pesticides in the Environment and its Bioremediation. *Eng. Life Sci.* **2005**, *5*, 497–526. [CrossRef]
16. Mittal, A.; Gupta, V.; Malviya, A.; Mittal, J. Process development for the batch and bulk removal and recovery of a hazardous, water-soluble azo dye (Metanil Yellow) by adsorption over waste materials (Bottom Ash and De-Oiled Soya). *J. Hazard. Mater.* **2008**, *151*, 821–832. [CrossRef]
17. Mondal, P.; Baksi, S.; Bose, D. Study of environmental issues in textile industries and recent wastewater treatment technology. *World Sci. News.* **2017**, *61*, 98–109.
18. Johnson, D.M.; Hokanson, D.R.; Zhang, Q.; Czupinski, K.D.; Tang, J. Feasibility of water purification technology in rural areas of developing countries. *J. Environ. Manag.* **2008**, *88*, 416–427. [CrossRef]
19. Eshelby, K. Dying for a drink. *BMJ* **2007**, *334*, 610–612. [CrossRef]
20. L'Vovich, M.I. *World Water Resources and Their Future*; American Geophysical Union: Washington, DC, USA, 1979.
21. Vörösmarty, C.J.; Green, P.; Salisbury, J.; Lammers, R.B. Global water resources: Vulnerability from climate change and population growth. *Science* **2000**, *289*, 284–288. [CrossRef]
22. Ritter, L.; Solomon, K.; Sibley, P.; Hall, K.; Keen, P.; Mattu, G.; Linton, B. Sources, pathways, and relative risks of contaminants in surface water and groundwater: A perspective prepared for the Walkerton inquiry. *J. Toxicol. Environ. Health Part A* **2002**, *65*, 1–142.
23. Pandiyan, S.; Arumugam, L.; Srirengan, S.P.; Pitchan, R.; Sevugan, P.; Kannan, K.; Pitchan, G.; Hegde, T.A.; Gandhirajan, V. Biocompatible Carbon Quantum Dots Derived from Sugarcane Industrial Wastes for Effective Nonlinear Optical Behavior and Antimicrobial Activity Applications. *ACS Omega* **2020**, *5*, 30363–30372. [CrossRef]
24. Bolong, N.; Ismail, A.F.; Salim, M.R.; Matsuura, T. A review of the effects of emerging contaminants in wastewater and options for their removal. *Desalination* **2009**, *239*, 229–246. [CrossRef]
25. Savage, N.; Diallo, M.S. Nanomaterials and Water Purification: Opportunities and Challenges. *J. Nanopart. Res.* **2005**, *7*, 331–342. [CrossRef]
26. Rickerby, D.; Morrison, M. Nanotechnology and the environment: A European perspective. *Sci. Technol. Adv. Mater.* **2007**, *8*, 19–24. [CrossRef]
27. Riu, J.; Maroto, A.; Rius, F.X. Nanosensors in environmental analysis. *Talanta* **2006**, *69*, 288–301. [CrossRef] [PubMed]
28. Theron, J.; Walker, J.A.; Cloete, T.E. Nanotechnology and Water Treatment: Applications and Emerging Opportunities. *Crit. Rev. Microbiol.* **2008**, *34*, 43–69. [CrossRef] [PubMed]
29. Fei, J.; Li, J. Metal Oxide Nanomaterials for Water Treatment. *Nanotechnol. Life Sci.* **2010**. [CrossRef]
30. Liu, H.; Ru, J.; Qu, J.; Dai, R.; Wang, Z.; Hu, C. Removal of persistent organic pollutants from micro-polluted drinking water by triolein embedded absorbent. *Bioresour. Technol.* **2009**, *100*, 2995–3002. [CrossRef]
31. Hoek, E.M.; Pendergast, M.T.M.; Ghosh, A.K. Nanotechnology-Based Membranes for Water Purification. In *Nanotechnology Applications for Clean Water*; William Andrew Publishing: Boston, USA, 2014; pp. 133–154. [CrossRef]
32. Nizam, N.U.M.; Hanafiah, M.M.; Mahmoudi, E.; Halim, A.A.; Mohammad, A.W. The removal of anionic and cationic dyes from an aqueous solution using biomass-based activated carbon. *Sci. Rep.* **2021**, *11*, 8623. [CrossRef]
33. Paul, J. Nanomaterials for the Decontamination of Waste Water Containing Pharmaceutical Drugs. 2015. Available online: https://www.diva-portal.org/smash/get/diva2:873252/FULLTEXT01.pdf(2015) (accessed on 17 March 2021).
34. Omole, M.A.; K'Owino, I.; Sadik, O.A. Nanostructured Materials for Improving Water Quality: Potentials and Risks. In *Nanotechnology Applications for Clean Water*; William Andrew: Boston, MA, USA, 2009; pp. 233–247. [CrossRef]
35. Van der Bruggen, B.; Vandecasteele, C. Removal of pollutants from surface water and groundwater by nanofiltration: Overview of possible applications in the drinking water industry. *Environ. Pollut.* **2003**, *122*, 435–445. [CrossRef]
36. Liu, J.; Chen, H.; Shi, X.; Nawar, S.; Werner, J.G.; Huang, G.; Ye, M.; Weitz, D.A.; Solovev, A.A.; Mei, Y. Hydrogel microcapsules with photocatalytic nanoparticles for removal of organic pollutants. *Environ. Sci. Nano* **2020**, *7*, 656–664. [CrossRef]
37. Safdar, M.; Simmchen, J.; Jänis, J. Correction: Light-driven micro- and nanomotors for environmental remediation. *Environ. Sci. Nano* **2017**, *4*, 2235. [CrossRef]

38. Esfahani, M.R.; Aktij, S.A.; Dabaghian, Z.; Firouzjaei, M.D.; Rahimpour, A.; Eke, J.; Escobar, I.C.; Abolhassani, M.; Greenlee, L.F.; Esfahani, A.R.; et al. Nanocomposite membranes for water separation and purification: Fabrication, modification, and applications. *Sep. Purif. Technol.* **2019**, *213*, 465–499. [CrossRef]
39. Sun, X.; Liu, Z.; Zhang, G.; Qiu, G.; Zhong, N.; Wu, L.; Cai, D.; Wu, Z. Reducing the pollution risk of pesticide using nano networks induced by irradiation and hydrothermal treatment. *J. Environ. Sci. Health Part B* **2015**, *50*, 901–907. [CrossRef]
40. Ray, P.C.; Yu, H.; Fu, P.P. Toxicity and Environmental Risks of Nanomaterials: Challenges and Future Needs. *J. Environ. Sci. Heal. Part C* **2009**, *27*, 1–35. [CrossRef]
41. Lu, F.; Astruc, D. Nanocatalysts and other nanomaterials for water remediation from organic pollutants. *Co-ord. Chem. Rev.* **2020**, *408*, 213180. [CrossRef]
42. Li, J.; Chen, C.; Zhang, R.; Wang, X. Nanoscale Zero-Valent Iron Particles Supported on Reduced Graphene Oxides by Using a Plasma Technique and Their Application for Removal of Heavy-Metal Ions. *Chem. Asian J.* **2015**, *10*, 1410–1417. [CrossRef]
43. Gasser, M.; Aly, M. Separation and recovery of rare earth elements from spent nickel–metal-hydride batteries using synthetic adsorbent. *Int. J. Miner. Process.* **2013**, *121*, 31–38. [CrossRef]
44. Ponou, J.; Wang, L.P.; Dodbiba, G.; Okaya, K.; Fujita, T.; Mitsuhashi, K.; Atarashi, T.; Satoh, G.; Noda, M. Recovery of rare earth elements from aqueous solution obtained from Vietnamese clay minerals using dried and carbonized parachlorella. *J. Environ. Chem. Eng.* **2014**, *2*, 1070–1081. [CrossRef]
45. Burton, F.L.; Stensel, H.D.; Tchobanoglous, G. *Metcalf & Eddy Wastewater Engineering: Treatment and Reuse*; McGraw Hill, Avenue of the Americas: New York, NY, USA, 2003; p. 10020.
46. Burkhard, R.; Deletic, A.; Craig, A. Techniques for water and wastewater management: A review of techniques and their integration in planning. *Urban Water* **2000**, *2*, 197–221. [CrossRef]
47. Kouras, A.; Zouboulis, A.; Samara, C.; Kouimtzis, T. Removal of pesticides from aqueous solutions by combined physicochemical processes—the behaviour of lindane. *Environ. Pollut.* **1998**, *103*, 193–202. [CrossRef]
48. Wu, D.; Niu, C.; Li, D.; Bai, Y. Solvent extraction of scandium (III), yttrium (III), lanthanum (III) and gadolinium (III) using Cyanex 302 in heptane from hydrochloric acid solutions. *J. Alloys Compd.* **2004**, *374*, 442–446. [CrossRef]
49. Kannan, K.; Radhika, D.; Nesaraj, A.; Sadasivuni, K.K.; Reddy, K.R.; Kasai, D.; Raghu, A.V. Photocatalytic, antibacterial and electrochemical properties of novel rare earth metal oxides-based nanohybrids. *Mater. Sci. Energy Technol.* **2020**, *3*, 853–861. [CrossRef]
50. Pearson, R.G. Hard and soft acids and bases—the evolution of a chemical concept. *Co-ord. Chem. Rev.* **1990**, *100*, 403–425. [CrossRef]
51. Al-Bastaki, N.M. Performance of advanced methods for treatment of wastewater: UV/TiO2, RO and UF. *Chem. Eng. Process. Process. Intensif.* **2004**, *43*, 935–940. [CrossRef]
52. Thanh, N.; Puntes, V.; Tung, L.; Fernig, D. Fifth international conference on fine particle magnetism. In *Journal of Physics: Conference Series*; IOP Publishing: Jakarta, Indonesia, 2005; pp. 70–76.
53. Kümmerer, K.; Erbe, T.; Gartiser, S.; Brinker, L. AOX—Emiissions from hospitals into municipal waste water. *Chemosphere* **1998**, *36*, 2437–2445. [CrossRef]
54. Peralta-Videa, J.R.; Zhao, L.; Lopez-Moreno, M.L.; de la Rosa, G.; Hong, J.; Gardea-Torresdey, J.L. Nanomaterials and the environment: A review for the biennium 2008–2010. *J. Hazard. Mater.* **2011**, *186*, 1–15. [CrossRef]
55. Mumma, R.O.; Raupach, D.C.; Waldman, J.P.; Tong, S.S.C.; Jacobs, M.L.; Babish, J.G.; Hotchkiss, J.H.; Wszolek, P.C.; Gutenman, W.H.; Bache, C.A.; et al. National survey of elements and other constituents in municipal sewage sludges. *Arch. Environ. Contam. Toxicol.* **1984**, *13*, 75–83. [CrossRef]
56. Chang, A.; Warneke, J.; Page, A.; Lund, L. *Accumulation of Heavy Metals in Sewage Sludge-Treated Soils*; 0047-2425; Wiley: Hoboken, NJ, USA, 1984.
57. Álvarez, E.; Mochón, M.; Sánchez, J.; Rodríguez, M. Heavy metal extractable forms in sludge from wastewater treatment plants. *Chemosphere* **2002**, *47*, 765–775. [CrossRef]
58. Sakaki, K.; Sugahara, H.; Kume, T.; Ohashi, M.; Naganawa, H.; Shimojo, K. Method for Synthesizing rare Earth Metal Extractant. U.S. Patent 8841482, 23 September 2014.
59. Ms, A.-R.; Arm, A.-R. Removal of Heavy Metals from Industrial Waste Water by Biomass-Based Materials: A Review. *J. Pollut. Eff. Control.* **2016**, *5*, 180. [CrossRef]
60. Ogata, T.; Narita, H.; Tanaka, M. Adsorption behavior of rare earth elements on silica gel modified with diglycolamic acid. *Hydrometallurgy* **2015**, *152*, 178–182. [CrossRef]
61. Ramakrishna, S. *An Introduction to Electrospinning and Nanofibers*; World Scientific: Singapore, 2005.
62. Rawat, D.; Mishra, V.; Sharma, R.S. Detoxification of azo dyes in the context of environmental processes. *Chemosphere* **2016**, *155*, 591–605. [CrossRef] [PubMed]
63. Santhosh, C.; Velmurugan, V.; Jacob, G.; Jeong, S.K.; Grace, A.N.; Bhatnagar, A. Role of nanomaterials in water treatment applications: A review. *Chem. Eng. J.* **2016**, *306*, 1116–1137. [CrossRef]
64. Adeleye, A.S.; Conway, J.R.; Garner, K.; Huang, Y.; Su, Y.; Keller, A.A. Engineered nanomaterials for water treatment and remediation: Costs, benefits, and applicability. *Chem. Eng. J.* **2016**, *286*, 640–662. [CrossRef]
65. Kobielska, P.A.; Howarth, A.J.; Farha, O.K.; Nayak, S. Metal–organic frameworks for heavy metal removal from water. *Co-ord. Chem. Rev.* **2018**, *358*, 92–107. [CrossRef]

66. Zhang, X.; Lin, B.; Zhao, K.; Wei, J.; Guo, J.; Cui, W.; Jiang, S.; Liu, D.; Li, J. A free-standing calcium alginate/polyacrylamide hydrogel nanofiltration membrane with high anti-fouling performance: Preparation and characterization. *Desalination* **2015**, *365*, 234–241. [CrossRef]
67. Yurekli, Y. Removal of heavy metals in wastewater by using zeolite nano-particles impregnated polysulfone membranes. *J. Hazard. Mater.* **2016**, *309*, 53–64. [CrossRef]
68. Tarabara, V.V. Multifunctional Nanomaterial-Enabled Membranes for Water Treatment. In *Nanotechnology Applications for Clean Water*; William Andrew Publishing: New York, NY, USA, 2009; pp. 59–75. [CrossRef]
69. Shahamat, Y.D.; Khani, M.R.; Mahdizadeh, H.; Kannan, K.; Kalankesh, L.R.; Kamarehei, B.; Baneshi, M.M. Olive Mill Wastewater (Omw) Treatment by Hybrid Processes of Electrocoagulation/Catalytic Ozonation And Biodegradation. *Environ. Eng. Manag. J.* **2020**, *19*, 1401–1410. [CrossRef]
70. Hunger, K. *Industrial Dyes: Chemistry, Properties, Applications*; Wiley-VCH Verlag GmbH & Co. KGaA: Weinheim, Germany, 2007.
71. Upadhyay, R.K.; Soin, N.; Roy, S.S. Role of graphene/metal oxide composites as photocatalysts, adsorbents and disinfectants in water treatment: A review. *RSC Adv.* **2014**, *4*, 3823–3851. [CrossRef]
72. Barros, Ó.; Costa, L.; Costa, F.; Lago, A.; Rocha, V.; Vipotnik, Z.; Silva, B.; Tavares, T. Recovery of Rare Earth Elements from Wastewater Towards a Circular Economy. *Molecules* **2019**, *24*, 1005. [CrossRef]

molecules

MDPI

Review

Treatment of Wastewaters with Zirconium Phosphate Based Materials: A Review on Efficient Systems for the Removal of Heavy Metal and Dye Water Pollutants

Monica Pica

Department of Pharmaceutical Sciences, University of Perugia, Vial del Liceo 1, 06123 Perugia, Italy; monica.pica@unipg.it

Abstract: Layered zirconium phosphate (ZrP) is a versatile material with phosphate (POH) groups able to exchange inorganic and organic cations or to intercalate basic molecules. The present review deals with the use of this material as a sorbent for heavy metal cations or dye molecules in wastewater treatments. The possibility to combine ZrP with polymers or other inorganic materials, in order to have suitable systems for real and large scale applications, was investigated, as well as the combination with photocatalytic materials to obtain hetrogeneous photocatalysts for the capture and photodegradation of organic dye molecules.

Keywords: zirconium phosphate; wastewater pollutants; ion exchange; heterogeneous photocatalysis

check for updates

Citation: Pica, M. Treatment of Wastewaters with Zirconium Phosphate Based Materials: A Review on Efficient Systems for the Removal of Heavy Metal and Dye Water Pollutants. *Molecules* **2021**, *26*, 2392. https://doi.org/10.3390/molecules26082392

Academic Editor: Farid Chemat

Received: 2 February 2021
Accepted: 19 April 2021
Published: 20 April 2021

1. Introduction

The progress of society led, on one hand, to a significant improvement in the quality of people's life, but, on the other hand, to the significant anthropogenic pollution of soil, water and air due to the intensive exploitation of natural resources. The scientific community is called to support society in this challenge, finding realistic and effective strategies to control, reduce and remedy environmental pollution. According to The Longman Dictionary of Environmental Science, environmental pollution is defined as "any harmful or undesirable change in the physical, chemical or biological quality of air, water or soil" [1].

The present review is focused on water pollution and, specifically, on the strategies to remove some categories of common pollutants from water.

Freshwater on Earth is only 3% of the total water, and about 20% of the population live in conditions of a lack of freshwater. According to the World Health Organization (WHO), more than one million people consume non-potable water, with letal effects causing about 30,000 daily deaths [2].

Inorganic compounds such as phosphates and nitrates, dyes and phenolic compounds, as well as pesticides, recalcitrant organic matter, sediments, heavy metal ions, and products derived from pharmaceutical preparations and industrial activities (plastic, leather, textile, paper, ceramics, glass, cosmetics, food, paints, soap, wax, biomedicine industry) are the most common water pollutants, which are often found in water resources as mixtures, leading to dangerous synergistic effects and complicating their detection, quantification and removal [2,3].

Dyes and heavy metal ions are among the main water pollutants derived from industrial activities. Synthetic dyes find application in paper printing, photography, and the pharmaceutical, cosmetic, food and textile industries, and their history started in 1856 when the first aniline dye, Mauveine, was discovered by William Henry Perkin. Dyes possess unsaturated groups, with conjugated chemical bonds, and often complex structures which are responsible for light absorption and emission in the visible region and for the difficulty of removal from water [2].

Heavy metals are defined as metals with a specific gravity above 5.0 and atomic weight between 63.5 and 200.6. Some of them are present in the human body and are fundamental for several biological processes, but when their concentration exceeds upper limit values, they become dangerous for human health. Their harmful effects on humans depend on the dosage, rate of emission and period of exposure [2,4].

The level of toxicity for the human body of some heavy metals follows the order Co < Al < Cr < Pb < Ni < Zn < Cu < Cd < Hg [5].

Table 1 reports the main harmful effect of some heavy metals, used in industrial activities, and their upper limit concentration in drinking water according to the the World Health Organization (WHO) [4].

Table 1. Upper limit concentrations, according WHO in drinking water for some heavy metals and their harmful effects on the human body [4].

Heavy Metal	Upper Limit Concentration in Drinking Water (mg/L)	Toxic Effects on Human Body
Lead	0.01	Harmful to heart, bones, intestines, reproductive and nervous systems.
Copper	2	Mucosal irritation, capillary damage, hepatic and renal damage, central nervous system damage.
Cadmium	0.003	Bone lesions, cancer, lung insufficiency, hypertension.
Zinc	3	Gastrointestinal effects.
Nickel	0.02	Cancer, skin allergy, lung fibrosis.
Mercury	0.001	Kidney disease, haemorrhagic gastritis and colitis, brain damage, cancer.
Chromium	0.05 (hexavalent chromium)	Cancer, healing ulcers.

Water treatments for the removal of dyes and heavy metals include biological processes (in aerobic or anaerobic conditions), chemical processes (such as ozonation, coagulation and precipitation, ion exchange, electro-coagulation) and physical processes (flotation, reverse osmosis, adsorption) [2]. Among these processes, adsorption is considered one of the most effective and competitive and is widely used in industrial applications. In an adsorption process, the substances of a fluid, liquid or gas bind to the external and interior surfaces of the adsorbent material. The main advantages are low cost, high efficiency, ease of operation and implementation, the possibility to use several solids as adsorbent materials, and the possibility to recover the adsorbent and the adsorbate [6]. It is interesting to compare the costs of the main technologies for wastewater treatment (reverse osmosis, ion exchange, electro-dialysis, electrolysis) with those of the adsorption process: the first range from 10 to 450 USD/m^3 treated water, while adsorption ranges from 5.0 to 200 USD/m^3 [7]. Depending on the nature of the adsorbent and on its textural properties, the adsorbent–adsorbate interaction can be physical, also called van der Waal's adsorption (adsorption into the adsorbent pores), and/or chemical (ion exchange or acid–base reactions). Generally speaking, besides low cost and availability, an adsorbent material should possess chemical and mechanical stability and good textural properties (high surface area and pore volume, suitable functional groups) in order to guarantee high adsorption efficiency, fast kinetics, recovery and reusage [2,7,8].

As far as the operation mode is concerned, adsorption processes are generally carried out by batch experiments at the lab scale, while a fixed-bed operation mode is suitable for large scale treatment applications, allowing the treatment of large water volumes in small physical areas, without requiring additional separation operations [2,9]. Moreover, fixed-bed columns for industrial processes have higher residence times and better heat and mass transfer characteristics than batch reactors [9].

Adsorption isotherm modeling is used to describe the interaction between adsorbate and adsorbent and how the adsorbate is distributed between the solution and the solid phase at the equilibrium state. Experimental data can be fitted by several adsorption isotherm models, among them Langmuir, Freundlich, and Sips [10–13]. The Langmuir model is based on the formation of an adsorbate monolayer at the outer surface of the adsorbent. According to this model, all the adsorption sites are energetically equivalent and identical.

Differently, the Freundlich isotherm is based on a heterogeneous adsorption with different adsorption energies.

The Sips model consists of the combination of the Langmuir and Freundlich isotherm models. The Sips model reduces to the Freundlich isotherm at low metal ion concentrations, while at high metal concentration it follows the Langmuir isotherm.

Kinetic studies also play an important role in the characterization of an adsorption process, since they provide information about the uptake rate of the adsorbate and, hence, on the efficiency of the adsorption process. Experimental data can be validated, among others, by pseudo-first order, pseudo-second order or intraparticle kinetic models [14,15].

According to Crini et al., adsorbent materials can be classified as conventional and unconventional materials [8,9,16].

Among conventional adsorbents, activated carbons, inorganic materials such as activated aluminas, silica gel, zeolites, and ion-exchange resins are included [17,18]. Unconventional adsorbents include materials from agricultural and industrial waste, natural materials such clays, biosorbents such as chitosan, and miscellaneous adsorbents such as alginates [8,9,19–24]. These sorbents have the advantage of being low-cost or free of cost materials. In addition to these, other unconventional materials have been investigated, among them graphene, carbon nanotubes, Metal Organic Frameworks, and layered zirconium phosphates [25–30].

Layered zirconium(IV) phosphates are a well known class of inorganic materials, whose intercalation and ion exchange properties are widely investigated [31]. Their chemical and thermal stability, the availability of several synthetic procedures, the possibility to easily tune its properties by changing the synthetic approach, the possibility to easily bind functional groups on the layer surface and their ability to uptake cations, both inorganic and organic, and basic molecules stimulated the interest in this class of compounds as potential and unconventional materials for the removal of heavy metals and dye from water. As proof of this, several papers reported their use for the removal of heavy metals and dyes from water. Nevertheless, to the best of our knowledge, specific reviews dedicated to the use of zirconium phosphates in pollution remediation were not found in the literature, except for the paper by Pandith et al., published in 2020 [32], that, however, was dedicated, among other things, just to the metal uptake by zirconium phosphate, while concerns about dye uptake were not addressed.

On the basis of these considerations, it seemed of interest for the scientific community to gather/collect the recent studies on water treatment by using zirconium phosphates for pollutants' removal, specifically heavy metal cations and dyes.

2. Structure of Zirconium Phosphate

The history of zirconium phosphate compounds started in the 1950s, when the first papers on their use as sorbents for metal cations were published [33–35].

Zirconium phosphates can be classified according to their crystal structure. One-dimensional, two-dimensional and three-dimensional structures are known [36] and, among these, α-type layered structures are the most studied as cation exchangers and intercalation hosts [31,37–39].

Zirconium phosphate of α-type ($Zr(HPO_4)_2\ H_2O$, ZrP) is made of layers in which the Zr atoms bond monohydrogenphosphate groups, with the P-OH groups pointing in the interlayer region, alternatively below and above the main plane. Six phosphates are coordinated to a Zr atom through oxygen atoms, and the oxygens of each phosphate are shared with three Zr atoms. Water molecules are located between the layers in six-sided cavities and form hydrogen bonds with the P–OH groups of the same layer [31].

The interlayer distance of ZrP is 7.56 Å and the presence of protogenic–OH groups is responsible for its cation exchange properties, and both inorganic and organic cations can be inserted in the interlayer region [31].

Microcrystalline ZrP has an ion exchange capacity of 6.6 meq/g. Generally, about half of the protons are exchanged at relatively low pH, with the formation of a half-exchanged phase which is converted in the fully exchanged phase at higher pH (neutral or slightly basic). Moreover, the protons of the P–OH groups can react with bases, leading to the intercalation of basic molecules [31].

ZrP is a highly versatile material, whose properties are strictly connected to its structure and degree of crystallinity which can be controlled by changing the synthetic conditions.

Amorphous, microcrystalline and nanocrystalline materials were obtained in different conditions. Amorphous gel materials were generally obtained by reaction between phosphoric acid and zirconyl chloride in water, without using any complexing agent for Zr(IV) [33–35,40], while post treatments of amorphous ZrP produced powders, granules and thin films [41,42]. Sol–gel and template methods were also developed in order to produce ZrP materials with different morphologies and porosity [43–50].

By refluxing amorphous ZrP in the presence of concentrated phosphoric acid, microcrystalline materials were obtained [51]. Alternatively, microcrystalline ZrP was prepared by Alberti et al. through hydrothermal synthesis by the slow decomposition of ZrF_6^{2-} complexes in the presence of phosphoric acid [52]. Chuah et al. proposed a modification of this hydrothermal method in the presence of small amounts of fluoride ions, leading to platelets and rods [53]. Alternatively to fluoride, oxalate anions were used as complexing anions for Zr(IV) [54].

Nanocrystalline ZrP was prepared by Pica et al. by using a quick procedure, consisting of mixing, at room temperature, zirconyl propionate and concentrated phosphoric acid in aliphatic alcohols [55]. A gel product was obtained, containing ZrP nanoparticles with an average planar size of tens of nanometers. By heating the gels to dryness, nanocrystalline ZrP, consisting of nanoplatelets with an average planar size of hundreds of nanometers, was obtained for phosphate to zirconium molar ratios higher than 4.

Amorphous, microcrystalline and nanocrystalline ZrP were investigated as sorbent materials for heavy metals and dyes, and in the following sections these applications are discussed.

3. Removal of Heavy Metals

Amorphous ZrP, prepared by the reaction of zirconyl chloride, dissolved in HCl solution, with phosphoric acid at room temperature, was employed by Pan et al. for the removal from water of heavy metal cations, specifically lead, zinc and cadmium [40]. The ion exchange properties of ZrP were evaluated, finding that about half of the total amount of protons, corresponding to 3.04 meq of H^+ per gram, is released into the solution at pH < 7, which is indispensable for the removal of heavy metals because most of heavy metals precipitate in solution at alkaline condition. Batch sorption tests, carried out at 30 °C by adding ZrP to a solution containing the selected heavy metals, proved that protons are stoichiometrically exchanged with the heavy metal cations, according to the following reaction:

$$\mathrm{Zr(HPO_4)_2} + \frac{1}{2}\mathrm{M^{2+}} \leftrightarrows \mathrm{ZrM_{\frac{1}{2}}H(PO_4)_2} + \mathrm{H^+} \quad (1)$$

Moreover, it was found that the sorption capacities of the three heavy metal ions decreased with increasing the pH solution, in agreement with the Le Chatelier Principle. The affinity order of ZrP for the cations, determined by sorption isotherms, was: $Pb^{2+} > Zn^{2+} \approx Cd^{2+}$. The authors speculated that, among other factors, this order could also be affected by the hard and soft properties of the involved species, according to the hard and soft acids and bases (HASB) theory [56–58]. Lead ion, as a Lewis acid, lies between hard and soft ones and preferably interacts with orthophosphate ion, a Lewis base also in borderline, while cadmium, a soft acid, presents weaker interaction with ZrP and consequently exhibited low selectivity.

Regeneration efficiencies, evaluated by the treatment of the cation-exchanged ZrP samples with HCl, were 95.3%, 99.6%, and 99.9% for ZrP loaded with Pb, Zn, and Cd, respectively. However, it should be pointed out that the synthetic procedure used for the synthesis of amorphous ZrP produced ultrafine particles that cannot be directly used at large scale in column operation, due to the unacceptably high pressure drop.

To overcome this problem, hybrid sorbents, made of ZrP particles immobilized onto porous materials or dispersed in a polymer matrix, were proposed for large-scale applications.

In the following, ZrP combined with polymers or other inorganic materials will be described.

3.1. ZrP–Polymer Composites

Several examples of sorbents based on polymer composites, consisting of ZrP particles dispersed in a polymer matrix, are reported in the literature. Charged and uncharged polymers were used, also offering, among other things, the advantage to in situ precipitate ZrP particles.

Pan et al. prepared a hybrid sorbent by loading ZrP particles onto a polystyrenesulfone cation exchanger D-001 [59]. D-001 was selected as the host material due to the presence of a negatively charged sulfonic acid group, which would improve the permeation of the targeted metal ions. As a matter of fact, D-001 is recommended by the US Environmental Protection Agency as the best available material for heavy metal contamination control [59].

The composite material (hereafter ZrP-001) was prepared by adding D-001 to a solution of $ZrOCl_2$ in HCl. After the evaporation of HCl, H_3PO_4 was added to precipitate ZrP. The ZrP wt% was 33% and its incorporation onto D-001 caused a decrease in the surface area and pore volume of the polymer matrix.

The ZrP-001 composite presented two different exchangeable sites for the uptake of heavy metals: the sulfonic group of D-001 and the phosphate groups of ZrP, according to Equations (1) and (2):

$$R - SO_3H + 1/2M^{2+} \leftrightarrows RSO_3M_{0.5} + H^+ \tag{2}$$

The ion-exchange capacity value of the composite, determined experimentally, was 3.20 meq/g, which was less than that calculated (4.07 meq/g) at neutral pH, but slightly higher that that found for ZrP in ref. [40].

Both batch and column sorption tests were carried out. From batch experiments, the uptake of the heavy metals at pH less than 0.5 was negligible, in agreement with the fact that the used ZrP-001 might be regenerated by strongly acidic solution, while at higher pH solution (under acidic or neutral conditions) the uptake was more favourable and higher than that of D-001; the same preference order found in ref. [40] was observed ($Pb^{2+} >> Zn^{2+} \approx Cd^{2+}$).

Fixed-bed column adsorption tests for the heavy metals were carried out in the presence of competing cations (Na^+, Ca^{2+}, and Mg^{2+}) and confirmed the more efficient sorption of heavy metals on ZrP-001 than on D-001, despite its lower ion exchange capacity (3.2 vs. 4.2 meq/g), and the higher selectivity of ZrP-001 for heavy metals with respect to innocuous cations. Moreover, less than 1% loss of ZrP in ZrP-001 beads was detected after five-column cycles.

A mesoporous polystyrene (MPS) matrix was also used to immobilize ZrP [60]. According to this procedure, the MPS host was first prepared by a flash freezing method, which allowed the acquisition of spherical beads of 1.5–2.0 mm in diameter, with abundant uniform pores of around 7.9 nm. MPS was then immersed into ethanol solutions containing $ZrOCl_2$, followed by evaporation to promote the diffusion of $[Zr(OH)_2(H_2O)_4]^{2+}$ cations inside the nanopores of the MPS host. The beads were then incubated with H_3PO_4 solution to induce the in situ growth of the ZrP nanoparticles. The ZrP loading in ZrP@MPS was about 7% in Zr mass and ZrP nanoparticles with an average diameter of 6.5 nm were well dispersed in ZrP@MPS. XRD and ^{31}P-MAS NMR analyses revealed the coexistence of α-ZrP and γ-ZrP and this was a quite surprising result, since γ-ZrP is normally obtained by boiling, under reflux conditions, an amorphous gel for several weeks [61].

The authors first evaluated the adsorption reactivity of both α-ZrP and ZrP@MPS toward Pb^{2+}. They found that pure MPS exhibited negligible adsorption toward heavy metals; moreover, the adsorption distribution coefficients K_d of ZrP@MPS were much higher (10–90 times higher) than those of α-ZrP. On the basis of the XPS studies, the authors proposed that ZrP@MPS adsorbed Pb^{II} mainly through the inner-sphere coordination rather than through an electrostatic interaction. They also examined the selective adsorption of the sample toward a series of heavy metal cations (Pd, Cd, Ni) in the presence of much higher concentrations of competing mineral cations, e.g., Ca^{2+}, in real contaminated water. With respect to other systems [59], the adsorption capacities toward heavy metal cations are less disturbed by Ca^{2+} (≈100%, >70%, and >40% for Pb^{2+}, Cd^{2+}, and Ni^{2+}, respectively), even at high Ca concentrations; the authors speculated that the higher affinity of the sorbent material toward heavy metals was mainly due to the kind of interaction between them: the specific inner-sphere coordination interaction between ZrP@MPS and heavy metal cations favoured the selectivity toward heavy metals more than nonspecific electrostatic attractions. ZrP@MPS was also used to treat simulated polluted water containing Pb^{2+} in a continuous column mode. A commercial D001 sample was used in parallel. The results showed that the D001 column produced clean water with a mass of ≈1500 times over the adsorbent mass, while the ZrP@MPS column was able to generate clean water with a mass of 7200 times over the adsorbent mass in each run. Finally, the ZrP@MPS absorbent material could be easily and effectively regenerated by treating the fully adsorbed sample with hydrochloric acid.

Polysulfide (PSF) capsules containing zirconium phosphate were prepared by Li et al. for the removal of Pb^{2+} ions from aqueous solutions [62]. Amorphous ZrP was prepared according to ref. [40], while PSF@ZrP capsules were prepared by the phase inversion precipitation technique. N-methylpyrrolidone was used as a solvent to dissolve PSF and disperse ZrP, while sodium dodecylsulfate in ethanol–water solution was used to form the PSF@ZrP capsules. Capsules with different PSF/ZrP mass ratios were prepared. The characterization of the PSF@ZrP capsules revealed the amorphous nature of ZrP and that their surface area was lower than that of the pure PSF capsules, due to the blockage of a fraction of the pores by the inorganic material. SEM images showed the presence of spherical capsules with a rough surface and with ZrP particles uniformly dispersed from the outer surface to the inner part of the capsules.

Adsorption experiments were carried out by using a batch method, and the effect of pH, contact time, initial concentration, temperature and competing ions was studied. Pure PSF capsules had an extremely low affinity to Pb^{2+}. The lead adsorption significantly increased from 12 to 62 mg g^{-1} with increasing the ZrP mass ratio in the capsules from 1:0 to 1:1 (PSF to ZrP). However, a further addition of ZrP up to 1:1.5 (PSF to ZrP) provoked a decrease in the adsorption capacity for Pb^{2+}. On the basis of these results, the authors selected the capsule with PSF/ZrP mass ratio = 1 to continue the study. The optimal pH condition for Pb^{2+} uptake was the original pH of the solution, that is, 5.75. Stronger acidic conditions did not favor the adsorption, suggesting that the PSF@ZrP capsules may be regenerated in strong acidic conditions. The equilibrium time was independent of the initial Pb^{2+} concentration. Moreover, the initial rate of adsorption increased with increasing the initial Pb^{2+} concentration because of the increased driving force, resulting from the difference between the Pb^{2+}concentration in the solution and onto the capsules. Although the amount of Pb^{2+} adsorbed at equilibrium increased from 56 to 102 mg g^{-1} with increasing the Pb^{2+} concentration from 100 to 300 mg L^{-1}, the removal percentage of Pb^{2+} from solution decreased from 56% to 34%.

PSF@ZrP capsules showed satisfactory affinity to bind Pb^{2+} with respect to other similar reported sorbents, such as alumina and modified alumina [63,64]. Furthermore, the maximum uptake amount of the PSF@ZrP capsules is over hundred times larger than that of ZrP modified silica [65].

The effects of competing cations such as Na^+ and K^+ on the Pb^{2+} uptake onto the PSF@ZrP capsules were also studied, finding that the amount of Pb^{2+} sorbed onto the PSF@ZrP capsules was slightly influenced when the concentration of the coexisting ions was low, while it decreased as the concentration of Na^+ and K^+ was hundreds of times more than that of Pb^{2+}. Finally, adsorption re-generation cycles were performed six times with no significant loss of adsorption capacity.

Macroporous polystyrene resins, with different functional groups, i.e., $-CH_2Cl$, $-SO_3H$, and $-CH_2N^+(CH_3)_3$, were also used to encapsulate ZrP nanoparticles, thus obtaining three nanocomposite adsorbents (denoted as ZrP–Cl, ZrP–S, and ZrP–N, respectively) for lead removal from water [66].

The host polymers with $-SO_3H$ and $-CH_2Cl$ functional groups were, respectively, prepared from polystyrene–divinylbenzene copolymer (St–DVB) by reaction with concentrated sulforic acid (hereafter S–St–DVB) and chloromethyl ether with zinc chloride as the catalyst (hereafter Cl–St–DVB), while $-CH_2N^+(CH_3)_3$ groups were introduced by the reaction of Cl–St–DVB with trimethylamine solution (hereafter N–St–DVB).

Polymer composites were prepared by immersing Cl–St–DVB, S–St–DVB, or N–St–DVB beads in ethanol solution containing different amounts of $ZrOCl_2$. Then, each $ZrOCl_2$-loaded polymer was treated with H_3PO_4 solution. The ZrP wt% in each composite was about 20 wt%.

TEM images of the three nanocomposites showed that both ZrP–S and ZrP–N consisted of ZrP nanoparticles with an average size of about 10–40 nm. Obvious particle agglomerates were observed within ZrP–Cl (20–140 nm). The authors suggested that the presence of negatively or positively charged functional groups in the polymeric supports was more favourable than the neutral chloromethyl group to form small ZrP nanoparticles. In other words, nanoparticle dispersion or aggregation is greatly associated with both van der Waals attraction and electrostatic double-layer repulsion interaction between the particles. The repulsion interactions between adjacent nanoparticles generally dominate the extent of their aggregation and particle size distribution. As proof of that, ZrP–N and ZrP–S had similar absolute values of zeta potential, which were higher than that of ZrP–Cl, in agreement with the improved ZrP dispersion of the two former nanocomposites. Sorption isotherms of Pb(II) by the three composites, in the presence of Ca(II) as a competing cation, showed that the sorption capacities were in the sequence of ZrP–S > ZrP–N > ZrP–Cl. The observed trend can be explained by considering both the ZrP particle size and the nature of electrical charge immobilized on the polymer surface.

The authors also evaluated the mechanical properties of the composites in light of their importance to improve their feasibility for practical application. On the basis of the sorption tests, S–St–DVB was selected as the host material to fabricate several composite adsorbents with different amounts of ZrP. Then, the mechanical strength of the resulting composites was examined in terms of compressive strength and compared with the pure S–St–DVB host polymer, finding that the maximum compressive strengths (MCS) of all the resulting nanocomposites were greatly improved with respect to that of the host polymer; the optimum nano-ZrP loading was about 5 wt%.

Alam et al. validated the ion exchange process on a composite cation exchanger based on nylon 6,6 and ZrP for a practical application in the wastewater treatment process [67]. ZrP was prepared by the reaction of zirconyl oxychloride with H_3PO_4 at pH = 1. Nylon 6,6 gel, obtained by treatment with concentrated formic acid, was added to ZrP and mixed thoroughly with constant stirring. Composite cation exchanger particles of mean radii of 125 nm in H^+ form were used to evaluate various kinetic parameters with heavy toxic metal pollutants such as lead, cadmium, zinc and copper. Kinetic studies were carried out at various temperatures under particle diffusion controlled phenomena. Various kinetic parameters such as self-diffusion coefficient (Do), energy of activation (E_a), and entropy of activation (ΔS^*) were evaluated to validate the ion exchange process. It was found that equilibrium is attained faster at a higher temperature with the particle diffusion controlled phenomenon. Moreover, the diffusion coefficient for the four heavy metal ions studied on ZrP composites followed the order $Cu^{2+} > Pb^{2+} > Zn^{2+} > Cd^{2+}$, also confirmed by the highest ΔS^* value for the H(I)–Cu(II) exchange.

Polyaniline (PANI) was used to prepare composite films loaded with ZrP platelets for potential-triggered adsorption of Pb^{2+} ions [68]. A highly crystalline α-ZrP was first synthesized by the hydrothermal method and then exfoliated by tetrabutylammonium hydroxide. ZrP nanosheets/PANI was deposited on a carbon nanotube modified Au electrode in aqueous solution by the electropolymerization of aniline, in the presence of ZrP, by the cyclic voltammetry (CV) method. The presence of CNTs had the advantage of increasing the roughness of the electrode surface, thus promoting the electrodeposition of ZrP/PANI. At the reduction state of CV, a quinonoid amine (=N-) of PANI seized the proton from the α-ZrP nanosheet to form benzenoid amine (NH) [69], leading to PO^- Pb^{2+} interactions. The incorporation and release of ions in the hybrid film was studied by the electrochemical quartz crystal microbalance (EQCM) technique, by in situ detection of the mass change of the film because of the charging and discharging associated with the ion exchange process between the film and electrolyte. It was found that, while the pure PANI film behaved as an anion exchanger, the α-ZrP/PANI hybrid film behaved as a cation exchanger. The mass increased when the composite film was reduced due to the insertion of Pb^{2+} ions into the film to neutralize the reduction centers. During the oxidation of the film, the Pb^{2+} ions were released into the solution to maintain charge balance and resulted in the mass decrease. A set of CV/EQCM experiments was conducted to quantify the preferential Pb^{2+} selectivity in the presence of Ni^{2+}, Cd^{2+}, Zn^{2+} and Co^{2+} ions. The ion adsorption selectivity follows the sequence of $Pb^{2+} > Ni^{2+} > Co^{2+} > Cd^{2+} > Zn^{2+}$, and the adsorption capacity towards Pb^{2+} ions was at least four times higher than that of other heavy metal ions.

Chen and coworkers fabricated a zirconium phosphate modified polyvinyl alcohol-polyvinylidene fluoride, (PVA)-PVDF, membrane for lead removal [70]. The zirconium ions and PVA were firstly coated onto a PVDF membrane through crosslinking reactions with glutaraldehyde, which was then modified by phosphate. Lead adsorption was studied by batch experiments. It was found that lead adsorption increased with an increase in pH up to pH = 5.5. The adsorption isotherms, studied at pH 5.5, showed that the experimental data were better described by the Langmuir equation than the Freundlich equation, suggesting that the adsorption sites onto the membrane are relatively homogenous. The maximum adsorption capacity was 121 mg-Pb/g at pH 5.5 for lead simulated water (prepared by DI water). A good selectivity in the adsorption towards lead with respect to zinc was also found, with a selectivity coefficient of lead/zinc around 10. Four cycles of adsorption and desorption were conducted to test the reusability of the modified PVDF membrane. After four cycles, the adsorption capacity was still as high as 95.6% of the virgin membrane in the first cycle.

An interesting paper by Hasan et al. reported the use of cellulose membranes coated with α-ZrP nanoparticles (α-ZrP-n) for the removal of heavy metals from wastewater [71]. The composites were prepared by spraying an aqueous dispersion of commercial ZrP nanoparticles, with an average particle size diameter of 100 nm, onto the surface of the pure cellulose membranes. Leaching tests proved that ZrP strongly interacted with cellulose, since no traces of Zr were found after the immersion of the membrane in water for 24 h. Moreover, mechanical tests showed that the presence of ZrP nanoparticles did not negatively affect the mechanical properties of cellulose fibers, while a decrease in the membrane porosity was observed with increasing the α-ZrP-n concentration, also leading to a reduction in water flux.

The synthetic wastewater sample containing heavy metals (Ni, Zn, Pb, Cu) was filtered using both the pristine cellulose and α-ZrP-n coated membranes using vacuum filtration.

At pH = 7, all composite membranes exhibited a removal percentage of all heavy metals higher than that of pure cellulose. The best results were obtained with Pb^{2+}, reaching a percentage of removal of about 60% with the membrane containing 1 wt% of α-ZrP-n.

3.2. Other ZrP Based Materials

In order to improve the textural properties ZrP, making it more suitable for large scale applications, several synthetic approaches were developed. In some cases, ZrP was combined with other inorganic materials and synergistic effects were observed.

Parida et al. proposed a titania pillared zirconium phosphate to remove hexavalent chromium from aqueous solutions by solar radiation [72]. Titania is a well known photocatalyst under UV light acting as a reducing agent able to reduce the harmful Cr(VI) to the less harmful Cr(III). Titania pillared ZrP was prepared from Na-exchanged ZrP following the procedure reported by Yamanaka et al. [73]. First, titania sol was prepared by the hydrolysis of titanium(IV) isopropoxide by HCl. Then, an aqueous suspension of sodium-exchanged ZrP was added to the sol. The suspension was filtered, washed and calcined at 500 °C. Composites with titania loadings in the range 1–10 wt% were prepared.

The photo-reduction of Cr(VI) was performed in batch. The solution was exposed to sunlight in closed flasks at room temperature with constant stirring. The effect of EDTA and 4-nitrophenol, as sacrificial electron donors, was also studied, keeping all other parameters fixed.

It was found that the photo-reduction of hexavalent chromium was strongly dependent on pH. The highest reaction rate was obtained at acidic pH (1–2). Moreover, by increasing the titania loading up to 2 wt%, the initial rate of the Cr(VI) photoreduction increased, and this could be due to the fact that an increase in titania resulted in higher surface area, which allowed more Cr(VI) species to be adsorbed on the surface, thus facilitating the photo-reduction process. Differently, the rate of photo-reduction of Cr(VI) decreased with increasing the initial Cr(VI) concentration. The percentage of photoreduction reached 100% at a lower Cr(VI) concentration (<10 mg/L). By increasing the catalyst dose or the irradiation time, the rate of photo-reduction of Cr(VI) initially increased; thereafter, it remained almost constant.

By comparing the effects of sacrificial electron donors such as EDTA and 4-nitrophenol, a significant effect was observed for EDTA. Moreover, experiments were also carried out by varying the atmosphere of the photo-reduction process by bubbling N_2, O_2 and air, finding that, among them, N_2 had the most significant effect on the photo-reduction, while the dissolved oxygen had no effect or very negligible effect on the photo-reduction process.

Graphene oxide–zirconium phosphate (GO–ZrP) nanocomposite was proposed by Pourbeyram as an adsorbent for the removal of Pb(II), Cd(II), Cu(II), and Zn(II) from aqueous solutions [74]. The GO–ZrP composite was prepared by the dropwise addition of zirconium chloride to a GO suspension under sonication, followed by the addition of sodium dihydrogen phosphate. The adsorption of phosphate on the surface of GO–Zr was represented schematically as follows (Figure 1):

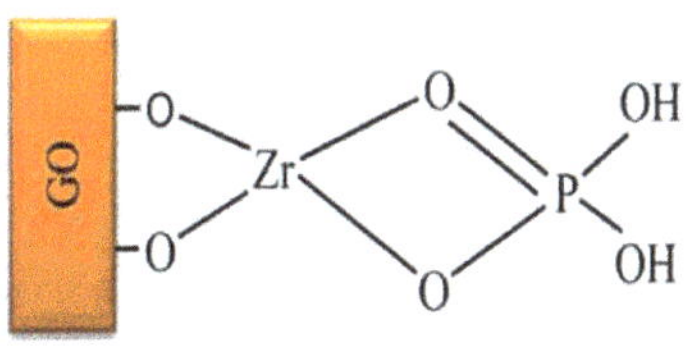

Figure 1. Schematic representation of ZrP bonded to the surface of graphene oxide (GO) ([74], with permission of ACS).

The transmission electron microscopy (TEM) images of the GO–ZrP nanocomposite revealed the presence of ZrP nanoparticles of ∼1–2 nm in diameter. The nanoparticles were well distributed, with a high density on the GO surface and interparticle distances in the range of ∼3–5 nm.

The adsorption of heavy metals was performed in a batch experiment. GO–ZrP nanocomposite was added to a solution containing Pb(II), Cd(II), Cu(II) and Zn(II), at desired initial concentrations. The effect of pH on the adsorption of heavy metals was studied in the pH range of 1–8. At the range of pH 3–6, high and relatively constant adsorption capacity was observed. It was found that, for all metals, the adsorption occurred in two different steps. During the first 10 min, the adsorption increased rapidly. After that, adsorption increased gradually and finally reached equilibrium after 20 min. After adsorption, a tendency of the nanocomposite to agglomerate and precipitate was observed. Moreover, the maximum adsorption capacity of GO was much lower than that for the GO–ZrP nanocomposite. On the other hand, on the GO–Zr nanocomposite, no adsorption of heavy metals was observed under the same conditions.

The amount of heavy metals adsorbed on the nanocomposite increased by increasing the initial concentration of the heavy metals in the range of 10–200 ppm, and then, the sorbent was finally saturated by a relatively constant amount of the heavy metals (∼200 ppm). The results of batch experiments indicated that the maximum adsorptions for Pb(II), Cd(II), Cu(II), and Zn(II) at pH 6 were 363, 232, 329, and 252 mg g^{-1}, respectively, corresponding to 3.5, 4.1, 10.4, and 7.7 meq g^{-1}, respectively. Moreover, the removal of heavy metals at lower amounts of sorbent took more time than that at higher amounts.

Adsorption isotherms showed that the process occurred at the functional groups/binding sites on the surface of the GO–ZrP nanocomposite, according to a monolayer adsorption. It was found that the adsorption capacity of the GO–ZrP nanocomposite was significantly higher than that of the most part of adsorbents. A possible configuration of the adsorption of metal ions (M) can be represented schematically as follows (Figure 2):

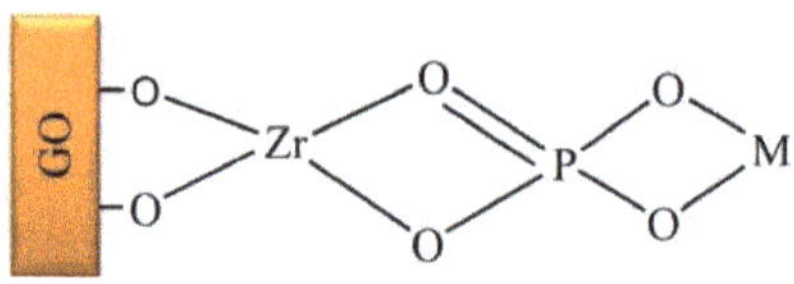

Figure 2. Schematic representation of the adsorption of a metal ion onto GO–ZrP ([74], with permission of ACS).

The desorption study showed that the GO–ZrP nanocomposite was effectively regenerated (∼100%) by the treatment of metal ion-loaded nanocomposite with 3 M HCl for ∼10 min. The nanocomposite was easily separated from the system via centrifugation. The high adsorption efficiency performance was maintained after being used for at least five cycles.

Besides inorganic compounds, organic molecules with high metal affinity were combined with ZrP to fabricate composite adsorbents with improved properties. On this regard, the ability of crown ethers to recognize cations in a selective fashion has been known since the 1970s [75], as well as their ability to interact with zirconium phosphates through covalent and uncovalent interactions [76,77]. Peng and coworkers fabricated a new organic–inorganic layer α-ZrP composite by the intercalation of 4-amino-benzo-18-crown-6 (AM–ZrP) to remove radioactive ^{90}Sr from solution [78]. This crown ether was chosen since it has a strong complexing ability for Sr^{2+}. α-ZrP was prepared by the fluorine reflux method. The intercalation of 4-amino-benzo-18-crown-6 was carried out on ZrP pre-intercalated with butylamine (BU–ZrP) in order to enlarge the interlayer region and promote the intercalation of crown ether. The authors speculated that the arrangement of 4-amino-benzo-18-crown-6 in the layer structure of α-ZrP comprised double inclined layers, which is one of the best modes that could load the maximum amount of 4-amino-benzo-18-crown-6.

The AM–ZrP composite exhibited excellent stability under acid and radioactive conditions: this is an important key factor for application in nuclear wastewater disposal, as most ^{90}Sr exists in acidic fission solution produced by dissolving uranium targets in strong HNO_3 solution. The intercalation of crown ether in ZrP also efficiently decreased its loss in the adsorption process.

The adsorption of Sr^{2+} by α-ZrP, BU–ZrP and AM–ZrP was measured in aqueous solution. Isotherm studies proved that the adsorption data were best fitted by the Langmuir model, indicating that Sr^{2+} adsorbed onto α-ZrP, BU–ZrP and AM–ZrP materials forms a monolayer. The maximum adsorption capacity of Sr^{2+} onto α-ZrP, BU–ZrP and AM–ZrP is 63.14 mg g^{-1}, 162.36 mg g^{-1} and 320.14 mg g^{-1}, respectively, at pH = 5. Moreover, adsorption kinetics of Sr^{2+} on α-ZrP, BU–ZrP and AM–ZrP showed that about 90% of Sr^{2+} ions were removed within the first 60 min, with equilibrium gradually reached in approximately 150 min. It is noteworthy that the adsorption capacity of AM–ZrP is higher than that of other similar zirconium adsorbents [78].

The study of Sr^{2+} uptake by AM–ZrP in the presence of other metal cations from Li, Na, K, Mg, Cs, Pd, Mo, Zr, Ca, and Ba proved its excellent selectivity for Sr^{2+}, which is mainly due to the complexing action, rather than ion exchange and surface physical adsorption. Unfortunately, the authors did not report information about the performances of AM–ZrP in column separation and nuclear wastewater disposal processes.

Melamine is a N-rich chelating molecule, which was investigated in heavy metal capture [79,80]. El-Shall et al. studied melamine zirconium phosphate (M–ZrP) adsorbent for the extraction of heavy metals from polluted water [81]. M–ZrP was prepared in two steps: first, melamine phosphate (MP) was prepared by the acid–base reaction between phosphoric acid as a proton donor and melamine as a proton acceptor. Then, zirconium tetrachloride was added to form amorphous ZrP. TEM images show that M–ZrP consisted of small particles connected together in mesoporous structures with a Brunauer–Emmet–Teller (BET) surface area of 320 m^2 g^{-1}.

The adsorption capacity of M–ZrP was tested for Pb(II), Hg(II) and Cd(II). It was found that the M–ZrP adsorbent shows exceptionally high adsorption affinity for Pb(II) with a capacity of 681 mg g^{-1} and 1000 mg g^{-1} using an adsorbent dose of 1 g L^{-1} and 2 g L^{-1}, respectively. The high adsorption capacity is also coupled with fast kinetics with an equilibrium time, required for the 100% removal of Pb(II), of the order of seconds and minutes, depending on the metal concentration.

In a mixture of six heavy metal ions, the removal efficiency was 100% for Pb(II), 99% for Hg(II), Cd(II) and Zn(II), 94% for Cu(II), and 90% for Ni(II) at a lower concentration, while at a higher concentration the removal efficiency for Pb(II) was 95% compared to 23% for Hg(II) and less than 10% for the other ions.

Despite the difficulties of using pure ZrP as a metal sorbent in a continuous operation mode on a large scale, it is noteworthy that researchers worked to develop strategies to optimize the particle morphology and textural properties of ZrP in order to improve, on one hand, its uptake capacity and, on the other hand, to make it more suitable for large scale applications.

Nakanishi et al. fabricated hierarchically porous ZrP monoliths combining micrometer order macropores and nanometer order mesopores [82]. Monoliths were prepared by a sol–gel process in which zirconium oxychloride reacted with phosphoric acid in the presence of suitable amounts of poly(ethylene glycol) (PEO) and polyacrylamide (PAAm) in order to induce phase separation during the sol–gel process. A macroporous co-continuous structure was obtained only when both polymers were added together into the starting solution. Mesopores, with an average mesopore size of 5 nm, were found in the supercritically dried ZrP monoliths. A syringe device with the tight-fit ZrP monolith was designed to efficiently treat contaminated water under a continuous flow condition. In each run, a metal salt solution was introduced through the syringe device to investigate the efficiency of ion adsorption, even at high ion concentrations. Altogether, eight kinds of metals were used, including Ag^+, Cs^+, Sr^{2+}, Cu^{2+}, Zn^{2+}, Pb^{2+}, Cd^{2+} and Fe^{3+}. The ZrP monolith showed the highest selectivity for Cu^{2+}, Pb^{2+} and Fe^{3+} and the lowest for Ag^+.

An egyptian zircon mineral was used by Ali for the synthesis of ZrP and its use for uptaking uranium was studied [83]. Zirconium chloride solution was obtained by Rosetta zircon and used for the synthesis of ZrP by reaction with sodium dihydrogen phosphate. The desired particle size (30–60 mesh) was selected by grinding and sieving. An amorphous ZrP material, with P:Zr molar ratio = 2:1, was obtained and used for U(VI) adsorption tests. It was found that U(VI) adsorption increased with increasing pH and reached a maximum (98.5%) at pH 5, and then declined sharply as pH was further increased. Moreover, 30 min was chosen as the optimum time where adsorption equilibrium reached about 98%. Kinetic studies showed that the adsorption process of U(VI) on ZrP can be expressed by a pseudo-second-order kinetic model. Thermodynamic studies revealed that the optimum temperature for U(VI) uptake was 298 K, and the positive value of ΔS reflected the good affinity of uranium ions towards the sorbent and the increasing randomness at the solid–solution interface during the adsorption process. U(VI) uptake was also studied in the presence of other metal ions (Cu, Ni, Fe, Th, and Pb). Moreover, it was found that the uranium uptake efficiency slightly decreased in the presence of competing ions, but it remained preferred with respect to the other cations. Adsorption isotherm studies revealed that the Langmuir isotherm described the system more adequately than the Freundlich isotherm and, according to that model, the adsorption occurred uniformly on the active sites of the sorbent, and once a sorbate occupied a site, no further sorption could take place at that site. A case study was carried out by using a Gattar leach liquor solution contacted with ZrP for 30 min at room temperature and the pH adjusted at 2–3. After pH adjustment and equilibration, the recovered solution was analyzed for the uranium concentration and it was found that the adsorption efficiency of ZrP was about 70%. It is also noteworthy that ZrP prepared from zircon mineral was more efficient than ZrP prepared from H_3PO_4 and $ZrOCl_2{\cdot}8H_2O$ (98% vs. 76.5%).

Pandith et al. prepared agglomerated spherical alpha zirconium phosphate nanoparticles by a facile and rapid microwave hydrothermal approach in the absence of any complexing or structure directing agent and tested their efficiency in the removal of radioactive $^{137}Cs^{+}$ and $^{90}Sr^{2+}$ ions from aqueous systems [84]. FTIR, X-Ray diffraction and BET analysis confirmed that the formation of α-ZrP occurred, with an average crystallite size of 3–4 nm, a mesoporous structure and a specific surface area of about 500 $m^2\ g^{-1}$. The removal of radioactive ions was tested by non-competitive batch measurements in acidic aqueous media (pH in the range 3.0–6.0, generally found in nuclear wastewater). It was found that about 98.3% of $^{90}Sr^{2+}$ was removed from the aqueous solution within 160 min of contact time. Although the adsorption of Cs^{+} reached equilibrium within 40 min, the removal % was only 76.5%. The removal of metal cations was observed to depend on the hydrated radii and on the charge on the exchanging ions.

Chuah et al. prepared a two-dimensional disodium zirconium phosphate, $Zr(NaPO_4)_2{\cdot}H_2O$ (hereafter indicated as α-Na_2ZrP), and investigated it as an ion exchanger for heavy metals [85]. Specifically, the materials were synthesized via a modified mechanochemistry-based method, involving only grinding and heating. Typically, $ZrOCl_2{\cdot}8H_2O$, Na_2HPO_4 and NaF (molar ratios 1:(2–4):0.3) were ground together in an agate mortar until a well-mixed paste was formed. Then, the mixture was heated to 120 °C for 24 h. For Na_2HPO_4:Zr > 2, the layered phase α-$Zr(NaPO_4)_2{\cdot}H_2O$ with an interlayer distance of 8.59 Å was formed. For a Na_2HPO_4: Zr = 4, nanoplatelets with regular size and shape were obtained. In addition, nanorods were also observed, probably formed due to the presence of the fluoride ion. The performance of α-Na_2ZrP as an ion exchanger was evaluated for several heavy metals (Pb, Cu, Zn, Co, Ni, Tl, Cd). It was found that the removal of Pb^{2+} was the highest, 99.9%, followed by Cu^{2+} (64.9%). In the absence of Pb^{2+}, the removal efficiency for Cu^{2+} increased sharply to 98%. When Pb^{2+}, Cu^{2+}, Cd^{2+}, and Tl^{+} were present together, the highest removal efficiency was still obtained for Pb^{2+}. It is noteworthy that α-Na_2ZrP was a much better ion-exchange material than the hydrogen form α-ZrP. Despite the high H^{+} concentration, the removal efficiency of Pb^{2+} and Cu^{2+} was ~92% at pH 2 and increased to almost 100% at pH 3–5. The uptake of Tl^+ was more susceptible to low pH, so that ~98% removal efficiency was obtained at pH 3–5. Low uptakes of Tl^+, Pb^{2+}, and Cu^{2+} were observed only at pH 1. At pH 3, it is reasonable to suppose that part of the sodium ions in the solid was exchanged by H^+, forming the monosodium-exchanged α-$Zr(NaPO_4)(HPO_4)\cdot 5H_2O$. However, despite these phase changes with pH, the ion-exchange efficiency was not compromised due to the large interlayer spacing in α-$Zr(NaPO_4)(HPO_4)\cdot 5H_2O$ (d = 11.8 Å) or in the more hydrated θ-$Zr(HPO_4)_2\cdot 6H_2O$ (d = 10.4 Å). Competitive studies showed that, in the presence of various interferent ions, the removal efficiency for Pb^{2+}, Cu^{2+}, and Cd^{2+} was >90% in the presence of Na^+, K^+, and Mg^{2+}, while the presence of an excess of Ca^{2+} resulted in reduced removal efficiencies for Pb^{2+}, Cu^{2+}, and Cd^{2+} by 10–30%.

Again, Chuah et al. synthesized α-$Zr(NH_4PO_4)_2\cdot H_2O$ by a single step minimalistic process, in which the reactants, $ZrOCl_2\cdot 8H_2O$ and $(NH_4)_2HPO_4$, were used in almost stoichiometric amounts in the solventless protocol where only the water of crystallization was present [86]. They found that by adding a small amount of NaF as a mineralizer, α-$Zr(NH_4PO_4)_2\cdot H_2O$ with good crystallinity and an interlayer distance of 9.44 Å was obtained. Scanning electron microscopy (SEM) images showed how the morphology of the α-$Zr(NH_4PO_4)_2\cdot H_2O$ particles depended on the amount of HF used for their preparation (Figure 3).

Figure 3. Scanning electron microscopy images of α-Zr-$(NH_4PO_4)_2\cdot H_2O$ prepared in the absence of HF (**a**), by using increasing amounts of HF (**b**,**c**) ([86], with permission of ACS).

HF also affected the surface area of the samples. Specifically, it was found that the samples prepared by using HF had a surface area significantly lower than that of the compound synthesized without HF.

The sample prepared with the minimum amount of HF was used for sorption studies. The uptake of Pb^{2+} and Cu^{2+} was studied under acidic conditions. At pH 1, no sorption of Pb^{2+} and Cu^{2+} was observed, mainly for two reasons: first, at high H^+ concentrations, there is the competition with heavy metal cations; secondly, at very low pH, ammonium ions are exchanged by protons forming α-Zr-$(NH_4PO_4)_2\cdot H_2O$, which has an interlayer distance lower than that of the ammonium exchanged form (7.6 Å vs. 9.4 Å). At pH 2 and above, ≥98.9% Pb^{2+} was removed, while for Cu^{2+}, the removal efficiency was close to 99.9% at pH ≥ 3. Sorption studies revealed a very strong affinity between the sorbent and the sorbate. Moreover, more than 60% of the target ions were removed from aqueous solutions in the first 10 min, and after 1 h the concentration became lower than 0.006 mmol/L. The high selectivity toward Pb^{2+} and Cu^{2+} was retained in the presence of other ions such as Na^+, K^+, Mg^{2+}, Ba^{2+} and Ca^{2+} despite their much higher concentrations. Furthermore, the large difference in the uptake of Pb^{2+} and Cu^{2+} over that for Zn^{2+}, Co^{2+} or Ni^{2+} also provides a method to separate these heavy metals.

Table 2 resumes the adsorption properties of some of the ZrP-based sorbent materials reported in the present review. The maximum ion uptake was referred to Pb^{2+}, the most studied heavy metal, since it is the most common metal that the human body can absorb in toxic amounts. For batch experiments, the maximum Pb^{2+} uptake, expressed in meq per gram of sorbent, was reported, while the adsorption percentage was used for continuous adsorption mode experiments.

Table 2. Maximum Pb^{2+} uptake, or adsorption percentage values, for ZrP-based sorbent materials.

Sorbent	Maximum Pb^{2+} Uptake (meq/g) or Adsorption Percentage (%)	Adsorption Mode	Ref.
Amorphous ZrP	≈3 at pH 5.5	Batch	[42]
ZrP monoliths	100%	Continuous	[82]
$Zr(NaPO_4)_2{\cdot}H_2O$	≈5 at pH 5	Batch	[85]
$Zr(NH_4PO_4)_2{\cdot}H_2O$	≈3.8 at pH ≈4.3–5	Batch	[86]
Amorphous ZrP/D001	100%	Continuous	[59]
α, γ-ZrP/mesoporous polystyrene	≈1.6 at pH = 5	Batch	[60]
Amorphous ZrP/polysulfone capsules	≈3 at pH 5.75	Batch	[62]
α-ZrP/polyaniline	≈1	Electrochemical quartz crystal microbalance	[68]
Amorphous ZrP/ (polyvinyl alcohol-polyvinylidene fluoride)	≈1.2 at pH 5.5	Batch	[70]
ZrP nanoparticles/cellulose	60%	Continuous	[71]
Amorphous ZrP/graphene oxide	≈3.5 at pH = 6	Batch	[74]
Amorphous ZrP/melamine	≈9.7 at pH 5.5	Batch	[81]

Table 2 shows that ZrP/melamine exhibited the highest metal uptake. This can be attributed to the complexing properties of N-rich molecules, similarly to crown ethers, as reported by Peng et al. [78]. Indeed, the metal uptake properties of ZrP intercalation compounds with nitrogen-containing molecules were already reported in 1985 by Ferragina et al., who studied the intercalation of 2,2′-bipyridyl into α-zirconium phosphate, and its coordination by Co^{2+}, Ni^{2+}, and Cu^{2+} [87].

Additionally, Takei et al. showed that γ-type zirconium phosphate intercalated with p-aminoazobenzene had an exceptional sorption ability towards rare earth elements, resulting about four times higher than that of unintercalated γ-ZrP [88]. They also proved that lanthanide uptake did not provoke the release of the azo-molecules.

4. Removal of Dyes

Organic dyes represent another important source of environmental contamination mainly coming from textile industries. Indeed, it was estimated that 1% to 15% of the dye is lost during dyeing processes and is released into wastewaters. Due to the large degree of aromaticity present in these molecules, dyes are difficult to degrade by traditional biological methods [89]. In the treatment of water contaminated by dyes, the adsorbent has a dual role: it should capture dyes from water and, at the same time, it should degrade them into environmentally compatible products, therefore also acting as a catalyst.

Organic dyes with positive charges or with protonable groups can be intercalated by zirconium phosphate through an ion-exchange process or an acid–base reaction involving POH groups [31].

In dye removal and degradation, zirconium phosphate is often employed as support for species with photocatalytic properties, with the aim to improve the surface area and, in turn, the number of available catalytic active sites.

Titania, silver nanoparticles, silver halides, graphene oxide, and carbon nitrides are some of the photocatalytic materials which were combined with zirconium phosphate [90]. The properties of the composites depend on the kind of photocatalyst and on its interaction with zirconium phosphate.

In 2007, Parida et al. tested the same material described in ref. [55], which was titania pillared zirconium phosphate, for the photocatalytic decolorisation of methylene blue (MB) [91]. The photo-oxidation of methylene blue was performed in batch by exposition to sunlight at room temperature. It was found that, with respect to pure ZrP, the presence of titania decreased the crystallite size which, in turn, increased the surface area and adsorption of methylene blue. This ultimately helped in increasing the initial rate of photodegradation/decolorisation of the dye.

Moreover, the photodecolorisation of methylene blue followed first-order kinetics, and at higher MB concentration the percentage of photodegradation decreased. On the other hand, with increasing the irradiation time, there was an increase in the percentage of photodecolorisation up to 240 min. It was also observed that the initial rate of photodegradation/decolorisation increased with increasing the H_2O_2 concentration, probably due to the formation of more $OH^{\bullet}$ radicals.

The effect of inorganic salts such as potassium persulphate and sodium chloride was also investigated. The effect of persulphate ion (electron scavenger) on the photocatalytic degradation of MB resulted in an increase in the initial rate of photodegradation of methylene blue with increasing the persulphate amount, attaining a 100% degradation in 4 h. Differently, with increasing the concentration of sodium chloride, the initial rate of photodegradation of MB decreased due to the hole scavenging properties of Cl^- ion.

The efficiency of titania pillared zirconium phosphate was comparable with that of other catalysts reported in the literature.

It is well known that noble metallic nanoparticles, such as gold and silver, exhibit unique spectral properties due to their surface plasmon resonance properties, which made them promising materials for several applications, such as photocatalysis, due to their ability to absorb visible light [92,93]. Moreover, the combination of silver nanoparticles with silver halides produces heterojunctions, in which the two components act so as to polarize the photo-induced charges, facilitating electron-hole separation, and in turn, the photocatalytic efficiency of the material [94].

Saffaj and coworkers studied the effect of silver loaded zirconium phosphate on the photodegradation of methylene blue [95]. Crystalline α-Zirconium phosphate was prepared by refluxing amorphous zirconium phosphate in a solution of phosphoric acid. For silver activation, amorphous α-ZrP was added to a solution of silver nitrate. Then, phosphoric acid was added and the mixture was refluxed for 2 days to yield a precipitate denoted Ag–ZrP. Both α-ZrP and Ag–ZrP exhibited an interlayer distance of 7.6 Å, but α -ZrP had a higher degree of crystallinity than Ag–ZrP. However, regarding the oxidation state of silver, the authors did not clarify if it is present as ion or metallic silver.

Methylene blue was selected as a model for the photocatalytic degradation experiments because it is a non-volatile and common contaminant in industrial wastewaters. A low-pressure mercury lamp UV was used as an artificial sunlight source. It was found that the photocatalytic efficiency of the dye degradation over Ag–ZrP and α-ZrP was highly pH dependent. Specifically, the photocatalytic efficiency increased with increasing pH values. In strong acidic conditions, the decolorization of the dye was negligible, while it dramatically increased at pH 8. Both α-ZrP and Ag–ZrP surfaces were positively charged below pH 4. Methylene blue is a cationic dye in aqueous solution and it can keep its cationic configuration in the pH range 3–11. At basic pH, electrostatic interactions between the negative catalyst surface and cationic dye led to strong adsorption of the latter on the inorganic support (despite the higher risk of hydrolysis of ZrP, occurring in basic conditions, note of the author).

Moreover, Ag–ZrP exhibited a higher degree of photocatalytic efficiency compared to α-ZrP. Specifically, a decrease in the concentration of MB in the solution around 90% was achieved with Ag–ZrP within 70 min, while it was less than 60% with α-ZrP in the same time range. Enhanced degradation with Ag–ZrP under UV irradiation was ascribed to the effects of Ag ions acting as electron traps leading to better electron excitation.

Composites based on zirconium phosphate and silver/silver halide heterojunctions demonstrated efficiency in the photodegradation of Rhodamine B (RhB) [96,97]. They were quickly and easily obtained by using nanosized silver(I) exchanged α-zirconium phosphate as a precipitating agent for silver halides. Specifically, a nanosized zirconium phosphate, prepared as reported by Pica et al. [55], first was reacted with different amounts of silver(I) acetate, then treated with HCl or HBr solutions in order to promote the precpitation of AgCl or AgBr. All composites contained AgX particles with size ≤ 1 μm, and

UV-Vis diffuse reflectance spectroscopy proved that ZP/AgX was effectively converted into ZP/x(Ag@AgX) after a few minutes of irradiation with a halogen lamp.

The photodegradation of RhB was studied in batch and the photocatalytic properties of the composites were compared with those of a pure AgX sample with particle size in the range of 0.5–2 μm. It was observed that, during irradiation, the absorbance decreased and the wavelength of the absorption maximum (λ_m) slowly shifted toward lower λ values. Watanabe et al., in 1977, proved that the ipsochromic shift of λ_m was due to de-ethylation occurring in a stepwise manner, and to cleavage processes [98]. It was found that, both for AgX and ZP/AgX or ZrP/AgCl, the degradation process started with RhB de-ethylation, followed by the cleavage of the de-ethylated species. However, the composites with 53 and 58 wt% of AgBr and AgCl, respectively, turned out to have better catalytic properties than the corresponding pure AgX samples. Specifically, the percentage of chromophore cleavage in the presence of ZP/AgBr and ZP/AgCl was around 90% after 6 min and 10 min, respectively, while in the presence of pure AgX, a 90% of percentage cleavage was achieved after 30 min. Moreover, reusage tests of the composite photocatalysts proved that they are stable for at least three cycles.

Ye et al. fabricated α-zirconium phosphate-pillared reduced graphene oxide (rGO–ZrP) composites for the removal of MB from aqueous solutions [99]. GO was synthesized from natural graphite powder by a modified Hummers method, while ZrP was synthesized by refluxing amorphous ZrP for 7 days in the presence of phosphoric acid. The GO–ZrP composite was prepared from a ZrP/methylamine intercalation compound, to which GO was slowly added. Then, hydrazine was added in order to reduce GO. BET specific surface areas of rGO–ZrP and rGO were 573 and 312 $m^2\ g^{-1}$, respectively. It is interesting that, with increasing the storage time in water, the BET specific surface area of rGO decreased to 44 $m^2\ g^{-1}$, while that of rGO–ZrP only decreased to 523 $m^2\ g^{-1}$, suggesting that the presence of ZrP hampered the rGO reaggregation to some extent.

The adsorption tests of MB onto the rGO–ZrP were conducted in a batch experiment at 30 °C. The maximum adsorption capacitiy of MB onto rGO–ZrP was about 90% higher than that onto rGO. With increasing the storage days in water, the maximum adsorption capacity of MB onto the rGO dramatically decreased, while that onto rGO–ZrP slightly decreased. Moreover, the maximum adsorption capacity of rGO–ZrP almost remained constant during the first six cycles of the adsorption–desorption process.

Chen et al. recently prepared α-zirconium phosphate/carbon nitride composites (ZrP/C_3N_4), with different ZrP/C_3N_4 mass ratios, and investigated their photocatalytic activity toward Rhodamine B under UV irradiation [100]. Zirconium phosphate was prepared by the hydrothermal reaction of zirconyl chloride and phosphoric acid, while carbon nitride was prepared by the calcination of urea. ZrP/C_3N_4 composites were prepared by mixing and sonicating different amounts of the two components in water. It was found that the particle size of C_3N_4 was in the range 200 nm–a few micrometers and higher amounts of C_3N_4 restricted the re-stacking process of α-ZrP nanoparticles. Photoluminescence spectra showed that when α-ZrP nanoparticles were composited with C_3N_4 nanosheets, a heterostructure was formed, and the high recombination rate of photogenerated carriers, which was the intrinsic property of C_3N_4, was efficiently reduced. Photocatalytic activity was tested under irradiation with UV-Vis light.

The photodegradation efficiency of RhB by ZrP under visible light was negligible since it cannot be excited by visible light irradiation. After incorporation with C_3N_4, the efficiency was improved slightly. In the condition of ultraviolet light irradiation, the photodegradation efficiency of ZrP/C_3N_4 composite material ($ZrP:C_3N_4$ mass ratio = 1:1) was much higher than that of α-ZrP and C_3N_4. After irradiation for 10 min, the photodegradation efficiency of α-ZrP and C_3N_4 was 64 and 67%, respectively, while the photodegradation efficiency of ZrP/C_3N_4 ($ZrP:C_3N_4$ mass ratio = 1:1) reached 98% in the same time range. The degradation efficiency of all composite systems reached above 90% after 18 min. Among those, the composite with ZrP: C_3N_4 mass ratio 1:2 had the highest degradation efficiency (≈100%).

The authors inferred that under UV irradiation, the photoinduced holes on the valence band of α-ZrP were transferred to the valence band of C_3N_4 through the heterostructure. The photogenerated electrons from the conduction band of C_3N_4 were transferred to the conduction band of α-ZrP, which achieved effective separation of photogenerated electron-hole pairs and reduced the electron-hole recombination. Moreover, an appropriate amount of C_3N_4 would benefit the separation of photogenerated electron-hole pairs, while an excess of C_3N_4 would block, to some extent, the ultraviolet light arriving at the surface of ZrP nanoparticles and then cause a decrease in the photodegradation rate.

Naushad et al. proposed a gelatin-Zr(IV) phosphate nanocomposite to remove toxic dyes from water [101]. Gelatin is a naturally occurring macromolecular and biodegradable protein produced by partial hydrolysis of collagen synthesized from the skins, white connective tissues and bones of animals.

Gelatin-Zr(IV) phosphate nanocomposite (GT/ZPNC) was synthesized using the sol–gel method. Specifically, $ZrOCl_2 \cdot 8H_2O$ and H_3PO_4 were mixed with continuous stirring at room temperature and the pH was kept between 0 and 1. Gelatin was then added in hot water to prepare gelatin gels, which were added to precipitates of Zr(IV) phosphate under continuous stirring (Figure 4).

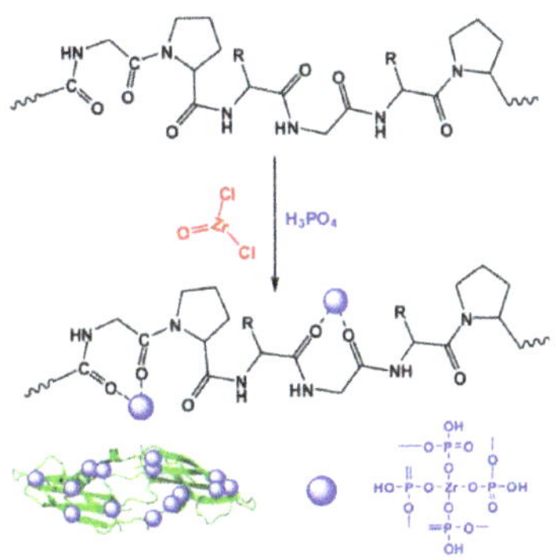

Figure 4. Proposed mechanism for the synthesis of gelatin-Zr(IV) phosphate nanocomposite (GT/ZPNC) ([101], with permission of Elsevier).

For the sake of clarity, the structure of $ZrOCl_2$ reported in Figure 4 is quite unusual and unprobable, since, as reported by Clearfield et al., zirconyl chloride does not contain ZrO^{2+} species but a tetrameric species $[Zr(OH)_2 \cdot 4H_2O]_4{}^{8+}$, which also exists in acidic solution of $ZrOCl_2 \cdot 8H_2O$. Chlorine is not bonded to Zr but it is present as a Cl^- ion [102].

The photocatalytic behaviour of GT/ZPNC particles was investigated against MB and FG (Fast green) dyes under solar light using batch process. The removal of FG dye in visible light reached about 90% within five hours, while in the dark it reached only 44%. Similar results were obtained for MB: about 38% of dye was removed in the dark and the removal increased up to 88% within five hours when exposed to sunlight. However, the authors did not carry out experiments in order to clarify if the higher dye removal in the presence of GT/ZPNC was due to a simple adsorption or to a photodegradation process.

5. Conclusions

Pollutant removal from water through adsorption is considered one of the most effective and competitive processes. The optimal adsorbents should be cheap, chemically stable, have a high surface area in order to make available a large number of active sites, and be easily regenerated and recoverable.

Among unconventional sorbents, layered zirconium phosphate has been investigated, both in batch and continuous conditions, as a potential material for the removal of heavy metals and dye due to its good chemical and thermal stability and, above all, its excellent cation exchange and intercalation properties.

Both amorphous and crystalline ZrP materials have been investigated as sorbent materials, often in the form of composites, in combination with organic polymers or other inorganic compounds, in order to have feasible hybrid sorbents for continuous and large-scale applications. Neutral and charged polymers, as well as inorganic compounds such as titania, graphene oxide, crown ethers, and N-rich molecules, have been used for the preparation of ZrP composites, which proved to be suitable and promising for a continuous adsorption mode. Among ZrP composites, those with N-rich molecules exhibited the highest metal uptake due to the chelating properties of these molecules toward metals.

Heavy metal uptake, occurring through ion exchange processes, has been studied as a function of different experimental conditions (pH, concentration of the sorbent and the sorbate, temperature, etc.), proving that the uptake is more efficient in moderate acidic conditions. The use of ZrP composites, which enable the achievement of higher surface areas, is beneficial for their adsorbent properties. Regardless of the characteristics of the sorbent, the highest affinity is observed toward lead and copper. Moreover, the affinity of zirconium phosphate-based materials for heavy metals is generally significantly higher than for alkaline and alkaline earth metals.

The uptake of organic dyes by ZrP can occur through both acid–base interactions and ion exchange processes. It is desirable that dyes should not be simply adsorbed on the solid material, but also degraded to nontoxic products, and photocatalytic degradation is often used. ZrP combined with photosensitive materials, among them titania, graphene oxide, carbon nitrides, silver nanoparticles and silver/silver halide heterojunctions, exhibits enhanced photocatalytic efficiency due to a more efficient light absorption and electron-hole separation.

The major part of the reported studies refer to lab scale experiments, proving that the use of zirconium phosphate as a sorbent material is still in the phase of fundametal research, and further experiments are undoubtedly necessary to prove the effective possibility to apply these materials on a large scale, where other aspects, such as economic advantages or disadvantages, should be evaluated.

Further developments of this topic are currently under investigation in my research group. New materials, based on layered zirconium phosphonates, bearing carboxylic and amino groups, covalently bonded to the inorganic layer, are promising potential sorbents for heavy metals and dyes. The idea was born from the synthesis of a novel mixed zirconium phosphate/phosphonate based on glyphosine, specifically a zirconium phosphate glycine-N, Nbismethylphosphonate, in which the presence of the carboxyl and amino groups, due to their coordinative capabilities, enhances the affinity of this material towards transition metals [103–105] and, at the same time, the presence of acidic groups could promote the adsorption of basic dyes.

Funding: This research received no external funding.

Acknowledgments: The author would like to acknowledge Professor Mario Casciola for scientific discussion and suggestions.

Conflicts of Interest: The author declares no conflict of interest.

References

1. Lawrence, E.; Jackson, J.M.; Jackson, A.R. *Longman Dictionary of Environmental Science*; Pearson Publications: Essex, UK, 1998.
2. Premkumar, M.P.; Thiruvengadaravi, K.V.; Kumar, P.S.; Nandagopal, J.; Sivanesan, S. Eco-Friendly Treatment Strategies for Wastewater Containing Dyes and Heavy Metals. In *Environmental Contaminants Measurement, Modelling and Control*; Gupta, T., Agarwal, A.K., Agarwal, R.A., Labhsetwar N., K., Eds.; Springer Nature: Singapore, Singapore, 2018; pp. 317–360.
3. Mirsal, I.A. Major Types of Soil Pollutants. In *Soil Pollution*; Springer: Berlin/Heidelberg, Germany, 2008; pp. 117–136.
4. Kinuthia, G.K.; Ngure, V.; Beti, D.; Lugalia, R.; Wangila, A.; Kamau, L. Levels of heavy metals in wastewater and soil samples from open drainage channels in Nairobi, Kenya: Community health implication. *Sci. Rep.* **2020**, *10*, 8434–8437. [CrossRef] [PubMed]
5. Mansourri, G.; Madani, M. Examination of the level of heavy metals in wastewater of Bandar Abbas Wastewater Treatment Plant. *Open J. Ecol.* **2016**, *6*, 55–61. [CrossRef]
6. Dotto, G.L.; McKay, G. Current scenario and challenges in adsorption for water treatment. *J. Environ. Chem. Eng.* **2020**, *8*, 103988. [CrossRef]

7. Ali, I.; Asim, M.; Khan, T.A. Low-Cost Adsorbents for the Removal of Organic Pollutants from Wastewater. *J. Environ. Manage* **2012**, *113*, 170–183. [CrossRef]
8. Crini, G. Non-conventional low-cost adsorbents for dye removal: A review. *Bioresour. Technol.* **2006**, *97*, 1061–1085. [CrossRef]
9. Crini, G.; Lichtfouse, E.; Wilson, L.; Morin-Crini, N. Conventional and non-conventional adsorbents for wastewater treatment. *Environ. Chem. Lett.* **2019**, *17*, 195–213. [CrossRef]
10. Al-Ghouti, M.A.; Da'ana, D.A. Guidelines for the use and interpretation of adsorption isotherm models: A review. *J. Hazard. Mater.* **2020**, *393*, 122383–122405. [CrossRef]
11. Nimibofa, A.; Ebelegi, A.N.; Wankasi, D. Modelling and interpretation of adsorption isotherms. *J. Chem.* **2017**, *2017*, 1–11.
12. Faust, S.D.; Osman, M.A. *Adsorption Processes for Water Treatment*; Butterworth Publishers: Waltham, MA, USA, 2013.
13. Bonilla-Petriciolet, A.; Mendoza-Castillo, D.I.; Reynel-Ávila, H.E. (Eds.) *Adsorption Processes for Water Treatment and Purification*; Springer International Publishing: Berlin, Germany, 2017.
14. Gupta, S.S.; Bhattacharyya, K.G. Kinetics of adsorption of metal ions on inorganic materials: A review. *Adv. Colloid Interface Sci.* **2011**, *162*, 39–58. [CrossRef]
15. Worch, E. *Adsorption Technology in Water Treatment: Fundamentals, Processes, and Modeling*; Walter de Gruyter: Berlin, Germany, 2012.
16. Crini, G.; Badot, P.M. (Eds.) *Sorption Processes and Pollution: Conventional and Non-conventional Sorbents for Pollutant Removal from Wastewaters*; Presses Universitaires de Franche-Comté,: Besançon, France, 2010.
17. Walcarius, A.; Mercier, L. Mesoporous organosilica adsorbents: Nanoengineered materials for removal of organic and inorganic pollutants. *J. Mater. Chem.* **2010**, *20*, 4478–4511. [CrossRef]
18. Cashin, V.B.; Eldridge, D.S.; Yu, A.; Zhao, D. Surface functionalization and manipulation of mesoporous silica adsorbents for improved removal of pollutants: A review. *Environ. Sci. Water Res.Technol.* **2018**, *4*, 110–128. [CrossRef]
19. Zhu, R.; Chen, Q.; Zhou, Q.; Xi, Y.; Zhu, J.; He, H. Adsorbents based on montmorillonite for contaminant removal from water: A review. *App. Clay Sci.* **2016**, *123*, 239–258. [CrossRef]
20. Desta, M.B. Batch Sorption Experiments: Langmuir and Freundlich Isotherm Studies for the Adsorption of Textile Metal Ions onto Teff Straw (*Eragrostis tef*) Agricultural Waste. *J. Thermodyn.* **2013**, *2013*, 1–6. [CrossRef]
21. Hameed, B.H. Spent tea leaves: A new non-conventional and low-cost adsorbent for removal of basic dye from aqueous solutions. *J. Hazard. Mater.* **2009**, *161*, 753–759. [CrossRef]
22. Agouborde, L.; Navia, R. Heavy metals retention capacity of a non-conventional sorbent developed from a mixture of industrial and agricultural wastes. *J. Hazard. Mater.* **2009**, *167*, 536–544. [CrossRef]
23. Collins, O.N.; Onu Chijioke, E. Adsorption of a dye (crystal violet) on an acid modified non-conventional adsorbent. *J. Chem. Technol. Metall.* **2019**, *54*, 95–110.
24. Da Costa, T.B.; da Silva, M.G.C.; Gurgel Adeodato Vieira, M. Recovery of rare-earth metals from aqueous solutions by bio/adsorption using non-conventional materials: A review with recent studies and promising approaches in column applications. *J. Rare Earths* **2020**, *38*, 339–355. [CrossRef]
25. Zhou, Y.; Lu, J.; Zhou, Y.; Liu, Y. Recent advances for dyes removal using novel adsorbents: A review. *Environ. Pollut.* **2019**, *252*, 352–365. [CrossRef]
26. Xu, J.; Cao, Z.; Zhang, Y.; Yuan, Z.; Lou, Z.; Xu, X.; Wang, X. A review of functionalized carbon nanotubes and graphene for heavy metal adsorption from water: Preparation, application, and mechanism. *Chemosphere* **2018**, *195*, 351–364. [CrossRef]
27. Cao, Y.; Li, X. Adsorption of graphene for the removal of inorganic pollutants in water purification: A review. *Adsorption* **2014**, *20*, 713–727. [CrossRef]
28. Baig, N.; Sajid, M.; Saleh, T.A. Graphene-based adsorbents for the removal of toxic organic pollutants: A review. *J. Environ. Manag.* **2019**, *244*, 370–382. [CrossRef]
29. Khan, N.A.; Hasan, Z.; Jhung, S.H. Adsorptive removal of hazardous materials using metal organic frameworks (MOFs): A review. *J. Hazard. Mater.* **2013**, *244*, 444–456. [CrossRef]
30. Liu, X.; Ma, R.; Wang, X.; Ma, Y.; Yang, Y.; Zhuang, L.; Zhang, S.; Jehan, R.; Chen, J.; Wang, X. Graphene oxide-based materials for efficient removal of heavy metal ions from aqueous solution: A review. *Environ. Pollut.* **2019**, *252*, 62–73. [CrossRef]
31. Clearfield, A.; Costantino, U. Layered Metal Phosphates and their Intercalation Chemistry. In *Comprehensive Supramolecular Chemistry, Solid-state Supramolecular Chemistry: Two-and Three-Dimensional Inorganic Networks*; Alberti, G., Bein, T., Eds.; Pergamon: Oxford, UK, 1996; Volume 7.
32. Bashir, A.; Ahad, S.; Malik, L.A.; Qureashi, A.; Manzoor, T.; Dar, G.N.; Pandith, A.H. Revisiting the Old and Golden Inorganic Material, Zirconium Phosphate: Synthesis, Intercalation, Surface Functionalization, and Metal Ion Uptake. *Ind. Eng. Chem. Res.* **2020**, *59*, 22353–22397. [CrossRef]
33. Kraus, K.A.; Phillips, H.O. Adsorption on inorganic materials. I. Cation exchange properties of zirconium phosphate[1]. *J. Am. Chem. Soc.* **1956**, *78*, 694. [CrossRef]
34. Amphlett, C.B.; McDonald, L.A.; Redman, M.J. Synthetic inorganic ionexchange materials—I zirconium phosphate. *J. Inorg. Nucl. Chem.* **1958**, *6*, 220–235. [CrossRef]
35. Amphlett, C.B.; McDonald, L.A.; Burgess, J.S.; Maynard, J.C. Synthetic inorganic ion-exchange materials—III. The separation of rubidium and caesium on zirconium phosphate. *J. Inorg. Nucl. Chem.* **1959**, *10*, 69–73. [CrossRef]
36. Vivani, R.; Alberti, G.; Costantino, F.; Nocchetti, M. New advances in zirconium phosphate and phosphonate chemistry: Structural archetypes. *Micropor. Mesopor. Mat.* **2008**, *107*, 58–70. [CrossRef]

37. Troup, J.M.; Clearfield, A. Mechanism of ion exchange in zirconium phosphates. 20. Refinement of the crystal structure of alpha-zirconium phosphate. *Inorg. Chem.* **1977**, *16*, 3311–3314. [CrossRef]
38. Albertsson, J.; Oskarsson, A.; Tellgren, R.; Thomas, J.O. Inorganic ion exchangers. 10. A neutron powder diffraction study of the hydrogen bond geometry in alpha-zirconium bis (monohydrogen orthophosphate) monohydrate. A model for the ion exchange. *J. Phys. Chem.* **1977**, *81*, 1574–1578. [CrossRef]
39. Pica, M.; Donnadio, A.; Casciola, M. From microcrystalline to nanosized α-zirconium phosphate: Synthetic approaches and applications of an old material with a bright future. *Coord. Chem. Rev.* **2018**, *374*, 218–235. [CrossRef]
40. Pan, B.; Zhang, Q.; Du, W.; Zhang, W.; Pan, B.; Zhang, Q.; Xu, Z.; Zhang, Q. Selective heavy metals removal from waters by amorphous zirconium phosphate: Behavior and mechanism. *Water Res.* **2007**, *41*, 3103–3111. [CrossRef]
41. Xiao, H.; Liu, S. Zirconium phosphate (ZrP)-based functional materials: Synthesis, properties and applications. *Mater. Des.* **2018**, *155*, 19–35. [CrossRef]
42. Khalameida, S.; Sydorchuk, V.; Skubiszewska-Zięba, J.; Charmas, B.; Skwarek, E.; Janusz, W. Hydrothermal, microwave and mechanochemical modification of amorphous zirconium phosphate structure. *J. Therm. Anal. Calorim.* **2017**, *128*, 795–806. [CrossRef]
43. Baran, V.; Caletka, R.; Tympl, M.; Urbanek, V. Application of the sol-gel method for the preparation of some inorganic ion-exchangers in spherical form. *J. Radioanal. Nucl. Chem.* **1975**, *24*, 353–359. [CrossRef]
44. Bogdanov, S.; Valiev, E.; Dorofeev, Y.A.; Pirogov, A.; Sharygin, L.; Moisseev, V.; Galkin, V. Structure of zirconium phosphate gels produced by the sol-gel method. *J. Phys. Condens. Matter* **1997**, *9*, 4031–4039. [CrossRef]
45. Benhamza, H.; Barboux, P.; Bouhaouss, A.; Josien, F.-A.; Livage, J. Sol–gel synthesis of $Zr(HPO_4)_2H_2O$. *J. Mater. Chem.* **1991**, *1*, 681–684. [CrossRef]
46. Jiménez-Jiménez, J.; Maireles-Torres, P.; Olivera-Pastor, P.; Rodríguez-Castellón, E.; Jiménez-López, A.; Jones, D.J.; Rozière, J. Surfactant-Assisted Synthesis of a Mesoporous Form of Zirconium Phosphate with Acidic Properties. *Adv. Mat.* **1998**, *10*, 812–815. [CrossRef]
47. Ferragina, C.; Cafarelli, P.; De Stefanis, A.; Di Rocco, R.; Petrilli, L. Synthesis and characterization of sol-gel zirconium phosphate with template surfactants by different methods. *J. Therm. Anal. Calorim.* **2003**, *71*, 1023–1034. [CrossRef]
48. Tarafdar, A.; Panda, A.B.; Pradhan, N.C.; Pramanik, P. Synthesis of spherical mesostructured zirconium phosphate with acidic properties. *Micropor. Mesopor. Mat.* **2006**, *95*, 360–365. [CrossRef]
49. Bellezza, F.; Cipiciani, A.; Costantino, U.; Marmottini, F. Adsorption of myoglobin onto porous zirconium phosphate and zirconium benzenephosphonate obtained with template synthesis. *Langmuir* **2006**, *22*, 5064–5069. [CrossRef] [PubMed]
50. Tian, X.; He, W.; Cui, J.; Zhang, X.; Zhou, W.; Yan, S.; Sun, X.; Han, X.; Han, S.; Yue, Y. Mesoporous zirconium phosphate from yeast biotemplate. *J. Colloid Interface Sci.* **2010**, *343*, 344–349. [CrossRef] [PubMed]
51. Sun, L.; Boo, W.J.; Sue, H.J.; Clearfield, A. Preparation of α-zirconium phosphate nanoplatelets with wide variations in aspect ratios. *New J. Chem.* **2007**, *31*, 39–43. [CrossRef]
52. Alberti, G.; Torracca, E. Crystalline insoluble salts of polybasic metals-II. Synthesis of crystalline zirconium or titanium phosphate by direct precipitation. *J. Inorg. Nucl. Chem.* **1968**, *30*, 317–318. [CrossRef]
53. Cheng, Y.; Wang, X.; Jaenicke, S.; Chuah, G.-K. Minimalistic liquid-assisted route to highly crystalline α-zirconium phosphate. *ChemSusChem* **2017**, *10*, 3235–3242. [CrossRef]
54. Capitani, D.; Casciola, M.; Donnadio, A.; Vivani, R. High yield precipitation of crystalline a-zirconium phosphate from oxalic acid solutions. *Inorg. Chem.* **2010**, *49*, 9409–9415. [CrossRef]
55. Pica, M.; Donnadio, A.; Capitani, D.; Vivani, R.; Troni, E.; Casciola, M. Advances in the chemistry of nanosized zirconium phosphates: A new mild and quick route to the synthesis of nanocrystals. *Inorg. Chem.* **2011**, *50*, 11623–11630. [CrossRef]
56. Alexandratos, S.D.; Zhu, X.P. Bifunctional coordinating polymers: Auxiliary groups as a means of tuning the ionic affinity of immobilized phosphate ligands. *Macromol.* **2005**, *38*, 5981–5986. [CrossRef]
57. Misono, M.; Ochiai, E.; Saito, Y.; Yoneda, Y. A new dual parameter scale for the strength of Lewis acids and bases with the evaluation of their softness. *J. Inorg. Nucl. Chem.* **1967**, *29*, 2685–2691. [CrossRef]
58. Pearson, R.G. Hard and soft acids and bases. *J. Am. Chem. Soc.* **1963**, *85*, 3534–3539. [CrossRef]
59. Zhang, Q.R.; Du, W.; Pan, B.C.; Pan, B.J.; Zhang, W.M.; Zhang, Q.J.; Xu, Z.W.; Zhang, Q.X. A comparative study on Pb^{2+}, Zn^{2+} and Cd^{2+} sorption onto zirconium phosphate supported by a cation exchanger. *J. Hazard. Mater.* **2008**, *152*, 469–475. [CrossRef]
60. Zhang, X.; Shen, J.; Pan, S.; Qian, J.; Pan, B. Metastable Zirconium Phosphate under Nanoconfinement with Superior Adsorption Capability for Water Treatment. *Adv. Funct. Mater.* **2020**, 1909014. [CrossRef]
61. Alberti, G.; Bartocci, M.; Santarelli, M.; Vivani, R. Zirconium Phosphate Chloride Dimethyl Sulfoxide, a Reactive Precursor of a Large Family of Layered Compounds. *Inorg. Chem.* **1997**, *36*, 3574–3575. [CrossRef]
62. Ma, X.; Li, Y.; Li, X.; Yang, L.; Wang, X. Preparation of novel polysulfone capsules containing zirconium phosphate and their properties for Pb^{2+} removal from aqueous solution. *J. Hazard. Mater.* **2011**, *188*, 296–303. [CrossRef]
63. Naiya, T.K.; Bhattacharya, A.K.; Das, S.K. Adsorption of Cd(II) and Pb(II) from aqueous solutions on activated alumina. *J. Colloid Interface Sci.* **2009**, *333*, 14–26. [CrossRef]
64. Mahmoud, M.E.; Osman, M.M.; Hafez, O.F.; Elmelegy, E. Removal and preconcentration of lead (II), copper (II), chromium (III) and iron (III) from wastewaters by surface developed alumina adsorbents with immobilized 1-nitroso-2- naphthol. *J. Hazard. Mater.* **2010**, *173*, 349–357. [CrossRef]

65. Nagata, N.; Kubota, L.T.; Bueno, M.I.M.S.; Peralta-Zamora, P.G. Adsorption parameters of Cd(II), Pb(II), and Hg(II) on Zirconium(IV) phosphate chemically grafted onto silica gel surface. *J. Colloid Interface Sci.* **1998**, *200*, 121–125. [CrossRef]
66. Zhang, Q.; Pan, B.; Zhang, S.; Wang, J.; Zhang, W.; Lv, L. New insights into nanocomposite adsorbents for water treatment: A case study of polystyrene-supported zirconium phosphate nanoparticles for lead removal. *J. Nanopart. Res.* **2011**, *13*, 5355–5364. [CrossRef]
67. AlOthman, Z.A.; Alam, M.M.; Naushad, M. Heavy toxic metal ion exchange kinetics: Validation of ion exchange process on composite cation exchanger nylon 6,6 Zr(IV) phosphate. *J. Ind. Eng. Chem.* **2013**, *19*, 956–960. [CrossRef]
68. Zhang, Q.; Du, X.; Ma, X.; Hao, X.; Guan, G.; Wang, Z.; Xue, C.; Zhang, Z.; Zuo, Z. Facile preparation of electroactive amorphous α-ZrP/PANI hybrid film for potential-triggered adsorption of Pb^{2+} ions. *J. Hazard. Mater.* **2015**, *289*, 91–100. [CrossRef]
69. Wang, Z.; Feng, Y.; Hao, X.; Huang, W.; Feng, X. A novel potential-responsive ion exchange film system for heavy metal removal. *J. Mater. Chem. A* **2014**, *2*, 10263–10272. [CrossRef]
70. Zhao, D.; Yu, Y.; Chen, J.P. Treatment of lead contaminated water by a PVDF membrane that is modified by zirconium, phosphate and PVA. *Water Res.* **2016**, *101*, 564–573. [CrossRef]
71. Ibrahim, Y.; Abdulkarem, E.; Naddeo, V.; Banat, F.; Hasan, S.W. Synthesis of super hydrophilic cellulose-alpha zirconium phosphate ion exchange membrane via surface coating for the removal of heavy metals from wastewater. *Sci. Total Environ.* **2019**, *690*, 167–180. [CrossRef] [PubMed]
72. Das, D.P.; Parida, K.; De, B.R. Photocatalytic reduction of hexavalent chromium in aqueous solution over titania pillared zirconium phosphate and titanium phosphate under solar radiation. *J. Mol. Cat. A: Chem.* **2006**, *245*, 217–224. [CrossRef]
73. Yoneyama, H.; Haga, S.; Yamanaka, S. Photocatalytic activities of microcrystalline titania incorporated in sheet silicates of clay. *J. Phys. Chem.* **1989**, *93*, 4833–4837. [CrossRef]
74. Pourbeyram, S. Effective Removal of Heavy Metals from Aqueous Solutions by Graphene Oxide–Zirconium Phosphate (GO–Zr-P) Nanocomposite. *Ind. Eng. Chem. Res.* **2016**, *55*, 5608–5617. [CrossRef]
75. Christensen, J.J.; Eatough, D.J.; Izatt, R.M. The Synthesis and Ion Bindings of Synthetic Multidentate Macrocyclic Compounds. *Chem. Rev.* **1974**, *74*, 351–384. [CrossRef]
76. Alberti, G.; Boccali, L.; Dionigi, C.; Vivani, R.; Kalchenko V., I.; Atamas L., I. Preparation and first characterization of a layered γ-zirconium phosphate derivative containing benzo 15-crown-5 groups covalently attached to inorganic layers. *Supramol. Chem.* **1996**, *7*, 129–135. [CrossRef]
77. Yamamoto, K.; Hasegawa, Y.; Nikki, K. Intercalation of Aminomethylcrowns into α-Zirconium Phosphate. *J. Incl. Phenom. Mol. Recognit. Chem.* **1998**, *31*, 289–303. [CrossRef]
78. Mu, W.; Yu, Q.; Gu, J.; Li, X.; Yang, Y.; Wei, H.; Peng, S. Bonding of crown ethers to α-zirconium phosphate—Novel layered adsorbent for radioactive strontium separation. *Sep. Purif. Technol.* **2020**, *240*, 116658–116667. [CrossRef]
79. Wang, L.; Guo, J.; Xiang, X.; Sang, Y.; Huang, J. Melamine-supported porous organic polymers for efficient CO_2 capture and Hg^{2+} removal. *Chem. Eng. J.* **2020**, *387*, 124070–124077. [CrossRef]
80. Li, H.; Li, Y.; Li, B.; Dai, Y.; Chen, X. Melamine-induced novel MSONs heterostructured framework: Controlled-switching between MOF and SOF via a self-assembling approach for rapid uranium sequestration. *Chem. Eng. J.* **2020**, *379*, 122279–122292. [CrossRef]
81. Bakry, A.M.; Awad, F.S.; Bobb, J.A.; Ibrahim, A.A.; El-Shall, M.S. Melamine-based functionalized graphene oxide and zirconium phosphate for high performance removal of mercury and lead ions from water. *RSC Adv.* **2020**, *10*, 37883–37890. [CrossRef]
82. Zhu, Y.; Shimizu, T.; Kitajima, T.; Morisato, K.; Moitra, N.; Brun, N.; Kiyomura, T.; Kanamori, K.; Takeda, K.; Kurata, H.; et al. Synthesis of robust hierarchically porous zirconium phosphate monolith for efficient ion adsorption. *New J. Chem.* **2015**, *39*, 2444–2450. [CrossRef]
83. Ali, A.H. Potentiality of zirconium phosphate synthesized from zircon mineral for uptaking uranium. *Separ. Sci. Tech.* **2018**, *53*, 2284–2296. [CrossRef]
84. Bashir, A.; Malik, L.A.; Dar, G.N.; Pandith, A.H. Microwave-Assisted Hydrothermal Synthesis of Agglomerated Spherical Zirconium Phosphate for Removal of C^{s+} and Sr^{2+} Ions from Aqueous System. In *Applications of Ion Exchange Materials in the Environment*; Inamuddin, A.M., Asiri, A., Eds.; Springer: Cham, Switzerland, 2019.
85. Cheng, Y.; Chui, S.S.-Y.; Wang, X.D.T.; Jaenicke, S.; Chuah, G.-K. One-Pot Synthesis of Layered Disodium Zirconium Phosphate: Crystal Structure and Application in the Remediation of Heavy-Metal-Contaminated Wastewater. *Inorg. Chem.* **2019**, *58*, 13020–13029. [CrossRef]
86. Cheng, Y.; Zhang, H.; Jaenicke, J.A.; Tan, E.C.P.; Chuah, G.-K. Minimalistic Synthesis of α-Zirconium Diammonium Phosphate and Zirconia for Applications in Ion Exchange and Catalysis. *ACS Sustain. Chem. Eng.* **2019**, *7*, 895–904. [CrossRef]
87. Ferragina, C.; La Ginestra, A.; Massucci, M.A.; Patrono, P.; Tomlinson, A.A.G. Intercalation of 2,2′-Bipyridyil into -Zirconlum Phosphate and in Situ Formation of Co^{2+}, Ni^{2+}, and Cu^{2+}/2,2′-Blpyridyl Complex Pillars. *J. Phys. Chem.* **1985**, *89*, 4762–4769. [CrossRef]
88. Fujimoto, S.; Takei, T.; Yanagida, S.; Kumada, N. Hybridization of layered zirconium phosphate with azo compounds and its photoresponsivity and adsorption of rare earth elements. *J. Ceram. Soc. Japan* **2019**, *127*, 830–836. [CrossRef]
89. Galindo, C.; Jacques, P.; Dalt, A. Photooxidation of the phenylazonaphthol AO20 on TIO_2: Kinetic and mechanistic investigations. *Chemosphere* **2001**, *45*, 997–1005. [CrossRef]
90. Pica, M. Zirconium phosphate catalysts in the XXI century: State of the art from 2010 to date. *Catalysts* **2017**, *7*, 190. [CrossRef]

91. Das, D.P.; Baliarsingh, N.; Parida, K.M. Photocatalytic decolorisation of methylene blue (MB) over titania pillared zirconium phosphate (ZrP) and titanium phosphate (TiP) under solar radiation. *J. Mol. Catal. A: Chem.* **2007**, *261*, 254–261. [CrossRef]
92. Feldheim, D.L.; Foss, C.A., Jr. *Metal Nanoparticles: Synthesis, Characterization, and Applications*; Marcel Dekker: New York, NY, USA, 2002.
93. Schmid, G. *Nanoparticles: From Theory to Application*; Wiley-VCH Verlag GmbH & Co. KGaA: Weinheim, Germany, 2004.
94. Pica, M. Silver halide-based composite photocatalysts: An updated account. *Rend. Fis. Acc. Lincei* **2019**, *30*, 453–467. [CrossRef]
95. Barhon, Z.; Saffaj, N.; Albizane, A.; Azzi, M.; Mamouni, R.; El Haddad, M. Effect of Modification of Zirconium Phosphate by Silver on Photodegradation of Methylene Blue. *J. Mater. Environ. Sci.* **2012**, *3*, 879–884.
96. Pica, M.; Nocchetti, M.; Ridolfi, B.; Donnadio, A.; Costantino, F.; Gentili, P.L.; Casciola, M. Nanosized zirconium phosphate/AgCl composite materials: A new synergy for efficient photocatalytic degradation of organic dye pollutants. *J. Mater. Chem. A.* **2015**, *3*, 5525–5534. [CrossRef]
97. Pica, M.; Calzuola, S.; Donnadio, A.; Gentili, P.L.; Nocchetti, M.; Casciola, M. De-ethylation and cleavage of rhodamine B by a zirconium phosphate/silver bromide composite photocatalyst. *Catalysts.* **2019**, *9*, 3. [CrossRef]
98. Watanabe, T.; Takirawa, T.; Honda, K. Photocatalysis through Excitation of Adsorbates. Rhodamine B Adsorbed to CdS. *J. Phys. Chem.* **1977**, *81*, 1845–1851. [CrossRef]
99. Wu, Z.; Zhang, L.; Guan, Q.; Ning, P.; Ye, D. Preparation of α-zirconium phosphate-pillared reduced graphene oxide with increased adsorption towards methylene blue. *Chem. Eng. Journal* **2014**, *258*, 77–84. [CrossRef]
100. Qing, S.; Chen, H.; Han, L.J.; Ye, Z.; Shi, L.; Shu, Z.; Chen, L.; Xu, L.; Xu, Q. Photocatalytic Activity Investigation of α-Zirconium Phosphate Nanoparticles Compositing with C_3N_4 under Ultraviolet Light. *ACS Omega* **2020**, *5*, 27873–27879. [CrossRef]
101. Thakur, M.; Sharma, G.; Ahamad, T.; Ghfar, A.A.; Pathania, D.; Naushad, M. Efficient photocatalytic degradation of toxic dyes from aqueous environment using gelatin-Zr(IV) phosphate nanocomposite and its antimicrobial activity. *Colloids Surf. B.* **2017**, *157*, 456–463. [CrossRef]
102. Clearfield, A.; Perry, H.P.; Gagnon, K.J. 5.08-Porous Pillared Clays and Layered Phosphates. In *Comprehensive Inorganic Chemistry II*, 2nd ed.; Reedijk, J., Poeppelmeier, K., Eds.; Elsevier: Amsterdam, The Netherlands, 2013; pp. 169–211. [CrossRef]
103. Costantino, F.; Vivani, R.; Bastianini, M.; Ortolani, L.; Piermatti, O.; Nocchetti, M.; Vaccaro, L. Accessing stable zirconium carboxyaminophosphonate nanosheets as support for highly active Pd nanoparticles. *Chem. Commun.* **2015**, *51*, 15990–15993. [CrossRef]
104. Kozell, V.; Giannoni, T.; Nocchetti, M.; Vivani, R.; Piermatti, O.; Vaccaro, L. Immobilized palladium nanoparticles on zirconium carboxy-aminophosphonatesnanosheets as an efficient recoverable heterogeneous catalyst for Suzuki-Miyaura and Heck coupling. *Catalysts* **2017**, *7*, 186. [CrossRef]
105. Costantino, F.; Nocchetti, M.; Bastianini, M.; Lavacchi, A.; Caporali, M.; Liguori, F. Robust zirconium phosphate-phosphonate nanosheets containing palladium nanoparticles as efficient catalyst for alkynes and nitroarenes hydrogenation reactions. *ACS Appl. Nano Mater.* **2018**, *1*, 1750–1757. [CrossRef]

molecules

Review

Application Research of Biochar for the Remediation of Soil Heavy Metals Contamination: A Review

Sheng Cheng [1,2,†], Tao Chen [1,2,*,†], Wenbin Xu [3], Jian Huang [1,2], Shaojun Jiang [1,2] and Bo Yan [1,2]

1 SCNU Environmental Research Institute, Guangdong Provincial Key Laboratory of Chemical Pollution and Environmental Safety & MOE Key Laboratory of Theoretical Chemistry of Environment, South China Normal University, Guangzhou 510006, China; ShengC@m.scnu.edu.cn (S.C.); 18371807641@163.com (J.H.); shaojunj93@163.com (S.J.); bo.yan@m.scnu.edu.cn (B.Y.)
2 School of Environment, South China Normal University, University Town, Guangzhou 510006, China
3 Dongjiang Environmental Company Limited, Nanshan District, Shenzhen 518057, China; xuwb@dongjiang.com.cn
* Correspondence: tao.chen@m.scnu.edn.cn; Tel./Fax: 1379-8199-406
† These authors contributed equally to this work.

Academic Editors: Chiara Bisio and Monica Pica
Received: 16 June 2020; Accepted: 8 July 2020; Published: 10 July 2020

Abstract: Soil contamination by heavy metals threatens the quality of agricultural products and human health, so it is necessary to choose certain economic and effective remediation techniques to control the continuous deterioration of land quality. This paper is intended to present an overview on the application of biochar as an addition to the remediation of heavy-metal-contaminated soil, in terms of its preparation technologies and performance characteristics, remediation mechanisms and effects, and impacts on heavy metal bioavailability. Biochar is a carbon-neutral or carbon-negative product produced by the thermochemical transformation of plant- and animal-based biomass. Biochar shows numerous advantages in increasing soil pH value and organic carbon content, improving soil water-holding capacity, reducing the available fraction of heavy metals, increasing agricultural crop yield and inhibiting the uptake and accumulation of heavy metals. Different conditions, such as biomass type, pyrolysis temperature, heating rate and residence time are the pivotal factors governing the performance characteristics of biochar. Affected by the pH value and dissolved organic carbon and ash content of biochar, the interaction mechanisms between biochar and heavy metals mainly includes complexation, reduction, cation exchange, electrostatic attraction and precipitation. Finally, the potential risks of in-situ remediation strategy of biochar are expounded upon, which provides the directions for future research to ensure the safe production and sustainable utilization of biochar.

Keywords: biochar; pyrolysis; heavy metals; soil remediation; bioavailability

1. Introduction

Soil is the final destination of heavy metals (HMs) whether they are from natural or anthropogenic sources. Mineral resource exploiting and smelting [1], metal electroplating [2], paint and coating processing [3], electronic equipment manufacturing [4], farmland sewage irrigating [5] and pesticide and chemical fertilizer nonstandard applying [6] are the primary anthropogenic activities that aggravate HMs contamination in soil. For instance, more than 30,000 tons of chromium and 800,000 tons of lead have been released into the environment globally in the past half century, most of which eventually accumulates in soils [7]. It is reported that approximately one-sixth of the total agricultural land area in China and about 600,000 hectares of brown field sites in America have been contaminated by HMs [8]. Hitherto, cadmium, lead and arsenic pollution, and associated ecological health risks in southeast China, are more severe than those in northwest China; similarly, those in industrial regions are worse than those in agricultural regions [7]. According to the results of a national soil survey [9] in 2014, the over-standard rates of cadmium, mercury, arsenic, copper, lead, chromium, zinc and nickel in soils of China are 7.0%, 1.6%, 2.7%, 2.1%, 1.5%, 1.1%, 0.9% and 4.8%, respectively. Soil HMs enter the food chain mainly through agricultural crops [10] and ultimately accumulate in organisms through diet, respiratory inhalation, skin contact and other exposure pathways, which directly or indirectly cause serious negative effects on human health [11,12]. The occurrence of cancer villages caused by HMs are the most immediate warning. Soil degradation and reduction of agricultural production land caused by HMs pollution brings an urgent need for the application of various efficient in-situ and ex-situ remediation techniques, to lessen the ecological risk of HMs and maximize the quality and security of agricultural land.

Over recent years, physical remediation (washing, thermal desorption, solidification and guest land methods), chemical remediation (vitrification, leaching, immobilization and electrokinetic methods), and biological remediation (microorganisms and plants) approaches have been applied to achieve this objective [13]. Nevertheless, these methods more or less exist with respective limitations i.e., complicated technique, inefficiency, poor feasibility, short duration, high economic cost, high secondary risk and so on [14]. At present, applying amendments to HM-contaminated soil is considered to be one of the most promising in-situ remediation techniques [15]. The frequently used soil additions include phosphate compounds, liming materials, clay minerals, coal fly ash, organic composts, metal oxides, and biochar [16,17]. In brief, the immobilization process of HMs can be achieved mainly through adsorption, complexation, reduction, and precipitation reactions, which cause the redistribution of HMs from soil liquid phases to solid phases, so as to reduce their mobility and bioavailability [16].

Biochar is a kind of carbon-rich, porous substance with abundant active organic functional groups and carbon aromatic structures with a neutral to alkaline pH value, relatively high cation exchange capacity, large specific surface area, and negative surface charge [18,19]. Numerous researchers report that the seed germination [20], plant growth [21–23], crop yields [23,24], and microbial activity and population [22,25,26] have been significantly increased in the HM-contaminated soil amended with biochars. Meanwhile, the effects of biochar on the immobilization/mobilization for different kinds of HMs have been confirmed in many pot experiments and field trials [21,27–29]. Furthermore, the production process of biochar is regarded as an efficient management method to dispose of a large number of organic wastes, which shows certain advantages in economic benefits and feasibility aspects. Remarkably, it cannot be completely ignored that there are still some potential risks in the field application of biochar, and these potential threats may hamper its further application. On this basis, the following aspects of biochar are reviewed: 1) preparation technologies, performance characteristics and influencing factors; 2) interaction mechanisms with different kinds of HMs; 3) effects on plant growth and HM bioavailability; and 4) potential risks in field engineering applications. In addition, the promotion and application of biochar in the future are also discussed.

2. Preparation of Biochar

The feedstock for biochar preparation mainly includes wood chips/branches [30–32], agricultural residues [21,33,34] and other woody biomass, as well as animal manure [35–38], sewage sludge [39–41] and other organic wastes. As shown in Table 1, thermochemical conversion technologies involved in the preparation process usually include fast pyrolysis, intermediate pyrolysis, slow pyrolysis, gasification, hydrothermal carbonization, torrefaction, etc. [42,43]. These technologies are mainly classified based on different heating rates, peak temperatures, residence times, reaction atmospheres and other parameters, where biochar, bio-oil and syngas are the main products. The yield of biochar is seriously affected by different operating conditions; several common thermochemical conversion technologies and their approximate product yields are discussed in Table 1.

Table 1. The reaction conditions and product distribution of various thermochemical conversion technologies.

Conversion Technologies	Temperature	Heating Rates	Reaction Atmosphere	Residence Time	Biochar	Bio-Oil	Syngas	Reference
Slow pyrolysis	300–650 °C	0.1–1 °C s^{-1}	Oxygen-free	1–24 h	25–35%	20–30%	25–35%	[44,45]
Intermediate pyrolysis	~500 °C	1.0–10 °C s^{-1}	Oxygen-free	10–20 s	20%	50%	30%	[46,47]
Fast pyrolysis	500–1000 °C	>200 K min^{-1}	Oxygen-free	<2 s	12–25%	50–75%	13–25%	[48–50]
Gasification	750–900 °C	50–100 °C s^{-1}	Air, steam, O_2, N_2, CO_2 or a mixture of these gases	10–20 s	10%	5%	85%	[47,50]
Hydrothermal carbonization	180–300 °C	5–10 °C min^{-1}	Confined system with a pressure of 2–6 MPa	1–16 h	50–80%	5–20%	2–5%	[19,47]
Torrefaction	250–300 °C	<50 °C min^{-1}	Inert atmosphere	10–60 min	60–80%	0%	20–40%	[44,51]

Fast pyrolysis can be defined as the thermochemical decomposition process of biomass with low energy density at a moderate pyrolysis temperature in the presence of little or no oxygen. Due to the characteristics of higher heating rate (>200 K min^{-1}) and shorter residence time (<2 s) in this process, bio-oil with high energy density, syngas with relatively low energy density and a small amount of biochar can be obtained (Table 1) [52]. On the contrary, slow pyrolysis is the common type of pyrolysis which is conducive to the formation of biochar rather than the generation of liquid and gaseous products. In this process, the biomass is pyrolyzed in a wide range of carbonization temperature with a heating rate of about 0.1~1 °C s^{-1} for a residence time between few hours and even days [52]. During slow pyrolysis, the fixed carbon content of biochar may increase with the rise of peak temperature, which is particularly prominent in the range of 400~500 °C [44]. The operating conditions of intermediate pyrolysis are between fast pyrolysis and slow pyrolysis, which better balances the distribution of solid–liquid product yield. The biochar produced by this pyrolysis mode has a brittle structure and does not contain a high quantity of reactive tar, which is suitable for the application of solid fuel, soil amendments and fertilizer [46].

Gasification is a process of direct contact oxidation of dry biomass with air, steam, oxygen, nitrogen, carbon dioxide or a mixture of these gases [43]. The primary product of gasification is a combustible gas which is packed with H_2, CO and CH_4, while the biochar with low yield contains a high amount of toxic substances such as polyaromatic hydrocarbons, and alkali and alkaline earth metals, that are attributed to the result of high-temperature reactions [51]. Hydrothermal carbonization is performed under a given pressure (2~6 MPa) and temperature (180~300 °C) that the feedstocks do not require for drying pretreatment, which is usually some wet biomass or dry biomass mixed with water. Compared to the pyrolysis and gasification processes, hydrothermal carbonization biochar (hydrochar) shows several advantages [51,53]. For instance, the hydrochar is characterized by high yield and high purity, and possesses a higher degree of aromatization and more surface functional groups. In addition, hydrochar contains a lower alkali and alkaline earth and heavy metal content, and a higher carbon content and heating value. Torrefaction, also referred to as mild pyrolysis, results in approximately 30% mass loss of biomass. The main products of torrefaction are organic carbon compounds with high specific energy density, but these cannot be referred to as a "biochar", because torrefaction is just the previous section of the pyrolysis process [43,51]. Consequently, the physicochemical properties of the torrefaction product is between that of biochar and biomass, and it also remains some volatile organic compounds.

Pyrolysis is a conventional process for the preparation of biochar. In this review paper, the properties of pyrolyzed biochar are preliminarily discussed, and the close connection between biochar performance characteristics and biomass feedstock species, pyrolysis temperature and residence time are clarified. According to the above factors (different preparation conditions), the performance characteristics of biochar are reviewed in detail in the next chapter, which provides a theoretical foundation for explaining the interaction mechanisms between biochar and soil HMs.

3. Performance Characteristics of Biochar

3.1. Elemental Composition

Element composition and content of biochar are a function of biomass species and pyrolysis temperature [54]. Generally, the content of total N, P, K, Ca, Mg and other nutrient elements in biochar prepared from poultry manure is higher than that of woody biomass, while the content of total C is the opposite [55]. Meanwhile, poultry manure is rich in mineral elements like K and P, which are important for plant growth, and thus the poultry-manure-derived biochar may be suitable as an ideal soil amendment instead of fertilizer [56].

The rise in pyrolysis temperature of biomass commonly results in the increase of ash and C content of biochar [50,57]. The N content of lignocellulosic type biochar increased slightly with the rise of pyrolysis temperature [58,59], while that of animal-manure- and sewage-sludge-derived biochar shows a downward trend [35,60–62]. Furthermore, the pyrolysis conditions of biomass such as a higher temperature and longer residence time are beneficial to the accumulation of total P and K [50,63], the release of Ca, Mg and Si, and the retention of Fe, Mn and S [47]. Correspondingly, with the increase in reaction time and temperature, other unstable substances (containing H and O elements) of biomass are removed by deoxygenation, dehydration and decarboxylation reactions progressively, which leads to the loss of volatile organic compounds and the reduction of H/C–O/C ratio [51,57,64,65]. These results indicate that the high-temperature biochar more easily forms a very stable crystal graphite-like structure and possesses a higher carbonization degree and more aromatic structures [66–68]. For example, Jindo et al. [66] reported that the O/C ratio of biochar pyrolyzed in the temperature range of 400~500 °C changed according to the following order: rice straw > rice husk > wood chips of apple tree > wood chips of oak tree. Therefore, these results indicate that there is a higher content of lignin and a slower mineralization rate in woody biomass (i.e., apple tree, oak tree), compared with herbaceous biochar (i.e., rice straw, rice husk) and sewage sludge biochar; furthermore, it has a lower O/C ratio, making woody biochar's structure more stable [67].

3.2. Functional Groups Abundance

As mentioned in Section 3.1, a high pyrolysis temperature commonly results in the decrease of the H/C, O/C and N/C ratios in biochar, which immediately indicates the decrease of its abundance of hydroxyl, carboxylic and amino functional groups [18]. Chen et al. [57] summarized the variation of FTIR characterization of pine wood shaving derived biochar with pyrolysis temperature: 1) for 150 °C, biochar is rich in -OH groups, CH_2 units, C=O, C=C, aromatic CO-, and phenolic-OH; 2) for 250 °C, C=O and C=C stretching vibrations were enhanced; 3) for 350 °C, the band CH_2 units disappeared completely, and aromatic ring and C=C stretching vibration of lignin strengthened; 4) for 500 °C, C=O and C=C stretching vibrations were significantly weakened; 5) for 700 °C, merely C=C of lignin and aromatic C-H vibrations were spotted. During the whole heating process, the band C-O-C of cellulose and hemicellulose reduced with the rise of pyrolysis temperature, until it disappeared. Hence, these results show that hemicellulose with short side chains, thermally stable cellulose, and lignin with phenolic structures contained in lignocellulosic are generally decomposed at 200~350 °C, 305~375 °C and 250~500 °C, respectively [69]. In addition, Ding et al. [70] indicated that compared with 250 °C pyrolyzed sugarcane-derived biochar, C≡C and C=O in 500 °C pyrolyzed biochar were relatively reduced, while C-O completely disappeared.

Moreover, the abundant surface functional groups such as C-O, C=O, -COOH and -OH in biochar possess high modifiability, which is the foundation for the preparation of various functionalized carbon materials [52]. For instance, Yang et al. [71] incubated walnut-shell-derived biochar with Al, Ca, Fe minerals or kaolinite, and the modified biochar's relative content of C-C, C=C and C-H increased from 63.8% to 72.5~81.8%, while C-O, C=O, and -COOH decreased from 36.3% to 16.6~26.5%. This result means that the interaction between biochar and minerals (Al, Ca, Fe, or kaolinite) has prevented the oxidation of C-C, C=C, C-H into C-O, C=O or -COOH, which enhanced the oxidation resistance of biochar surface. This is related to the modification process of biochar, which has been reviewed by previous scholars in the following papers [18,31,72], and will not be further analyzed in this paper.

3.3. Cation Exchange Capacity (CEC) and Specific Surface Area (SSA)

With the pyrolysis processes conducting, the SSA (specific surface area) value of biochar has significantly increased compared to the feedstock, while the relatively low temperature pyrolyzed biochar has the highest CEC (cation exchange capacity) value [73]. Although high-temperature biochar cannot possess the highest CEC and SSA values simultaneously, there are adequate functional groups remaining in the biochar structures to provide negative charges. The low O/C atomic ratio of high-temperature biochar results in a decrease in the CEC value, which is mainly manifested by the reduction of volatile organic compounds and acidic functional groups [74]. In other words, the high SSA and pH value of biochar pyrolyzed at higher temperatures (>600 °C) may compensate for the low CEC value to supply greater CEC provision to soil [54]. Thereby, biochar is a combination of charged surface functional groups and specific surface area, which can combine with HMs by adsorption and complexation reactions. For example, Yuan et al. [75] raised the pyrolysis temperature of sewage sludge from 300 °C to 700 °C, and found that the O/C ratio of biochar decreased from 0.33 to 0.05 and the volatile matter content reduced from 27.4% to 5.5%, but the SSA value increased from 14.37 $m^2\ g^{-1}$ to 26.70 $m^2\ g^{-1}$. Similarly, Heitkötter et al. [76] used corn digestate (derived from maize silage) as feedstock, as the temperature increased from 400 °C to 600 °C, the CEC value decreased by 29.9%, but the SSA value increased by 50.7% which exactly made up for the deficiencies in the CEC value reduction. The cation exchange capacity of biochar was enhanced by the tendency of attracting positive charges through its surface functional groups, which is an important feature for the remediation of HM-contaminated soil [77]. Furthermore, the porosity and pore size of biochar still depend on pyrolysis temperature, because the release of volatile organic compounds at higher pyrolysis temperatures may promote the formation of micropores.

3.4. pH Value

The pH value of biochar is mostly alkaline, and normally increases with the rise of pyrolysis temperature, which means biochar possesses the abilities to improve soil pH and CEC value, and to reduce soil acidity and bioavailability of certain HMs [78–80]. Fidel et al. [81] summarized four broad categories of biochar alkalinity, including: (1) surface organic functional groups; (2) soluble organic compounds; (3) carbonates/bicarbonates; and (4) other inorganic alkalis such as oxides, hydroxides, sulfates, sulfides, and orthophosphates. Surface organic functional groups have a long-term effect on the amelioration of soil properties (e.g., pH and CEC value), while soluble organic and inorganic alkalis contribute to the short-term improvement of soil acidification [54,81]. The functional groups separated from the pyrolysis of biomass are predominantly acidic in essence, such as the carboxyl group, hydroxyl group, or formyl group [73]. For others, the alkalinity of remaining solid (includes ash) raised with the increase number of functional groups released by biomass, therefore, the increase of pH value of biochar is the direct result of the increasing degree of carbonization [73]. The carbonates formed by mineral elements were considered to be the primary alkaline substances in biochar, and especially the biochar pyrolyzed at high temperatures possesses a higher content of carbonates a stronger buffer capacity [80]. Shen et al. [34] reported that the pH value (7.94) of rice-straw-derived biochar pyrolyzed at 300 °C is alkalescence, while the biochars with higher pH values (10.40 and 10.68) can be obtained at higher pyrolysis temperatures (500 °C and 700 °C). The results are attributed to the decomposition of acidic functional groups such as the carboxyl group and phenolic hydroxyl group, and the formation of alkaline minerals like K_2O in high-temperature pyrolyzed biochar. In addition, there is a high ash content in poultry manure and algae biomass, so the pH value of biochar pyrolyzed at the same temperature is higher than that of other woody biochar [82]. However, in some studies, the biochar produced by hydrothermal carbonization is typically acidic [83]. For example, the pH value of *Miscanthus*-derived hydrochar prepared by Gronwald et al. [84] at 200 °C is 3.8, and Cui et al. [85] found that the pH values of *Hydrocotyle verticillata*-, *Myriophyllum spicatum*- and *Canna indica*-derived hydrochar (200 °C) are 5.07, 4.97 and 6.48. Liu et al. [86] adjusted the pH value of the initial solution (pH = 2, 3, 5, 7, 9, 11, 12) of the hydrothermal carbonization process, so that the pH value of the prepared sewage-sludge-derived hydrochar was still weakly acidic or neutral (corresponding pH = 5.05, 6.11, 7.24, 6.60, 6.62, 6.74, 6.94). The presence of carboxyl functional groups on the surface of hydrochar as a result of formation of acetic and formic acids during the hydrothermal carbonization process could be the reason for the low pH value [87].

4. Remediation of Soil HMs Contamination by Biochar

4.1. Interaction Mechanisms of Biochar and HMs in Soil

The different sources of biomass feedstocks, and the diverse pyrolysis conditions applied in the preparation processes, lead to various biochar performance characteristics, which may in turn affect the interaction mechanisms between biochar and HMs. On the other hand, the greatest concerns of HMs have been focused on copper, arsenic, cadmium, lead, mercury, and chromium. Table 2 summarizes the research progress on biochar applications for the remediation of HM-contaminated soil, which included different types of biochar, different application conditions, and different HM treatment efficiencies. The various mechanisms proposed for the interaction of biochar with HMs are summarized in Figure 1. It shows that the abundant surface functional groups, mineral substances, alkaline metal ions, π-electrons, organic matters, and pore structures of micropores provided by biochar are the effective binding sites of HMs. Biochar is able to absorb or combine soil HMs through complexation, reduction, cation exchange, electrostatic attraction, and precipitation functions, or convert HMs from inorganic states into organic states, which changes HM mobility and bioavailability [14,88,89], and then improves soil agronomic benefits. Therefore, the interaction mechanisms between biochar and HMs are critical for the soil remediation and are discussed in detail in the following sections.

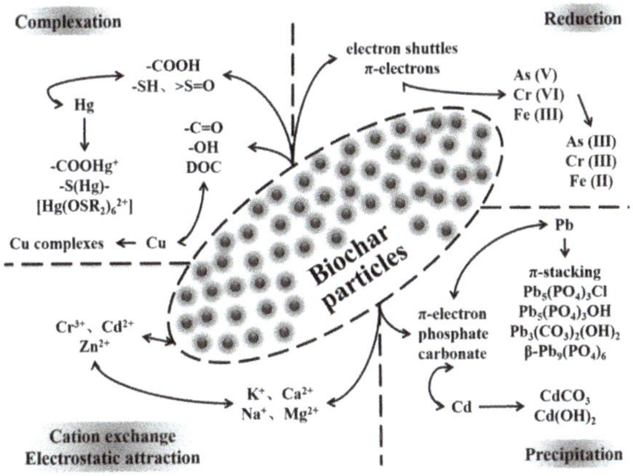

Figure 1. Interaction mechanism between biochar particles and HMs in soil.

Table 2. Remediation efficiency of biochar on HM (heavy metal)-contaminated soil.

Metals	Types of Biomass Feedstock	Pyrolysis Temperature	Dosage	Duration [a]	Soil Types	Total Metal Content	Immobilization Efficiency (Evaluation Method)	Reference
Cu	Chicken manure	500 °C	5%	14 d	Sedimentary alfisol	800 mg kg^{-1}	73% (NH_4NO_3-extractable)	[36]
	Orange bagasse	500 °C	60 t ha^{-1}	24 m	Fallow field soil	100 mg kg^{-1}	28% (citric acid-extractable)	[90]
	Chicken manure	550 °C	5%	14 d	Mine soil	1805 mg kg^{-1}	79% (NH_4NO_3-extractable)	[91]
	Orange bagasse	500 °C	60 t ha^{-1}	6 m	Fallow field soil	100 mg kg^{-1}	41% (citric acid-extractable)	[92]
	Bamboo	600 °C	15%	20 d	Sediment soil	134.6 mg kg^{-1}	79.7% (HOAc-soluble)	[93]
	Oat hull	300 °C	5%	24 m	Sedimentary alfisol	338 mg kg^{-1}	68% (exchangeable fraction)	[22]
	Chicken manure	550 °C	5%	14 d	Hills soil	160 mg kg^{-1}	−45% (NH_4NO_3-extractable)	[91]
	Sewage sludge	500 °C	30 t ha^{-1}	6 m	Fallow field soil	100 mg kg^{-1}	−18% (citric acid-extractable)	[92]
As	Rice straw	500 °C	3%	30 d	Paddy soil	120 mg kg^{-1}	As concentration increased by 234.5% in soil solution	[26]
	Rice straw	450 °C	1–3%	96 d	Paddy field soil	212 mg kg^{-1}	As concentration increased in soil porewater	[94]
	Soybean stover	700 °C	10%	90 d	Agricultural soil	1945 mg kg^{-1}	As mobility increased greatly in soil	[25]
	Rice straw	300 °C	10%	35 d	Paddy field soil	92.3 mg kg^{-1}	As concentration increased in soil pore solution	[95]
	Oil palm fibers	700 °C	3%	20 d	Paddy field soil	0.3 mg kg^{-1}	As concentration increased in soil solution	[96]
	Sewage sludge	200 °C	3%	6 d	Agricultural soil	98.7 mg kg^{-1}	−81.9% (water-soluble)	[97]
	Sewage sludge	350 °C	3%	6 d	Agricultural soil	98.7 mg kg^{-1}	42.2% (water-soluble)	[97]
	Corn straw	600 °C	0.5–2%	100 d	Paddy soil	73 mg kg^{-1}	As(V): 11.7–28.5% (phosphate-extractable) As(III): 54.0–81.6% (phosphate-extractable)	[23]
Cd	Wheat straw	350–550 °C	40 t ha^{-1}	3 y	Paddy soil	5 mg kg^{-1}	59% ($CaCl_2$-extractable) 24% (DTPA-extractable)	[98]
	Bamboo	600 °C	15%	20 d	Sediment soil	3.8 mg kg^{-1}	31.2% (HOAc-soluble)	[93]
	Soybean straw	350 °C	3%	6 d	Agricultural soil	1.36 mg kg^{-1}	65.7% (water-soluble)	[97]
	Maize straw	550 °C	30 t ha^{-1}	~6 m	Paddy soil	2.04 mg kg^{-1}	50.4% (DTPA-extractable)	[29]
	Rice straw	450 °C	1–2%	96 d	Paddy field soil	10.8 mg kg^{-1}	Cd concentration decreased in soil porewater	[94]
	Sugarcane bagasse	500 °C	1.5%	4 m	Agricultural soil	50 mg kg^{-1}	40.4% (DTPA-extractable)	[21]
	Corn stalk	550 °C	2%	30 d	Arable land soil	2.0 mg kg^{-1}	91% ($CaCl_2$-extractable)	[99]
	Hickory nut shell	500 °C	30 t ha^{-1}	~6 m	Paddy soil	2.04 mg kg^{-1}	53.6% (DTPA-extractable)	[29]

Table 2. *Cont.*

Metals	Types of Biomass Feedstock	Pyrolysis Temperature	Dosage	Duration [a]	Soil Types	Total Metal Content	Immobilization Efficiency (Evaluation Method)	Reference
Pb	Soybean stover	700 °C	10%	90 d	Agricultural soil	1945 mg kg^{-1}	95% (NH_4OAc-extractable) 95% (TCLP-extractable)	[25]
	Vegetable waste	500 °C	5%	45 d	Agricultural soil	1445 mg kg^{-1}	87% (NH_4OAc-extractable)	[100]
	Bamboo sawdust	600 °C	37.5%	30 d	Sediment soil	589.7 mg kg^{-1}	100% (TCLP-extractable)	[101]
	Red pepper stalk	650 °C	2.5%	45 d	Agricultural soil	1445 mg kg^{-1}	65% (NH_4OAc-extractable)	[100]
	Wheat straw	350–550 °C	40 t ha^{-1}	3 y	Paddy soil	100 mg kg^{-1}	59% ($CaCl_2$-extractable) 27% (DTPA-extractable)	[98]
	Bamboo	600 °C	15%	20 d	Sediment soil	44.3 mg kg^{-1}	73.2% (HOAc-soluble)	[93]
Hg	Rice husk	550 °C	1–5%	10 d	Field soil	1000 mg kg^{-1}	>94% (TCLP-extractable)	[102]
	Rice hull	480–660 °C	24 t ha^{-1}	118 d	Farmland soil	129 mg kg^{-1}	Hg concentration decreased by 44% in soil porewater	[28]
	Wheat straw	350–450 °C	72 t ha^{-1}	118 d	Farmland soil	129 mg kg^{-1}	Hg concentration decreased by 26% in soil porewater	[28]
	Sewage sludge	600 °C	5%	17 w	Paddy field soil	2.1 mg kg^{-1}	MeHg concentration increased by 67% in soil	[41]
	Sewage sludge	600 °C	5%	17 w	Paddy field soil	65.3 mg kg^{-1}	MeHg concentration increased by 29% in soil	[41]
Cr	Waste wood	900 °C	1–5%	11 w	Tannery waste soil	12285 mg kg^{-1}	28–68% ($CaCl_2$-extractable)	[103]
	Sugarcane bagasse	500 °C	1.5%	4 m	Agricultural soil	50 mg kg^{-1}	49.6% (DTPA-extractable)	[21]
	Rice straw	500 °C	40 t ha^{-1}	~4 m	Paddy field soil	432.8 mg kg^{-1}	Cr(total): 48.1% (HNO_3/H_2SO_4-extractable) Cr(VI): 22.3% (HNO_3/H_2SO_4-extractable)	[104]
	Wheat straw	600 °C	0.25%	180 d	Cr-spiked soil	308 mg kg^{-1}	Cr(VI): 47.1% (TCLP-extractable) Cr(VI): 65.5% ($CaCl_2$-extractable)	[105]

[a] Units explanation: d for days, w for weeks, m for months and y for years.

4.1.1. Copper (Cu)

As described by Meier et al. [36], the functional groups (especially for -OH) and negative ζ-potential existing in chicken-manure-derived biochar has been proved with high affinity for Cu. The immobilization process of Cu is achieved by increasing soil pH value and inducing the liming effect, to stimulate the complexation of Cu(II) with biochar surface functional groups (e.g., C=O and phenolic-OH). Additionally, Rechberger et al. [30] found that carbonates and hydroxides in the ash of woodchip-derived biochar are the important adsorbents for Cu(II), which are able to promote the formation of $CuCO_3$ and $Cu(OH)_2$, and this is also illustrated by the bamboo-derived biochar prepared by Zhang et al. [93]. Therefore, the essential point of Cu immobilization is to use the alkalinity of biochar to improve soil pH value [92]. On the other side of the shield, the mobility of Cu is highly affected by the content of soil-dissolved organic carbon (DOC). For instance, Park et al. [91] introduced chicken-manure-derived biochar to Cu-spiked soil, which led to the increase of soil DOC content and provoked the conversion of Cu(II) into Cu complexes with higher solubility. Wagner et al. [106] also reported that the *Miscanthus*-derived biochar can increase the Cu concentration in soil solution. In other words, the increase of Cu concentration in soil pore water is the proximate consequence of the Cu(II) desorption from soil with the form of organic complex. Hence, the immobilization/mobilization of Cu by biochar should be further studied according to the actual types of biochar.

4.1.2. Arsenic (As)

Phosphorus (P) and As have similar chemical properties and the soil P content is a critical factor in controlling the mobility of As [25]. There is a significant positive correlation between phosphate content and arsenate content in As-contaminated soils. After the soils are treated with P-containing biochar, the competition between soluble phosphate and arsenate that occurred on the adsorption sites of soil particles promoted the desorption of As from soil solid phase, and increased the As concentration in pore water [45], such as soybean-stover-, pine-needle- [25], and rice-straw-derived biochar [94]. The role of soil DOC is similar to P [95], but DOC has another effect. For example, Wang et al. [26] indicated that the application of rice-straw-derived biochar under the anaerobic conditions increased the abundance of Fe-reducing bacteria (e.g., *Clostridum*, *Bacillus* and *Caloramator*) in paddy soil, promoted the reduction of As(V) adsorbed on the amorphous Fe/Al oxides. To be brief, the increase of DOC content in soils enhanced the microbial reduction effect of As(V), and finally stimulated the release of As (III) from paddy soil. Under the anoxic conditions, similar results were obtained in the application of oil palm fiber derived biochar prepared by Qiao et al. [96]. Biochar, with high aromaticity and alkalinity, and which can act as an electron shuttle, likes humus to promote the microbial reduction of Fe(III) and As(V) simultaneously [26,96]. In addition, Choppala et al. [27] reported that the π-electrons provided by the functional groups on the surface of chicken-manure-derived biochar could promote the reduction of As(V), which is another important factor to enhance the mobility of As. However, the Fe-biochar prepared by Yin et al. [94] and Mn-biochar prepared by Yu et al. [23] are able to adsorb As to the Fe/Mn oxides of biochar surface, thus commendably limiting the migration of As into soil solution, which proved the effectiveness of As immobilization by the modified biochar (not listed in Table 2).

4.1.3. Cadmium (Cd)

The activity of Cd in soils strongly depends on soil pH value [99]. The alkaline substances such as CO_3^{2-}, PO_4^{3-} and OH^- contained in biochar commonly have strong adsorption and binding capacity to Cd in soils, which makes the free Cd(II) transform into $Cd(OH)_2$, $Cd_3(PO_4)_2$ and $CdCO_3$ precipitates [30]. The high adsorption affinity produced by the cation exchange effects of soil calcite ($CaCO_3$) with Cd(II) is the main factor to reduce the bioavailability of Cd; of course, the abundant functional groups and large specific surface area in biochar are also critical for Cd immobilization [29]. For instance, Yin et al. [94] used 1~2% rice-straw-derived biochar to treat farmland soil in a mining area, and found that the Cd concentration in pore water of soil rhizosphere was significantly reduced, and the corn-straw-derived biochar prepared by Gao et al. [99] decreased 91% of the $CaCl_2$-extractable Cd content in the farmland soil. Besides, Bian et al. [98] applied 40 t ha^{-1} wheat-straw-derived biochar to paddy soil, and the $CaCl_2$-extractable Cd and DTPA-extractable Cd was reduced by 59% and 24%, respectively, over the past three years. Thereby, the result of biochar in increasing soil pH value is extremely effective for the immobilization of Cd. It is worth mentioning that the basic substances in biochar can also stimulate the deprotonation of acid functional groups of biochar, and further enhance the adsorption capacity of Cd [107].

4.1.4. Lead (Pb)

The immobilization process of Pb in soils by biochar is relatively simple. For example, Ahmad et al. [25] considered that the immobilization process of Pb stimulated by soybean-stover -derived biochar is attributed to the π-cation electron donor–acceptor interaction, which occurs by the π-electron-rich biochar graphene surface and π-electron-deficient positively charged Pb(II) ion. Furthermore, the cation exchange and precipitation reactions between basic substances (CO_3^{2-}, OH^- and other alkaline earth Ca^{2+}, Mg^{2+}) in biochar and Pb(II) are able to achieve a significant immobilization effect of Pb, such as the formation of $Pb_3(CO_3)_2(OH)_2$ precipitation. In particular, the poultry-manure-derived biochar [38] and the sewage-sludge-derived biochar [39] contain abundant phosphates, which are able to form insoluble compounds, such as $Pb_5(PO_4)_3Cl$, $Pb_5(PO_4)_3OH$ and β-$Pb_9(PO_4)_6$, with Pb(II) to reduce the mobility of Pb [25,101]. The functional groups also have certain effects on the immobilization of Pb. Igalavithana et al. [100] reported that the vegetable-waste-derived biochar not only improved soil pH value, but also promoted the immobilization of Pb by the strong covalent bonding action of N-containing functional groups (especially for -NH_2) on the biochar surface, thereby effectively decreasing the concentration of NH_4OAc-extractable Pb.

4.1.5. Mercury (Hg)

Hg, as a special metal, has great toxicity in soil, but biochar is an effective tool for soil remediation. As Wang et al. [31] remarked, the carboxyl group in a hardwood-derived biochar surface and soil Hg(II) ion develop a coordination reaction generating a complex of -$COOHg^+$ precipitate, thereby reducing the mobility of Hg and the thiol functionalities and sulfoxide groups are also able to react with Hg(II) ion to form -S(Hg)- and [$Hg(OSR_2)_6^{2+}$] precipitates. Additionally, Xing et al. [28] found that rice-husk-derived biochar possesses a higher sulfate concentration compared with wheat-straw-derived biochar, which is more effective to promote the mercury–sulfur coordination reaction and to produce sulfides precipitation. It is extremely important to know that methylation of Hg is a special environmental biogeochemical behavior, and once the inorganic-Hg in soils is converted into methylmercury (MeHg), its toxicity and bioaccumulation will be enormously enhanced [108]. The release of rice root exudates reduced soil pH value, thereby enhancing the methylation of Hg, while the application of alkaline biochar increases soil pH value and effectively inhibits methylation. Zhang et al. [41] reported that sewage-sludge-derived biochar with high organic matter content can stimulate the growth and activity of heterotrophic microorganisms in Hg-contaminated acidic farmland soil, thus promoting the formation of MeHg. However, the utilization rate of MeHg in rice has been significantly decreased, which led to the inhibition of MeHg accumulation in rice and effectively reduced the bioavailability of organic-Hg [41].

4.1.6. Chromium (Cr)

Firstly, it should be pointed out that Cr exists in soils in the form of two valence states, namely Cr(VI) and Cr(III). Therefore, the immobilization of Cr in soils is a complicated combination process of adsorption–reduction–precipitation. The interaction mechanism between biochar and Cr is mainly manifested in the surface adsorption effect of Cr(VI) by the oxygen-containing functional groups such as C-O, C=O, -COOH and -OH in biochar, as well as the reduction reaction of electrons provided by biochar [105]. Specifically, biochar transforms Cr(VI) into Cr(III) with lower toxicity and solubility by adsorption–reduction effects, and participates in the formation of Cr_2O_3 and/or $Cr(OH)_3$ precipitation, so as to achieve the purpose of Cr immobilization [109,110]. On the other hand, Choppala et al. [27] applied chicken-manure-derived biochar to Cr-spiked soil with low organic matter content, which significantly increased the supply of soil organic carbon and nutrients, enhancing the soil respiration and microbial activity effectively and finally showing a superior microbial Cr(VI) reduction effect and reduced the Cr biotoxicity. Moreover, Mandal et al. [111] found that the Cr(VI) reduction effect of animal-manure-derived biochar in acidic soil was significantly higher than that in alkaline soil.

In summary, many HMs can be continuously immobilized in soil by biochar through specific or non-specific surface adsorption. It is noteworthy that not only the activity of Cu and As are affected by DOC content in soil, but also Pb and Cd. This phenomenon is attributed to the fact that the high-dose application of biochar in soil results in the sharp increase of DOC content, which leads to the complexation of Pb and Cd with DOC, and finally increases the mobility and bioavailability of Pb and Cd. In addition, the adsorption and cation exchange reactions between HMs and biochar ash (e.g., K^+, Na^+, Ca^{2+}, Mg^{2+} and other alkali metal ions) are also important factors for the immobilization of HMs.

4.2. Effect of Biochar on Bioavailability of HMs in Soil

Plants mainly absorb, transport, and accumulate HMs from contaminated soil by roots. Thus, the primary objectives of biochar soil remediation are limiting the migration and transformation rates of HMs in soil and reducing their bioavailability, so as to prevent HMs from entering organisms through the food chain and to eliminate their toxic effects. The ultimate goal of HM-contaminated soils remediation is to increase crop yields on the premise of ensuring food production safety.

On the one hand, the introduction of biochar provides a source of organic matter, N, P, K, Ca, Mg and other nutrients to the soil, which enhances soil enzyme and microbial activities. On the other hand, the plant root environment, soil water retention and saturated hydraulic conductivity [112] can be improved with the presence of biochar, and plant growth and nutrient absorption can be promoted. Finally, it increases the plants biomass, and dilutes the content of HMs in plant tissues to reduce their phytotoxicity. Meier et al. [36] indicated that 5% chicken-manure-derived biochar can reduce the uptake of Cu from 66.9 mg kg^{-1} to 36.6 mg kg^{-1} in the aboveground part of *Oenothera picensis* plants in copper-mine-polluted soil, and increase the biomass of shoots and roots by 3.5 times and 3.1 times respectively. Xing et al. [28] reported that after applying 24 t ha^{-1} and 72 t ha^{-1} rice-husk-derived biochar to mercury-contaminated farmland soils, the Hg content in rice grains reduced to 10 ng g^{-1} and 7.2 ng g^{-1}, which significantly inhibited the transportation of Hg from soil to rice grains, and successfully reached the national standard (below 20 ng g^{-1}). Similarly, Li et al. [97] applied 3% of soybean-straw-derived biochar (hydrothermal carbon at 350°C) to the arsenic and cadmium co-contaminated farmland soil, which reduced the bioaccumulation of As in rice plants by 88%. Besides, compared with the control group, As(III) content decreased from 3.47 mg kg^{-1} to 0.29 mg kg^{-1}, As(V) decreased from 715 μg kg^{-1} to 150 μg kg^{-1}, and the treatment effects on Cd were similar [97]. As for cadmium, a field trial studied by Zheng et al. [113] showed that, when the application rates of soybean-straw-derived biochar and rice-straw-derived biochar were 20 t ha^{-1}, the content of Cd in rice roots, rice shoots, rice husks and rice grains decreased by 25.0~44.1% and 19.9~44.2%, and 46.2%~70.6% and 25.8%~70.9%, respectively. Furthermore, the effectiveness of rice-straw-derived biochar in reducing Cd accumulation in rice grains was also confirmed in the long-term field effects studied by Zhang et al. [29]. Cd contamination of farmland soil is particularly prominent in China, but it happens that China is a large agricultural country, which has a large output of agricultural waste with low cost. Considering the economic value of biochar application, using rice straw, wheat straw and other agricultural wastes to produce biochar and returning it to the field is the best choice.

In a word, the immobilization effects of soil HMs, the limitation of HM uptake and accumulation by plants, the enhancement of plant biomass, and the dilution effect of HMs in plant tissues are the four pivotal performances of biochar to reduce the bioavailability of HMs [114,115]. Although certain

HMs can be combined with soil DOC and converted into organic complex forms, which are activated finally, these organic complexes commonly possess stability and are not readily or directly absorbed by plants, which has little impact on HM content in plant tissues.

5. Potential Risks

Applying biochar as an addition to soil in-situ remediation should not only consider the HM immobilization/mobilization effects, but also take into account the long-term stability and potential ecological risks of biochar, which similarly depend on the type and performance characteristics of biochar. Figure 2 briefly introduced the advantages and disadvantages of biochar in the remediation of soil HM contamination. Firstly, it is reported that biochar may be the carrier of HMs [116], volatile organic compounds (VOCs) [117], dioxin (PCDD/Fs) and polycyclic aromatic hydrocarbons (PAHs) [118,119] and other toxic substances, and the demands for remediation of HM contamination in soil varies from 1.5 to 72 t ha^{-1} or even higher [24,28,98,113]. Therefore, the toxic substances may be released into the soil/air/water environment with the application of the double-edged biochar, which will pose a secondary pollution and ecological risk. So far, from the perspective of soil, biochar plays an important role in the carbon utilization and non-CO_2 greenhouse gases emission reduction [104]. However, the research on increasing greenhouse gas emissions by biochar has also been reported, i.e., under specific conditions, the application of biochar can promote the emissions of CO_2 [120], N_2O [121], CH_4 [122] to a certain extent. For example, although biochar can reduce CH_4 emissions, it may also promote N_2O emissions, and vice versa, and this depends on the diversity of biochar application conditions. Therefore, the emission reduction effect of biochar cannot be reflected in all types of greenhouse gases, and blind application may cause negative effects, so the biochar–greenhouse gas interaction should be considered in the field application of biochar. Meanwhile, other research has indicated that biochar can inhibit the efficacy of soil pesticides and their biodegradation effects [123], which makes the ability of agricultural weeding and insecticides unable to achieve the expected effects. The residue of pesticides may be related to the strong adsorption and binding capacity of biochar. Furthermore, although biochar can improve the biological activity of bacteria (e.g., *Geobacter*, *Anaeromyxobacter* and *Clostridium*) [26,96], it may bring about a negative impact on the survival, growth and diversity of, for example, acidophilic earthworm and fungi biological communities [124,125].

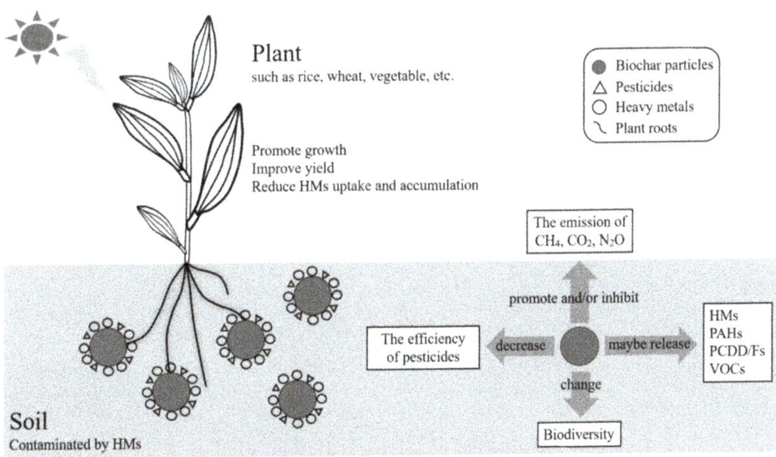

Figure 2. Advantages and disadvantages of biochar in the remediation of soil HM contamination.

6. Summary and Future Perspectives

It is a feasible strategy of green, economic and environmental protection to apply biochar to remediate the HM-contaminated soil. Several thermochemical conversion technologies have been used to prepare biochar, and the performance characteristics of biochar are highly affected by the type of feedstock materials, pyrolysis temperature, and residence time. The immobilization or mobilization mechanisms of HMs includes complexation, reduction, cation exchange, electrostatic attraction and precipitation reactions. The HM uptake and accumulation of plants can be suppressed by biochar through the variation in soil pH value, DOC content, and other alkaline mineral substances content, and achieve the goal of promoting plant growth, reducing HM bioavailability and improving soil quality in unison.

Modern soil environmental management and remediation technologies are beginning to pay attention to the long-term stability of biochar and the ecological responses of contaminants with the most toxicological fractions. Consequently, in order to meet demand and application-based results, the future application of biochar is predicted as follows:

(1) It is expected that the secondary ecological risks in the process of biochar production will still need to be focused on, i.e., the formation of HMs, VOCs, PCDD/Fs and PAHs. These pollutants are essentially derived from biomass, so it is necessary to screen and pretreat raw biomass materials to remove HMs. In addition, we must adjust the pyrolysis temperature, residence time, pyrolysis atmosphere, pressure conditions and other process parameters to minimize the generation of associated pollutants. Furthermore, it is necessary to establish and perfect the regulatory systems related to the application of biochar in practical engineering.

(2) With the passage of time, soil components will occur varying degrees of physicochemical and biochemical changes, including the properties of biochar (aged biochar) in the meantime. Although the research on soil HM immobilization by biochar has developed to laboratory pot experiments and short-term field trials, the ultimate goal is to extend it to large-scale engineering applications. Hence, carrying out long-term positioning tests in different types of HM-contaminated soil must be conducted, so as to further verify the long-term stability of biochar and its long-term effects on soil environment (such as aggregation, surface potential and density magnitude), and lastly to ensure the quality and safety of agricultural land.

(3) Biochar cannot completely remediate the heavy metals contaminated soil. Therefore, in order to improve the remediation/improvement effects of multi-heavy metals contaminated soil, the multifunctional biochar materials (e.g., biochar inoculated with microorganisms and biochar modified by chemicals and minerals) should be gradually put on the stage of engineering application.

(4) The advantages and disadvantages between the economic cost (production) and benefit value (application) of biochar need to be carefully measured. In order to enhance economic availability, easier production processes and cheaper sources of raw biomass materials need to be discovered, which could provide a platform for improving production efficiency and reducing economic burdens, i.e., to achieve the purpose of commercial practicality.

Author Contributions: Conceptualization, S.C. and T.C.; methodology, S.C.; software, S.C.; validation, J.H.; W.X. and S.J.; formal analysis, W.X.; investigation, S.C.; resources, T.C.; data curation, S.C.; writing—original draft preparation, S.C.; writing—review and editing, T.C.; visualization, S.J.; supervision, T.C.; project administration, T.C. and B.Y.; funding acquisition, B.Y. All authors have read and agreed to the published version of the manuscript.

Funding: This work was financially supported by the National key research and development plan (2018YFC1802800), the Guangdong provincial science and technology program (2015B020237003), the Provincial Science and Technology Plan Project of Guangdong Province, Shao Guan (2018SG00118), and the Guangdong Provincial Key Laboratory of Chemical Pollution and Environmental Safety (2019B030301008).

Acknowledgments: The authors are very grateful to the anonymous reviewers for their revising suggestions.

Conflicts of Interest: The authors declare no conflict of interest.

References

1. Xiao, R.; Wang, S.; Li, R.H.; Wang, J.J.; Zhang, Z.Q. Soil heavy metal contamination and health risks associated with artisanal gold mining in Tongguan, Shaanxi, China. *Ecotoxicol. Environ. Saf.* **2017**, *141*, 17–24. [CrossRef]
2. Liu, J.; Zhang, X.H.; Henry, T.; Wang, D.Q.; Zhu, Y.N. Heavy metal contamination and risk assessment in water, paddy soil, and rice around an electroplating plant. *Environ. Sci. Pollut. Res.* **2011**, *18*, 1623–1632. [CrossRef]
3. Udosen, E.; Akpan, E.; Sam, S. Levels of some heavy Metals in Cocoyam (Colocasia esculentum) grown on Soil receiving Effluent from a Paint Industry. *J. Appl. Sci. Environ. Manag.* **2016**, *20*, 215–218. [CrossRef]
4. Kaya, M. Recovery of metals and nonmetals from electronic waste by physical and chemical recycling processes. *Waste Manag.* **2016**, *57*, 64–90. [CrossRef]
5. Rattan, R.; Datta, S.; Chhonkar, P.; Suribabu, K.; Singh, A. Long-term impact of irrigation with sewage effluents on heavy metal content in soils, crops and groundwater—A case study. *Agric. Ecosyst. Environ.* **2005**, *109*, 310–322. [CrossRef]
6. Gimeno-García, E.; Andreu, V.; Boluda, R. Heavy metals incidence in the application of inorganic fertilizers and pesticides to rice farming soils. *Environ. Pollut.* **1996**, *92*, 19–25. [CrossRef]
7. Yang, Q.Q.; Li, Z.Y.; Lu, X.N.; Duan, Q.N.; Huang, L.; Bi, J. A review of soil heavy metal pollution from industrial and agricultural regions in China: Pollution and risk assessment. *Sci. Total Environ.* **2018**, *642*, 690–700. [CrossRef]
8. Mahar, A.; Wang, P.; Ali, A.; Awasthi, M.K.; Lahori, A.H.; Wang, Q.; Li, R.; Zhang, Z. Challenges and opportunities in the phytoremediation of heavy metals contaminated soils: A review. *Ecotoxicol. Environ. Saf.* **2016**, *126*, 111–121. [CrossRef]
9. Ministry of Environmental Protection of China. National Soil Pollution Survey Bulletin. 2014. Available online: http://www.mee.gov.cn/gkml/sthjbgw/qt/201404/t20140417_270670.htm (accessed on 16 June 2020).
10. Zheng, S.N.; Wang, Q.; Yuan, Y.Z.; Sun, W.M. Human health risk assessment of heavy metals in soil and food crops in the Pearl River Delta urban agglomeration of China. *Food Chem.* **2020**, *316*, 126213. [CrossRef]
11. He, Y.; Li, B.B.; Zhang, K.N.; Li, Z.; Chen, Y.G.; Ye, W.M. Experimental and numerical study on heavy metal contaminant migration and retention behavior of engineered barrier in tailings pond. *Environ. Pollut.* **2019**, *252*, 1010–1018. [CrossRef]
12. Wu, H.W.; Liu, Q.Y.; Ma, J.; Liu, L.L.; Qu, Y.J.; Gong, Y.W.; Yang, S.H.; Luo, T. Heavy Metal(loids) in typical Chinese tobacco-growing soils: Concentrations, influence factors and potential health risks. *Chemosphere* **2020**, *245*, 125591. [CrossRef]
13. Dhaliwal, S.S.; Singh, J.; Taneja, P.K.; Mandal, A. Remediation techniques for removal of heavy metals from the soil contaminated through different sources: A review. *Environ. Sci. Pollut. Res.* **2019**, *27*, 1–15. [CrossRef]
14. Lahori, A.H.; Guo, Z.Y.; Zhang, Z.Q.; Li, R.H.; Mahar, A.; Awasthi, M.K.; Shen, F.; Sial, T.A.; Kumbhar, F.; Wang, P. Use of biochar as an amendment for remediation of heavy metal-contaminated soils: Prospects and challenges. *Pedosphere* **2017**, *27*, 991–1014. [CrossRef]
15. Karna, R.R.; Luxton, T.; Bronstein, K.E.; Hoponick Redmon, J.; Scheckel, K.G. State of the science review: Potential for beneficial use of waste by-products for in situ remediation of metal-contaminated soil and sediment. *Crit. Rev. Environ. Sci. Technol.* **2017**, *47*, 65–129. [CrossRef]
16. Bolan, N.; Kunhikrishnan, A.; Thangarajan, R.; Kumpiene, J.; Park, J.; Makino, T.; Kirkham, M.B.; Scheckel, K. Remediation of heavy metal(loid)s contaminated soils–to mobilize or to immobilize? *J. Hazard. Mater.* **2014**, *266*, 141–166. [CrossRef]
17. Palansooriya, K.N.; Shaheen, S.M.; Chen, S.S.; Tsang, D.C.W.; Hashimoto, Y.; Hou, D.; Bolan, N.S.; Rinklebe, J.; Ok, Y.S. Soil amendments for immobilization of potentially toxic elements in contaminated soils: A critical review. *Environ. Int.* **2020**, *134*, 105046. [CrossRef]
18. Li, H.B.; Dong, X.L.; Silva, E.B.D.; Oliveira, L.M.D.; Chen, Y.S.; Ma, L.Q. Mechanisms of metal sorption by biochars: Biochar characteristics and modifications. *Chemosphere* **2017**, *178*, 466–478. [CrossRef]
19. Tan, X.F.; Liu, Y.G.; Zeng, G.M.; Wang, X.; Hu, X.J.; Gu, Y.L.; Yang, Z.Z. Application of biochar for the removal of pollutants from aqueous solutions. *Chemosphere* **2015**, *125*, 70–85. [CrossRef]
20. Martins, G.C.; Penido, E.S.; Alvarenga, I.F.S.; Teodoro, J.C.; Bianchi, M.L.; Guilherme, L.R.G. Amending potential of organic and industrial by-products applied to heavy metal-rich mining soils. *Ecotoxicol. Environ. Saf.* **2018**, *162*, 581–590. [CrossRef]

21. Bashir, S.; Hussain, Q.; Akmal, M.; Riaz, M.; Hu, H.; Ijaz, S.S.; Iqbal, M.; Abro, S.; Mehmood, S.; Ahmad, M. Sugarcane bagasse-derived biochar reduces the cadmium and chromium bioavailability to mash bean and enhances the microbial activity in contaminated soil. *J. Soils Sediments* **2017**, *18*, 874–886. [CrossRef]
22. Moore, F.; González, M.E.; Khan, N.; Curaqueo, G.; Sanchez-Monedero, M.; Rilling, J.; Morales, E.; Panichini, M.; Mutis, A.; Jorquera, M. Copper immobilization by biochar and microbial community abundance in metal-contaminated soils. *Sci. Total Environ.* **2018**, *616*, 960–969. [CrossRef]
23. Yu, Z.H.; Qiu, W.W.; Wang, F.; Lei, M.; Wang, D.; Song, Z.G. Effects of manganese oxide-modified biochar composites on arsenic speciation and accumulation in an indica rice (Oryza sativa L.) cultivar. *Chemosphere* **2017**, *168*, 341–349. [CrossRef] [PubMed]
24. Nie, C.L.; Yang, X.; Niazi, N.K.; Xu, X.Y.; Wen, Y.H.; Rinklebe, J.; Ok, Y.S.; Xu, S.; Wang, H.L. Impact of sugarcane bagasse-derived biochar on heavy metal availability and microbial activity: A field study. *Chemosphere* **2018**, *200*, 274–282. [CrossRef]
25. Ahmad, M.; Ok, Y.S.; Kim, B.Y.; Ahn, J.H.; Lee, Y.H.; Zhang, M.; Moon, D.H.; Al-Wabel, M.I.; Lee, S.S. Impact of soybean stover-and pine needle-derived biochars on Pb and As mobility, microbial community, and carbon stability in a contaminated agricultural soil. *J. Environ. Manag.* **2016**, *166*, 131–139. [CrossRef]
26. Wang, N.; Xue, X.M.; Juhasz, A.L.; Chang, Z.Z.; Li, H.B. Biochar increases arsenic release from an anaerobic paddy soil due to enhanced microbial reduction of iron and arsenic. *Environ. Pollut.* **2017**, *220*, 514–522. [CrossRef]
27. Choppala, G.; Bolan, N.; Kunhikrishnan, A.; Bush, R. Differential effect of biochar upon reduction-induced mobility and bioavailability of arsenate and chromate. *Chemosphere* **2016**, *144*, 374–381. [CrossRef]
28. Xing, Y.; Wang, J.X.; Xia, J.C.; Liu, Z.M.; Zhang, Y.H.; Du, Y.; Wei, W.L. A pilot study on using biochars as sustainable amendments to inhibit rice uptake of Hg from a historically polluted soil in a Karst region of China. *Ecotoxicol. Environ. Saf.* **2019**, *170*, 18–24. [CrossRef]
29. Zhang, M.; Shan, S.D.; Chen, Y.G.; Wang, F.; Yang, D.Y.; Ren, J.K.; Lu, H.Y.; Ping, L.F.; Chai, Y.J. Biochar reduces cadmium accumulation in rice grains in a tungsten mining area-field experiment: Effects of biochar type and dosage, rice variety, and pollution level. *Environ. Geochem. Health.* **2019**, *41*, 43–52. [CrossRef]
30. Rechberger, M.V.; Kloss, S.; Wang, S.L.; Lehmann, J.; Rennhofer, H.; Ottner, F.; Wriessnig, K.; Daudin, G.; Lichtenegger, H.; Soja, G. Enhanced Cu and Cd sorption after soil aging of woodchip-derived biochar: What were the driving factors? *Chemosphere* **2019**, *216*, 463–471. [CrossRef]
31. Wang, A.O.; Ptacek, C.J.; Blowes, D.W.; Gibson, B.D.; Landis, R.C.; Dyer, J.A.; Ma, J. Application of hardwood biochar as a reactive capping mat to stabilize mercury derived from contaminated floodplain soil and riverbank sediments. *Sci. Total Environ.* **2019**, *652*, 549–561. [CrossRef]
32. Wang, S.S.; Gao, B.; Zimmerman, A.R.; Li, Y.C.; Ma, L.; Harris, W.G.; Migliaccio, K.W. Removal of arsenic by magnetic biochar prepared from pinewood and natural hematite. *Bioresour. Technol.* **2015**, *175*, 391–395. [CrossRef]
33. Li, B.; Yang, L.; Wang, C.Q.; Zhang, Q.P.; Liu, Q.C.; Li, Y.D.; Xiao, R. Adsorption of Cd(II) from aqueous solutions by rape straw biochar derived from different modification processes. *Chemosphere* **2017**, *175*, 332–340. [CrossRef]
34. Shen, Z.T.; Hou, D.Y.; Jin, F.; Shi, J.X.; Fan, X.L.; Tsang, D.C.; Alessi, D.S. Effect of production temperature on lead removal mechanisms by rice straw biochars. *Sci. Total Environ.* **2019**, *655*, 751–758. [CrossRef]
35. Cao, X.D.; Harris, W. Properties of dairy-manure-derived biochar pertinent to its potential use in remediation. *Bioresour. Technol.* **2010**, *101*, 5222–5228. [CrossRef]
36. Meier, S.; Curaqueo, G.; Khan, N.; Bolan, N.; Cea, M.; Eugenia, G.M.; Cornejo, P.; Ok, Y.S.; Borie, F. Chicken-manure-derived biochar reduced bioavailability of copper in a contaminated soil. *J. Soils Sediments* **2017**, *17*, 741–750. [CrossRef]
37. Sahin, O.; Taskin, M.; Kaya, E.; Atakol, O.; Emir, E.; Inal, A.; Gunes, A. Effect of acid modification of biochar on nutrient availability and maize growth in a calcareous soil. *Soil Use Manag.* **2017**, *33*, 447–456. [CrossRef]
38. Wang, Y.; Lin, Y.X.; Chiu, P.C.; Imhoff, P.T.; Guo, M.X. Phosphorus release behaviors of poultry litter biochar as a soil amendment. *Sci. Total Environ.* **2015**, *512*, 454–463. [CrossRef]
39. Ho, S.H.; Yang, Z.K.; Nagarajan, D.; Chang, J.S.; Ren, N.Q. High-efficiency removal of lead from wastewater by biochar derived from anaerobic digestion sludge. *Bioresour. Technol.* **2017**, *246*, 142–149. [CrossRef]

40. Jin, J.W.; Li, Y.A.; Zhang, J.Y.; Wu, S.C.; Cao, Y.C.; Liang, P.; Zhang, J.; Wong, M.H.; Wang, M.Y.; Shan, S.D. Influence of pyrolysis temperature on properties and environmental safety of heavy metals in biochars derived from municipal sewage sludge. *J. Hazard. Mater.* **2016**, *320*, 417–426. [CrossRef]
41. Zhang, J.; Wu, S.C.; Xu, Z.T.; Wang, M.Y.; Man, Y.B.; Christie, P.; Liang, P.; Shan, S.D.; Wong, M.H. The role of sewage sludge biochar in methylmercury formation and accumulation in rice. *Chemosphere* **2019**, *218*, 527–533. [CrossRef]
42. Rangabhashiyam, S.; Balasubramanian, P. The potential of lignocellulosic biomass precursors for biochar production: Performance, mechanism and wastewater application—A review. *Ind. Crop. Prod.* **2019**, *128*, 405–423.
43. Zhang, Z.K.; Zhu, Z.Y.; Shen, B.X.; Liu, L.N. Insights into biochar and hydrochar production and applications: A review. *Energy* **2019**, *171*, 581–598. [CrossRef]
44. Shaheen, S.M.; Niazi, N.K.; Hassan, N.E.; Bibi, I.; Wang, H.L.; Tsang, D.C.; Ok, Y.S.; Bolan, N.; Rinklebe, J. Wood-based biochar for the removal of potentially toxic elements in water and wastewater: A critical review. *Int. Mater. Rev.* **2019**, *64*, 216–247. [CrossRef]
45. Ahmad, M.; Rajapaksha, A.U.; Lim, J.E.; Zhang, M.; Bolan, N.; Mohan, D.; Vithanage, M.; Lee, S.S.; Ok, Y.S. Biochar as a sorbent for contaminant management in soil and water: A review. *Chemosphere* **2014**, *99*, 19–33. [CrossRef] [PubMed]
46. Tripathi, M.; Sahu, J.N.; Ganesan, P. Effect of process parameters on production of biochar from biomass waste through pyrolysis: A review. *Renew. Sustain. Energy Rev.* **2016**, *55*, 467–481. [CrossRef]
47. Qambrani, N.A.; Rahman, M.M.; Won, S.; Shim, S.; Ra, C. Biochar properties and eco-friendly applications for climate change mitigation, waste management, and wastewater treatment: A review. *Renew. Sustain. Energy Rev.* **2017**, *79*, 255–273. [CrossRef]
48. Kumar, A.; Saini, K.; Bhaskar, T. Advances in design strategies for preparation of biochar based catalytic system for production of high value chemicals. *Bioresour. Technol.* **2019**, *299*, 122564. [CrossRef]
49. Manyà, J.J. Pyrolysis for biochar purposes: A review to establish current knowledge gaps and research needs. *Environ. Sci. Technol.* **2012**, *46*, 7939–7954. [CrossRef]
50. Xie, T.; Reddy, K.R.; Wang, C.W.; Yargicoglu, E.; Spokas, K. Characteristics and applications of biochar for environmental remediation: A review. *Crit. Rev. Environ. Sci. Technol.* **2015**, *45*, 939–969. [CrossRef]
51. Kambo, H.S.; Dutta, A. A comparative review of biochar and hydrochar in terms of production, physico-chemical properties and applications. *Renew. Sustain. Energy Rev.* **2015**, *45*, 359–378. [CrossRef]
52. Liu, W.J.; Jiang, H.; Yu, H.Q. Development of biochar-based functional materials: Toward a sustainable platform carbon material. *Chem. Rev.* **2015**, *115*, 12251–12285. [CrossRef] [PubMed]
53. Cheng, F.; Li, X.W. Preparation and application of biochar-based catalysts for biofuel production. *Catalysts* **2018**, *8*, 346. [CrossRef]
54. Gul, S.; Whalen, J.K.; Thomas, B.W.; Sachdeva, V.; Deng, H. Physico-chemical properties and microbial responses in biochar-amended soils: Mechanisms and future directions. *Agric. Ecosyst. Environ.* **2015**, *206*, 46–59. [CrossRef]
55. Singh, B.; Singh, B.P.; Cowie, A.L. Characterisation and evaluation of biochars for their application as a soil amendment. *Soil Res.* **2010**, *48*, 516–525. [CrossRef]
56. Cantrell, K.B.; Hunt, P.G.; Uchimiya, M.; Novak, J.M.; Ro, K.S. Impact of pyrolysis temperature and manure source on physicochemical characteristics of biochar. *Bioresour. Technol.* **2012**, *107*, 419–428. [CrossRef]
57. Chen, Z.M.; Chen, B.L.; Chiou, C.T. Fast and slow rates of naphthalene sorption to biochars produced at different temperatures. *Environ. Sci. Technol.* **2012**, *46*, 11104–11111. [CrossRef]
58. Keiluweit, M.; Nico, P.S.; Johnson, M.G.; Kleber, M. Dynamic molecular structure of plant biomass-derived black carbon (biochar). *Environ. Sci. Technol.* **2010**, *44*, 1247–1253. [CrossRef]
59. Suliman, W.; Harsh, J.B.; Abu-Lail, N.I.; Fortuna, A.M.; Dallmeyer, I.; Garcia-Perez, M. Influence of feedstock source and pyrolysis temperature on biochar bulk and surface properties. *Biomass Bioenergy* **2016**, *84*, 37–48. [CrossRef]
60. Hossain, M.K.; Strezov, V.; Chan, K.Y.; Ziolkowski, A.; Nelson, P.F. Influence of pyrolysis temperature on production and nutrient properties of wastewater sludge biochar. *J. Environ. Manag.* **2011**, *92*, 223–228. [CrossRef]

61. Hussain, M.; Farooq, M.; Nawaz, A.; Al-Sadi, A.M.; Solaiman, Z.M.; Alghamdi, S.S.; Ammara, U.; Ok, Y.S.; Siddique, K.H. Biochar for crop production: Potential benefits and risks. *J. Soils Sediments* **2017**, *17*, 685–716. [CrossRef]
62. Yuan, H.R.; Lu, T.; Wang, Y.Z.; Chen, Y.; Lei, T.Z. Sewage sludge biochar: Nutrient composition and its effect on the leaching of soil nutrients. *Geoderma* **2016**, *267*, 17–23. [CrossRef]
63. Peng, X.; Ye, L.L.; Wang, C.H.; Zhou, H.; Sun, B. Temperature-and duration-dependent rice straw-derived biochar: Characteristics and its effects on soil properties of an Ultisol in southern China. *Soil Tillage Res.* **2011**, *112*, 159–166. [CrossRef]
64. Wu, W.X.; Yang, M.; Feng, Q.B.; McGrouther, K.; Wang, H.L.; Lu, H.H.; Chen, Y.X. Chemical characterization of rice straw-derived biochar for soil amendment. *Biomass Bioenergy* **2012**, *47*, 268–276. [CrossRef]
65. Zhao, B.; O'Connor, D.; Zhang, J.L.; Peng, T.Y.; Shen, Z.T.; Tsang, D.C.; Hou, D.Y. Effect of pyrolysis temperature, heating rate, and residence time on rapeseed stem derived biochar. *J. Clean. Prod.* **2018**, *174*, 977–987. [CrossRef]
66. Jindo, K.; Mizumoto, H.; Sawada, Y.; Sanchez-Monedero, M.A.; Sonoki, T. Physical and chemical characterization of biochars derived from different agricultural residues. *Biogeosciences* **2014**, *11*, 6613–6621. [CrossRef]
67. Leng, L.J.; Huang, H.J. An overview of the effect of pyrolysis process parameters on biochar stability. *Bioresour. Technol.* **2018**, *270*, 627–642. [CrossRef] [PubMed]
68. Wang, Z.Y.; Liu, G.C.; Zheng, H.; Li, F.M.; Ngo, H.H.; Guo, W.S.; Liu, C.; Chen, L.; Xing, B.S. Investigating the mechanisms of biochar's removal of lead from solution. *Bioresour. Technol.* **2015**, *177*, 308–317. [CrossRef] [PubMed]
69. Yaashikaa, P.R.; Kumar, P.S.; Varjani, S.J.; Saravanan, A. Advances in production and application of biochar from lignocellulosic feedstocks for remediation of environmental pollutants. *Bioresour. Technol.* **2019**, *292*, 122030. [CrossRef]
70. Ding, W.C.; Dong, X.L.; Ime, I.M.; Gao, B.; Ma, L.Q. Pyrolytic temperatures impact lead sorption mechanisms by bagasse biochars. *Chemosphere* **2014**, *105*, 68–74. [CrossRef]
71. Yang, F.; Zhao, L.; Gao, B.; Xu, X.Y.; Cao, X.D. The interfacial behavior between biochar and soil minerals and its effect on biochar stability. *Environ. Sci. Technol.* **2016**, *50*, 2264–2271. [CrossRef]
72. Rajapaksha, A.U.; Chen, S.S.; Tsang, D.C.; Zhang, M.; Vithanage, M.; Mandal, S.; Gao, B.; Bolan, N.S.; Ok, Y.S. Engineered/designer biochar for contaminant removal/immobilization from soil and water: Potential and implication of biochar modification. *Chemosphere* **2016**, *148*, 276–291. [CrossRef]
73. Weber, K.; Quicker, P. Properties of biochar. *Fuel* **2018**, *217*, 240–261. [CrossRef]
74. Lee, J.W.; Kidder, M.; Evans, B.R.; Paik, S.; Buchanan Iii, A.; Garten, C.T.; Brown, R.C. Characterization of biochars produced from cornstovers for soil amendment. *Environ. Sci. Technol.* **2010**, *44*, 7970–7974. [CrossRef] [PubMed]
75. Yuan, H.R.; Lu, T.; Huang, H.Y.; Zhao, D.D.; Kobayashi, N.; Chen, Y. Influence of pyrolysis temperature on physical and chemical properties of biochar made from sewage sludge. *J. Anal. Appl. Pyrolysis.* **2015**, *112*, 284–289. [CrossRef]
76. Heitkötter, J.; Marschner, B. Interactive effects of biochar ageing in soils related to feedstock, pyrolysis temperature, and historic charcoal production. *Geoderma* **2015**, *245*, 56–64. [CrossRef]
77. Kavitha, B.; Reddy, P.V.L.; Kim, B.; Lee, S.S.; Pandey, S.K.; Kim, K.H. Benefits and limitations of biochar amendment in agricultural soils: A review. *J. Environ. Manag.* **2018**, *227*, 146–154. [CrossRef]
78. Gai, X.P.; Wang, H.Y.; Liu, J.; Zhai, L.M.; Liu, S.; Ren, T.Z.; Liu, H.B. Effects of feedstock and pyrolysis temperature on biochar adsorption of ammonium and nitrate. *PLoS ONE* **2014**, *9*, e113888. [CrossRef]
79. Maroušek, J.; Vochozka, M.; Plachý, J.; Žák, J. Glory and misery of biochar. *Clean Technol. Environ. Policy* **2017**, *19*, 311–317. [CrossRef]
80. Yuan, J.H.; Xu, R.K.; Zhang, H. The forms of alkalis in the biochar produced from crop residues at different temperatures. *Bioresour. Technol.* **2011**, *102*, 3488–3497. [CrossRef]
81. Fidel, R.B.; Laird, D.A.; Thompson, M.L.; Lawrinenko, M. Characterization and quantification of biochar alkalinity. *Chemosphere* **2017**, *167*, 367–373. [CrossRef]
82. Tag, A.T.; Duman, G.; Ucar, S.; Yanik, J. Effects of feedstock type and pyrolysis temperature on potential applications of biochar. *J. Anal. Appl. Pyrolysis.* **2016**, *120*, 200–206. [CrossRef]

83. Magdziarz, A.; Wilk, M.; Wądrzyk, M. Pyrolysis of hydrochar derived from biomass–Experimental investigation. *Fuel* **2020**, *267*, 117246. [CrossRef]
84. Gronwald, M.; Helfrich, M.; Don, A.; Fuß, R.; Well, R.; Flessa, H. Application of hydrochar and pyrochar to manure is not effective for mitigation of ammonia emissions from cattle slurry and poultry manure. *Biol. Fertil. Soils.* **2018**, *54*, 451–465. [CrossRef]
85. Cui, X.Q.; Lu, M.; Khan, M.B.; Lai, C.Y.; Yang, X.E.; He, Z.L.; Chen, G.Y.; Yan, B.B. Hydrothermal carbonization of different wetland biomass wastes: Phosphorus reclamation and hydrochar production. *Waste Manag.* **2020**, *102*, 106–113. [CrossRef]
86. Liu, X.M.; Zhai, Y.B.; Li, S.H.; Wang, B.; Wang, T.F.; Liu, Y.L.; Qiu, Z.Z.; Li, C.T. Hydrothermal carbonization of sewage sludge: Effect of feed-water pH on hydrochar's physicochemical properties, organic component and thermal behavior. *J. Hazard. Mater.* **2020**, *388*, 122084. [CrossRef]
87. Garlapalli, R.K.; Wirth, B.; Reza, M.T. Pyrolysis of hydrochar from digestate: Effect of hydrothermal carbonization and pyrolysis temperatures on pyrochar formation. *Bioresour. Technol.* **2016**, *220*, 168–174. [CrossRef]
88. Wang, M.M.; Zhu, Y.; Cheng, L.R.; Andserson, B.; Zhao, X.H.; Wang, D.Y.; Ding, A.Z. Review on utilization of biochar for metal-contaminated soil and sediment remediation. *J. Environ. Sci.* **2018**, *63*, 156–173. [CrossRef]
89. Xu, M.; Wu, J.; Luo, L.; Yang, G.; Zhang, X.H.; Peng, H.; Yu, X.Y.; Wang, L.L. The factors affecting biochar application in restoring heavy metal-polluted soil and its potential applications. *Chem. Ecol.* **2018**, *34*, 177–197. [CrossRef]
90. Gonzaga, M.I.S.; Matias, M.I.D.A.S.; Andrade, K.R.; Jesus, A.N.D.; Cunha, G.D.C.; Andrade, R.S.D.; Santos, J.C.D.J. Aged biochar changed copper availability and distribution among soil fractions and influenced corn seed germination in a copper-contaminated soil. *Chemosphere* **2020**, *240*, 124828. [CrossRef]
91. Park, J.H.; Choppala, G.K.; Bolan, N.S.; Chung, J.W.; Chuasavathi, T. Biochar reduces the bioavailability and phytotoxicity of heavy metals. *Plant Soil* **2011**, *348*, 439. [CrossRef]
92. Gonzaga, M.I.S.; Mackowiak, C.; de Almeida, A.Q.; Wisniewski Jr, A.; de Souza, D.F.; da Silva Lima, I.; de Jesus, A.N. Assessing biochar applications and repeated Brassica juncea L. production cycles to remediate Cu contaminated soil. *Chemosphere* **2018**, *201*, 278–285. [CrossRef] [PubMed]
93. Zhang, C.; Shan, B.Q.; Zhu, Y.Y.; Tang, W.Z. Remediation effectiveness of Phyllostachys pubescens biochar in reducing the bioavailability and bioaccumulation of metals in sediments. *Environ. Pollut.* **2018**, *242*, 1768–1776. [CrossRef] [PubMed]
94. Yin, D.X.; Wang, X.; Peng, B.; Tan, C.Y.; Ma, L.Q. Effect of biochar and Fe-biochar on Cd and As mobility and transfer in soil-rice system. *Chemosphere* **2017**, *186*, 928–937. [CrossRef] [PubMed]
95. Kim, H.B.; Kim, S.H.; Jeon, E.K.; Kim, D.H.; Tsang, D.C.; Alessi, D.S.; Kwon, E.E.; Baek, K. Effect of dissolved organic carbon from sludge, Rice straw and spent coffee ground biochar on the mobility of arsenic in soil. *Sci. Total Environ.* **2018**, *636*, 1241–1248. [CrossRef]
96. Qiao, J.T.; Li, X.M.; Li, F.B. Roles of different active metal-reducing bacteria in arsenic release from arsenic-contaminated paddy soil amended with biochar. *J. Hazard. Mater.* **2018**, *344*, 958–967. [CrossRef]
97. Li, G.; Khan, S.; Ibrahim, M.; Sun, T.R.; Tang, J.F.; Cotner, J.B.; Xu, Y.Y. Biochars induced modification of dissolved organic matter (DOM) in soil and its impact on mobility and bioaccumulation of arsenic and cadmium. *J. Hazard. Mater.* **2018**, *348*, 100–108. [CrossRef] [PubMed]
98. Bian, R.J.; Joseph, S.; Cui, L.Q.; Pan, G.X.; Li, L.Q.; Liu, X.Y.; Zhang, A.F.; Rutlidge, H.; Wong, S.; Chia, C. A three-year experiment confirms continuous immobilization of cadmium and lead in contaminated paddy field with biochar amendment. *J. Hazard. Mater.* **2014**, *272*, 121–128. [CrossRef]
99. Gao, X.; Peng, Y.T.; Zhou, Y.Y.; Adeel, M.; Chen, Q. Effects of magnesium ferrite biochar on the cadmium passivation in acidic soil and bioavailability for packoi (Brassica chinensis L.). *J. Environ. Manag.* **2019**, *251*, 109610. [CrossRef]
100. Igalavithana, A.D.; Kwon, E.E.; Vithanage, M.; Rinklebe, J.; Moon, D.H.; Meers, E.; Tsang, D.C.; Ok, Y.S. Soil lead immobilization by biochars in short-term laboratory incubation studies. *Environ. Int.* **2019**, *127*, 190–198. [CrossRef]
101. Huang, D.L.; Deng, R.; Wan, J.; Zeng, G.M.; Xue, W.J.; Wen, X.F.; Zhou, C.Y.; Hu, L.; Liu, X.G.; Xu, P. Remediation of lead-contaminated sediment by biochar-supported nano-chlorapatite: Accompanied with the change of available phosphorus and organic matters. *J. Hazard. Mater.* **2018**, *348*, 109–116. [CrossRef]

102. O'Connor, D.; Peng, T.Y.; Li, G.H.; Wang, S.X.; Duan, L.; Mulder, J.; Cornelissen, G.; Cheng, Z.L.; Yang, S.; Hou, D.Y. Sulfur-modified rice husk biochar: A green method for the remediation of mercury contaminated soil. *Sci. Total Environ.* **2018**, *621*, 819–826. [CrossRef] [PubMed]
103. Herath, I.; Iqbal, M.; Al-Wabel, M.I.; Abduljabbar, A.; Ahmad, M.; Usman, A.R.; Ok, Y.S.; Vithanage, M. Bioenergy-derived waste biochar for reducing mobility, bioavailability, and phytotoxicity of chromium in anthropized tannery soil. *J. Soils Sediments* **2017**, *17*, 731–740. [CrossRef]
104. Zhou, J.M.; Chen, H.L.; Tao, Y.L.; Thring, R.W.; Mao, J.L. Biochar amendment of chromium-polluted paddy soil suppresses greenhouse gas emissions and decreases chromium uptake by rice grain. *J. Soils Sediments* **2019**, *19*, 1756–1766. [CrossRef]
105. Lyu, H.H.; Zhao, H.; Tang, J.C.; Gong, Y.Y.; Huang, Y.; Wu, Q.H.; Gao, B. Immobilization of hexavalent chromium in contaminated soils using biochar supported nanoscale iron sulfide composite. *Chemosphere* **2017**, *194*, 360–369. [CrossRef] [PubMed]
106. Wagner, A.; Kaupenjohann, M. Biochar addition enhanced growth of D actylis glomerata L. and immobilized Zn and Cd but mobilized Cu and Pb on a former sewage field soil. *Eur. J. Soil Sci.* **2015**, *66*, 505–515. [CrossRef]
107. Qian, T.T.; Wu, P.; Qin, Q.Y.; Huang, Y.N.; Wang, Y.J.; Zhou, D.M. Screening of wheat straw biochars for the remediation of soils polluted with Zn (II) and Cd (II). *J. Hazard. Mater.* **2019**, *362*, 311–317. [CrossRef]
108. Zhang, H.; Feng, X.B.; Larssen, T.; Shang, L.H.; Li, P. Bioaccumulation of methylmercury versus inorganic mercury in rice (Oryza sativa L.) grain. *Environ. Sci. Technol.* **2010**, *44*, 4499–4504. [CrossRef]
109. Xia, S.P.; Song, Z.L.; Jeyakumar, P.; Bolan, N.; Wang, H.L. Characteristics and applications of biochar for remediating Cr (VI)-contaminated soils and wastewater. *Environ. Geochem. Health* **2019**, 1–25. [CrossRef]
110. Xu, Z.B.; Xu, X.Y.; Tao, X.Y.; Yao, C.B.; Tsang, D.C.; Cao, X.D. Interaction with low molecular weight organic acids affects the electron shuttling of biochar for Cr (VI) reduction. *J. Hazard. Mater.* **2019**, *378*, 120705. [CrossRef]
111. Mandal, S.; Sarkar, B.; Bolan, N.; Ok, Y.S.; Naidu, R. Enhancement of chromate reduction in soils by surface modified biochar. *J. Environ. Manag.* **2017**, *186*, 277–284. [CrossRef]
112. Ni, J.J.; Bordoloi, S.; Shao, W.; Garg, A.; Xu, G.Z.; Sarmah, A.K. Two-year evaluation of hydraulic properties of biochar-amended vegetated soil for application in landfill cover system. *Sci. Total Environ.* **2020**, *712*, 136486. [CrossRef] [PubMed]
113. Zheng, R.L.; Chen, Z.; Cai, C.; Tie, B.Q.; Liu, X.L.; Reid, B.J.; Huang, Q.; Lei, M.; Sun, G.X.; Baltrėnaitė, E. Mitigating heavy metal accumulation into rice (Oryza sativa L.) using biochar amendment—A field experiment in Hunan, China. *Environ. Sci. Pollut. Res.* **2015**, *22*, 11097–11108. [CrossRef] [PubMed]
114. Glaser, B.; Wiedner, K.; Seelig, S.; Schmidt, H.P.; Gerber, H. Biochar organic fertilizers from natural resources as substitute for mineral fertilizers. *Agron. Sustain. Dev.* **2015**, *35*, 667–678. [CrossRef]
115. Xu, P.; Sun, C.X.; Ye, X.Z.; Xiao, W.D.; Zhang, Q.; Wang, Q. The effect of biochar and crop straws on heavy metal bioavailability and plant accumulation in a Cd and Pb polluted soil. *Ecotoxicol. Environ. Saf.* **2016**, *132*, 94–100. [CrossRef]
116. Xu, Y.G.; Qi, F.J.; Bai, T.X.; Yan, Y.B.; Wu, C.C.; An, Z.R.; Luo, S.; Huang, Z.; Xie, P. A further inquiry into co-pyrolysis of straws with manures for heavy metal immobilization in manure-derived biochars. *J. Hazard. Mater.* **2019**, *38*, 120870. [CrossRef]
117. Buss, W.; Mašek, O. High-VOC biochar—Effectiveness of post-treatment measures and potential health risks related to handling and storage. *Environ. Sci. Pollut. Res.* **2016**, *23*, 19580–19589. [CrossRef]
118. Zielińska, A.; Oleszczuk, P. Effect of pyrolysis temperatures on freely dissolved polycyclic aromatic hydrocarbon (PAH) concentrations in sewage sludge-derived biochars. *Chemosphere* **2016**, *153*, 68–74. [CrossRef]
119. Lyu, H.H.; He, Y.H.; Tang, J.C.; Hecker, M.; Liu, Q.L.; Jones, P.D.; Codling, G.; Giesy, J.P. Effect of pyrolysis temperature on potential toxicity of biochar if applied to the environment. *Environ. Pollut.* **2016**, *218*, 1–7. [CrossRef]
120. Yang, X.; Meng, J.; Lan, Y.; Chen, W.F.; Yang, T.X.; Yuan, J.; Liu, S.N.; Han, J. Effects of maize stover and its biochar on soil CO_2 emissions and labile organic carbon fractions in Northeast China. *Agric. Ecosyst. Environ.* **2017**, *240*, 24–31. [CrossRef]

121. Liu, J.Y.; Shen, J.L.; Li, Y.; Su, Y.R.; Ge, T.D.; Jones, D.L.; Wu, J.S. Effects of biochar amendment on the net greenhouse gas emission and greenhouse gas intensity in a Chinese double rice cropping system. *Eur. J. Soil Biol.* **2014**, *65*, 30–39. [CrossRef]
122. Ribas, A.; Mattana, S.; Llurba, R.; Debouk, H.; Sebastià, M.; Domene, X. Biochar application and summer temperatures reduce N_2O and enhance CH_4 emissions in a Mediterranean agroecosystem: Role of biologically-induced anoxic microsites. *Sci. Total Environ.* **2019**, *685*, 1075–1086. [CrossRef] [PubMed]
123. Cheng, H.; Jones, D.L.; Hill, P.; Bastami, M.S. Biochar concomitantly increases simazine sorption in sandy loam soil and lowers its dissipation. *Arch. Agron. Soil Sci.* **2017**, *63*, 1082–1092. [CrossRef]
124. Chen, J.H.; Liu, X.Y.; Zheng, J.W.; Zhang, B.; Lu, H.; Chi, Z.Z.; Pan, G.X.; Li, L.Q.; Zheng, J.F.; Zhang, X.H. Biochar soil amendment increased bacterial but decreased fungal gene abundance with shifts in community structure in a slightly acid rice paddy from Southwest China. *Appl. Soil Ecol.* **2013**, *71*, 33–44. [CrossRef]
125. Anyanwu, I.N.; Alo, M.N.; Onyekwere, A.M.; Crosse, J.D.; Nworie, O.; Chamba, E.B. Influence of biochar aged in acidic soil on ecosystem engineers and two tropical agricultural plants. *Ecotoxicol. Environ. Saf.* **2018**, *153*, 116–126. [CrossRef]

molecules

MDPI

Review

Recent Developments in Chitosan-Based Adsorbents for the Removal of Pollutants from Aqueous Environments

Daniele C. da Silva Alves [1,2], Bronach Healy [1], Luiz A. de Almeida Pinto [2], Tito R. Sant'Anna Cadaval, Jr. [2] and Carmel B. Breslin [1,*]

1 Department of Chemistry, Maynooth University, W23 F2H6 Maynooth, Co. Kildare, Ireland; DANIELE.COSTADASILVAALVES.2021@MUMAIL.IE (D.C.d.S.A.); bronach.healy.2017@mumail.ie (B.H.)
2 School of Chemistry and Food, Federal University of Rio Grande, Rio Grande, RS 96203-900, Brazil; dqmpinto@furg.br (L.A.d.A.P.); titoeq@gmail.com (T.R.S.C.J.)
* Correspondence: Carmel.Breslin@mu.ie

Abstract: The quality of water is continuously under threat as increasing concentrations of pollutants escape into the aquatic environment. However, these issues can be alleviated by adsorbing pollutants onto adsorbents. Chitosan and its composites are attracting considerable interest as environmentally acceptable adsorbents and have the potential to remove many of these contaminants. In this review the development of chitosan-based adsorbents is described and discussed. Following a short introduction to the extraction of chitin from seafood wastes, followed by its conversion to chitosan, the properties of chitosan are described. Then, the emerging chitosan/carbon-based materials, including magnetic chitosan and chitosan combined with graphene oxide, carbon nanotubes, biochar, and activated carbon and also chitosan-silica composites are introduced. The applications of these materials in the removal of various heavy metal ions, including Cr(VI), Pb(II), Cd(II), Cu(II), and different cationic and anionic dyes, phenol and other organic molecules, such as antibiotics, are reviewed, compared and discussed. Adsorption isotherms and adsorption kinetics are then highlighted and followed by details on the mechanisms of adsorption and the role of the chitosan and the carbon or silica supports. Based on the reviewed papers, it is clear, that while some challenges remain, chitosan-based materials are emerging as promising adsorbents.

Keywords: chitosan; adsorbent; carbon; graphene oxide; silica; magnetic separation; dyes; heavy metals; adsorption; Langmuir isotherm

Citation: da Silva Alves, D.C.; Healy, B.; Pinto, L.A.d.A.; Cadaval, T.R.S., Jr.; Breslin, C.B. Recent Developments in Chitosan-Based Adsorbents for the Removal of Pollutants from Aqueous Environments. *Molecules* **2021**, *26*, 594. https://doi.org/10.3390/molecules26030594

Academic Editors: Chiara Bisio and Monica Pica

Received: 7 December 2020
Accepted: 21 January 2021
Published: 23 January 2021

1. Introduction

Improving water quality is one of the major environmental challenges worldwide to be solved, since water resources are increasingly scarce due to population growth, climate change and increased demand for water in industrial and agricultural activities [1]. In addition, the inappropriate disposal of organic and inorganic contaminants combined with disinformation and neglect in the treatment of these compounds can result in irreversible damage to the aquatic environment and, consequently, to humans [2,3]. Dyes, phenolic compounds, metallic ions and micropollutants, such as pesticides and drugs, have all been detected in wastewaters, surface and even drinking water, indicating that the conventional methods used in treatment plants are not optimised for their removal [4]. Consequently, the removal of these pollutants with high toxicity, even when present at low concentrations, has been increasingly studied in the scientific world [5,6].

Several techniques have been developed based on hybrid systems [7], membrane filtration [8] and biological degradation [9] to reduce the content of pollutants in water. However, the slow response, sensitivity and high energy demand are some of the disadvantages of such techniques. In addition, they are not very efficient when the effluent has a low content of suspended colloidal particle and a high load of organic matter. A promising alternative to the treatments mentioned is adsorption, due to its simplicity of operation and effectiveness [10]. Therefore, the search for new adsorbent materials that can be used to remedy aquatic contamination has been stimulated. The biopolymer chitosan is attracting considerable interest as a matrix for adsorbent material development, since this biopolymer has a high density of hydroxyl groups (–OH) and primary amines (–NH_2) that act as active adsorption sites, making it an efficient adsorbent [11].

Chitosan (β–(1–4)–D–glucosamine) is a polysaccharide that possesses two types of monomers, one containing an acetamido group (2–acetamido–2–deoxy–β–D–glucopyranose residues), and another containing an amino group (2–amino–2–deoxy–β–D–glucopyranose residues). Chitosan is not available directly in the environment, it is obtained from chitin (β–(1–4)–*N*–acetyl–D–glucosamine), usually by the alkaline or enzymatic *N*-deacetylation of chitin [12–14]. The characteristic that differentiates the structure of chitin and chitosan is the substitution of the acetamide group at position 2. This directly influences the solubility properties of these compounds, with chitin being insoluble and inert, and chitosan soluble in weak acids [13,15].

Currently, chitin and chitosan are produced commercially in Japan, United States, India, Poland, Australia, and Norway, and to a lesser extent, in Canada, Italy, Chile and Brazil [16,17]. The annual production of chitin in nature has been estimated to be as high as 1×10^{10} to 1×10^{12} tonnes [18], which makes this biopolymer a cheap and available resource. However, as the biodegradation rate of chitin is slow, the production of high amounts through shrimp and seafood processing creates an environmental problem, and the conversion of this waste material into high-value products, such as chitosan, can be an attractive solution [19].

During the past few decades, many researchers have focused on the development of chitosan-based materials to solve problems in environmental and biomedical engineering fields, and in the development of innovative products for other different applications [20–22]. From 2010 to 2015, more than 15,000 articles and about 20 books on chitosan were published worldwide. This ever increasing interest is connected with the attractive properties of chitosan, such as biodegradability, low toxicity and biocompatibility, coupled with the availability of natural resources required for its chemical and enzymatic modifications for specific end uses [19,23]. Chitosan has been used in many fields such as food, medicine, cosmetics and wastewater treatment [24]. In addition, the efficient utilisation of marine biomass resources has become an environmental priority that leads to intensified research on chitosan production and its applications [25].

In this review, the applications of chitosan-based materials as adsorbents for the removal of pollutants from aqueous environments are reviewed. Although there are a number of review articles that describe the environmental applications of chitosan [19,24,26,27], in this review we focus initially on the sources, properties and chemical modifications of chitosan and discuss the various factors that influence its properties as a material for the removal of pollutants. Next, we review the support materials utilised and combined with chitosan, including the emerging carbon and silicon supports, providing a more comprehensive account than previously published.

2. Chitosan

Chitin is the second most abundant natural polysaccharide (biopolymer) on Earth, following cellulose [26]. In 1811, the first chitin was extracted by Henri Bracon from mushrooms and was named initially as "fungine". In 1823, Odier [28] found the same material in the insect exoskeleton and called it chitin. Later, chitin was found in crab shells, confirming that it can be found in the crustacean exoskeleton. In 1859, Charles Roughet discovered that chitin could be transformed into a water-soluble form after chemical modification [19]. In 1894, this chemically modified chitin was called chitosan by Hoppe-Seyler [29,30]. Although chitin was discovered 30 years before cellulose, most of the research was focused on cellulose, due to the high investments made by the textile industries. Thus, chitin and chitosan remained restricted only to basic research during this period. Around 1970, interest in natural products increased and investigations directed at exploring the potential applications of chitosan began to attract considerable attention [13]. In 1977, the "1st International Conference on Chitin and Chitosan" was organised in Boston, USA, where the scientific and industrial communities leveraged the world's interest in these biopolymers [23].

2.1. Source and Production

Chitosan is derived from chitin, usually obtained from natural sources such as the residues of shrimp, crab and lobster, fungal mycelia and green algae [15,31,32]. This biopolymer can also be found in the exoskeletons of insects and fungal cell walls, as shown in Figure 1a. Chitin content in fungi varies between 19% and 42%, compared to the exoskeleton, where chitin can reach up to 75% [29,33]. Not all shellfish wastes are good sources of chitin. The blue crab (*Callinectes*) contains 14% of chitin, while oyster and clam shells have chitin content in the range of 4–6% [29,34]. Chitin content can vary with the proportion of minerals, proteins, and carotenoids, depending on the species, reproductive cycle phase, nutritional status, age, and also the peeling conditions during the processing [13,17]. Chitosan can be obtained by deacetylation of chitin through enzymatic or alkaline methods (Figure 1b). However, the enzymatic method has been limited only to the laboratory scale, while the alkaline method has been more widely used on an industrial scale, due to its short processing time, simplicity, and low operational costs [29].

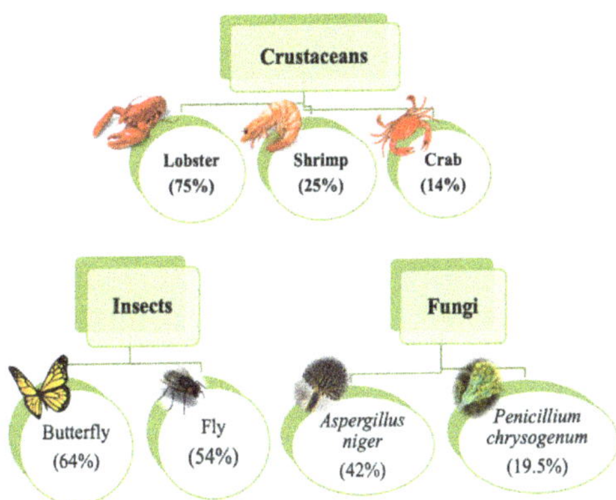

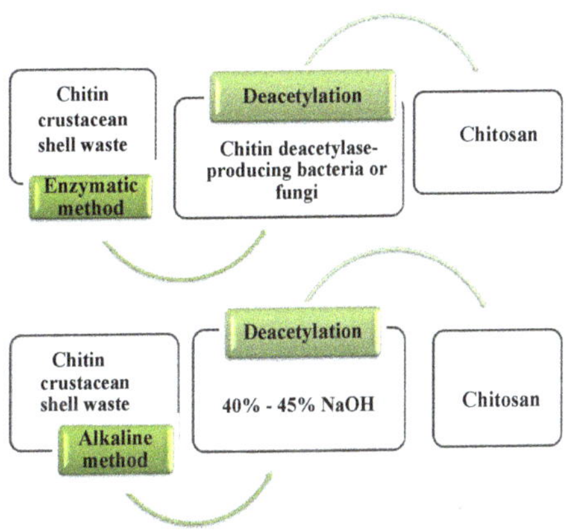

Figure 1. (**a**) Examples of chitin content from different sources, (**b**) Chitin deacetylation methods to produce chitosan.

2.1.1. Enzymatic Methods

The first step in the production of chitosan is the extraction of chitin from the seafood waste and this can be achieved using both biological and chemical approaches. The biological process involves the use of bacteria, that produce organic acids, and enzymes for the demineralisation and deproteinisation of crustacean shells [35]. In the demineralisation step, the lactic acid produced by bacteria reacts with the calcium carbonate component in the crustacean biomass waste resulting in the formation of calcium lactate, which can be precipitated and removed by washing, while proteases from the bacteria eliminate proteins [35]. *Bacillus cereus* A21 has shown high activity with both the demineralisation and deproteinisation steps from shrimp shell wastes, 91% and 80%, respectively [36]. Likewise, the deproteinisation and demineralisation from natural crab shell wastes by *Bacillus pumilus*A1 were 94% and 80%, respectively [37]. These results suggest that the chitin production process by the enzymatic method of the seafood wastes could be applicable and become a friendly environmental alternative.

The deacetylation step that gives rise to the production of chitosan can be promoted by chitin deacetylase. This enzyme was first found in *Mucorrouxii* (*Zygomycetes*) in 1974, by Araki and Ito [38]. In 1984, Davis and Bartnicki-Garcia found evidence that changes in the culture medium during the fungi growth phase can directly influence the production of chitosan [39]. The reaction mechanism of the chitin deacetylase from *Mucorrouxii* is considered as a multipoint attack mode; specifically, the enzyme systematically hydrolyses acetyls from the non-reducing end of the binding site after it binds to a substrate chain, and then leaves the substrate and binds to another one [40]. Strains that produce a large amount of extracellular deacetylase with high activity are very valuable in the production of chitin deacetylase and the production of chitosan. Nevertheless, there are still some problems such as low enzyme activity and low yields of deacetylase-producing strains. Moreover, natural chitins are crystals, not a good substrate for deacetylase. Hence, many preparations still need to be carried out before the chitin deacetylase method can be used in the industrial production of chitosan [41].

Chitin deacetylase-producing bacteria, such as *Serritia* sp. and *Bacillus*sp, may replace the current fungal strains. This microorganism culture method is another possibility to obtain chitosan, removing the acetyl groups by catalysing the substrate with the enzymes produced by these microorganisms. Moreover, bacteria grow faster than fungi in large-scale fermentation processes. Recently, research has concentrated on the breeding of microorganism strains and optimisation of the culture medium. Chitosan formed by this method has shown good ion adsorption capacity [41], making it suitable for environmental applications.

2.1.2. Alkaline Methods

The traditional chemical methods to extract chitin from crustacean shell wastes involve three steps: demineralisation, deproteinisation and decolourisation. In the first step, crustacean shells are washed, dried, and grounded to smaller sizes [42]. In order to remove mineral constituents, mainly calcium carbonate, the powdered raw material is treated with dilute hydrochloric acid followed by the precipitation of calcium chloride. Alkali treatment is used for deproteinisation of the demineralised shells. Proteins are eliminated through solubilising with dilute aqueous sodium hydroxide and in the process, N-acetyl groups within the polymer backbone are hydrolysed. The recovery of protein may be obtained by lowering the pH to about 4.0. An additional decolourisation step can be incorporated when a colourless product is required. Acetone or organic solvent mixtures are used to remove the pigments such as carotenoids [15].

Chitosan is obtained by deacetylation of chitin in 40–45% sodium hydroxide, as shown in the sketch of Figure 2. The alkaline treatments hydrolyse the acetyl groups and transform the *N*–acetyl–D–glucosamine units into D–glucosamine units with free NH_2 groups [15]. Chitosan, with different degrees of deacetylation, is generated depending on the reaction temperature, time, and concentration of the alkaline solution [43]. An additional purification step can be performed to obtain high purity chitosan. For this, the deacetylated product is dissolved in acid, centrifuged or filtered, and the chitosan is precipitated through the addition of alkali generating pure chitosan (90–95%) in paste form. These steps can be repeated to obtain higher purity chitosan (purity >99.9%) [15,23].

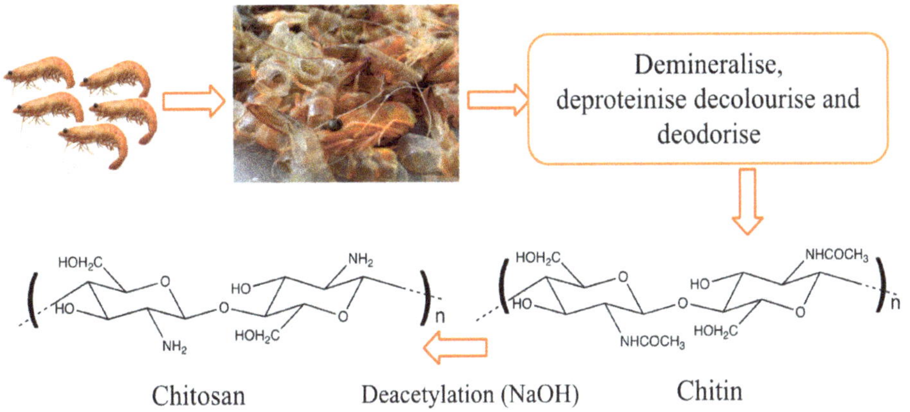

Figure 2. Process of obtaining chitosan by the deacetylation alkaline treatment of chitin from shrimp shell wastes.

The drying of chitosan is an important step in its production. In general, after drying, the desired product should contain a moisture content lower than 10% (wet basis), to ensure good physicochemical and microbiological aspects during prolonged storage. Polymerisation and Maillard reactions are the main alterations that should be avoided during the drying operation [44]. Chitosan is composed mainly of carbohydrate monomer units that, at high temperatures, are capable of undergoing caramelisation of the polymer. Therefore, in this process, one of the key parameters is temperature [45,46]. Different techniques have been used to obtain good quality dried chitosan, such as spouted bed drying [47], spray drying [48], convective tray drying [49], oven drying and infrared drying [50], lyophilization [51] and low-pressure superheated steam drying [52]. All these techniques have shown that the most important factors are temperature and residence time, which must be controlled to obtain a high-quality product. For example, Dotto et al. [47] have shown that an increase in the temperature (from 90 to 100 and 110 °C), using a spouted bed drying technique, causes an increase in powder darkening, an increase in molecular weight (from 147 to 25 kDa) and increased particle size (from 100 to 200 μm). Hence, the best powder quality was obtained at 90 °C, which resulted in the final humidity content being within the commercial range (10%).

2.2. Structure and Properties

Chitosan, a partially deacetylated product of chitin, is a biopolymer composed of β-(1–4)–D–glucosamine, as shown in Figure 3. This biopolymer is a glycosaminoglycan and consists of two common sugars, β-(1–4)–2–acetamido–D–glucose and β-(1–4)–2–amino–D–glucose, glucosamine, and *N*–acetylglucosamin, respectively. The proportion of each depends on the alkaline treatment, and usually deviates from an equal contribution. In terms of structure, chitosan is analogous to cellulose, in which the hydroxyl (in cellulose) has been replaced by acetamido or amino groups (in chitosan) at carbon-2 [53]. Thus, unlike other polysaccharides abundant in carbon, oxygen, and hydrogen; chitin and chitosan contain additional nitrogen (6.89%), making them interesting commercially [54].

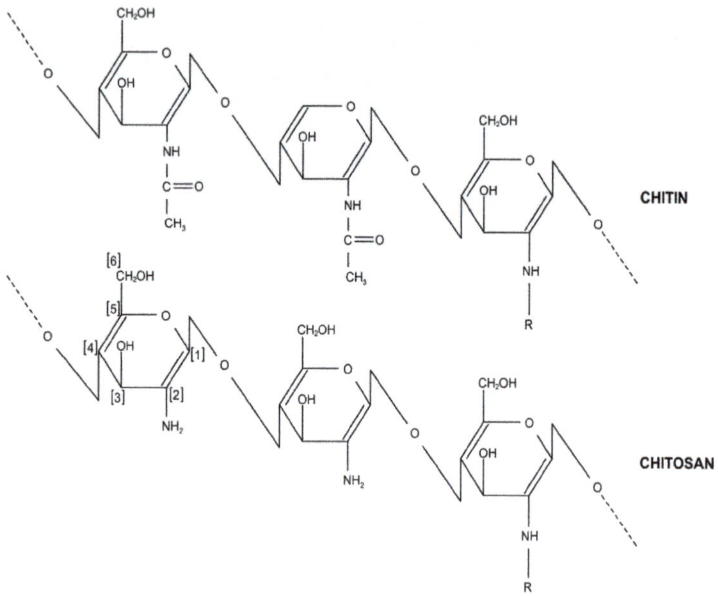

Figure 3. Chemical structure of chitin and chitosan.

The physicochemical properties of chitin and chitosan strongly depend on molecular chain orientation and regular packing. The abundance of hydroxyl groups and highly reactive amino groups in chitosan or its N-acetyl counterpart with a strong tendency for intra- and inter-molecular hydrogen bonding, results in the formation of linear aggregates and rigid crystalline domains. However, chitosan is usually less crystalline than chitin, which presumably makes chitosan more accessible to reagents and, consequently, more soluble. Most of the aqueous acids dissolve chitosan whereas chitin is soluble in very few solvents. The protonation of amino groups by acids along the chitosan chain creates a multitude of cationic sites, which increases its solubility by increasing the polarity. This unique property expands the potential applications of chitosan, including its ability to adsorb different pollutants. Amine groups, for example, are strongly attracted to metal ions due to the lone pair of electrons on the nitrogen atoms [55,56]. The protonation of these amine groups may lead to the electrostatic attraction of anionic compounds, such as anionic dyes [57] and halogens [58]. Moreover, the existence of these free $-NH_2$ and $-OH$ active groups allows the adsorption of other pollutants, such as, phenol [59,60], antibiotics [61,62] and pesticides [63]. Hence, the chitosan adsorption capacity depends on its crystallinity, affinity to water, and deacetylation percentage [64].

The main properties of chitosan are summarised in Table 1. Some of the intrinsic properties of chitosan, such as its polycationic character in acid media, its ability to form hydrogen bonds, van der Walls and electrostatic interactions, make it an efficient adsorbent material. Other characteristics, such as the degree of deacetylation (DD), crystallinity, molecular weight (MW), solubility, surface area and particle size will all influence the properties of the final chitosan-based material and its adsorption potential [23]. Therefore, these properties and their optimisation are central in the formation of efficient adsorbent materials.

Table 1. Main chitosan properties, according to the information reported in [13,27].

Physicochemical properties	• Linear aminopolysaccharide with a high nitrogen content • Ionic conductive • Cationic biopolymer with high charge density • Flocculating agent • Complexing and chelating properties • Existence of reactive groups for chemical activation and cross-linking • Adsorption properties • pKa varies from 6.5 and 6.7 • Enable to form intermolecular hydrogen bonds
Biological properties	• Bioadhesivity • Non-toxic • Bioactivity • Adsorbable • Biodegradable • Antimicrobial activity • Blood anticoagulants • Antiacid, antiulcer and antitumoral properties • Hypolipidermic activity

2.2.1. Deacetylation Degree

The deacetylation degree (DD) of chitosan is one of the most important parameters as it defines the acetyl content in the biopolymer and it can be increased by repeating or prolonging the alkaline treatment step in the chitin deacetylation process [15]. It can be easily determined using several analytical tools, including UV spectrophotometry [65], X-ray diffraction [66], FTIR-spectroscopy [67,68] and titration methods [69–71]. Increases in DD lead to an increase in the number of free amino groups on the chitosan polymeric chain. These amino groups are responsible for differences in the physicochemical properties and structure of chitosan, due to intra- and inter-molecular hydrogen bonds. As a consequence, the chitosan solubility and polycationic character are increased, expanding the applications of chitosan [72].

The control and manipulation of the physicochemical properties of chitosan, such as the mechanical properties, crystallinity, swelling and thermal degradation, have been shown to correlate with the distribution of the acetyl groups along the main chain [73,74]. With an increase in the DD, the charge density along the chain increases and the chitosan chain becomes more flexible, tending to form a random coil with more inter- and intra-molecular hydrogen bonds within the chain. In Table 2, the influence of DD on the physicochemical properties is shown, where it is seen that the DD % has a significant impact.

Table 2. Influence of deacetylation degree (DD) on physicochemical properties of chitosan in different forms.

Form	DD (%)	Properties	Results	Ref.
Films	82, 80–85, 100	Crystallinity, tensile strength, % elongation, swelling index	Crystallinity and tensile strength increased, while % elongation and swelling index decreased with increase in DD (highest MW)	[75]
Powder	76–92	Thermal degradation	Thermal stability decreased with increase in DD	[76]
Films	70–95	Crystallinity, tensile strength, % elongation, swelling index	Crystallinity and tensile strength increased with increase in DD, swelling index decreased with increase in DD; % elongation increased when the DD increased from 70% to 80% and decreased when the DD increased from 80% to 95%	[77]
Microspheres	48, 62, 75	Swelling index	Swelling index increased with increase in DD	[78]
Membranes	75, 87, 96	Crystallinity, swelling index, tensile strength, % elongation	87% DD presented lower crystallinity and mechanical properties, but higher swelling index than 75% or 96% DD	[79]
Nanofibers	59, 76, 85	Thermal degradation	Thermal stability decreased with increase in DD	[80]
Films	15–70	Tensile strength, % elongation	Tensile strength, and % elongation increased with increase in DD	[81]
Sponge	58, 73, 82, 88, 91	Swelling index, tensile strength	Swelling index increased and tensile strength decreased with increase in DD	[82]
Films	72–85	Crystallinity, tensile strength, % elongation, degradation rate	Crystallinity increased and tensile strength decreased with increase in DD, % elongation decreased when the DD was increased from 72% to 75% and decreased when it increased from 75% to 85%. No difference in degradation rates	[83]
Beads	83,94,96	Tensile strength, thermal degradation	Tensile strength, thermal stability increased with increased in DD	[84]

In most cases, an increase in DD was shown to result in an increase in tensile strength and crystallinity and a decrease in the percentage of elongation of the materials. This effect of increasing tensile strength with increasing DD is usually attributed to an increase in crystallinity of the chitosan. Chitosan chains with higher DD have fewer acetyl side groups leading to a more efficient and regular packing of the polymer chains, which in turn, promotes crystallinity in the chitosan [77,85,86]. On the other hand, chitosan with lower DD presents more acetyl side groups that prevent regular packing of the chains due to steric hindrances leading to a reduced crystalline or an amorphous structure [77,86]. Despite the improved tensile strength and stiffness of the chitosan which is observed on increasing the crystallinity this also leads to an increase in the brittleness and a decrease in the percentage of elongation [86]. For example, Zhuang et al. [86] evaluated chitosan films with different DD (81.0%, 88.1% and 95.2%). They reported that the tensile strengths of chitosan films increased from 28.86 to 32.96 MPa and the elongation decreased from 54.31% to 41.66% as DD increased from 81.0% to 95.2%. In applications where film-formation properties of chitosan are important, chitosan with improved tensile strength is an advantage [81,87]. Liu et al. [88] developed composite films of gelatin and chitosan of different MW and DD, and evaluated the interactions between the two polymers in order to improve the films produced. It was verified that the tensile strengths of gelatine films were improved, especially when using chitosan of higher DD and MW. On the other hand, Moura et al. [89] reported a decrease in tensile strength and in elongation of chitosan films with increasing DD, while Nunthanide et al. [75] found both an increase and decrease in tensile strength and elongation, depending on the molecular weight, on increasing DD. These studies which account for the role of MW in the observed results are interesting and highlight the role of both MW and DD.

DD also influences the swelling and thermal degradation characteristics of chitosan [77,78,90,91]. These studies have shown that chitosan with higher DD exhibits faster thermal degradation rates and reduced swelling, as compared to lower DD chitosan. These characteristics may also depend on crystallinity. Chitosan with higher DD and crystallinity are expected to have a close-packed microstructure, which limits water permeation and thus reduced swelling [77]. Moreover, a decrease in the N-acetyl content results in a decrease in the thermal stability as the N-acetyl domains are more thermally stable than the deacetylated ones [92]. Khoulenjani et al. [90], in using chitosan with different DDs (56%, 64% and 74%), showed that the swelling index decreased from 216% to 115% with an increase in DD from 56% to 74%. Nunthanide et al. [75] also reported that the films become more brittle with a lower swelling index with an increase in DD. Wanjun et al. [93] verified that an increase in DD resulted in a decrease in the thermal stability of chitosan due to the decreased acetyl content. This relationship between thermal effects and the DD of chitosan has also been confirmed by Kittur et al. [76]. In another study, Tavares et al. [84] demonstrated that the DD of chitosan had a positive influence on the thermal degradation behaviour. They prepared genipin-crosslinked chitosan beads and evaluated the effect of the chitosan DD (83%, 94% and 96%) on their characteristics. It was verified that the chemical interactions between chitosan and genipin result in a material more thermally stable, especially when a higher chitosan DD (96%) was used. This behaviour was attributed to the decrease in the hydrophilic groups available to form hydrogen bonds with water molecules, resulting in a material more thermally stable.

However, the highly hydrophilic character of chitosan with high DD might be a disadvantage for its surface modification and hence limit the development of chitosan-based materials. Iamsamai et al. [94] have shown that the DD of chitosan plays a critical role in the dispersion of multiwall carbon nanotubes (MWCNTs) and their stability. They confirmed that the chitosan surface coverage on the MWCNTs was twice as high when modifying the surface of the nanotubes with the 61% DD than when using the 93% DD chitosan; suggesting that the dispersion of MWCNTs with chitosan might be improved when using chitosan having a lower DD level.

In addition to the above properties, DD also affects the adsorption properties of chitosan-based materials, since it is linked directly to its cationic properties. Piccin et al. [95] studied the adsorption of FD&C Red 40 dye by chitosan powder with different deacetylation degrees. It was shown that an increase in the DD from 42% to 84% caused an increase in the adsorption capacity from 266 to 373 mg g^{-1}. Habiba et al. [96] prepared a chitosan/polyvinyl alcohol/TiO_2 composite with different DD for methyl orange adsorption. They have shown that the adsorption capacity was higher for the composite containing chitosan with higher DD. Furthermore, the composite containing chitosan with higher DD was more reusable and stable with good adsorption capacity even after 15 regenerations. Józwiaket al. [97] have developed chitosan adsorbents with different forms (flakes and hydrogel granules) and different DD (75%, 85% and 90%) to remove Reactive Black 5 (RB5) from aqueous solutions. The highest adsorption capacity (1559.7 mg g^{-1}) was obtained for the chitosan-hydrogel granules formed with 90% DD. Chitosan hydrogel granules reached up to 224% higher adsorption capacity ($q_{DD75\%}$ = 1307.5 mg g^{-1}) than chitosan in the form of flakes ($q_{DD75\%}$ = 403.4 mg g^{-1}), which indicated that the chitosan form is also important to the adsorption operation. Besides, the DD of chitosan had a particularly large impact on the RB5 adsorption effectiveness of chitosan in the form of flakes. The adsorption on the flakes with a 90% DD was 1049.6 mg g^{-1} and was higher by 260% than on the flakes with 75% DD. All the authors of these studies concluded that the DD had influenced the physicochemical properties as well as the interactions of chitosan with the pollutants in the adsorption process. Therefore, DD is a major factor in matching chitosan to other materials to develop potential adsorbent materials [98].

2.2.2. Molecular Weight

The molecular weight (MW) of chitosan is a characteristic associated with the number of monomeric units per polymer molecule (n). The deacetylation process brings about a change in MW and depending on the source and preparation procedure, the average MW of chitosan may range from 50 to 2000 kDa [99]. The MW of chitosan can be measured by light scattering, high-performance liquid chromatography (HPLC) and viscosimetric methods [23]. Chitosan can be classified as low molecular weight (LMW), medium molecular weight (MMW) and high molecular weight (HMW) [99]. Generally, the MW of chitosan can be modified by using depolymerisation techniques where the high MW chitosan is converted to a lower MW. These MW modifications are important as they can preserve the integral structure of chitosan [99,100].

The control, evaluation and modifications of this characteristic are fundamental since MW affects many of the physicochemical properties, including solubility, viscosity, crystallinity, tensile strength, adsorption and elasticity. Consequently, MW has a significant effect on the applications of chitosan. Zhou et al. [73] prepared hydrogels with different MW of chitosan and verified that the viscosity of the hydrogels increased with MW, increasing from 88 to 1360 kDa at 37 °C. Moreover, the increase of MW was favourable for sol-to-gel transition and high molecular chitosan was optimal for hydrogel preparation. In Table 3, the various molecular weights employed in chitosan-based materials are summarised. In all cases, the MW is a key factor that influences the tensile strength (TS) and elongation-at-break (EB) properties, as well as the different physical forms of chitosan.

Zhong et al. [101] studied the effect of MW on the properties of chitosan films and found that the conductivity, viscosity, surface tension, and crystallinity of the chitosan film were raised with increasing MW due to an increase in the proportion of amine-groups and degrees of chitosan chain entanglements. Moura et al. [89] verified that the tensile strength, elongation-at-break and water barrier properties of chitosan films were improved with an increase in MW. On the other hand, Ziani et al. [102] showed that the low MW films exhibited greater tensile strengths and percentage of elongation compared to the high MW films despite the high DD of the low MW chitosan. In this study, they verified that the MW had more influence on the mechanical properties than DD. These characteristics were attributed to the number of hetero-monomers, which form stronger films than the character of the acetylated or deacetylated monomers. In general, according to Tables 2 and 3, several studies have demonstrated that DD and MW can be used to manipulate the physical-mechanical and the thermal degradation properties of chitosan materials. However, these studies also highlight that there is a significant and complex interaction between DD and MW and this interaction can lead to conflicting results, e.g., tensile strengths have been shown to increase and decrease with an increase in DD, according to the MW, and increasing MW can both increase and decrease the percentage of elongation, depending on DD. It is also noted that the type or mode of fabrication of the chitosan (e.g., films, gels, membranes, etc.) may be further influenced by the DD and MW properties.

Table 3. Tensile strength (TS) and elongation-at-break (EB) properties of chitosan-materials with different molecular weight (MW) chitosan.

Material	Chitosan MW (kDa)	TS (MPa)	EB (%)	Ref.
Chitosan film	6.55	8.67 ± 1.72	32.53 ± 4.78	[101]
	12.93	12.05 ± 2.24	35.52 ± 6.32	
	47.70	11.51 ± 2.25	25.74 ± 3.69	
Chitosan coated cellulose paper	25.00	9.70 ± 1.50	6.7 ± 1.3	[103]
	2100	13.40 ± 1.50	6.9 ± 0.5	
Chitosan film	101.0	22.30 ± 0.2	8.7 ± 0.2	[89]
	153.6	29.50 ± 0.1	11.4 ± 0.2	
	201.7	39.80 ± 0.1	15.7 ± 0.2	
Chitosan-starch composite film	LMW	5.77 ± 0.62	9.04 ± 1.42	[104]
	MMW	20.90 ± 3.52	4.67 ± 0.58	
	HMW	22.30 ± 2.21	9.09 ± 0.42	

2.2.3. Solubility

The solubility of chitosan is a fundamental property that is particularly important in the fabrication of chitosan-based materials [17,26,105]. The main factors that affect this property are DD and MW. It is known that due to the high degree of acetylation, chitin is hydrophobic making it insoluble in water and most organic solvents, decreasing its applications [13]. On the other hand, with higher DD levels, more amino groups in the molecular chain become protonated to give higher degrees of solubility [106,107]. However, an increase in the MW brings about an increase in the intra- and inter-molecular hydrogen bonds within the chains, giving rise to entanglement of the chains and a reduction in solubility [108].Chitosan is soluble in weak acids but insoluble above a pH of 7. The pH has a significant influence on the charged state and properties of chitosan due to the presence of the amino groups [74]. At low pH, the amino groups of chitosan are protonated and become positively charged which leads to a soluble cationic polyelectrolyte. However, as the pH increases to above 6, the amino groups of chitosan are deprotonated, the biopolymer loses its charge, and this gives an insoluble structure. The soluble-insoluble transition occurs at about a pH of 6.5 (p*Ka* of the amino group). This characteristic makes chitosan a cationic polyelectrolyte (p*Ka* ≈ 6.5), one of the few found in nature [19,27].

In addition to the properties mentioned previously, solubility depends also on the type of acid used [13]. Formic acid is one of the best solvents when aqueous solutions of chitosan are required and the formic acid concentrations can range from 0.2–100% [14]. Acetic acid (1%) has been the most used solvent for the solubilisation of chitosan [13]. However, acetic acid solutions with high concentrations and at elevated temperatures can give rise to the depolymerisation of chitosan [109]. Rinaudo et al. [109,110] observed that for acetic and hydrochloric acid, the chitosan solubility was entirely related to the pH and to the ionic strength, while Kurita et al. [111] verified that it was dependent on chain flexibility, degree of ionisation, crystallinity, solvation of the chain, and the presence of acetyl-glucosamine blocks. Shamov et al. [112] have observed that chitosan solubility is also influenced by interactions between the hydrocarbon chains of the carboxylic acids. There are many other factors that have vital effects on chitosan solubility. These factors can include alkali concentration, temperature, time of deacetylation, prior treatments applied to chitin isolation, particle size, etc. [113]. In addition, these studies also highlight that solubility in acidic solution imparts the chitosan with excellent gel-forming properties and can expand the potential applications of chitosan-composite materials.

2.2.4. Surface Area and Particle Size

Chitosan surface area and particle size are important characteristics which are related to the porosity, pore volume and pore size distribution of the chitosan. Surface area and particle size are fundamental for adsorption applications, since accessible sites and a porous structure are required [114,115]. It is known that chitosan powders or flakes are non-porous materials which present a low surface area (lower than 10 $m^2\ g^{-1}$) [23]. Thus, chemical and physical modifications of chitosan have been performed to increase the surface area and improve potential applications [17,27,58,59,116,117]. Phongying et al. [118] obtained chitosan directly from chitin and prepared chitosan nanoscaffolds in order to improve the surface area, particle size and pore volume. They verified that the surface area of their chitosan scaffolds (55.75 $m^2\ g^{-1}$) was approximately seven times higher than the commercial chitosan flakes (7.70 $m^2\ g^{-1}$). Moreover, the pore volume and pore size of the chitosan nanoscaffolds were higher. Esquerdo et al. [119] developed chitosan scaffolds and verified that the new material had a specific surface area, porosity and pore volume of 1135 $m^2\ g^{-1}$, 92.2% and 0.0079 $m^3\ kg^{-1}$, respectively. These values are higher that other chitosan-based materials, such as chitosan powders (surface area of 4.2 $m^2\ g^{-1}$ and pore volume of 9.5×10^{-6} $m^3\ kg^{-1}$) [120], chitosan flakes (surface area range of 4–6 $m^2\ g^{-1}$), chitosan beads (surface area range of 30–40 $m^2\ g^{-1}$) [121], chitosan hydrogel beads (porosity of 85%) [122], and chitosan–graphene mesostructures (surface area of 603.2 $m^2\ g^{-1}$) [123]. These studies confirm that modification of chitosan leads to an improvement in the surface area and, consequently, in the porosity and pore volume.

Moreover, the particle size of the adsorbents has a significant effect on the final solute concentration, and hence on the overall performance of the adsorption process. Larger particle sizes reduce the uptake due to the lower specific surface area. Thus, an increase in surface area of adsorbent results in new active sites formed, thus allowing more binding of solute molecules [124]. Piccin et al. [120] investigated the effects of particle size, surface area and pore volume of chitosan on the adsorption of FD&C Red 40. The particle sizes used were 0.10, 0.18 and 0.26 mm, with surface areas of 4.2, 3.4 and 1.6 $m^2\ g^{-1}$, respectively. The results showed that an increase in the surface area and a decrease in particle size doubled the adsorption capacity. Dotto et al. [57] evaluated chitin and chitosan as adsorbents for tartrazine dye. They verified that chitosan showed better adsorbent properties than chitin due to its higher deacetylation degree and higher surface area, pore volume and pore size. These characteristics are particularly important for adsorption applications because it provides access to large pollutant molecules, enabling them to reach the internal adsorption sites.

3. Chitosan Supports

Although chitosan is an effective adsorbent for a variety of pollutants (as illustrated in Section 2), it nevertheless suffers from poor mechanical properties and thermal stability combined with a relatively low surface area. Therefore, it is not surprising that it has been modified with a variety of other additives to form composites or hybrids. These additives include cellulose [125,126], starch [127], other biopolymers such as alginate [128], gelatin [129], clays, such as bentonite [130], zeolites [131], metal organic frameworks (MOFs) [132], conducting polymers, such as polypyrrole [133] and other polymeric systems comprising methacrylamide [134],polyacrylamide [135], polyurethane [136], poly(vinyl alcohol) [137] and lignosulfonate [138]. These additives are interesting because they can form interpenetrated polymers with chitosan.

3.1. Chitosan Combined with Carbon-Based Materials

In more recent times, there has been considerable interest in combining chitosan (CS) with carbon-based materials as many carbon-based materials have very good adsorption qualities and these materials can also be employed to enhance the surface area of the adsorbents. Shown in Figure 4 is a summary of the number of papers published in 2019 and 2020 that have employed chitosan combined with various carbon-based materials as adsorbents. It is clearly evident from this analysis that it is graphene and especially graphene oxide (GO) that is dominating the carbon-based materials, with a somewhat lower number of papers describing the use of activated carbon. In the following sections, these CS/carbon-based materials are introduced, highlighting their properties and abilities to facilitate adsorption.

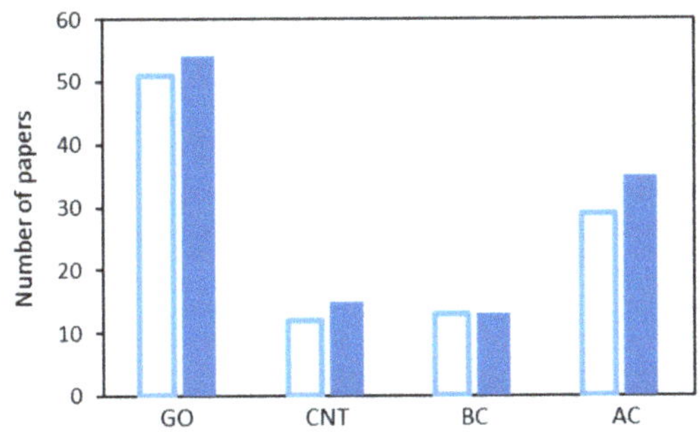

Figure 4. Number of papers published in 2019 (open) and 2020 (solid) focussed on various CS/carbon-based materials, where the carbon materials are graphene oxide (GO), carbon nanotubes (CNT), biochar (BC) and activated carbon (AC). All data taken from Scopus.

3.1.1. Chitosan/Graphene Composites

Since its discovery, graphene has been used in a wide variety of applications, ranging from sensors [139], batteries [140], electro-Fenton [141] to electronics [142]. It has also been recognised as an adsorbent material, as it possesses a large surface area and there is considerable evidence to show that π-π interactions occur between the aromatic rings of various organic pollutants and the basal planes of graphene [143]. These π-π interactions occur between aromatic pollutants and pristine graphene, but fortunately graphene oxide, which is considerably easier to synthesise and is more cost effective, is an especially promising adsorbent [144]. GO contains a number of oxygen containing functional groups, such as epoxides (C–O–C), hydroxyl (–OH), carboxylic (–COOH) and carbonyl groups (>C=O) [145], while other oxygen containing groups, such as ketones and quinones, have also been detected [146]. These functional groups can facilitate the binding of positively charged molecules through electrostatic interactions [147]. Indeed, numerous studies have demonstrated the excellent ability of GO to adsorb various planar aromatic molecules, such as dyes, through a combination of π–π stacking, electrostatic interaction and hydrogen bonding [148,149].

Graphene oxide (GO) is normally synthesised by oxidising graphite using the well-known modified Hummers method [150]. The interlayer spacing increases as the graphite is oxidised to give GO sheets that can be exfoliated through a relatively simple liquid-phase exfoliation and/or ultrasonication. The GO sheets are stable in colloidal solutions and are easily combined with chitosan to give CS/GO composites. Typically, the chitosan is dissolved in acetic acid and the GO is added to form a homogeneous mixture. The CS/GO hydrogel can be easily formed, by a combination of violent shaking and sonication [144], adding NaOH [151], freeze drying [152], or by employing solvothermal reactions [153]. Chitosan is a positively charged polysaccharide at near neutral pH due to protonation of the amino groups and therefore it attracts the negatively charged GO sheets. These electrostatic interactions combined with hydrogen bonding facilitates the formation of the CS/GO hydrogel to give stable composites with excellent thermal and mechanical properties [154], as illustrated in the schematic provided in Figure 5. Indeed, it has been shown by Fan et al. [155] using FTIR measurements, that the –NH groups on the chitosan chains react with the –COOH groups of GO to form a linking –NHCO– group. Using these approaches, CS/GO composites have been formed as beads [156], membranes [157,158] and columns [144,159] and employed successfully as adsorbents for the removal of pollutants from aqueous media.

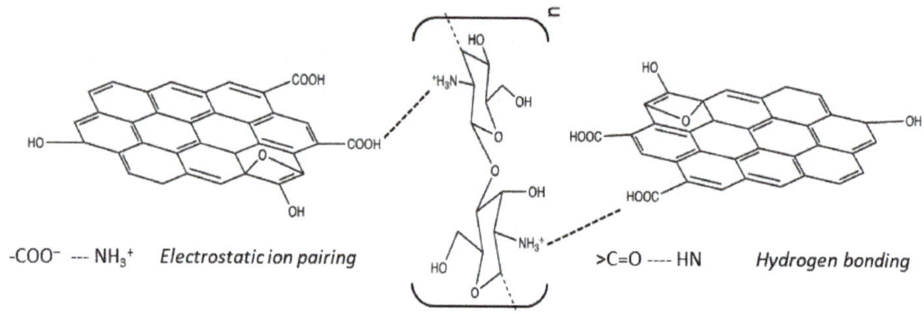

Figure 5. Schematic representation of the interactions between chitosan and GO.

Several studies have been reported using CS/GO composites and these hydrogels have been employed to adsorb and remove various dyes from water [160], heavy metal ions [161], phenolic compounds [162] and pharmaceutical and personal care products [163]. In more recent years, other components have been added in an attempt to further enhance the adsorption capacity of the CS/GO composites, while three-dimensional GO and graphene based aerogels have also been developed and these are now described in turn.

Magnetic Chitosan/GO

Magnetic chitosan has emerged as an exciting new material in environmental applications and recently there has been much interest in the applications of magnetic CS/GO [164,165]. The introduction of magnetism facilitates the separation of the adsorbent from the aqueous medium through a simple magnetic process [166]. It is normally difficult to separate chitosan-based adsorbents, and indeed other adsorbents, from aqueous environments through conventional filtration and sedimentation techniques, as these adsorbents can block filters and are often lost, contributing to secondary pollution. The Fe_3O_4, a ferromagnetic black iron oxide, is the most widely employed, as it possesses good compatibility, low toxicity and also has high magnetic properties [167]. Furthermore, it contains both Fe(II) and Fe(III), and with the presence of Fe(II), which has the potential to act as an electron donor, oxidation of the pollutants can be achieved. Fe_3O_4 can also be formed as rods, spheres, wires and nanoparticles and these can be combined with CS/GO. There has also been a report where FeO(OH) was utilised with CS/GO [168], while γ-Fe_2O_3 has been combined with chitosan and employed as a magnetic adsorbent [169].

Magnetic CS/GO can be easily formed through both in-situ [170] and ex-situ methods [171] and variations of these two approaches. The GO/Fe_3O_4 can be initially formed before being combined with chitosan [172], or the CS/Fe_3O_4 can be firstly formed [173]. For example, Singh et al. [165] used the reactions between the carboxyl and epoxy groups on GO and the amine groups on chitosan to form amide and hydroxyl functionalised groups that facilitated the conversion of the iron ions to the iron oxide, enabling the in-situ preparation of the magnetic CS/GO. Alternatively, the Fe_3O_4 nanoparticles can be initially synthesised using simple methods, such as co-precipitation using ferric and ferrous salts, as illustrated in the schematic provided in Figure 6a. The Fe_3O_4 nanoparticles are then combined with the CS/GO hydrogel [174]. Using these approaches, Tran et al. [175] showed that a large number of the Fe_3O_4 nanoparticles were immobilised onto the GO sheets, Figure 6b, while Rebekah et al. [164] also concluded that the Fe_3O_4 nanoparticles became attached to the edges and basal planes of GO.

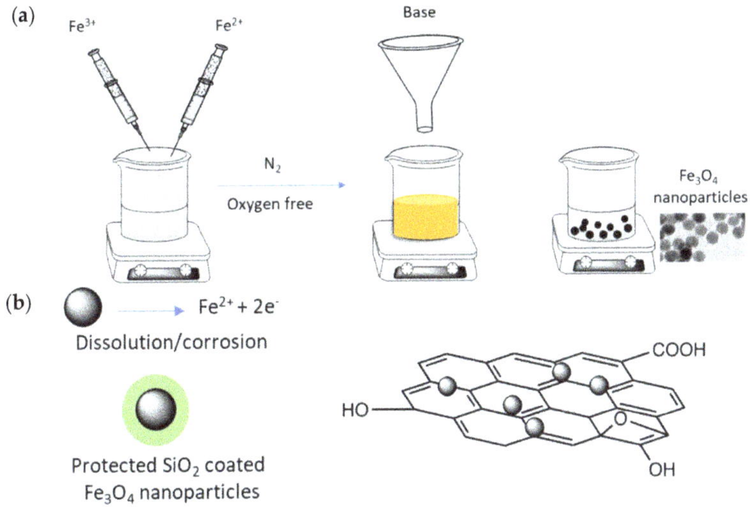

Figure 6. (**a**) Schematic representation of the co-precipitation method used to prepare Fe_3O_4 (**b**) SiO_2 coated Fe_3O_4 nanoparticles deposited and dispersed on GO.

The dispersion and aggregation, size dispersion and shape of the Fe_3O_4 nanoparticles within the hydrogels are all important characteristics in terms of their performance as adsorbents. In general, the Fe_3O_4 nanoparticles appear aggregated, due to their magnetic nature [173]. Some authors have estimated the particle sizes or have observed some isolated particles among the clusters. Spherical Fe_3O_4 clustered particles were observed by Gul et al. [167] with some isolated particles of approximately 90 nm. Shafaati et al. [176] have prepared spherical Fe_3O_4 particles with an average size of 45 nm with evidence of some agglomeration, but when they were combined with chitosan an increase in the particle size was observed, indicating more extensive agglomeration during the reaction with chitosan or as the authors suggested, the chitosan polymer chains may provide links between the neighbouring Fe_3O_4 particles. Again, Jiang et al. [177] have shown that the Fe_3O_4 particles can become severely aggregated, but when the Fe_3O_4 particles were coated with silica the aggregation was markedly reduced. TEM micrographs indicated that the GO sheets were decorated with the silica coated Fe_3O_4 particles with the more wrinkled GO sheets providing more adsorption sites for the particles. A similar finding, highlighting the role of silica in reducing aggregation of the Fe_3O_4 particles, was reported by Tang et al. [178]. This reduction in the aggregation was attributed to a decrease in the dipole-dipole interactions between the silica modified Fe_3O_4 nanoparticles. Furthermore, the inert silica coating layers can protect the magnetic cores as the Fe_3O_4 particles are susceptible to dissolution and corrosion in acidic solutions, which lead to the loss of magnetism [179], as illustrated in Figure 6b.

Chitosan/rGO

While GO is the main form of graphene employed with chitosan, there is also evidence to show that reduced GO, designated as rGO, can be employed to give CS/rGO hydrogel adsorbent materials. The rGO is formed through the reduction of GO and this can be achieved using various thermal approaches, where the GO is heated to high temperatures to transform the oxygen-containing groups to gaseous CO or CO_2 [180], reducing agents, such as borohydride or ascorbic acid [181], or through the electrochemical reduction of GO [182–184]. However, it is very difficult to completely reduce GO and maintain it in the fully reduced form and therefore rGO will always contain some oxygen-containing functional groups. The rGO is considerably more conducting compared to GO, and therefore it can be easily decorated with various metals or metal oxide particles or single atoms. Indeed, Pradeep and co-workers [185] employed the conducting nature and properties of rGO to form well dispersed and uncapped silver, gold, platinum, palladium and manganese oxide decorated rGO, which was then supported within a chitosan hydrogel. The redox reaction between the metal ion precursors and rGO leads to the progressive oxidation of rGO back to GO, providing the metal decorated graphene sheets with functional groups, facilitating its incorporation within chitosan.

CS/rGO has also been combined with Fe_3O_4 to give magnetic CS/rGO adsorbents and employed to give the effective adsorption of an antibiotic [186] and dyes [187]. While the conducting rGO can be beneficial in depositing well dispersed metal/metal oxide particles through reduction, there is evidence to show that CS/GO composites have a higher adsorption capacity when compared with the reduced GO counterparts. For example, Gu et al. [188] compared the performance of chitosan combined with GO and rGO in adsorbing and removing a dye from aqueous solutions and found that while adsorption was evident with both systems, the CS/GO was the more efficient adsorbent. This appears to be related to the presence of the functional groups providing a combination of π–π stacking, electrostatic interaction and hydrogen bonding with the pollutants [148,149].

Chitosan with 3D Graphene, Graphene Aerogels, Foams and Sponges

Although GO sheets can be well dispersed within chitosan, restacking of these sheets can occur over time to give GO aggregates and this, in turn, will reduce the surface area of the adsorbent, reducing its adsorption capacity. Consequently, there has been increasing interest in using three-dimensional (3D) GO or rGO hierarchical macrostructures for environmental applications [189,190]. The 3D GO structures can be fabricated as foams, sponges and as porous or macro-porous aerogels [191] and are based on the bending and wrinkling of the GO sheets to give a low mass density and very high specific surface areas [192]. These 3D materials have the potential to act as scaffolds with very good mechanical strength and a high specific surface area, facilitating adsorption. Moreover, they are easily recovered from the liquid phase following adsorption. However, 3D GO and rGO structures without any other additives can have relatively poor stability in water, but this stability can be enhanced considerably by combining the 3D GO network with biopolymers such as chitosan. Indeed, it was shown by Ma et al. [193], in studying the adsorption and removal of methylene blue, that the GO foam was susceptible to collapse, but its macroscopic morphology could be maintained over three repeated uses when combined with chitin. Similarly, 3D GO combined with high molecular weight chitosan was successfully applied in five repeated cycles of adsorption followed by regeneration, achieving a 90% adsorption capacity [152]. Very good stability and recyclability was also achieved with layered chitosan/GO sponges, with a regeneration efficiency greater than 80% over five cycles [194]. A number of CS/aerogel composites have been formed and these have been employed successfully in the removal of Cu(II) [195], tetracycline [196], azo dyes [197], anionic and cationic dyes [198], hexavalent chromium [199] and 4-nonylphenol [151].

Chitosan/GO with Other Additives

Other additives have been combined with CS/GO adsorbents and these have included β-cyclodextrins exploiting the hydrophobic properties of the β-cyclodextrin to enhance the adsorption of dyes. These β-cyclodextrin modified CS/GO composite materials have been fabricated and employed to adsorb methylene blue [200]. In this case the authors clearly showed that the extent of adsorption was enhanced on going from GO to CS/GO to CS/GO/β-cyclodextrin, illustrating the beneficial effects of incorporating the β-cyclodextrin. Yan et al. [201] employed a similar CS/GO/β-cyclodextrin composite to adsorb Mn(II), while Li et al. [202] found that the added β-cyclodextrin improved the adsorption of Cr(VI). Similar findings were reported in studying the adsorption of hydroquinone [203] and dye molecules [204].

Polypyrrole, a well-known conducting polymer, has also been combined with CS/GO by polymerising the corresponding pyrrole monomer within the CS/GO dispersion. This gives ternary hydrogel composites with a conducting polymer that has the ability to bind anionic and cationic species as dopants and these materials have been shown to give efficient adsorbents [205,206]. Moreover these hybrids can be further decorated with magnetic nanoparticles, enabling the removal of the adsorbent from water following the adsorption process [207]. Other polymeric systems that have been combined with CS/GO include polyacrylamide [208] and polyacrylate [209]. These high molecular weight polymers can improve the swelling and adsorption behaviour of the CS/GO hydrogels.

Although chitosan has a number of binding sites for metal ions, some of these are consumed in the crosslinking with the GO sheets. Consequently, additives that have additional binding sites have been added with the aim of enhancing the adsorption capacity. Particularly interesting additives include polydopamine, a mussel adhesive, that is easily formed through the oxidation and polymerisation of dopamine in slightly alkaline solutions [210]. It is a promising adsorbent material [211]. Polydopamine has a high density of amine and catechol groups and the combination of chitosan and polydopamine gives more binding groups and has been used to adsorb Cr(VI) [212] and Cu(II), Pb(II) and Cd(II) [213]. Other interesting materials are layered double hydroxides (LDHs) that have the general formula $[M^{2+}{}_{1-x}M^{3+}{}_{x}(OH)_2]^{x+}[(A^{n-})_{x/n}mH_2O]$ where M^{2+} and M^{3+} are the divalent and trivalent cations, respectively, such as Fe^{2+} and Al^{3+}, while A^{n-1} represents the intercalating anions. These layered materials have very good adsorption properties for metal ions and have been used extensively for the removal of heavy metal ions [214]. It is not surprising that LDHs have recently been combined with chitosan and GO to give efficient adsorbents with enhanced adsorption performances [215,216]. Recently, metal-organic frameworks (MOFs) have also been combined with CS/GO [217,218] to give good adsorption properties. MOFs have received considerable interest in environmental science and chemistry as these materials have high porosity and high specific surface areas, with tunable pore structures. Indeed they have been used for heavy metal adsorption and are attracting applications in wastewater treatment [219]. However, MOFs, which are typically powders, are difficult to separate from aqueous environments and this is limiting their environmental applications. The CS/GO hydrogel provides a matrix for encapsulating these powdered materials and as detailed earlier the GO sheets can be easily decorated with magnetic iron to introduce magnetic separation.

A number of other additives has been combined with CS/GO, such as kaolin as a filler to enhance the mechanical strength of the hydrogel composite [220], lignosulfonate for additional binding sites [221,222], triethylenetetramine providing amine groups to enhance adsorption [223], hydroxyapatite to enhance strength and adsorption capacity [224] and silica as it contains a number of silanol groups (Si–OH) [225] and it can be furthermore employed to aid the dispersion of GO within chitosan to give effective adsorbents [226]. Moreover, other biopolymers have been combined with chitosan to form blends which are then combined with GO to give high performing adsorbents. These comprise CS/GO/gelatin [227], CS/GO/alginate [228], CS/GO/heparin [229] and CS/GO/cellulose blends [230].

3.1.2. Chitosan/Carbon Nanotubes

Carbon nanotubes (CNTs), like GO sheets, have high surface areas and excellent stability. Therefore, there has been considerable interest in combining these carbon-based materials with chitosan to give adsorbent materials. CNTs are now readily synthesised as single-walled (SWCNT) and multi-walled nanotubes (MWCNT), distinguished by the number or graphitic layers folded over to form the tubes, with very high aspect ratios. They can be well dispersed within chitosan minimising their agglomeration. For good dispersion, the CNTs are normally treated in nitric acid to generate –COOH groups [231] and these groups can also bind with the chitosan. More recently, the CNTs have been functionalised with valine and starch to aid their dispersion within chitosan and enhance their affinity for the adsorption of heavy metal ions [232]. In addition, they have been coated with polydopamine thin films to aid dispersion and minimise aggregation within chitosan [233]. Similar to that employed in the formation of CS/GO, the CS/CNTs are formed by initially dissolving the chitosan in acetic acid, then the CNTs are added, dispersed and normally a crosslinking agent, such as glutaraldehyde [215], is used. These CS/CNT composites have been employed as adsorbents and used in the removal of Cr(VI) [234], V(V), Cr(VI), Cu(II), As(V) and Ag(I) from biological and environmental samples [235], Cu(II) [236], U(VI) [237], Pb(II) [238], phosphate [239], phenol [60], fluoride [58], diazinon [240], food dyes [241] and dyes [242].

Magnetic separation has also been developed and this provides a convenient method to remove the CS/CNT adsorbents from the aquatic environment. This is especially important for CNTs as there is considerable concern over the environmental and ecological risks associate with the release of CNTs into the environment [243,244]. For example, Zhou et al. [245] decorated CNTs with $-NH_2$ functionalised super paramagnetic $CoFe_2O_4$ nanoparticles and combined these magnetic CNTs with chitosan and employed the resulting composites for the removal of Pb(II) and tetrabromobisphenol A. Magnetic Fe_3O_4 nanoparticles have also been used to form magnetic CS/CNTs composites and employed to remove Pb(II) [246].

Multicomponent and multifunctional CS/CNTs have also been formed. For example, CS/CNT has been further modified with poly(acrylic acid) and poly(4–aminodiphenylamine). The resulting adsorbent enabled the removal of Cr(VI) through adsorption and reduction to the Cr(III) species. The partially oxidised poly(4–aminodiphenylamine) was transformed in the presence of Cr(VI) into its fully oxidised form with the corresponding reduction of Cr(VI) to Cr(III) [247]. Alsabagh et al. [248] have fabricated a multifunctional nanocomposite comprising chitosan, well dispersed silver and copper nanoparticles and CNTs for the adsorption of Cu(II), Cd(II) and Pb(II). Other components have been added to CS/CNT and these include a prussian blue analogue [249], while a cellulose acetate (CA) and chitosan solution were used as an electrospinning solution and employed to form multicomponent electrospun CA/CS/CNTs/Fe_3O_4/TiO_2 nanofibers [250].

CS/CNTs have also been formulated to give selective adsorption. While many adsorbents can give relatively high adsorption capacity, it is more challenging to obtain selective adsorption. One avenue that can be employed is imprinting technology. This has been used successfully with ion imprinted polymers, whereby the polymer is formed with a template molecule through a copolymerisation process. The template molecule is then removed leaving behind cavities in the polymer matrix with an affinity for that template, facilitating its rebinding. Li et al. [251] have used this approach to form CS/CNTs composites for the selective capture of Gd(III) by imprinting the chitosan with the Gd(III).

3.1.3. Chitosan/Biochar

Biochar (BC) is a porous carbon rich material which is obtained through the pyrolysis of organic matter, in the presence of a limited concentration of oxygen. It has attracted much attention in environmental applications as it has a porous structure [252,253]. Moreover, it is a cost-effective material as it is fabricated from wastes, mainly agricultural and forestry waste materials. However, the adsorption capacity of biochar is limited and the density of the functional groups on its surface depend on the pyrolytic temperature with a general loss in these functional groups as the pyrolytic temperature is increased [253]. Accordingly, much attention has been paid to the modification of the biochar production process and modification of the surface through oxidation and/or functionalisation, to give more effective adsorbents [254,255]. Treatment of the BC with H_2O_2 is an interesting modification that gives rise to an increase in the concentration of the oxygen-containing functional groups and aids the removal of heavy metal ions from water [254].

Chitosan has been coated onto biochar surfaces [256] and employed as a dispersing and stabilising reagent to form CS/BC composites [257,258]. The BC powders are difficult to retrieve from aqueous solutions, but when the BC is incorporated within the chitosan

hydrogel, it is more readily separated from the solution phase. Separation can be further facilitated by forming magnetic CS/BC hydrogels [259,260]. In addition, by using the chitosan solution phase, it is possible to add a number of other additives or reagents in addition to the BC, giving the composite more functional properties. For example, while CS/BC composites have a number of functional groups, such as amine and hydroxyl groups, other additives that increase the number of functional groups can be introduced within the hydrogel matrix. Using this approach, pyromellitic dianhydride (PMDA) has been employed as it can react with the amine groups of chitosan to give additional amides and carboxyl groups and this facilitates electrostatic interactions and complexation with heavy metal ions [261]. Indeed, it was found that the CS/PMDA modified BC exhibited selective adsorption for Cu(II) and this was attributed to the *N*-containing functional groups and carbonyl groups. Moreover, poly(acrylic acid), with carboxylate groups, was grafted to the chitosan modified BC to give not only additional functional groups, but also enhance chemical stability with stronger intermolecular forces [262]. Supramolecules, such as cyclodextrins, which have hydrophobic cavities and hydrophilic exteriors, and can form inclusion complexes with a wide range of organic molecules, have also been combined with CS/BC to give higher performing adsorbents [263]. These CS/BC composites have been employed as adsorbents in a number of studies to remove heavy metal ions from water [264], including Cr(VI) [265]. In addition, they have been utilised in the removal of phosphates [266], nitrates and phosphates [267], fluorides [268], benzoates [269] and various antibiotic and pharmaceutical molecules, such as diclofenac, ibuprofen and naproxen [270], ciprofloxacin [271,272] and ofloxacin [273].

3.1.4. Chitosan/Activated Carbon

Activated carbon (AC) is well-known as an adsorbent material. It has been used in a number of environmental applications [274]. However, its relatively high cost is limiting its more widespread applications. One of the more commonly used starting materials in the synthesis of AC is coal [275], but given the depleted amounts of coal now available, this gives rise to an increase in the price of coal-based AC. Consequently, there is a recent focus on developing more environmentally acceptable synthesis and fabrication methods, using starting materials such as mandarin peel [276] and coconut shell [274]. Another avenue being exploited is the fabrication of multifunctional adsorbent materials that contain relatively small amounts of AC. Chitosan, with its high density of functional groups and good dispersion properties, is an ideal companion material. Indeed, there is evidence to suggest that this combination is effective as an adsorbent material. On comparing the maximum adsorption capacities of AC, chitosan and CS/AC for Cd(II), Hydari et al. [277] observed values of 10.3, 10.0, and 52.63 mg g^{-1} for AC, chitosan and CS/AC, respectively. Likewise, Auta and Hameed [278] observed synergistic effects between AC and chitosan in the removal of cationic and anionic dyes, while Fatombi et al. [279] also concluded that the best performance was achieved with a CS/AC composite.

CS/AC composites have been formed using commercially available activated carbon, coconut shell charcoal/carbon [280,281], renewable waste tea [278], sapotaceae seed shells [282] *Typhalatifolia* leaves [283] and olive stones as the carbon source [284]. These composites can be formed through a surface modification process, where the surface of the AC is modified by chitosan [285]. Babel et al. [280] concluded that surface modification of coconut shell charcoal with chitosan significantly improved the adsorption of Cr(VI). They also found that the pre-treatment of the AC with acids gave rise to enhanced adsorption. Amuda et al. [281] arrived at a similar conclusion, and showed that chitosan coated acid treated coconut shell carbon was very effective in the removal of Zn(II). Alternatively, the chitosan can be dissolved in acid and then the AC can be added in the form of a powder to generate CS/AC [286,287]. Crosslinking agents, such as glutaraldehyde [288], genipin [289] or epichlorohydrin [290], can be added to generate the composite hydrogels. With this latter approach, the ratio of activated carbon to chitosan can be easily varied [291], while other additives can be introduced. For example, a CS/AC was formed with SiO_2/Fe_3O_4

to give magnetic CS/AC [288], while CS/AC was combined with an anionic surfactant, sodium dodecyl sulphate (SDS), to adsorb a cationic dye [292]. In addition, chitosan has been blended with polyvinyl alcohol [293,294], and polyethylene glycol [295] and then combined with activated carbon, while CS/CA has also been combined with alginate to form CS/CA/alginate adsorbent beads [289].

Activated carbon fibres, regarded as the third generation of carbonaceous adsorbents, have also been employed with chitosan. These have been utilised as membranes [296] and have been decorated with iron oxides and modified with chitosan to remove arsenic, phenol and humic acid from water, with high adsorption capacity for As(V) [297]. Magnetic activated carbon nanofibers based on chitosan and cellulose acetate have also been fabricated for the adsorption of Cr(VI), Ni(II) and phenol from aqueous solutions [298]. In addition, a number of magnetic CS/AC composites has been formed with the majority involving Fe_3O_4 [299,300], while others have employed $CoFe_2O_4$ [301] and barium ferrite [302].

These CS/AC composites have been evaluated for the removal of phenols [303], parabens [304], dyes [305], food dyes [306], anti-inflammatory drugs [307], acetaminophen [308], organic molecules, such as aniline [309], and various heavy metal ions [289]. Generally, there is good agreement that the combination of chitosan and AC gives rise to enhanced adsorption, when compared to the individual chitosan and AC counterparts.

3.2. *Chitosan Combined with Inorganic Adsorbent Materials*

While chitosan has been combined with various carbon based materials, as illustrated in Section 3.1, there is growing interest in the use of inorganic components, such as activated alumina [310], mesoporous alumina [311], silica and ordered mesoporous silica-based materials, as the chitosan support materials [312,313]. Silica has very good physical, mechanical and thermal stability and can be easily functionalised due to its hydroxyl groups. In particular, mesoporous silica is a fascinating material, which first gained prominence in the 1990s with a regular mesostructure, with uniform pore distribution and tunable pore sizes, very high specific surface areas, combined with thermal and mechanical stability [314]. It is attracting considerable interest as an adsorbent material [315]. These materials can be formed by a simple sol–gel synthesis route comprising hydrolysis, condensation and polycondensation reactions using various templates or surfactant molecules [316]. In particular, the template-assisted mesoporous silica synthesis using surfactants is gaining considerable attention. Typically, liquid silicon alkoxide precursors, such as tetramethyoxysilane or tetraethoxysilane are used. The successive polymerisation, gelation, drying and aging steps can be tailored to control the microstructure of the final materials. The surfactant–silica assembly occurs simultaneously with condensation of the inorganic species to produce the mesoporous silica composite.

CS/silica composites have been formed using a variety of methods which can be broadly grouped into two main approaches, comprising silica supported chitosan, where the chitosan is coated or adsorbed onto the silica support, and secondly a CS/silica hybrid that is fabricated using the sol-gel methodology. Several reports have focussed on SiO_2 as a bead, particle, nanoparticle or powder, where the SiO_2 particles are added to the chitosan solution phase to give a chitosan coated particle [317]. The SiO_2 particles can also be functionalised with amine and carboxylic groups to give more efficient binding with the chitosan [318]. Silica layers have also been added to previously formed chitosan-based beads to give organic-inorganic (CS/silica) layered structures, with greater stability [319] and sol-gel synthesis has been employed to immobilise chitosan onto silica particles [320]. Sol-gel synthesis is more commonly used to form a CS/silica hybrid layer on silica bead/particle supports [321,322]. For example, Xu et al. [323] covalently linked chitosan with an epoxide containing siloxane through the sol-gel process to give a hybrid chitosan layer on silica particles. The CS/silica hybrid has been further modified with EDTA (ethylenediaminetetraacetic acid), which is very well known to form stable chelates with a number of metal ions [324], to give adsorbents for heavy metal ions [325]. While the sol-gel synthesis is very versatile, Blachnio et al. [326], on comparing three CS/silica composites formed by the adsorption of chitosan on silica gel and fumed silica and by the sol-gel process, concluded that the adsorbed chitosan had a higher adsorption capacity for dye molecules, although the CS/silica fabricated using the sol-gel synthesis had a high surface area of 600 $m^2\ g^{-1}$.

There has been considerable interest in combining mesoporous silica with chitosan to combine the good adsorption properties of chitosan with the large surface area and adjustable pore size of silica. Likewise, magnetic mesoporous silica, which has a magnetic Fe_3O_4 core surrounded by the mesoporous silica is attracting a lot of attention in environmental applications [327]. These magnetic materials are environmentally acceptable with no toxicity, are biocompatible, have high surface area, very good stability and the outer mesoporous silica can be functionalised and modified by chitosan to add functional groups. The cross-linking method can be employed to decorate the mesoporous silica with the chitosan. Cross-linking agents, such as glutaraldehyde [328], formaldehyde [329] and epoxides [327], can be used, while in a recent study, He et al. [330] used thiol-ene click chemistry to achieve binding between chitosan and magnetic mesoporous silica. The surface areas, pore sizes and volumes of a number of these materials are summarised in Table 4. In general, the surface area, pore size and volume of the mesoporous silica are reduced as higher amounts of chitosan are added and partially fill the pores. However, these chitosan and mesoporous silica composites possess good surface areas with a high density of functional groups and with the potential to give magnetic separation.

Table 4. Surface area, pore diameter and pore volume of CS/mesoporous silica composites and hybrids.

System	Surface Area/$m^2\ g^{-1}$	Pore Size /nm	Pore Volume/ $cm^3\ g^{-1}$	Ref
CS	130.2	3.98	0.482	[331]
CS/silica (electrospun)	272.3	3.52	0.431	
SBA-15	809.4	6.6	1.10	[332]
SBA-15 (10% CS)	653.9	6.6	0.90	
SBA-15 (20% CS)	461.9	6.7	0.80	
CS	150	-	0.753	[312]
CS/silica (43% CS)	342	-	1.092	
Silica	739	-	3.645	
SBA-15	876	7.8	1.30	[333]
SBA-15/CS/Fe_2O_3	446	6.7	0.90	
CS/silica (81.3% Si)	357.3	8.18	0.730	[325]
CS/silica (74.4% Si)	309.7	6.19	0.479	
CS/silica (59.9% Si)	268.1	6.08	0.407	

4. Adsorption and Removal of Pollutants

The removal of pollutants from aquatic environments through adsorption is one of the more popular approaches in environmental applications. The aim in these technologies is to remove the maximum amount of pollutant and therefore adsorption isotherms have been used extensively to develop an understanding of the adsorption equilibria. In this section, these adsorption models are briefly introduced, followed by adsorption kinetics and finally a comparison of the performance of the various chitosan composites is made.

4.1. Adsorption Models and Adsorption Kinetics

Adsorption isotherms are frequently employed in the study of adsorption, facilitating a quantitative comparison of different adsorbent materials. In addition, they are often used to optimise the use of adsorbents, by observing the adsorption capacity as a function of the experimental conditions. Several different isotherm models have been employed to analyse experimental adsorption data and these include the Langmuir, Freundlich, Temkin, Frumkin, Redlich-Peterson (R-P), Halsey, Henderson and Dubinin-Radushkevich isotherms. However, the two most frequently used models with chitosan and chitosan-based composite materials are the Langmuir [223,334] and to a lesser extent the Freundlich isotherms [335]. The Langmuir adsorption model is described in Equation (1) and the linear form commonly employed in fitting data in Equation (2). Here q_e is the equilibrium concentration of the adsorbate, q_m is the monolayer adsorption capacity, C_e is the concentration of the adsorbate in the aqueous phase and K is a constant. In this model all sites are considered as energetically equivalent, to give monolayer adsorption with no interactions between adjacent adsorbates. In this analysis, the adsorbent has a finite capacity for the adsorbate and a saturation point is reached where no further adsorption occurs. The BiLangmuir model can also be applied with chitosan-based composites [317] and, in this case, the relationship is given in Equation (3), where q_{m1} and q_{m2} represent the maximum adsorption capacities of two different adsorption sites and K_{L1} and K_{L2} correspond to these two sites. The Freundlich model assumes multilayer adsorption on a heterogeneous surface and can be described by Equations (4) and (5), where q_e represents the amount of adsorbent adsorbed at the surface, C_e is the equilibrium concentration, and n and K_F are the Freundlich constant and Freundlich exponent, respectively. The Freundlich constant, K_F, provides a measure of the adsorption capacity and the magnitude of n is related to the extent of adsorption with $n > 1$, indicating favourable adsorption. An adsorption plot, using simulated data, is illustrated in Figure 7, where a schematic of monolayer and multilayer adsorption is also shown. In this example, the experimental data are more consistent with the Freundlich isotherm.

$$q_e = \frac{q_m K_L C_e}{1 + K_L C_e} \tag{1}$$

$$\frac{C_e}{q_e} = \frac{1}{K_L q_m} + \frac{C_e}{q_m} \tag{2}$$

$$q_e = \frac{q_{m1} K_{L1} C_e}{1 + K_{L1} C_e} + \frac{q_{m2} K_{L2} C_e}{1 + K_{L2} C_e} \tag{3}$$

$$q_e = K_F C_e^{1/n} \tag{4}$$

$$\log q_e = \log K_F + \frac{1}{n} \log C_e \tag{5}$$

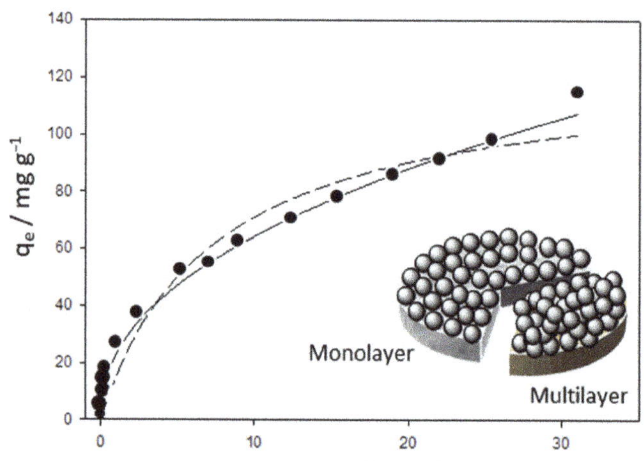

Figure 7. Schematic of an adsorption plot showing experimental data (symbols), with the —— Freundlich isotherm and – – – Langmuir isotherm fitting and the inset shows monolayer and multilayer adsorption processes.

The kinetics of the adsorption process are important as these studies provide information on the rate of adsorption which is relevant in terms of the contact time required to remove the maximum amount of adsorbate. The Lagergren rate equation is one of the most widely used adsorption rate equations for the adsorption of adsorbates from a solution phase and this has been used with various chitosan composites [336]. The pseudo-first order (PFO) and pseudo second-order models (PSO) are described in Equations (6) and (7) where q_t and q_e represent the mass of the adsorbing molecule per unit mass of adsorbent at time t and at equilibrium, while k_1 and k_2 correspond to the first- and second-order rate constants and t is the time. Other kinetic models have been employed in the study of chitosan composite materials and these include a double-exponential kinetic model [337], and a generalised fractal kinetic model (Brouers-Sotolongo model) [338]. Generally, the adsorption kinetics are controlled by the (i) rate of diffusion of the adsorbate from the bulk solution to the adsorbent-solution boundary, (ii) diffusion form the boundary layer to the adsorbent surface, (iii) diffusion of the adsorbate within the adsorbent material, i.e., intraparticle diffusion and (iv) the rate of the adsorption step. Normally, the diffusion process in the bulk solution can be eliminated through agitation, while the adsorption is fast and the rate-determining step is typically intraparticle diffusion [200].

$$\ln(q_e - q_t) = lnq_e - k_1 t \tag{6}$$

$$\frac{t}{q_t} = \frac{1}{q_t} + \frac{1}{k_2 q_e^2} \tag{7}$$

4.2. A Comparison of the Chitosan Supported Composites in the Adsorption of Pollutants

The performance of the chitosan composites in the removal of heavy metal ions is summarised and illustrated in Tables 5–7, where the chitosan is combined with GO, CNTs, BC, AC and silica as support materials.

The Langmuir isotherm is based on the assumption that the surface of the absorbent is homogenous, and every adsorption site is equal. Although these chitosan-composites have various functional groups, which in turn, give rise to different affinities with the adsorbates, the Langmuir model and to a lesser extent, the Freundlich model, correlate well with most of the experimental studies. It is also evident in Tables 5–7, that many of the 3D or porous supports give higher adsorption capacity values, highlighting the influence of the more porous materials, while the magnetic chitosan materials, (MSC), with Fe_3O_4 particles/nanoparticles, also perform well. It is difficult to make a direct comparison between the carbon-based materials and silica-based supports, as the adsorption capacity depends on the nature of the chitosan. However, some of the highest adsorption capacities are seen when chitosan is combined with GO.

It can be seen from these tables that a large variety of heavy metal ions have been adsorbed and removed. This is not surprising as these ions are toxic and pose a significant threat to both human beings and aquatic life. As illustrated, the adsorption capacity varies considerably from relatively low values of 9.4 mg g^{-1} to much higher values in the vicinity of 957 mg g^{-1}. These variations appear to be somewhat related to the nature of the heavy metal ions, with relatively high adsorption values for Pb(II) and lower values for Cr(VI). However, the nature and properties of the chitosan, including its DD levels, MW, porosity, particle size, solubility (Section 2), crosslinking agents and the ratio of chitosan to the carbon-based or silica supports are also important elements that will influence the extent of adsorption.

Table 5. Adsorption performance of CS/GO composites in the removal of heavy metal ions.

Adsorbent	Adsorbate	pH	Ads. Cap./mg g^{-1}	Kinetic Model	Isotherm Model	Ref
CS/GO-SH	Cd(II)	5.0	177	PSO	Freundlich	[339]
	Pb(II)	5.0	447	PSO	Freundlich	
	Cu(II)	5.0	425	PSO	Freundlich	
CS/GO	Cu(II)	6.0	254	PSO	Langmuir	[195]
CS/GO aerogel	Cu(II)	6.0	407	PSO	Langmuir	
CS/GOnano-fibrous	Pb(II)	6.0	461	Double-exp	R–P	[337]
	Cu(II)	6.0	423	Double-exp	R–P	
	Cr(VI)	3.0	310	Double-exp	R–P	
MCS/GO with EDTA	Pb(II)	5.0	206	PSO	Langmuir	[336]
	Cu(II)	5.5	207	PSO	Langmuir	
	As (III)	8.0	43	PSO	Freundlich	
MCS/GO	Cu(II)	7.0	217	PSO	Langmuir	[340]
MCS/GO	As(III)	7.3	2.3	PSO	Langmuir	[341]
CS/GO/MOF	Cr(VI)	3.0	144	PSO	Langmuir	[217]
MCS/GO	Cr(VI)	2.0	270	PSO	Langmuir	[342]
MSC/GO with IL	Pb(II)	5.0	85	PSO	Langmuir	[343]
MSC/GO gel beads	Cd(II)	6.0	86	PSO	Langmuir	[228]
	Pb(II)	5.0	189	PSO	Langmuir	
	Cu(II)	5.0	55	PSO	Langmuir	
MSC/3D-GO	Pb(II)	8.5	957	-	-	[344]
CS-GO/CMC aerogel	Cr(VI)	-	127	-	Langmuir	[230]
MCS/GO	Cr(VI)	2.0	100	PSO	Freundlich	[345]
CS/GO	U(VI)	6.0	78	PSO	Freundlich	[346]
MCS/3D graphene	Pb(II)	8.5	947	-	-	[344]
CS/GO-PVA	Cd(II)	8.0	172	PSO	Langmuir	
CS/GO-PVA	Ni(II)	8.0	71	PSO	Langmuir	[347]
MCS/GO-EDTA	Pb(II)	8.3	666	PFO	Langmuir	[172]
CS/GO gel	Pb(II)	6.0	470	PSO	Langmuir	[348]
CS/GO-silica	Pb(II)	6.0	256	PSO	Langmuir	[225]

In terms of the experimental conditions employed in these studies, it is well documented that the pH plays a significant role in the adsorption process. It is well known that the $-NH_2$ and –OH groups on the chitosan chains have a strong association with metal cations and this facilitates the adsorption of various heavy metal ions at slightly acidic or near neutral pH values. This is clearly evident in Tables 5–7, where several of the studies are performed in slightly acidic solutions. As the chitosan becomes protonated, with the formation of NH_3^+ at lower pH values, this protonated group repels the cationic heavy metal ions. Moreover, the chitosan becomes less stable and more soluble in highly acidic environments. Indeed, the pH of zero charge (pH_{pzc}) has been determined as 6.0 for CS/GO composites, indicating that the surface of CS/GO is positively charged for pH values < 6.0 but for pH values > 6.0, the surface adopts a negative charge [365]. However, the speciation of the metal ions is also important, and this is illustrated in Figure 8, where the Pourbaix diagrams for various heavy metals, designated as M, and Cr are shown. In general, for heavy metals, such as Cu(II), the adsorption capacity increases gradually as the pH increases from about 2 to 7 and then it decreases rapidly as the pH is further increased. The rapid decrease at higher pH values is due to the formation of insoluble hydroxide species. While higher concentrations of the metal cations are present in the solution phase at low pH values, the adsorption capacity is poor, which can be attributed to the protonation of chitosan. As the pH is increased, the$-NH_2$ chelating groups become available and at these conditions the concentrations of the metal cations is still sufficiently high to facilitate chelation with the chitosan. The nature of the –COOH functional groups on GO and the other carbon-based materials is also pH dependent with the generation of $-COO^-$ at higher pH values and again this anionic group will bind with metal cations. Therefore, the maximum adsorption is seen at pH values from about 5 to 7, as illustrated in Tables 5–7, for a number of heavy metal ions. The Pourbaix diagram of Cr is somewhat different with the generation of the anionic dichromate ($Cr_2O_7^{2-}$) and chromate ($HCrO_4^-$) ions at the lower more acidic pH values and the insoluble oxide phases at higher pH values. In this case, the $HCrO_4^-$ ions can be adsorbed at the chitosan through electrostatic interactions with the protonated chitosan, to give more favourable adsorption at pH values between approximately 2.0 and 4.0.

Table 6. Adsorption performance of CS/CNT, CS/BC and CS/AC composites in the removal heavy metal ions.

Adsorbent	Adsorbate	pH	Ads. Cap./ mg g^{-1}	Kinetic Model	Isotherm Model	Ref
CS/CNT/ $CoFe_2O_4$	Pb(II)	6.0	140	PSO	Langmuir	[245]
MCS/CNT	Pb(II)	5.0	101	PSO	Sips	[246]
CS/CNT/PDA	Cu(II)	7.0	112	PSO	Langmuir	[233]
CS/CNT	Cu(II)	7.0	115	PSO	Langmuir	[349]
CS/CNT at 293 K	Cr(VI)	2.0	142	PSO	Langmuir	[350]
at 303 K	Cr(VI)	2.0	151	PSO	Langmuir	
at 313 K	Cr(VI)	2.0	164	PSO	Langmuir	
CS/CNT/PB	Cs(I)	6.0	219	PSO	Freundlich	[249]
	Sr(II)	6.0	205	PSO	Freundlich	
CS/CNT	U(VI)	4.0	126	PSO	Langmuir	[351]
MCS/CNT	Cr(III)	4.0	66	PSO	Langmuir	[352]
	Cr(VI)	4.0	449	PSO	Langmuir	
CS/BC/ PMDA	Cu(II)	5.0	96	PSO	Langmuir	[261]
	Pb(II)	5.0	13	PSO	Langmuir	
	Cd(II)	5.0	38	PSO	Langmuir	
CS/BC/β-CD	Cr(VI)	2.0	206	PSO	Freundlich	[263]
MCS/BC	Cr(VI)	3.0	30	PSO	Freundlich	[353]
	Cu(II)	5.8	54	PSO	Freundlich	
CS/BC	Hg(II)	3.0	594	PSO	Langmuir	[258]
	Pb(II)	5.0	210	PSO	Langmuir	
CS/BC/PAA	Mn(II)	3–7	139	PSO	Langmuir	[262]
	Co(II)	3–7	135	PSO	Langmuir	
	Pb(II)	3–7	476	PSO	Langmuir	
CS/BC/Clay	Cu(II)	5.0	121	Elovich	Freundlich	[354]
	Pb(II)	5.0	336	PSO	Temkin	
	Zn(II)	5.0	134	Elovich	Freundlich	
CS/AC	Zn(II)	6.0	60	-	Langmuir	[281]
CS/AC	Cr(VI)	5.0	84	PSO	Langmuir	[282]
CS/AC	Pb(II)	5.0	125	PFO	Freundlich	[287]
	Cd(II)	5.0	69	PFO	Freundlich	
CS/PEO/AC	Fe(III)	3.0	217	-	Langmuir/	[355]
	Cu(II)	5.0	195	-	Freundlich	
CS/AC	Hg(II)	7.0	576	-	Langmuir	[289]
CS/AC/PVA	Cr(VI)	2.0	109	PSO	Langmuir	[293]

Abbreviations: PMDA: pyromellitic dianhydride; PAA: poly(acrylic acid); PVA: poly(vinyl alcohol); PEO: poly(ethylene oxide); PB: Prussian blue.

Table 7. Adsorption performance of CS/silica composites in the removal heavy metal ions from water.

Adsorbent	Adsorbate	pH	Ads. Capacity /mg g^{-1}	Kinetic Model	Isotherm Model	Ref
CS/Silica Mesoporous	V(V)	6.5	16	-	-	[329]
	Cu(II)	6.5	21	-	-	
	Pb(II)	6.5	22	-	-	
	Cd(II)	6.5	12	-	-	
	Hg(II)	6.5	13	-	-	
MCS/Silica/ PAM	Cu(II)	5.0	43	PSO	Langmuir	[356]
	Pb(II)	5.0	63	PSO	Langmuir	
	Hg(II)	5.0	263	PSO	Langmuir	
CS/SBA-15	Pb(II)	5.0	57	PSO	Langmuir	[357]
MCS/SBA-15	Cu(II)	6.0	107	PSO	Langmuir	[333]
	Zn(II)	6.0	100	PSO	Langmuir	
MCS/SBA-15	Zn(II)	6.0	107	PSO	Langmuir	[358]
CS/Silica	Hg(II)	6.0	204		Langmuir	[359]
	As(V)	6.0	198		Freundlich	
CS/KCC-1	Pb(II)	9.0	168	PSO	Langmuir	[360]
MSC/Silica	Cu(II)	5.0	73	PSO	Langmuir	[361]
SC/Silica	Cr(VI)	3.0	9	PSO	Langmuir	[362]
MSC/Silica	Cu(II)	6.0	350	-	Freundlich	[363]
MSC/Silica aerogel	Cd(II)	8.0	71	PSO	Langmuir	[364]
CS/GO/Silica	Pb(II)	6.0	256	PSO	Langmuir	[225]
MCS/Silica	Cr(III)	4.0	39	PSO	Bi-langmuir	[317]

Abbreviations: polyacrylamide (PAM).

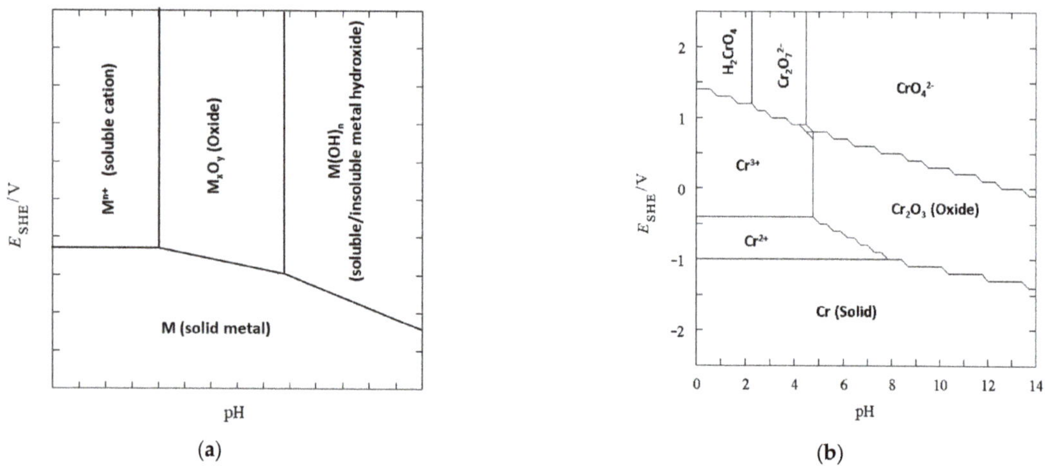

Figure 8. Pourbaix diagrams illustrating the speciation of (**a**) heavy metals, M, and (**b**) Cr in aqueous solutions.

Other conditions that can alter the adsorption capacity of the heavy metal ions are temperature and ionic strength. Generally, the adsorption capacity increases with higher temperatures [195], as ΔG becomes more negative, implying that the adsorption process becomes more favourable at higher temperatures, with $\Delta G° < 0$, $\Delta H° > 0$ and $\Delta S° > 0$ for several adsorbents, however, adsorption can also be exothermic [337]. There have been relatively few studies devoted to the selectivity of the adsorption process at these chitosan composites. This is especially important in terms of the potential applications of the adsorbents, as the water samples or industrial effluents are likely to contain other cations and anions, such as Cl^-, NO_3^-, SO_4^{2-}, Mg^{2+} and Na^+. For example, it was found that an increase in the ionic strength inhibited the adsorption of Cu(II) at CS/GO [195]. One approach that can be employed to enhance the binding of a particular heavy metal ion is templating [251]. This is generally successful provided the target metal ion is reasonably different to the size of the co-existing ions.

The adsorption capacity of the various supported chitosan composites in the removal of dye molecules and other organic molecules, including antibiotics, is summarised in Table 8. Various cationic and anionic dyes have been employed as model compounds and very impressive adsorption capacities have been obtained in a number of studies. In particular, the adsorption of methylene blue (MB), a cationic dye molecule, is very high at CS/GO composites reaching values > 1000 mg g^{-1} in a number of studies, as illustrated in Table 8. This good adsorption is attributed mainly to the $\pi-\pi$ interactions between the MB and GO layers. The electrostatic interactions between the $-COO^-$ groups on GO and the cationic MB can also facilitate adsorption, provided the solution pH is not acidic giving rise to the formation of the unionised –COOH groups. Therefore, it is the GO that is largely responsible for the adsorption of cationic dyes [144]. On the other hand, the chitosan plays a more significant role in the adsorption of anionic dye molecules. Although $\pi-\pi$ interactions will exist between GO and the anionic dyes, the electrostatic repulsion between the $-COO^-$ groups and the anionic dye will inhibit its adsorption at higher pH values, where the COOH groups are ionised. The electrostatic interactions between the cationic centres in the chitosan chains and the anionic dyes will have a significant effect at low pH values, while a combination of electrostatic forces, van der Waals interactions and hydrogen bonding are likely to occur at higher pH values [366]. This explains the good adsorption of anionic dyes observed at near neutral pH values in Table 8. The concentrations of the dye molecules can also influence the adsorption process. Many of these dye molecules can form dimers and aggregates and this becomes more relevant as the concentrations of the dye molecules increase, with aggregates forming in solution and at the surface. Indeed, the impressive adsorption of rhodamine B [367] and its adsorption kinetics were questioned as the aggregation of rhodamine B in water was not taken into account in the original study [368].

Various antibiotics have also been adsorbed at the chitosan-based composites, as shown in Table 8. Again, many of these molecules have benzene rings which facilitate their adsorption onto the carbon surface through π-π electron donor-acceptor interactions. Furthermore, many of these organic molecules have –OH, >C=O and $-NH_2$ groups which can be involved in hydrogen bonding with the oxygen groups on the carbon surfaces in the chitosan composites [369]. Hydrophobic interactions may also be relevant [370]. These organic molecules tend to be hydrophobic making hydrophobic interaction between the antibiotics and the carbon surfaces possible. However, high numbers of oxygen-containing groups, such as –OH and –COOH on the carbon surfaces tend to make the surface more hydrophilic.

The CS/silica composites are only emerging as potential adsorbents and compared with the chitosan-carbon based systems, there are much fewer reports focused on the removal of dyes and organic molecules with these adsorbents. This may be due in part to the silanol groups, which are hydrophilic, and easily form hydrogen bonds with water, thus limiting the adsorption process. However, the mesoporous silica surfaces can be functionalised, and this provides the opportunity to design more hydrophobic surfaces that can be tailored to adsorb organic molecules. Indeed, there is clear evidence in Table 8 that the CS/silica composites can be employed in the removal of dyes.

Table 8. Adsorption performance of chitosan composites in the removal pharmaceuticals, organics and drug molecules from water.

Adsorbent	Adsorbate	pH	Ads. Capacity	Kinetic Model	Isotherm Model	Ref
CS/GO	MB	5.3	95	PSO	Langmuir	[155]
MCS/GO β-CD	MB	-	84	PSO	Langmuir	[204]
CS/GO monoliths	MO	4.2	567	PSO	No fit	[371]
MCS/GO	Acid orange	3.0	42	PSO	Langmuir	[372]
MCS/GO	MO	4.0	398	PSO	Langmuir	[373]
CS/GO aerogel	MO	4.0	686	PSO	-	[197]
	Amido black	4.0	573	PSO	-	
CS/GOβ-CD	MB	12.0	1134	PSO	Freundlich	[200]
CS/GO aerogel	Metanil yellow	6.8	430	PSO	Langmuir	[365]
CS/GO aerogels	Indigo carmine	6.8	534	-	Langmuir	[198]
	MB	7.0	168	-	Langmuir	
CS/GO/Ligno-sulfonate aerogel	MB	7.0	1023	PSO	Langmuir	[221]
CS/GO/ Cellulose	MB	6.0	3190	PSO	Langmuir	[230]
MCS/GOCS-EDTA	Rhodamine B	7.5	1085	PSO	Langmuir	[367]
CS/GO aerogel	Congo red	7.0	384	PSO	Langmuir	[374]
MCS/GO	MB	8.5	2478	PFO	Sips	[170]
MCS/CNT	Congo red	6.0	262	PSO	Langmuir	[375]
MCS/CNT	Acid red	3.0	809	PSO	R−P. Freundlich	[376]
MSC/CNT/ SiO_2	DB 71	6.8	61	PSO	Langmuir	[377]
	RB 19	2.0	97	PSO	Langmuir	
CS/CNT	DB 71	6.2	29	PSO	Langmuir	[378]
	FdR17	3.0	1508	Avrami	Langmuir	[241]
	FdB1	3.0	1480	Avrami	Langmuir	
CS/CNT/GO	Rhodamine B	-	9.6	PSO	-	[215]
CS/CNT	Phenol	6.5	404	PSO	Dubinin-Radushkevic	[60]
CS/AC	Crystal violet	9.0	12.5 (323K)	PSO	Langmuir and Freundlich	[283]
CS/AC	FBL2	3.0	155	Avrami	-	[286]
CS/AC	FR17	3.0	133	Avrami	-	
CS/AC/PVA	MB	6.0	468	PFO	Langmuir and Freundlich	[294]

Table 8. *Cont.*

Adsorbent	Adsorbate	pH	Ads. Capacity	Kinetic Model	Isotherm Model	Ref
CS/AC/PEG	MB	7.0	424	PSO	Langmuir	[295]
CS/AC mesoporous	Thionine	3–11	61	PSO	Freundlich	[305]
CS/AC	Indigo carmine	3.0	208	PSO	Langmuir and Freundlich	[279]
CS/AC with Carica papaya seeds	MB	8.0	302	PSO Elovich	Langmuir	[379]
CS/AC/hexa-decylamine	RB 5	4.0	666	PSO	Freundlich	[380]
CS/AC/ Cellulose	Tylosin	7.0	59	PSO	Langmuir	[381]
CS/Silica	Acid Red 88	7.0	25	PSO	Langmuir	[320]
MCS/Silica/ Glu	MB	5.0	185	PSO	Langmuir	[382]
	Crystal violet	5.0	390	PSO	Langmuir	
	Light yellow	5.0	228	PSO	Langmuir	
MCS/Silica	Rhodamine B	-	191	-	Langmuir	[383]
CS/Silica/ ZnO	MB	7.0	293	-	Langmuir	[384]
CS/Silica/ PVA	Direct Red 80	2.0	322	PSO	Langmuir	[385]
CS/GO	Tetracycline	9.0	1130	PFO	Temkin	[196]
MCS/GO	Rifampicin	5.0	102	PSO	Langmuir	[176]
CS/GO	Ciprofloxacin	7.0	5.3	-	Langmuir	[227]
	Ofloxacin	7.0	8.3	-	Langmuir	
MCS/rGO	Cefixime	6.4	30	-	Freundlich	[187]
CS/CNT	Diazinon	5.5	222	PSO	Sips	[240]
CS/BC beads	Ciprofloxacin	3.0	76	PSO	Langmuir	[386]
CS/BC	Ciprofloxacin	7.0	106	PSO	Langmuir	[272]
	Enrofloxacin	7.0	100	PSO	Langmuir	
MCS/AC	Ciprofloxacin	-	90	PSO	Langmuir	[299]
	Erythromycin	-	178	PSO	Langmuir	
	Amoxicillin	-	526	PSO	Langmuir	

5. Conclusions and Future Perspectives

It is clear from the reports reviewed and the growing number of publications, where chitosan and chitosan-based materials are employed as adsorbents, that these materials are emerging as interesting candidates in the formulation of adsorbents for environmental applications. Chitosan can be easily combined with different support materials, and while earlier studies were devoted to blending chitosan with other polymeric materials, many of the more recent reports are focused on combing chitosan with carbon-based materials with GO, and to a lesser extent activated carbon, attracting considerable attention. Likewise, there is increasing attention being focussed on merging chitosan and mesoporous silica. The CS/GO composites have shown impressive adsorption capacity with both dyes and heavy metal ions.

However, this research field is still in its infancy and a number of challenges exist and must be addressed before these chitosan-carbon based or chitosan-silica based materials can be employed as adsorbents for the removal of a variety of pollutants. One of the more challenging aspects, that has direct implications in terms of costs, is the regeneration of the adsorbents. Ideally, adsorbents should have the capacity to be regenerated and used multiple times. Regeneration is normally achieved using NaOH or acid treatments, where, for example, heavy metal ions are released from the chitosan. However, these treatments lead to a progressive hydrolysis of the polysaccharide on the chitosan. Consequently, the adsorption capacity decreases with each adsorption-regeneration cycle. New regeneration processes are required to give more longer lasting and cost-effective chitosan-based adsorbents. Other challenges are the introduction of selectivity in the adsorption process. Real water samples contain a number of ions that will compete with the removal of heavy metal ions, consuming the adsorption sites and reducing the uptake of the targeted pollutants. Furthermore, the removal of neutral pollutant molecules using these chitosan-based materials is more difficult to achieve, although the addition of GO provides two dimensional sheets that facilitate the adsorption of aromatic ring structures. In addition, the adsorbents need to be removed from the aquatic environment or employed in a continuous flow system. While the development of magnetic chitosan-based materials provides the opportunity to remove the adsorbents using magnetic separation, these magnetic materials are only emerging and it is not entirely clear if they can be sufficiently anchored within the chitosan composites to prevent their leaching over longer terms. However, the development of silica coated and protected magnetic iron-containing particles is promising. There are added concerns over the environmental impact of GO and CNTs, which if leached from the chitosan composites, can enter the aquatic system and have adverse effects on the aquatic ecosystem. Therefore, the CS-carbon and CS-silica composites must be stable and not prone to leaching of GO flakes, CNTs, or the magnetic iron oxide particles. Consequently, studies that monitor the leaching of the various carbon, silica and iron species from the chitosan composites are needed from an environmental perspective.

Fundamental studies on kinetics and intraparticle diffusion require further study. Most of the kinetic models employed are relatively simple pseudo-second order models while the impermeable nature of the GO sheets, CNTs, carbon and silica particles on the internal diffusion of the pollutants are not well developed. While the physical and chemical properties of chitosan can be tailored by varying its molecular weight, DD levels, particle sizes, etc, the complex relationships between some of these parameters and how they control the adsorption process and capacity are difficult to establish.

Nevertheless, these chitosan-based materials, and especially the emerging chitosan-carbon and chitosan-silica based composites, have a promising future as adsorbent materials. While chitosan is currently unable to compete with activated carbon in commercial and industrial settings, it is nevertheless an attractive and viable material as it is derived from chitin, which can be found in abundance and extracted from seafood wastes. With further developments aimed at strengthening the mechanical properties of chitosan, the development of recovery protocols, scale-up of production using green solvents and implementation of nonthermal technologies, industrial exploitation can become a reality.

Moreover, there has been a recent explosion in the development of new two-dimensional materials. Some of these 2D layered materials, for example MXenes, MoS_2 and $MoSe_2$, are potential adsorbents and could be easily combined with chitosan to generate a new family of high-performance materials.

Author Contributions: Conceptualization, D.C.d.S.A. and C.B.B.; writing—original draft preparation, D.C.d.S.A., B.H., L.A.d.A.P., T.R.S.C.J. and C.B.B.; writing—review and editing, D.C.d.S.A., B.H., L.A.d.A.P., T.R.S.C.J. and C.B.B.; supervision, L.A.d.A.P., T.R.S.C.J. and C.B.B. All authors have read and agreed to the published version of the manuscript.

Funding: This research was funded by Coordenação de Aperfeiçoamento de Pessoal de Nível Superior (CAPES)/Brazil, grant number 001 and by Conselho Nacional de Desenvolvimento Científico e Tecnológico (CNPq)/Brazil. This research also was funded by Secretaria de Desenvolvimento, Ciência e Tecnologia/RS/Brazil, grant numbers DCIT 70/2015 and DCIT 77/2016.

Conflicts of Interest: The authors declare no conflict of interest.

References

1. Brindha, K.; Schneider, M. Impact of urbanization on groundwater quality. In *GIS and Geostatistical Techniques for Groundwater Science*; Elsevier Inc.: Amsterdam, The Netherlands, 2019; pp. 179–196. ISBN 9780128154137.
2. Tran, N.H.; Reinhard, M.; Gin, K.Y.H. Occurrence and fate of emerging contaminants in municipal wastewater treatment plants from different geographical regions-a review. *Water Res.* **2018**, *133*, 182–207. [CrossRef] [PubMed]
3. Brindha, K.; Paul, R.; Walter, J.; Tan, M.L.; Singh, M.K. Trace metals contamination in groundwater and implications on human health: Comprehensive assessment using hydrogeochemical and geostatistical methods. *Environ. Geochem. Health* **2020**, *42*, 3819–3839. [CrossRef] [PubMed]
4. Burant, A.; Selbig, W.; Furlong, E.T.; Higgins, C.P. Trace organic contaminants in urban runoff: Associations with urban land-use. *Environ. Pollut.* **2018**, *242*, 2068–2077. [CrossRef] [PubMed]
5. Pai, S.M.; Kini, S.; Selvaraj, R.; Pugazhendhi, A. A review on the synthesis of hydroxyapatite, its composites and adsorptive removal of pollutants from wastewater. *J. Water Process Eng.* **2020**, *38*, 101574. [CrossRef]
6. Soylak, M.; Ozalp, O.; Furkan, U. Magnetic nanomaterials for the removal, separation and preconcentration of organic and inorganic pollutants at trace levels and their practical applications: A review. *Trends Environ. Anal. Chem.* **2020**, 106192. [CrossRef]
7. Reyes-Contreras, C.; Matamoros, V.; Ruiz, I.; Soto, M.; Bayona, J.M. Evaluation of PPCPs removal in a combined anaerobic digester-constructed wetland pilot plant treating urban wastewater. *Chemosphere* **2011**, *84*, 1200–1207. [CrossRef] [PubMed]
8. Zhang, Y.; Li, J.L.; Cai, T.; Cheng, Z.L.; Li, X.; Chung, T.S. Sulfonated hyperbranched polyglycerol grafted membranes with antifouling properties for sustainable osmotic power generation using municipal wastewater. *J. Memb. Sci.* **2018**, *563*, 521–530. [CrossRef]
9. Zhou, G.J.; Ying, G.G.; Liu, S.; Zhou, L.J.; Chen, Z.F.; Peng, F.Q. Simultaneous removal of inorganic and organic compounds in wastewater by freshwater green microalgae. *Environ. Sci. Process. Impacts* **2014**, *16*, 2018–2027. [CrossRef] [PubMed]
10. Zhang, N.; Ishag, A.; Li, Y.; Wang, H.; Guo, H.; Mei, P.; Meng, Q.; Sun, Y. Recent investigations and progress in environmental remediation by using covalent organic framework-based adsorption method: A review. *J. Clean. Prod.* **2020**, *277*, 123360. [CrossRef]
11. Wan Ngah, W.S.; Teong, L.C.; Hanafiah, M.A.K.M. Adsorption of dyes and heavy metal ions by chitosan composites: A review. *Carbohydr. Polym.* **2011**, *83*, 1446–1456. [CrossRef]
12. Crini, G.; Badot, P.M. Application of chitosan, a natural aminopolysaccharide, for dye removal from aqueous solutions by adsorption processes using batch studies: A review of recent literature. *Prog. Polym. Sci.* **2008**, *33*, 399–447. [CrossRef]
13. Rinaudo, M. Chitin and chitosan: Properties and applications. *Prog. Polym. Sci.* **2006**, *31*, 603. [CrossRef]
14. Yi, H.; Wu, L.Q.; Bentley, W.E.; Ghodssi, R.; Rubloff, G.W.; Culver, J.N.; Payne, G.F. Biofabrication with chitosan. *Biomacromolecules* **2005**, *6*, 2881–2894. [CrossRef]
15. Weska, R.F.; Moura, J.M.; Batista, L.M.; Rizzi, J.; Pinto, L.A.A. Optimization of deacetylation in the production of chitosan from shrimp wastes: Use of response surface methodology. *J. Food Eng.* **2007**, *80*, 749–753. [CrossRef]
16. FAO. *Fishery and Aquaculture Statistics*; FAO: Rome, Italy, 2018.
17. Gonçalves, J.O.; Esquerdo, V.M.; Cadaval, T.R.S.; Pinto, L.A.A. Chitosan-based hydrogels. In *Sustainable Agriculture Reviews 36*; Springer: Aix-en-Provence, France, 2019; pp. 49–123. ISBN 9783540228608.
18. Varun, T.K.; Senani, S.; Jayapal, N.; Chikkerur, J.; Roy, S.; Tekulapally, V.B. Extraction of chitosan and its oligomers from shrimp shell waste, their characterization and antimicrobial effect. *Vet. World* **2017**, *10*, 170–175. [CrossRef] [PubMed]
19. Bakshi, P.S.; Selvakumar, D.; Kadirvelu, K.; Kumar, N.S. Chitosan as an environment friendly biomaterial—A review on recent modifications and applications. *Int. J. Biol. Macromol.* **2020**, *150*, 1072–1083. [CrossRef] [PubMed]
20. Maity, J.; Ray, S.K. Chitosan based nano composite adsorbent—synthesis, characterization and application for adsorption of binary mixtures of Pb(II) and Cd(II) from water. *Carbohydr. Polym.* **2017**, *182*, 159–171. [CrossRef] [PubMed]

21. Qu, B.; Luo, Y. Chitosan-based hydrogel beads: Preparations, modifications and applications in food and agriculture sectors—A review. *Int. J. Biol. Macromol.* **2020**, *152*, 437–448. [CrossRef]
22. Shokrgozar, M.A.; Mottaghitalab, F.; Mottaghitalab, V.; Farokhi, M. Fabrication of porous chitosan/poly (vinyl alcohol) reinforced single-walled carbon nanotube nanocomposites for neural tissue engineering. *J. Biomed. Nanotechnol.* **2011**, *7*, 276–284. [CrossRef]
23. Dotto, G.L.; Pinto, L.A.A. General considerations about chitosan. In *Frontiers in Biomaterials*; Bentham Science: Sharjah, UAE, 2017; pp. 3–33.
24. Kumar, D.; Gihar, S.; Shrivash, M.K.; Kumar, P.; Kundu, P.P. A review on the synthesis of graft copolymers of chitosan and their potential applications. *Int. J. Biol. Macromol.* **2020**, *163*, 2097–2112. [CrossRef]
25. Muñoz, I.; Rodríguez, C.; Gillet, D.; Moerschbacher, B.M. Life cycle assessment of chitosan production in India and Europe. *Int. J. Life Cycle Assess.* **2018**, *23*, 1151–1160. [CrossRef]
26. Pakdel, P.M.; Peighambardoust, S.J. Review on recent progress in chitosan-based hydrogels for wastewater treatment application. *Carbohydr. Polym.* **2018**, *201*, 264–279. [CrossRef] [PubMed]
27. Jiménez-Gómez, C.P.; Cecilia, J.A. Chitosan: A Natural biopolymer with a wide and varied range of applications. *Molecules* **2020**, *25*, 3981. [CrossRef] [PubMed]
28. Odier, A. Mémoir sur la composition chimique des parties cornées des insectes. *Mémoirs la Soc. d'Histoire Nat.* **1823**, *1*, 29–42.
29. Hamed, I.; Özogul, F.; Regenstein, J.M. Industrial applications of crustacean by-products (chitin, chitosan, and chitooligosaccharides): A review. *Trends Food Sci. Technol.* **2016**, *48*, 40–50. [CrossRef]
30. Hoppe-Seyler, F. Ueber chitin un cellulose. *Berichte der Dtsch. Chem. Gesellschaft* **1894**, *27*, 3329–3331. [CrossRef]
31. Muslim, S.; Al, I.M.S.; Naseer, A.; Ali, M.; Kadham, B.; Ahmad, M.; Saadoon, S.; Hammood, N.; Nsayef, S. Chitosan extracted from Aspergillus flavus shows synergistic effect, eases quorum sensing mediated virulence factors and biofilm against nosocomial pathogen *Pseudomonas aeruginosa*. *Int. J. Biol. Macromol.* **2018**, *107*, 52–58. [CrossRef]
32. Berger, L.R.R.; Stamford, T.C.M.; De Oliveira, K.; Árabe, R.; Pessoa, A.D.M.P.; De Lima, M.A.B.; Pintado, M.M.E.; Câmara, M.P.S.; Franco, L.D.O.; Magnani, M.; et al. Chitosan produced from Mucorales fungi using agroindustrial by-products and its efficacy to inhibit Colletotrichum species. *Int. J. Biol. Macromol.* **2018**, *108*, 635–641. [CrossRef]
33. Campana-Filho, S.P.; Signini, R. Efeito de aditivos na desacetilação de quitina. *Polímeros: Ciência e Tecnologia* **2001**, *11*, 169–173. [CrossRef]
34. Kaur, S.; Dhillon, G.S. Recent trends in biological extraction of chitin from marine shell wastes: A review. *Crit. Rev. Biotechnol.* **2015**, *35*, 44–61. [CrossRef]
35. Arbia, W.; Arbia, L.; Adour, L.; Amrane, A. Chitin extraction from crustacean shells using biological methods—A review. *Food Technol. Biotechnol.* **2013**, *51*, 12–25.
36. Ghorbel-bellaaj, O.; Younes, I.; Maâlej, H.; Hajji, S.; Nasri, M. Chitin extraction from shrimp shell waste using *Bacillus bacteria*. *Int. J. Biol. Macromol.* **2012**, *51*, 1196–1201. [CrossRef] [PubMed]
37. Hajji, S.; Ghorbel-bellaaj, O.; Younes, I.; Jellouli, K.; Nasri, M. Chitin extraction from crab shells by *Bacillus bacteria*. Biological activities of fermented crab supernatants. *Int. J. Biol. Macromol.* **2015**, *79*, 167–173. [CrossRef] [PubMed]
38. Araki, Y.; Ito, E. A pathway of chitosan formation in Mucor rouxii. *Biochem. Biophys. Res. Commun.* **1974**, *56*, 669–675. [CrossRef]
39. Davis, L.L.; Bartnicki-Garcia, S. Chitosan synthesis by the tandem action of chitin synthetase and chitin deacetylase from mucor rouxifi. *Biochemistry* **1984**, 1065–1073. [CrossRef]
40. Tsigos, I.; Zydowicz, N.; Martinou, A.; Domard, A.; Bouriotis, V. Mode of action of chitin deacetylase from Mucor rouxii on N -acetylchitooligosaccharides. *Eur. J. Biochem.* **1999**, *705*, 698–705. [CrossRef]
41. Teng, D. From chitin to chitosan. In *Chitosan-Based Hydrogels: Functions and Applications*; Taylor & Francis: Boca Raton, FL, USA, 2012; pp. 5–8.
42. Yuan, Y.; Chen, L.; Shi, W.; Wang, Z.; Zhang, H. Insight into physicochemical, rheological, and antibacterial properties of chitosan extracted from antarctic krill: A comparative study. *Molecules* **2020**, *25*, 4074. [CrossRef]
43. De Moura, C.M.; De Moura, J.M.; Soares, N.M.; Antonio, L.; Pinto, D.A. Evaluation of molar weight and deacetylation degree of chitosan during chitin deacetylation reaction: Used to produce biofilm. *Chem. Eng. Process. Process Intensif.* **2011**, *50*, 351–355. [CrossRef]
44. García-Bermejo, A.B.; Cardelle-Cobas, A.; Ruiz-Matute, A.I.; Montañés, F.; Olano, A.; Corzo, N. Effect of drying methods on the reactivity of chitosan towards Maillard reaction. *Food Hydrocoll.* **2012**, *29*, 27–34. [CrossRef]
45. Dotto, G.L.; de Souza, V.C.; de Moura, J.M.; de Moura, C.M.; de Almeida Pinto, L.A. Influence of drying techniques on the characteristics of chitosan and the quality of biopolymer films. *Dry. Technol.* **2011**, *29*, 1784–1791. [CrossRef]
46. Thakhiew, W.; Devahastin, S.; Soponronnarit, S. Physical and mechanical properties of chitosan films as affected by drying methods and addition of antimicrobial agent. *J. Food Eng.* **2013**, *119*, 140–149. [CrossRef]
47. Dotto, G.L.; Souza, V.C.; Pinto, L.A.A. Drying of chitosan in a spouted bed: The influences of temperature and equipment geometry in powder quality. *LWT Food Sci. Technol.* **2011**, *44*, 1786–1792. [CrossRef]
48. Estevinho, B.N.; Rocha, F.; Santos, L.; Alves, A. Microencapsulation with chitosan by spray drying for industry applications—A review. *Trends Food Sci. Technol.* **2013**, *31*, 138–155. [CrossRef]
49. Batista, L.M.; da Rosa, C.A.; Pinto, L.A.A. Diffusive model with variable effective diffusivity considering shrinkage in thin layer drying of chitosan. *J. Food Eng.* **2007**, *81*, 127–132. [CrossRef]

50. Srinivasa, P.C.; Ramesh, M.N.; Kumar, K.R.; Tharanathan, R.N. Properties of chitosan films prepared under different drying conditions. *J. Food Eng.* **2004**, *63*, 79–85. [CrossRef]
51. Kim, M.Y.; Lee, J. Chitosan fibrous 3D networks prepared by freeze drying. *Carbohydr. Polym.* **2011**, *84*, 1329–1336. [CrossRef]
52. Mayachiew, P.; Devahastin, S. Effects of drying methods and conditions on release characteristics of edible chitosan films enriched with Indian gooseberry extract. *Food Chem.* **2010**, *118*, 594–601. [CrossRef]
53. Ibrahim, M.; Osman, O.; Mahmoud, A.A. Spectroscopic analyses of cellulose and chitosan: FTIR and modeling approach. *J. Comput. Theor. Nanosci.* **2011**, *8*, 117–123. [CrossRef]
54. Xia, W.; Liu, P.; Zhang, J.; Chen, J. Biological activities of chitosan and chitooligosaccharides. *Food Hydrocoll.* **2011**, *25*, 170–179. [CrossRef]
55. Gerhardt, R.; Farias, B.S.; Moura, J.M.; De Almeida, L.S.; Adriano, R.; Dias, D.; Cadaval, T.R.S.; Pinto, L.A.A. Development of chitosan/*Spirulina* sp. blend films as biosorbents for Cr^{6+} and Pb^{2+} removal. *Int. J. Biol. Macromol.* **2020**, *155*, 142–152. [CrossRef]
56. Yang, Z.; Chai, Y.; Zeng, L.; Gao, Z.; Zhang, J.; Ji, H. Efficient removal of copper ion from wastewater using a stable chitosan gel material. *Molecules* **2019**, *24*, 4205. [CrossRef] [PubMed]
57. Dotto, G.L.; Vieira, M.L.G.; Pinto, L.A.A. Kinetics and mechanism of tartrazine adsorption onto chitin and chitosan. *Ind. Eng. Chem. Res.* **2012**, *51*, 6862–6868. [CrossRef]
58. Affonso, L.N.; Marques, J.L.; Lima, V.V.C.; Gonçalves, J.O.; Barbosa, S.C.; Primel, E.G.; Burgo, T.A.L.; Dotto, G.L.; Pinto, L.A.A.; Cadaval, T.R.S. Removal of fluoride from fertilizer industry effluent using carbon nanotubes stabilized in chitosan sponge. *J. Hazard. Mater.* **2020**, *388*, 122042. [CrossRef] [PubMed]
59. Alves, D.C.S.; Coseglio, B.B.; Pinto, L.A.A.; Cadaval, T.R.S. Development of *Spirulina*/chitosan foam adsorbent for phenol adsorption. *J. Mol. Liq.* **2020**, *309*, 113256. [CrossRef]
60. Alves, D.C.S.; Gonçalves, J.O.; Coseglio, B.B.; Burgo, T.A.L.; Dotto, G.L.; Pinto, L.A.A.; Cadaval, T.R.S. Adsorption of phenol onto chitosan hydrogel scaffold modified with carbon nanotubes. *J. Environ. Chem. Eng.* **2019**, *7*, 103460. [CrossRef]
61. Soares, S.F.; Fernandes, T.; Trindade, T.; Daniel-da-silva, A.L. Trimethyl chitosan/siloxane-hybrid coated Fe_3O_4 nanoparticles for the uptake of sulfamethoxazole from water. *Molecules* **2019**, *24*, 1958. [CrossRef]
62. Phasuphan, W.; Praphairaksit, N.; Imyim, A. Removal of ibuprofen, diclofenac, and naproxen from water using chitosan-modified waste tire crumb rubber Warintorn. *J. Mol. Liq.* **2019**, 111554. [CrossRef]
63. Sanchez, L.M.; Ollier, R.P.; Pereira, A.E.S.; Fraceto, L.F.; Alvarez, V.A. *Pesticide Removal from Industrial Effluents Using Biopolymeric Materials*; Elsevier Inc.: Amsterdam, The Netherlands, 2020; ISBN 9780128181348.
64. Gonçalves, J.O.; Duarte, D.A.; Dotto, G.L.; Pinto, L.A.A. Use of chitosan with different deacetylation degrees for the adsorption of food dyes in a binary system. *Clean–Soil Air Water* **2013**, *9*, 767–774. [CrossRef]
65. Liu, D.; Wei, Y.; Yao, P.; Jiang, L. Determination of the degree of acetylation of chitosan by UV spectrophotometry using dual standards. *Carbohydr. Res.* **2006**, *341*, 782–785. [CrossRef]
66. Kumirska, J.; Czerwicka, M.; Kaczyński, Z.; Bychowska, A.; Brzozowski, K.; Thöming, J.; Stepnowski, P. Application of spectroscopic methods for structural analysis of chitin and chitosan. *Mar. Drugs* **2010**, *8*, 1567–1636. [CrossRef]
67. Duarte, M.L.; Ferreira, M.C.; Marvão, M.R.; Rocha, J. An optimised method to determine the degree of acetylation of chitin and chitosan by FTIR spectroscopy. *Int. J. Biol. Macromol.* **2002**, *31*, 1–8. [CrossRef]
68. Beil, S.; Schamberger, A.; Naumann, W.; MacHill, S.; Van Pée, K.H. Determination of the degree of N-acetylation (DA) of chitin and chitosan in the presence of water by first derivative ATR FTIR spectroscopy. *Carbohydr. Polym.* **2012**, *87*, 117–122. [CrossRef]
69. Wang, C.; Yuan, F.; Jin, L.; Du, X.; Cai, H. Evaluation of the deacetylation degree of chitosan with two-abrupt-change coulometric titration. *Electroanalysis* **2016**, *28*, 401–406. [CrossRef]
70. Zhang, Y.; Zhang, X.; Ding, R.; Zhang, J.; Liu, J. Determination of the degree of deacetylation of chitosan by potentiometric titration preceded by enzymatic pretreatment. *Carbohydr. Polym.* **2011**, *83*, 813–817. [CrossRef]
71. Kong, X. Simultaneous determination of degree of deacetylation, degree of substitution and distribution fraction of -COONa in carboxymethyl chitosan by potentiometric titration. *Carbohydr. Polym.* **2012**, *88*, 336–341. [CrossRef]
72. Gupta, V.K.S. Application of low-cost adsorbents for dye removal—A review. *J. Environ. Manag.* **2009**, *90*, 2313–2342. [CrossRef]
73. Zhou, H.Y.; Guang, X.; Kong, M.; Sheng, C.; Su, D.; Kennedy, J.F. Effect of molecular weight and degree of chitosan deacetylation on the preparation and characteristics of chitosan thermosensitive hydrogel as a delivery system. *Carbohydrate* **2008**, *73*, 265–273. [CrossRef]
74. Pillai, C.K.S.; Paul, W.; Sharma, C.P. Chitin and chitosan polymers: Chemistry, solubility and fiber formation. *Prog. Polym. Sci.* **2009**, *34*, 641–678. [CrossRef]
75. Nunthanid, J.; Puttipipatkhachorn, S.; Yamamoto, K.; Peck, G.E. Physical properties and molecular behavior of chitosan films. *Drug Dev. Ind. Pharm.* **2001**, *27*, 143–157. [CrossRef]
76. Kittur, F.S.; Harish Prashanth, K.V.; Udaya Sankar, K.; Tharanathan, R.N. Characterization of chitin, chitosan and their carboxymethyl derivatives by differential scanning calorimetry. *Carbohydr. Polym.* **2002**, *49*, 185–193. [CrossRef]
77. Wenling, C.; Duohui, J.; Jiamou, L.; Yandao, G.; Nanming, Z.; Xiufang, Z. Effects of the degree of deacetylation on the physicochemical properties and schwann cell affinity of chitosan films. *J. Biomater. Appl.* **2005**, *20*, 157–177. [CrossRef] [PubMed]
78. Gupta, K.C.; Jabrail, F.H. Effects of degree of deacetylation and cross-linking on physical characteristics, swelling and release behavior of chitosan microspheres. *Carbohydr. Polym.* **2006**, *66*, 43–54. [CrossRef]

79. Trung, T.S.; Thein-han, W.W.; Qui, N.T.; Ng, C.-H.; Stevens, W.F. Functional characteristics of shrimp chitosan and its membranes as affected by the degree of deacetylation. *Bioresour. Technol.* **2006**, *97*, 659–663. [CrossRef] [PubMed]
80. Nam, Y.S.; Park, W.H.; Ihm, D.; Hudson, S.M. Effect of the degree of deacetylation on the thermal decomposition of chitin and chitosan nanofibers. *Carbohydr. Polym.* **2010**, *80*, 291–295. [CrossRef]
81. Potivas, T.; Laokuldilok, T. Deacetylation of chitin and the properties of chitosan films with various deacetylation degrees. *Food Appl. Biosci.* **2014**, *13*, 559–568. [CrossRef]
82. Sukul, M.; Sahariah, P.; Reseland, J.E.; Lauzon, L.; Borges, J.; Másson, M.; Mano, J.F.; Haugen, H.J.; Reseland, J.E. In vitro biological response of human osteoblasts in 3D chitosan sponges with controlled degree of deacetylation and molecular weight. *Carbohydr. Polym.* **2020**, 117434. [CrossRef]
83. Foster, L.J.R.; Ho, S.; Hook, J.; Basuki, M.; Marçal, H. Chitosan as a biomaterial: Influence of degree of deacetylation on its physiochemical, material and biological properties. *PLoS ONE* **2015**, *10*, e0135153. [CrossRef]
84. Tavares, L.; Emanuel, E.; Flores, E.; Rodrigues, R.C.; Hertz, P.F. Effect of deacetylation degree of chitosan on rheological properties and physical chemical characteristics of genipin-crosslinked chitosan beads. *Food Hydrocoll.* **2020**, *106*. [CrossRef]
85. Li, J.; Revol, J.F.; Marchessault, R.H. Effect of degree of deacetylation of chitin on the properties of chitin crystallites. *J. Appl. Polym. Sci.* **1997**, *65*, 373–380. [CrossRef]
86. Zhuang, C.; Zhong, Y.; Zhao, Y. Effect of deacetylation degree on properties of chitosan films using electrostatic spraying technique. *Food Control* **2019**, *97*, 25–31. [CrossRef]
87. Sivashankari, P.R.; Prabaharan, M. Deacetylation modification techniques of chitin and chitosan. In *Chitosan Based Biomaterials: Fundamentals*; Elsevier: Amsterdam, The Netherlands, 2017; Volume 1, pp. 117–133. ISBN 9780081002308.
88. Liu, Z.; Ge, X.; Lu, Y.; Dong, S.; Zhao, Y.; Zeng, M. Effects of chitosan molecular weight and degree of deacetylation on the properties of gelatine-based films. *Food Hydrocoll.* **2012**, *26*, 311–317. [CrossRef]
89. Moura, J.M.; Farias, B.S.; Rodrigues, D.A.S. Preparation of chitosan with different characteristics and its application for biofilms production. *J. Polym. Environ.* **2015**, *23*, 470–477. [CrossRef]
90. Bagheri-Khoulenjani, S.; Taghizadeh, S.M.; Mirzadeh, H. An investigation on the short-term biodegradability of chitosan with various molecular weights and degrees of deacetylation. *Carbohydr. Polym.* **2009**, *78*, 773–778. [CrossRef]
91. Barbosa, H.F.G.; Francisco, D.S.; Ferreira, A.P.G.; Cavalheiro, É.T.G. A new look towards the thermal decomposition of chitins and chitosans with different degrees of deacetylation by coupled TG-FTIR. *Carbohydr. Polym.* **2019**, *225*, 115232. [CrossRef] [PubMed]
92. Gámiz-gonzález, M.A.; Correia, D.M.; Lanceros-mendez, S.; Sencadas, V. Kinetic study of thermal degradation of chitosan as a function of deacetylation degree. *Carbohydr. Polym.* **2017**, *167*, 52–58. [CrossRef]
93. Wanjun, T.; Cunxin, W.; Donghua, C. Kinetic studies on the pyrolysis of chitin and chitosan. *Polym. Degrad. Stab.* **2005**, *87*, 389–394. [CrossRef]
94. Iamsamai, C.; Hannongbua, S.; Ruktanonchai, U. The effect of the degree of deacetylation of chitosan on its dispersion of carbon nanotubes. *Carbon* **2010**, *48*, 25–30. [CrossRef]
95. Piccin, J.S.; Vieira, M.L.G.; Gonçalves, J.O.; Dotto, G.L.; Pinto, L.A.A. Adsorption of FD & C Red No. 40 by chitosan: Isotherms analysis. *J. Food Eng.* **2009**, *95*, 16–20. [CrossRef]
96. Habiba, U.; Chin, T.; Siddique, T.A.; Salleh, A.; Chin, B.; Afifi, A.M. Effect of degree of deacetylation of chitosan on adsorption capacity and reusability of chitosan/polyvinyl alcohol/TiO_2 nano composite. *Int. J. Biol. Macromol.* **2017**, *104*, 1133–1142. [CrossRef]
97. Józwiak, T.; Filipkowska, U.; Szymczyk, P.; Zysk, M. Effect of the form and deacetylation degree of chitosan sorbents on sorption effectiveness of Reactive Black 5 from aqueous solutions. *Int. J. Biol. Macromol.* **2017**, *95*, 1169–1178. [CrossRef]
98. Cadaval, T.R.S.; Camara, A.S.; Dotto, G.L.; Pinto, L.A. de A. Adsorption of Cr(VI) by chitosan with different deacetylation degrees. *Desalin. Water Treat.* **2013**, *51*, 7690–7699. [CrossRef]
99. Tsai, M.; Chen, R.H. Modifying the molecular weight of chitosan. In *Chitosan Based Biomaterials: Fundamentals*; Elsevier: Duxford, UK, 2017; Volume 1, pp. 135–158. ISBN 9780081002308.
100. De Farias, B.S.; Grundmann, D.D.R.; Rizzi, F.Z.; Martins, N.S.S.; Cadaval, T.R.S.A.; Pinto, L.A.D.A. Production of low molecular weight chitosan by acid and oxidative pathways: Effect on physicochemical properties. *Food Res. Int.* **2019**, *123*, 88–94. [CrossRef] [PubMed]
101. Zhong, Y.; Zhuang, C.; Gu, W.; Zhao, Y. Effect of molecular weight on the properties of chitosan films prepared using electrostatic spraying technique. *Carbohydr. Polym.* **2019**, *212*, 197–205. [CrossRef] [PubMed]
102. Ziani, K.; Oses, J.; Coma, V.; Mate, J.I. Effect of the presence of glycerol and Tween 20 on the chemical and physical properties of films based on chitosan with different degree of deacetylation. *Food Sci. Technol.* **2008**, *41*, 2159–2165. [CrossRef]
103. Tanpichai, S.; Witayakran, S.; Wootthikanokkhan, J.; Srimarut, Y. Mechanical and antibacterial properties of the chitosan coated cellulose paper for packaging applications: Effects of molecular weight types and concentrations of chitosan. *Int. J. Biol. Macromol.* **2020**, *155*, 1510–1519. [CrossRef]
104. Bof, M.J.; Bordadgaray, V.C.; Locaso, D.E.; García, M.A. Chitosan molecular weight effect on starch-composite film properties. *Food Hydrocoll.* **2015**, *51*, 281–294. [CrossRef]
105. Akpan, E.I.; Gbenebor, O.P.; Adeosun, S.O.; Cletus, O. Solubility, degree of acetylation, and distribution of acetyl groups in chitosan. In *Handbook of Chitin and Chitosan*; Elsevier: Amsterdam, The Netherlands, 2020; pp. 131–164. ISBN 9780128179703.

106. Kubota, N.; Eguchi, Y. Facile preparation of water-soluble N-acetylated chitosan and molecular weight dependence of its water-solubility. *Polym. J.* **1997**, *29*, 123–127. [CrossRef]
107. Bajaj, M.; Winter, J.; Gallert, C. Effect of deproteination and deacetylation conditions on viscosity of chitin and chitosan extracted from Crangon crangon shrimp waste. *Biochem. Eng. J.* **2011**, *56*, 51–62. [CrossRef]
108. Chang, S.H.; Lin, H.T.V.; Wu, G.J.; Tsai, G.J. pH Effects on solubility, zeta potential, and correlation between antibacterial activity and molecular weight of chitosan. *Carbohydr. Polym.* **2015**, *134*, 74–81. [CrossRef]
109. Rinaudo, M.; Pavlov, G.; Desbrieres, J. Influence of acetic acid concentration on the solubilization of chitosan. *Int. J. Polym.* **1999**, *40*, 7029–7032. [CrossRef]
110. Rinaudo, M.; Pavlov, G.; Desbrières, J. Solubilization of chitosan in strong acid medium. *Int. J. Polym. Anal. Charact.* **1999**, *5*, 267–276. [CrossRef]
111. Kurita, K.; Kamiya, M.; Nishimura, S.I. Solubilization of a rigid polysaccharide: Controlled partial N-Acetylation of chitosan to develop solubility. *Carbohydr. Polym.* **1991**, *16*, 83–92. [CrossRef]
112. Shamov, M.V.; Bratskaya, S.Y.; Avramenko, V.A. Interaction of carboxylic acids with chitosan: Effect of pK and hydrocarbon chain length. *J. Colloid Interface Sci.* **2002**, *249*, 316–321. [CrossRef] [PubMed]
113. Lu, S.; Song, X.; Cao, D.; Chen, Y.; Yao, K. Preparation of water-soluble chitosan. *J. Appl. Polym. Sci.* **2004**, *91*, 3497–3503. [CrossRef]
114. Vieira, M.L.G.; Pinheiro, C.P.; Silva, K.A.; Lutke, S.F.; Cadaval, T.R.S.A.; Dotto, G.; de Pinto, L.A.A. Chitosan and cyanoguanidine-crosslinked chitosan coated glass beads and its application in fixed bed adsorption. *Chem. Eng. Commun.* **2019**, *206*, 1474–1486. [CrossRef]
115. Jayasuriya, A.C. Production of micro- and nanoscale chitosan particles for biomedical applications. In *Chitosan Based Biomaterials: Fundamentals*; Elsevier: Duxford, UK, 2017; Volume 1, pp. 185–209. ISBN 9780081002308.
116. Santos, J.C.C.; Mansur, A.A.P.; Mansur, H.S. One-step biofunctionalization of quantum dots with chitosan and n-palmitoyl chitosan for potential biomedical applications. *Molecules* **2013**, *18*, 6550–6572. [CrossRef]
117. Gonçalves, J.O.; Dotto, G.L.; Pinto, L.A.A. Cyanoguanidine-crosslinked chitosan to adsorption of food dyes in the aqueous binary system. *J. Mol. Liq.* **2015**, *211*, 425–430. [CrossRef]
118. Phongying, S.; Aiba, S.; Chirachanchai, S. Direct chitosan nanoscaffold formation via chitin whiskers. *Polymer* **2007**, *48*, 393–400. [CrossRef]
119. Esquerdo, V.; Jr, T.R.S.C.; Dotto, G.L.; Pinto, L.A. Chitosan scaffold as an alternative adsorbent for the removal of hazardous food dyes from aqueous solutions. *J. Colloid Interface Sci.* **2014**, *424*, 7–15. [CrossRef]
120. Piccin, J.S.; Dotto, G.L.; Vieira, M.L.G.; Pinto, L.A.A. Kinetics and mechanism of the food dye FD & C Red 40 adsorption onto chitosan. *J. Chem. Eng. Data* **2011**, *56*, 3759–3765.
121. Wu, F.C.; Tseng, R.L.; Juang, R.S. Comparative adsorption of metal and dye on flake- and bead-types of chitosans prepared from fishery wastes. *J. Hazard. Mater.* **2000**, *73*, 63–75. [CrossRef]
122. Chatterjee, S.; Lee, D.S.; Lee, M.W.; Woo, S.H. Enhanced adsorption of congo red from aqueous solutions by chitosan hydrogel beads impregnated with cetyl trimethyl ammonium bromide. *Bioresour. Technol.* **2009**, *100*, 2803–2809. [CrossRef] [PubMed]
123. Cheng, J.S.; Du, J.; Zhu, W. Facile synthesis of three-dimensional chitosan-graphene mesostructures for reactive black 5 removal. *Carbohydr. Polym.* **2012**, *88*, 61–67. [CrossRef]
124. Karimi, S.; Tavakkoli Yaraki, M.; Karri, R.R. A comprehensive review of the adsorption mechanisms and factors influencing the adsorption process from the perspective of bioethanol dehydration. *Renew. Sustain. Energy Rev.* **2019**, *107*, 535–553. [CrossRef]
125. Wittmar, A.S.M.; Böhler, H.; Kayali, A.L.; Ulbricht, M. One-step preparation of porous cellulose/chitosan macro-spheres from ionic liquid-based solutions. *Cellulose* **2020**, *27*, 5689–5705. [CrossRef]
126. Li, N.; Bai, R. Copper adsorption on chitosan-cellulose hydrogel beads: Behaviors and mechanisms. *Sep. Purif. Technol.* **2005**, *42*, 237–247. [CrossRef]
127. Chen, L.; Hao, H.; Zhang, W.; Shao, Z. Adsorption mechanism of copper ions in aqueous solution by chitosan–carboxymethyl starch composites. *J. Appl. Polym. Sci.* **2020**, *137*. [CrossRef]
128. Zhao, X.; Wang, X.; Lou, T. Preparation of fibrous chitosan/sodium alginate composite foams for the adsorption of cationic and anionic dyes. *J. Hazard. Mater.* **2021**, *403*, 124054. [CrossRef]
129. Preethi, J.; Karthikeyan, P.; Vigneshwaran, S.; Meenakshi, S. Facile synthesis of Zr^{4+} incorporated chitosan/gelatin composite for the sequestration of Chromium(VI) and fluoride from water. *Chemosphere* **2021**, *262*, 128317. [CrossRef]
130. Da Silva, J.C.S.; França, D.B.; Rodrigues, F.; Oliveira, D.M.; Trigueiro, P.; Silva Filho, E.C.; Fonseca, M.G. What happens when chitosan meets bentonite under microwave-assisted conditions? Clay-based hybrid nanocomposites for dye adsorption. *Colloids Surf. A Physicochem. Eng. Asp.* **2021**, *609*, 125584. [CrossRef]
131. Lin, J.; Zhan, Y. Adsorption of humic acid from aqueous solution onto unmodified and surfactant-modified chitosan/zeolite composites. *Chem. Eng. J.* **2012**, *200*, 202–213. [CrossRef]
132. Zhao, R.; Ma, T.; Zhao, S.; Rong, H.; Tian, Y.; Zhu, G. Uniform and stable immobilization of metal-organic frameworks into chitosan matrix for enhanced tetracycline removal from water. *Chem. Eng. J.* **2020**, *382*. [CrossRef]
133. Karthik, R.; Meenakshi, S. Removal of hexavalent chromium ions from aqueous solution using chitosan/polypyrrole composite. *Desalin. Water Treat.* **2015**, *56*, 1587–1600. [CrossRef]

134. Kundakci, S. Synthesis of Methacrylamide/chitosan polymeric cryogels and swelling/dye sorption properties. *Polym. Sci. Ser. A* **2020**, *62*, 481–493. [CrossRef]
135. Da Silva, R.C.; de Aguiar, S.B.; da Cunha, P.L.R.; de Paula, R.C.M.; Feitosa, J.P.A. Effect of microwave on the synthesis of polyacrylamide-g-chitosan gel for azo dye removal. *React. Funct. Polym.* **2020**, *148*. [CrossRef]
136. Da Rosa Schio, R.; da Rosa, B.C.; Gonçalves, J.O.; Pinto, L.A.A.; Mallmann, E.S.; Dotto, G.L. Synthesis of a bio–based polyurethane/chitosan composite foam using ricinoleic acid for the adsorption of Food Red 17 dye. *Int. J. Biol. Macromol.* **2019**, *121*, 373–380. [CrossRef]
137. Zhu, H.-Y.; Fu, Y.-Q.; Jiang, R.; Yao, J.; Xiao, L.; Zeng, G.-M. Novel magnetic chitosan/poly(vinyl alcohol) hydrogel beads: Preparation, characterization and application for adsorption of dye from aqueous solution. *Bioresour. Technol.* **2012**, *105*, 24–30. [CrossRef]
138. Gu, F.; Geng, J.; Li, M.; Chang, J.; Cui, Y. Synthesis of chitosan-lignosulfonate composite as an adsorbent for dyes and metal ions removal from wastewater. *ACS Omega* **2019**, *4*, 21421–21430. [CrossRef]
139. Lee, C.W.; Suh, J.M.; Jang, H.W. Chemical sensors based on two-dimensional (2D) materials for selective detection of ions and molecules in liquid. *Front. Chem.* **2019**, *7*. [CrossRef]
140. Chang, Y.-M.; Lin, H.-W.; Li, L.-J.; Chen, H.-Y. Two-dimensional materials as anodes for sodium-ion batteries. *Mater. Today Adv.* **2020**, *6*. [CrossRef]
141. Yu, T.; Breslin, C.B. Graphene-modified composites and electrodes and their potential applications in the electro-fenton process. *Materials* **2020**, *13*, 2254. [CrossRef] [PubMed]
142. Li, D.; Lai, W.-Y.; Zhang, Y.-Z.; Huang, W. Printable transparent conductive films for flexible electronics. *Adv. Mater.* **2018**, *30*. [CrossRef] [PubMed]
143. Rochefort, A.; Wuest, J.D. Interaction of substituted aromatic compounds with graphene. *Langmuir* **2009**, *25*, 210–215. [CrossRef] [PubMed]
144. Chen, Y.; Chen, L.; Bai, H.; Li, L. Graphene oxide-chitosan composite hydrogels as broad-spectrum adsorbents for water purification. *J. Mater. Chem. A* **2013**, *1*, 1992–2001. [CrossRef]
145. Liu, Z.; Rios-Carvajal, T.; Ceccato, M.; Hassenkam, T. Nanoscale chemical mapping of oxygen functional groups on graphene oxide using atomic force microscopy-coupled infrared spectroscopy. *J. Colloid Interface Sci.* **2019**, *556*, 458–465. [CrossRef]
146. Aliyev, E.; Filiz, V.; Khan, M.M.; Lee, Y.J.; Abetz, C.; Abetz, V. Structural characterization of graphene oxide: Surface functional groups and fractionated oxidative debris. *Nanomaterials* **2019**, *9*, 1180. [CrossRef]
147. Şinoforoğlu, M.; Gür, B.; Arik, M.; Onganer, Y.; Meral, K. Graphene oxide sheets as a template for dye assembly: Graphene oxide sheets induce H-aggregates of pyronin (Y) dye. *RSC Adv.* **2013**, *3*, 11832–11838. [CrossRef]
148. Ramesha, G.K.; Vijaya Kumara, A.; Muralidhara, H.B.; Sampath, S. Graphene and graphene oxide as effective adsorbents toward anionic and cationic dyes. *J. Colloid Interface Sci.* **2011**, *361*, 270–277. [CrossRef]
149. Chowdhury, S.; Balasubramanian, R. Recent advances in the use of graphene-family nanoadsorbents for removal of toxic pollutants from wastewater. *Adv. Colloid Interface Sci.* **2014**, *204*, 35–56. [CrossRef]
150. Eda, G.; Chhowalla, M. Chemically derived graphene oxide: Towards large-area thin-film electronics and optoelectronics. *Adv. Mater.* **2010**, *22*, 2392–2415. [CrossRef]
151. Javadi, E.; Baghdadi, M.; Taghavi, L.; Ahmad Panahi, H. Removal of 4-nonylphenol from surface water and municipal wastewater effluent using three-dimensional graphene oxide–chitosan aerogel beads. *Int. J. Environ. Res.* **2020**, *14*, 513–526. [CrossRef]
152. Lai, K.C.; Lee, L.Y.; Hiew, B.Y.Z.; Yang, T.C.-K.; Pan, G.-T.; Thangalazhy-Gopakumar, S.; Gan, S. Utilisation of eco-friendly and low cost 3D graphene-based composite for treatment of aqueous Reactive Black 5 dye: Characterisation, adsorption mechanism and recyclability studies. *J. Taiwan Inst. Chem. Eng.* **2020**, *114*, 57–66. [CrossRef]
153. Zhang, M.; Ma, G.; Zhang, L.; Chen, H.; Zhu, L.; Wang, C.; Liu, X. Chitosan-reduced graphene oxide composites with 3D structures as effective reverse dispersed solid phase extraction adsorbents for pesticides analysis. *Analyst* **2019**, *144*, 5164–5171. [CrossRef] [PubMed]
154. Yang, X.; Tu, Y.; Li, L.; Shang, S.; Tao, X.-M. Well-dispersed chitosan/graphene oxide nanocomposites. *ACS Appl. Mater. Interfaces* **2010**, *2*, 1707–1713. [CrossRef] [PubMed]
155. Fan, L.; Luo, C.; Li, X.; Lu, F.; Qiu, H.; Sun, M. Fabrication of novel magnetic chitosan grafted with graphene oxide to enhance adsorption properties for methyl blue. *J. Hazard. Mater.* **2012**, *215–216*, 272–279. [CrossRef]
156. Li, K.; Zhao, B.; Yu, Q.; Xu, J.; Li, X.; Wei, D.; Qian, L.; Liu, G.; Wang, W. Porous graphene oxide/chitosan beads with honeycomb-biomimetic microchannels as hydrophilic adsorbent for the selective capture of glycopeptides. *Microchim. Acta* **2020**, *187*, 324. [CrossRef]
157. Zhang, J.; Chen, N.; Su, P.; Li, M.; Feng, C. Fluoride removal from aqueous solution by zirconium-chitosan/graphene oxide membrane. *React. Funct. Polym.* **2017**, *114*, 127–135. [CrossRef]
158. Li, C.; Ashraf, M.A. Preparation of a new graphite oxide/chitosan composite membrane and its application in environmental protection engineering. *Funct. Mater.* **2019**, *26*, 441–445. [CrossRef]
159. Vo, T.S.; Vo, T.T.B.C.; Suk, J.W.; Kim, K. Recycling performance of graphene oxide-chitosan hybrid hydrogels for removal of cationic and anionic dyes. *Nano Converg.* **2020**, *7*, 4. [CrossRef]
160. Du, Q.; Li, Y.; Li, J.; Zhang, Z.; Qiao, B.; Sui, K.; Wang, D.; Wang, C.; Li, H.; Xia, Y. Preparation of graphene oxide/chitosan pellets and their adsorption properties for Congo Red. *Int. J. Nanosci.* **2019**, *18*, 1850030. [CrossRef]

161. Liu, L.; Li, C.; Bao, C.; Jia, Q.; Xiao, P.; Liu, X.; Zhang, Q. Preparation and characterization of chitosan/graphene oxide composites for the adsorption of Au(III) and Pd(II). *Talanta* **2012**, *93*, 350–357. [CrossRef] [PubMed]
162. Mohseni Kafshgari, M.; Tahermansouri, H. Development of a graphene oxide/chitosan nanocomposite for the removal of picric acid from aqueous solutions: Study of sorption parameters. *Colloids Surf. B Biointerfaces* **2017**, *160*, 671–681. [CrossRef] [PubMed]
163. Delhiraja, K.; Vellingiri, K.; Boukhvalov, D.W.; Philip, L. Development of highly water stable graphene oxide-based composites for the removal of pharmaceuticals and personal care products. *Ind. Eng. Chem. Res.* **2019**, *58*, 2899–2913. [CrossRef]
164. Rebekah, A.; Bharath, G.; Naushad, M.; Viswanathan, C.; Ponpandian, N. Magnetic graphene/chitosan nanocomposite: A promising nano-adsorbent for the removal of 2-naphthol from aqueous solution and their kinetic studies. *Int. J. Biol. Macromol.* **2020**, *159*, 530–538. [CrossRef] [PubMed]
165. Singh, N.; Riyajuddin, S.; Ghosh, K.; Mehta, S.K.; Dan, A. Chitosan-graphene oxide hydrogels with embedded magnetic iron oxide nanoparticles for dye removal. *ACS Appl. Nano Mater.* **2019**, *2*, 7379–7392. [CrossRef]
166. Kharissova, O.V.; Dias, H.V.R.; Kharisov, B.I. Magnetic adsorbents based on micro- and nano-structured materials. *RSC Adv.* **2015**, *5*, 6695–6719. [CrossRef]
167. Gul, K.; Sohni, S.; Waqar, M.; Ahmad, F.; Norulaini, N.A.N. Functionalization of magnetic chitosan with graphene oxide for removal of cationic and anionic dyes from aqueous solution. *Carbohydr. Polym.* **2016**, *152*, 520–531. [CrossRef]
168. Shan, H.; Peng, S.; Zhao, C.; Zhan, H.; Zeng, C. Highly efficient removal of As(III) from aqueous solutions using goethite/graphene oxide/chitosan nanocomposite. *Int. J. Biol. Macromol.* **2020**, *164*, 13–26. [CrossRef]
169. Zhu, H.-Y.; Jiang, R.; Xiao, L.; Li, W. A novel magnetically separable γ-Fe_2O_3/crosslinked chitosan adsorbent: Preparation, characterization and adsorption application for removal of hazardous azo dye. *J. Hazard. Mater.* **2010**, *179*, 251–257. [CrossRef]
170. Sarkar, A.K.; Bediako, J.K.; Choi, J.-W.; Yun, Y.-S. Functionalized magnetic biopolymeric graphene oxide with outstanding performance in water purification. *NPG Asia Mater.* **2019**, *11*, 4. [CrossRef]
171. Xu, L.; Suo, H.; Wang, J.; Cheng, F.; Liu, H.; Qiu, H. Magnetic graphene oxide decorated with chitosan and Au nanoparticles: Synthesis, characterization and application for detection of trace Rhodamine B. *Anal. Methods* **2019**, *11*, 3837–3843. [CrossRef]
172. Foroughi, M.; Azqhandi, M.H.A. A biological-based adsorbent for a non-biodegradable pollutant: Modeling and optimization of Pb(II) remediation using GO-CS-Fe_3O_4-EDTA nanocomposite. *J. Mol. Liq.* **2020**, *318*, 114077. [CrossRef]
173. Tasmia Shah, J.; Jan, M.R. Microextraction of selected endocrine disrupting phenolic compounds using magnetic chitosan biopolymer graphene oxide nanocomposite. *J. Polym. Environ.* **2020**, *28*, 1673–1683. [CrossRef]
174. Anush, S.M.; Chandan, H.R.; Gayathri, B.H.; Asma Manju, N.; Vishalakshi, B.; Kalluraya, B. Graphene oxide functionalized chitosan-magnetite nanocomposite for removal of Cu(II) and Cr(VI) from waste water. *Int. J. Biol. Macromol.* **2020**, *164*, 4391–4402. [CrossRef] [PubMed]
175. Tran, H.V.; Hoang, L.T.; Huynh, C.D. An investigation on kinetic and thermodynamic parameters of methylene blue adsorption onto graphene-based nanocomposite. *Chem. Phys.* **2020**, *535*, 110793. [CrossRef]
176. Shafaati, M.; Miralinaghi, M.; Shirazi, R.H.S.M.; Moniri, E. The use of chitosan/Fe_3O_4 grafted graphene oxide for effective adsorption of rifampicin from water samples. *Res. Chem. Intermed.* **2020**. [CrossRef]
177. Jiang, X.; Pan, W.; Chen, M.; Yuan, Y.; Zhao, L. The fabrication of a thiol-modified chitosan magnetic graphene oxide nanocomposite and its adsorption performance towards the illegal drug clenbuterol in pork samples. *Dalt. Trans.* **2020**, *49*, 6097–6107. [CrossRef]
178. Tang, T.; Cao, S.; Xi, C.; Li, X.; Zhang, L.; Wang, G.; Chen, Z. Chitosan functionalized magnetic graphene oxide nanocomposite for the sensitive and effective determination of alkaloids in hotpot. *Int. J. Biol. Macromol.* **2020**, *146*, 343–352. [CrossRef]
179. Huang, Y.; Fulton, A.N.; Keller, A.A. Simultaneous removal of PAHs and metal contaminants from water using magnetic nanoparticle adsorbents. *Sci. Total Environ.* **2016**, *571*, 1029–1036. [CrossRef]
180. Pei, S.; Cheng, H.-M. The reduction of graphene oxide. *Carbon* **2012**, *50*, 3210–3228. [CrossRef]
181. Zhang, J.; Yang, H.; Shen, G.; Cheng, P.; Zhang, J.; Guo, S. Reduction of graphene oxide vial-ascorbic acid. *Chem. Commun.* **2010**, *46*, 1112–1114. [CrossRef] [PubMed]
182. Wang, Z.; Zhou, X.; Zhang, J.; Boey, F.; Zhang, H. Direct electrochemical reduction of single-layer graphene oxide and subsequent functionalization with glucose oxidase. *J. Phys. Chem. C* **2009**, *113*, 14071–14075. [CrossRef]
183. Guo, H.-L.; Wang, X.-F.; Qian, Q.-Y.; Wang, F.-B.; Xia, X.-H. A green approach to the synthesis of graphene nanosheets. *ACS Nano* **2009**, *3*, 2653–2659. [CrossRef]
184. Shao, Y.; Wang, J.; Engelhard, M.; Wang, C.; Lin, Y. Facile and controllable electrochemical reduction of graphene oxide and its applications. *J. Mater. Chem.* **2010**, *20*, 743–748. [CrossRef]
185. Sreeprasad, T.S.; Maliyekkal, S.M.; Lisha, K.P.; Pradeep, T. Reduced graphene oxide-metal/metal oxide composites: Facile synthesis and application in water purification. *J. Hazard. Mater.* **2011**, *186*, 921–931. [CrossRef]
186. Sheshmani, S.; Mashhadi, S. Potential of magnetite reduced graphene oxide/chitosan nanocomposite as biosorbent for the removal of dyes from aqueous solutions. *Polym. Compos.* **2018**, *39*, E457–E462. [CrossRef]
187. Ciğeroğlu, Z.; Küçükyıldız, G.; Erim, B.; Alp, E. Easy preparation of magnetic nanoparticles-rGO-chitosan composite beads: Optimization study on cefixime removal based on RSM and ANN by using Genetic Algorithm Approach. *J. Mol. Struct.* **2021**, *1224*, 129182. [CrossRef]
188. Guo, X.; Qu, L.; Tian, M.; Zhu, S.; Zhang, X.; Tang, X.; Sun, K. Chitosan/graphene oxide composite as an effective adsorbent for reactive red dye removal. *Water Environ. Res.* **2016**, *88*, 579–588. [CrossRef]

189. Liu, F.; Chung, S.; Oh, G.; Seo, T.S. Three-dimensional graphene oxide nanostructure for fast and efficient water-soluble dye removal. *ACS Appl. Mater. Interfaces* **2012**, *4*, 922–927. [CrossRef]
190. Cheng, C.; Deng, J.; Lei, B.; He, A.; Zhang, X.; Ma, L.; Li, S.; Zhao, C. Toward 3D graphene oxide gels based adsorbents for high-efficient water treatment via the promotion of biopolymers. *J. Hazard. Mater.* **2013**, *263*, 467–478. [CrossRef]
191. Ma, Y.; Chen, Y. Three-dimensional graphene networks: Synthesis, properties and applications. *Natl. Sci. Rev.* **2015**, *2*, 40–53. [CrossRef]
192. Sun, Z.; Fang, S.; Hu, Y.H. 3D Graphene materials: From understanding to design and synthesis control. *Chem. Rev.* **2020**, *120*, 10336–10453. [CrossRef] [PubMed]
193. Ma, Z.; Liu, D.; Zhu, Y.; Li, Z.; Li, Z.; Tian, H.; Liu, H. Graphene oxide/chitin nanofibril composite foams as column adsorbents for aqueous pollutants. *Carbohydr. Polym.* **2016**, *144*, 230–237. [CrossRef] [PubMed]
194. Zhang, D.; Li, N.; Cao, S.; Liu, X.; Qiao, M.; Zhang, P.; Zhao, Q.; Song, L.; Huang, X. A layered chitosan/graphene oxide sponge as reusable adsorbent for removal of heavy metal ions. *Chem. Res. Chin. Univ.* **2019**, *35*, 463–470. [CrossRef]
195. Yu, B.; Xu, J.; Liu, J.-H.; Yang, S.-T.; Luo, J.; Zhou, Q.; Wan, J.; Liao, R.; Wang, H.; Liu, Y. Adsorption behavior of copper ions on graphene oxide-chitosan aerogel. *J. Environ. Chem. Eng.* **2013**, *1*, 1044–1050. [CrossRef]
196. Zhao, L.; Dong, P.; Xie, J.; Li, J.; Wu, L.; Yang, S.-T.; Luo, J. Porous graphene oxide-chitosan aerogel for tetracycline removal. *Mater. Res. Express* **2014**, *1*, 015601. [CrossRef]
197. Wang, Y.; Xia, G.; Wu, C.; Sun, J.; Song, R.; Huang, W. Porous chitosan doped with graphene oxide as highly effective adsorbent for methyl orange and amido black 10B. *Carbohydr. Polym.* **2015**, *115*, 686–693. [CrossRef]
198. Salzano de Luna, M.; Ascione, C.; Santillo, C.; Verdolotti, L.; Lavorgna, M.; Buonocore, G.G.; Castaldo, R.; Filippone, G.; Xia, H.; Ambrosio, L. Optimization of dye adsorption capacity and mechanical strength of chitosan aerogels through crosslinking strategy and graphene oxide addition. *Carbohydr. Polym.* **2019**, *211*, 195–203. [CrossRef]
199. Mei, J.; Zhang, H.; Mo, S.; Zhang, Y.; Li, Z.; Ou, H. Prominent adsorption of Cr(VI) with graphene oxide aerogel twined with creeper-like polymer based on chitosan oligosaccharide. *Carbohydr. Polym.* **2020**, *247*, 116733. [CrossRef]
200. Liu, Y.; Huang, S.; Zhao, X.; Zhang, Y. Fabrication of three-dimensional porous β-cyclodextrin/chitosan functionalized graphene oxide hydrogel for methylene blue removal from aqueous solution. *Colloids Surf. A Physicochem. Eng. Asp.* **2018**, *539*, 1–10. [CrossRef]
201. Yan, J.; Li, X.; Qiu, F.; Zhao, H.; Yang, D.; Wang, J.; Wen, W. Synthesis of β-cyclodextrin-chitosan-graphene oxide composite and its application for adsorption of manganese ion (II). *Mater. Technol.* **2016**, *31*, 406–415. [CrossRef]
202. Li, L.; Fan, L.; Sun, M.; Qiu, H.; Li, X.; Duan, H.; Luo, C. Adsorbent for chromium removal based on graphene oxide functionalized with magnetic cyclodextrin-chitosan. *Colloids Surf. B Biointerface* **2013**, *107*, 76–83. [CrossRef] [PubMed]
203. Li, L.; Fan, L.; Sun, M.; Qiu, H.; Li, X.; Duan, H.; Luo, C. Adsorbent for hydroquinone removal based on graphene oxide functionalized with magnetic cyclodextrin-chitosan. *Int. J. Biol. Macromol.* **2013**, *58*, 169–175. [CrossRef] [PubMed]
204. Fan, L.; Luo, C.; Sun, M.; Qiu, H.; Li, X. Synthesis of magnetic β-cyclodextrin-chitosan/graphene oxide as nanoadsorbent and its application in dye adsorption and removal. *Colloids Surf. B Biointerface* **2013**, *103*, 601–607. [CrossRef] [PubMed]
205. Salahuddin, N.; EL-Daly, H.; El Sharkawy, R.G.; Nasr, B.T. Synthesis and efficacy of PPy/CS/GO nanocomposites for adsorption of ponceau 4R dye. *Polymer* **2018**, *146*, 291–303. [CrossRef]
206. Kamal, S.; Khan, F.; Kausar, H.; Khan, M.S.; Ahmad, A.; Ishraque Ahmad, S.; Asim, M.; Alshitari, W.; Nami, S.A.A. Synthesis, characterization, morphology, and adsorption studies of ternary nanocomposite comprising graphene oxide, chitosan, and polypyrrole. *Polym. Compos.* **2020**, *41*, 3758–3767. [CrossRef]
207. Salahuddin, N.A.; EL-Daly, H.A.; El Sharkawy, R.G.; Nasr, B.T. Nano-hybrid based on polypyrrole/chitosan/grapheneoxide magnetite decoration for dual function in water remediation and its application to form fashionable colored product. *Adv. Powder Technol.* **2020**, *31*, 1587–1596. [CrossRef]
208. Wang, Z.; Zhang, G.; Li, Y. Preparation of chitosan/polyacrylamide/graphene oxide composite membranes and study of their methylene blue adsorption properties. *Materials* **2020**, *13*, 4407. [CrossRef]
209. Chang, Z.; Chen, Y.; Tang, S.; Yang, J.; Chen, Y.; Chen, S.; Li, P.; Yang, Z. Construction of chitosan/polyacrylate/graphene oxide composite physical hydrogel by semi-dissolution/acidification/sol-gel transition method and its simultaneous cationic and anionic dye adsorption properties. *Carbohydr. Polym.* **2020**, *229*, 115431. [CrossRef]
210. Hong, S.; Na, Y.S.; Choi, S.; Song, I.T.; Kim, W.Y.; Lee, H. Non-covalent self-assembly and covalent polymerization co-contribute to polydopamine formation. *Adv. Funct. Mater.* **2012**, *22*, 4711–4717. [CrossRef]
211. Fu, J.; Xin, Q.; Wu, X.; Chen, Z.; Yan, Y.; Liu, S.; Wang, M.; Xu, Q. Selective adsorption and separation of organic dyes from aqueous solution on polydopamine microspheres. *J. Colloid Interface Sci.* **2016**, *461*, 292–304. [CrossRef] [PubMed]
212. Li, L.; Wei, Z.; Liu, X.; Yang, Y.; Deng, C.; Yu, Z.; Guo, Z.; Shi, J.; Zhu, C.; Guo, W.; et al. Biomaterials cross-linked graphene oxide composite aerogel with a macro–nanoporous network structure for efficient Cr (VI) removal. *Int. J. Biol. Macromol.* **2020**, *156*, 1337–1346. [CrossRef] [PubMed]
213. Li, T.; Liu, X.; Li, L.; Wang, Y.; Ma, P.; Chen, M.; Dong, W. Polydopamine-functionalized graphene oxide compounded with polyvinyl alcohol/chitosan hydrogels on the recyclable adsorption of Cu(II), Pb(II) and Cd(II) from aqueous solution. *J. Polym. Res.* **2019**, *26*, 281. [CrossRef]
214. Ma, L.; Wang, Q.; Islam, S.M.; Liu, Y.; Ma, S.; Kanatzidis, M.G. Highly selective and efficient removal of heavy metals by layered double hydroxide intercalated with the MoS_4^{2-} ion. *J. Am. Chem. Soc.* **2016**, *138*, 2858–2866. [CrossRef] [PubMed]

215. Wang, R.; Zhang, X.; Zhu, J.; Bai, J.; Gao, L.; Liu, S.; Jiao, T. Facile preparation of self-assembled chitosan-based composite hydrogels with enhanced adsorption performances. *Colloids Surf. A Physicochem. Eng. Asp.* **2020**, *598*, 124860. [CrossRef]
216. Priya, V.N.; Rajkumar, M.; Magesh, G.; Mobika, J.; Sibi, S.P.L. Chitosan assisted Fe-Al double layered hydroxide/reduced graphene oxide composites for As(V) removal. *Mater. Chem. Phys.* **2020**, *251*, 123108. [CrossRef]
217. Samuel, M.S.; Subramaniyan, V.; Bhattacharya, J.; Parthiban, C.; Chand, S.; Singh, N.D.P. A GO-CS@MOF [Zn(BDC)(DMF)] material for the adsorption of chromium(VI) ions from aqueous solution. *Compos. Part B Eng.* **2018**, *152*, 116–125. [CrossRef]
218. Guo, X.; Yang, H.; Liu, Q.; Liu, J.; Chen, R.; Zhang, H.; Yu, J.; Zhang, M.; Li, R.; Wang, J. A chitosan-graphene oxide/ZIF foam with anti-biofouling ability for uranium recovery from seawater. *Chem. Eng. J.* **2020**, *382*, 122850. [CrossRef]
219. Li, J.; Wang, X.; Zhao, G.; Chen, C.; Chai, Z.; Alsaedi, A.; Hayat, T.; Wang, X. Metal-organic framework-based materials: Superior adsorbents for the capture of toxic and radioactive metal ions. *Chem. Soc. Rev.* **2018**, *47*, 2322–2356. [CrossRef]
220. Zhang, Y.; Chen, S.; Feng, X.; Yu, J.; Jiang, X. Self-assembly of sponge-like kaolin/chitosan/reduced graphene oxide composite hydrogels for adsorption of Cr(VI) and AYR. *Environ. Sci. Pollut. Res.* **2019**, *26*, 28898–28908. [CrossRef]
221. Yan, M.; Huang, W.; Li, Z. Chitosan cross-linked graphene oxide/lignosulfonate composite aerogel for enhanced adsorption of methylene blue in water. *Int. J. Biol. Macromol.* **2019**, *136*, 927–935. [CrossRef] [PubMed]
222. Zeng, W.; Liu, Y.-G.; Hu, X.-J.; Liu, S.-B.; Zeng, G.-M.; Zheng, B.-H.; Jiang, L.-H.; Guo, F.-Y.; Ding, Y.; Xu, Y. Decontamination of methylene blue from aqueous solution by magnetic chitosan lignosulfonate grafted with graphene oxide: Effects of environmental conditions and surfactant. *RSC Adv.* **2016**, *6*, 19298–19307. [CrossRef]
223. Ge, H.; Ma, Z. Microwave preparation of triethylenetetramine modified graphene oxide/chitosan composite for adsorption of Cr(VI). *Carbohydr. Polym.* **2015**, *131*, 280–287. [CrossRef] [PubMed]
224. Sirajudheen, P.; Karthikeyan, P.; Ramkumar, K.; Meenakshi, S. Effective removal of organic pollutants by adsorption onto chitosan supported graphene oxide-hydroxyapatite composite: A novel reusable adsorbent. *J. Mol. Liq.* **2020**, *318*, 114200. [CrossRef]
225. Azizkhani, S.; Mahmoudi, E.; Abdullah, N.; Ismail, M.H.S.; Mohammad, A.W.; Hussain, S.A. Synthesis and characterisation of graphene oxide-silica-chitosan for eliminating the Pb(II) from aqueous solution. *Polymers* **2020**, *12*, 1922. [CrossRef] [PubMed]
226. Chen, J.; Ma, Y.; Wang, L.; Han, W.; Chai, Y.; Wang, T.; Li, J.; Ou, L. Preparation of chitosan/SiO_2-loaded graphene composite beads for efficient removal of bilirubin. *Carbon N. Y.* **2019**, *143*, 352–361. [CrossRef]
227. Kovtun, A.; Campodoni, E.; Favaretto, L.; Zambianchi, M.; Salatino, A.; Amalfitano, S.; Navacchia, M.L.; Casentini, B.; Palermo, V.; Sandri, M.; et al. Multifunctional graphene oxide/biopolymer composite aerogels for microcontaminants removal from drinking water. *Chemosphere* **2020**, *259*. [CrossRef]
228. Wu, Z.; Deng, W.; Zhou, W.; Luo, J. Novel magnetic polysaccharide/graphene oxide @Fe_3O_4 gel beads for adsorbing heavy metal ions. *Carbohydr. Polym.* **2019**, *216*, 119–128. [CrossRef]
229. Wei, H.; Han, L.; Tang, Y.; Ren, J.; Zhao, Z.; Jia, L. Highly flexible heparin-modified chitosan/graphene oxide hybrid hydrogel as a super bilirubin adsorbent with excellent hemocompatibility. *J. Mater. Chem. B* **2015**, *3*, 1646–1654. [CrossRef]
230. Huang, T.; Shao, Y.-W.; Zhang, Q.; Deng, Y.-F.; Liang, Z.-X.; Guo, F.-Z.; Li, P.-C.; Wang, Y. Chitosan-cross-linked graphene oxide/carboxymethyl cellulose aerogel globules with high structure stability in liquid and extremely high adsorption ability. *ACS Sustain. Chem. Eng.* **2019**, *7*, 8775–8788. [CrossRef]
231. Tofighy, M.A.; Mohammadi, T. Copper ions removal from aqueous solutions using acid-chitosan functionalized carbon nanotubes sheets. *Desalin. Water Treat.* **2016**, *57*, 15384–15396. [CrossRef]
232. Mallakpour, S.; Nezamzadeh Ezhieh, A. Preparation and characterization of chitosan-poly(vinyl alcohol) nanocomposite films embedded with functionalized multi-walled carbon nanotube. *Carbohydr. Polym.* **2017**, *166*, 377–386. [CrossRef] [PubMed]
233. Zeng, G.; Liu, X.; Liu, M.; Huang, Q.; Xu, D.; Wan, Q.; Huang, H.; Deng, F.; Zhang, X.; Wei, Y. Facile preparation of carbon nanotubes based carboxymethyl chitosan nanocomposites through combination of mussel inspired chemistry and Michael addition reaction: Characterization and improved Cu^{2+} removal capability. *J. Taiwan Inst. Chem. Eng.* **2016**, *68*, 446–454. [CrossRef]
234. Jung, C.; Heo, J.; Han, J.; Her, N.; Lee, S.-J.; Oh, J.; Ryu, J.; Yoon, Y. Hexavalent chromium removal by various adsorbents: Powdered activated carbon, chitosan, and single/multi-walled carbon nanotubes. *Sep. Purif. Technol.* **2013**, *106*, 63–71. [CrossRef]
235. Dai, B.; Cao, M.; Fang, G.; Liu, B.; Dong, X.; Pan, M.; Wang, S. Schiff base-chitosan grafted multiwal led carbon nanotubes as a novel solid-phase extraction adsorbent for determination of heavy metal by ICP-MS. *J. Hazard. Mater.* **2012**, *219–220*, 103–110. [CrossRef]
236. Popuri, S.R.; Frederick, R.; Chang, C.-Y.; Fang, S.-S.; Wang, C.-C.; Lee, L.-C. Removal of copper (II) ions from aqueous solutions onto chitosan/carbon nanotubes composite sorbent. *Desalin. Water Treat.* **2014**, *52*, 691–701. [CrossRef]
237. Chen, J.-H.; Lu, D.-Q.; Chen, B.; Ouyang, P.-K. Removal of U(VI) from aqueous solutions by using MWCNTs and chitosan modified MWCNTs. *J. Radioanal. Nucl. Chem.* **2013**, *295*, 2233–2241. [CrossRef]
238. Nitayaphat, W.; Jintakosol, T. *Removal of Silver(I) from Aqueous Solutions by Chitosan/Carbon Nanotube Nanocomposite Beads*; Trans Tech Publications Ltd.: Bäch, Switzerland, 2014; Volume 893, ISBN 9783038350255.
239. Huang, Y.; Lee, X.; Grattieri, M.; Macazo, F.C.; Cai, R.; Minteer, S.D. A sustainable adsorbent for phosphate removal: Modifying multi-walled carbon nanotubes with chitosan. *J. Mater. Sci.* **2018**, *53*, 12641–12649. [CrossRef]
240. Firozjaee, T.T.; Mehrdadi, N.; Baghdadi, M.; Nabi Bidhendi, G.R.N. The removal of diazinon from aqueous solution by chitosan/carbon nanotube adsorbent. *Desalin. Water Treat.* **2017**, *79*, 291–300. [CrossRef]

241. Gonçalves, J.O.; da Silva, K.A.; Rios, E.C.; Crispim, M.M.; Dotto, G.L.; de Almeida Pinto, L.A. Chitosan hydrogel scaffold modified with carbon nanotubes and its application for food dyes removal in single and binary aqueous systems. *Int. J. Biol. Macromol.* **2020**, *142*, 85–93. [CrossRef]
242. Wang, W.; Du, Y.; Liu, M.; Yang, M.; Han, R. Adsorption of congo red from solution using chitosan modified carbon nanotubes. *Desalin. Water Treat.* **2019**, *156*, 96–105. [CrossRef]
243. Chen, M.; Zhou, S.; Zhu, Y.; Sun, Y.; Zeng, G.; Yang, C.; Xu, P.; Yan, M.; Liu, Z.; Zhang, W. Toxicity of carbon nanomaterials to plants, animals and microbes: Recent progress from 2015-present. *Chemosphere* **2018**, *206*, 255–264. [CrossRef] [PubMed]
244. Song, B.; Chen, M.; Ye, S.; Xu, P.; Zeng, G.; Gong, J.; Li, J.; Zhang, P.; Cao, W. Effects of multi-walled carbon nanotubes on metabolic function of the microbial community in riverine sediment contaminated with phenanthrene. *Carbon* **2019**, *144*, 1–7. [CrossRef]
245. Zhou, L.; Ji, L.; Ma, P.-C.; Shao, Y.; Zhang, H.; Gao, W.; Li, Y. Development of carbon nanotubes/$CoFe_2O_4$ magnetic hybrid material for removal of tetrabromobisphenol A and Pb(II). *J. Hazard. Mater.* **2014**, *265*, 104–114. [CrossRef] [PubMed]
246. Wang, Y.; Shi, L.; Gao, L.; Wei, Q.; Cui, L.; Hu, L.; Yan, L.; Du, B. The removal of lead ions from aqueous solution by using magnetic hydroxypropyl chitosan/oxidized multiwalled carbon nanotubes composites. *J. Colloid Interface Sci.* **2015**, *451*, 7–14. [CrossRef] [PubMed]
247. Kim, M.K.; Shanmuga Sundaram, K.; Anantha Iyengar, G.; Lee, K.-P. A novel chitosan functional gel included with multiwall carbon nanotube and substituted polyaniline as adsorbent for efficient removal of chromium ion. *Chem. Eng. J.* **2015**, *267*, 51–64. [CrossRef]
248. Alsabagh, A.M.; Fathy, M.; Morsi, R.E. Preparation and characterization of chitosan/silver nanoparticle/copper nanoparticle/carbon nanotube multifunctional nano-composite for water treatment: Heavy metals removal; kinetics, isotherms and competitive studies. *RSC Adv.* **2015**, *5*, 55774–55783. [CrossRef]
249. Li, T.; He, F.; Dai, Y.D. Prussian blue analog caged in chitosan surface-decorated carbon nanotubes for removal cesium and strontium. *J. Radioanal. Nucl. Chem.* **2016**, *310*, 1139–1145. [CrossRef]
250. ZabihiSahebi, A.; Koushkbaghi, S.; Pishnamazi, M.; Askari, A.; Khosravi, R.; Irani, M. Synthesis of cellulose acetate/chitosan/SWCNT/Fe_3O_4/TiO_2 composite nanofibers for the removal of Cr(VI), As(V), methylene blue and congo red from aqueous solutions. *Int. J. Biol. Macromol.* **2019**, *140*, 1296–1304. [CrossRef]
251. Li, K.; Gao, Q.; Yadavalli, G.; Shen, X.; Lei, H.; Han, B.; Xia, K.; Zhou, C. Selective adsorption of Gd^{3+} on a magnetically retrievable imprinted chitosan/carbon nanotube composite with high capacity. *ACS Appl. Mater. Interfaces* **2015**, *7*, 21047–21055. [CrossRef]
252. Mohan, D.; Sarswat, A.; Ok, Y.S.; Pittman, C.U. Organic and inorganic contaminants removal from water with biochar, a renewable, low cost and sustainable adsorbent—A critical review. *Bioresour. Technol.* **2014**, *160*, 191–202. [CrossRef] [PubMed]
253. Tan, X.; Liu, Y.; Zeng, G.; Wang, X.; Hu, X.; Gu, Y.; Yang, Z. Application of biochar for the removal of pollutants from aqueous solutions. *Chemosphere* **2015**, *125*, 70–85. [CrossRef] [PubMed]
254. Xue, Y.; Gao, B.; Yao, Y.; Inyang, M.; Zhang, M.; Zimmerman, A.R.; Ro, K.S. Hydrogen peroxide modification enhances the ability of biochar (hydrochar) produced from hydrothermal carbonization of peanut hull to remove aqueous heavy metals: Batch and column tests. *Chem. Eng. J.* **2012**, *200–202*, 673–680. [CrossRef]
255. Fernandez, M.E.; Ledesma, B.; Román, S.; Bonelli, P.R.; Cukierman, A.L. Development and characterization of activated hydrochars from orange peels as potential adsorbents for emerging organic contaminants. *Bioresour. Technol.* **2015**, *183*, 221–228. [CrossRef]
256. Zhou, Y.; Gao, B.; Zimmerman, A.R.; Fang, J.; Sun, Y.; Cao, X. Sorption of heavy metals on chitosan-modified biochars and its biological effects. *Chem. Eng. J.* **2013**, *231*, 512–518. [CrossRef]
257. Zhou, Y.; Gao, B.; Zimmerman, A.R.; Chen, H.; Zhang, M.; Cao, X. Biochar-supported zerovalent iron for removal of various contaminants from aqueous solutions. *Bioresour. Technol.* **2014**, *152*, 538–542. [CrossRef]
258. Ifthikar, J.; Jiao, X.; Ngambia, A.; Wang, T.; Khan, A.; Jawad, A.; Xue, Q.; Liu, L.; Chen, Z. Facile one-pot snthesis of sustainable carboxymethyl chitosan—Sewage sludge biochar for effective heavy metal chelation and regeneration. *Bioresour. Technol.* **2018**, *262*, 22–31. [CrossRef]
259. Liu, S.; Huang, B.; Chai, L.; Liu, Y.; Zeng, G.; Wang, X.; Zeng, W.; Shang, M.; Deng, J.; Zhou, Z. Enhancement of As(V) adsorption from aqueous solution by a magnetic chitosan/biochar composite. *RSC Adv.* **2017**, *7*, 10891–10900. [CrossRef]
260. Son, E.-B.; Poo, K.-M.; Mohamed, H.O.; Choi, Y.-J.; Cho, W.-C.; Chae, K.-J. A novel approach to developing a reusable marine macro-algae adsorbent with chitosan and ferric oxide for simultaneous efficient heavy metal removal and easy magnetic separation. *Bioresour. Technol.* **2018**, *259*, 381–387. [CrossRef]
261. Deng, J.; Liu, Y.; Liu, S.; Zeng, G.; Tan, X.; Huang, B.; Tang, X.; Wang, S.; Hua, Q.; Yan, Z. Competitive adsorption of Pb(II), Cd(II) and Cu(II) onto chitosan-pyromellitic dianhydride modified biochar. *J. Colloid Interface Sci.* **2017**, *506*, 355–364. [CrossRef]
262. Zhang, L.; Tang, S.; He, F.; Liu, Y.; Mao, W.; Guan, Y. Highly efficient and selective capture of heavy metals by poly(acrylic acid) grafted chitosan and biochar composite for wastewater treatment. *Chem. Eng. J.* **2019**, *378*, 122215. [CrossRef]
263. Huang, X.; Liu, Y.; Liu, S.; Tan, X.; Ding, Y.; Zeng, G.; Zhou, Y.; Zhang, M.; Wang, S.; Zheng, B. Effective removal of Cr(VI) using β-cyclodextrin-chitosan modified biochars with adsorption/reduction bifuctional roles. *RSC Adv.* **2015**, *6*, 94–104. [CrossRef]
264. Li, R.; Liang, W.; Huang, H.; Jiang, S.; Guo, D.; Li, M.; Zhang, Z.; Ali, A.; Wang, J.J. Removal of cadmium(II) cations from an aqueous solution with aminothiourea chitosan strengthened magnetic biochar. *J. Appl. Polym. Sci.* **2018**, *135*, 46239. [CrossRef]

265. Zhou, D.; Xie, G.; Hu, X.; Cai, X.; Zhao, Y.; Hu, X.; Jin, Q.; Fu, X.; Tan, X.; Liang, C.; et al. Coupling of Kenaf biochar and magnetic $BiFeO_3$ onto cross-linked chitosan for enhancing separation performance and Cr(VI) ions removal efficiency. *Int. J. Environ. Res. Public Health* **2020**, *17*, 788. [CrossRef] [PubMed]
266. Huang, Y.; Lee, X.; Grattieri, M.; Yuan, M.; Cai, R.; Macazo, F.C.; Minteer, S.D. Modified biochar for phosphate adsorption in environmentally relevant conditions. *Chem. Eng. J.* **2020**, *380*, 122375. [CrossRef]
267. Song, Q.; Yang, B.; Wang, H.; Xu, S.; Cao, Y. Effective removal of copper (II) and cadmium (II) by adsorbent prepared from chitosan-modified magnetic biochar. *J. Resid. Sci. Technol.* **2016**, *13*, 197–205. [CrossRef]
268. Jiang, X.; Xiang, X.; Hu, H.; Meng, X.; Hou, L. Facile Fabrication of Biochar/Al_2O_3 adsorbent and its application for fuoride removal from aqueous solution. *J. Chem. Eng. Data* **2019**, *64*, 83–89. [CrossRef]
269. Liu, D.; Zhu, Y.; Li, Z.; Xiao, M.; Jiang, C.; Chen, M.; Chen, Y. Microfibrillar polysaccharide-derived biochars as sodium benzoate adsorbents. *ACS Omega* **2017**, *2*, 2959–2966. [CrossRef]
270. Mojiri, A.; Kazeroon, R.A.; Gholami, A. Cross-linked magnetic chitosan/activated biochar for removal of emerging micropollutants from water: Optimization by the artificial neural network. *Water* **2019**, *11*, 551. [CrossRef]
271. Afzal, M.Z.; Yue, R.; Sun, X.-F.; Song, C.; Wang, S.-G. Enhanced removal of ciprofloxacin using humic acid modified hydrogel beads. *J. Colloid Interface Sci.* **2019**, *543*, 76–83. [CrossRef]
272. Nguyen, H.T.; Phuong, V.N.; Van, T.N.; Thi, P.N.; Dinh Thi Lan, P.; Pham, H.T.; Cao, H.T. Low-cost hydrogel derived from agro-waste for veterinary antibiotic removal: Optimization, kinetics, and toxicity evaluation. *Environ. Technol. Innov.* **2020**, *20*, 101098. [CrossRef]
273. Zhu, C.; Lang, Y.; Liu, B.; Zhao, H. Ofloxacin adsorption on chitosan/biochar composite: Kinetics, isotherms, and effects of solution chemistry. *Polycycl. Aromat. Compd.* **2019**, *39*, 287–297. [CrossRef]
274. Kannan, N.; Sundaram, M.M. Kinetics and mechanism of removal of methylene blue by adsorption on various carbons—A comparative study. *Dye. Pigment.* **2001**, *51*, 25–40. [CrossRef]
275. Elkady, M.; Shokry, H.; Hamad, H. New activated carbon from mine coal for adsorption of dye in simulated water or multiple heavy metals in real wastewater. *Materials* **2020**, *13*, 2498. [CrossRef] [PubMed]
276. Unugul, T.; Nigiz, F.U. Preparation and characterization an active carbon adsorbent from waste mandarin peel and determination of adsorption behavior on removal of synthetic dye solutions. *Water. Air. Soil Pollut.* **2020**, *231*, 538. [CrossRef]
277. Hydari, S.; Sharififard, H.; Nabavinia, M.; Parvizi, M.R. A comparative investigation on removal performances of commercial activated carbon, chitosan biosorbent and chitosan/activated carbon composite for cadmium. *Chem. Eng. J.* **2012**, *193–194*, 276–282. [CrossRef]
278. Auta, M.; Hameed, B.H. Coalesced chitosan activated carbon composite for batch and fixed-bed adsorption of cationic and anionic dyes. *Colloids Surf. B Biointerface* **2013**, *105*, 199–206. [CrossRef]
279. Fatombi, J.K.; Idohou, E.A.; Osseni, S.A.; Agani, I.; Neumeyer, D.; Verelst, M.; Mauricot, R.; Aminou, T. Adsorption of indigo carmine from aqueous solution by chitosan and chitosan/activated carbon composite: Kinetics, isotherms and thermodynamics studies. *Fib. Polym.* **2019**, *20*, 1820–1832. [CrossRef]
280. Babel, S.; Kurniawan, T.A. Cr(VI) removal from synthetic wastewater using coconut shell charcoal and commercial activated carbon modified with oxidizing agents and/or chitosan. *Chemosphere* **2004**, *54*, 951–967. [CrossRef]
281. Amuda, O.S.; Giwa, A.A.; Bello, I.A. Removal of heavy metal from industrial wastewater using modified activated coconut shell carbon. *Biochem. Eng. J.* **2007**, *36*, 174–181. [CrossRef]
282. Amuda, O.S.; Adelowo, F.E.; Ologunde, M.O. Kinetics and equilibrium studies of adsorption of chromium(VI) ion from industrial wastewater using Chrysophyllum albidum (Sapotaceae) seed shells. *Colloids Surf. B Biointerface* **2009**, *68*, 184–192. [CrossRef] [PubMed]
283. Jayasantha Kumari, H.; Krishnamoorthy, P.; Arumugam, T.K.; Radhakrishnan, S.; Vasudevan, D. An efficient removal of crystal violet dye from waste water by adsorption onto TLAC/Chitosan composite: A novel low cost adsorbent. *Int. J. Biol. Macromol.* **2017**, *96*, 324–333. [CrossRef] [PubMed]
284. Ba, S.; Alagui, A.; Hajjaji, M. Retention and release of hexavalent and trivalent chromium by chitosan, olive stone activated carbon, and their blend. *Environ. Sci. Pollut. Res.* **2018**, *25*, 19585–19604. [CrossRef] [PubMed]
285. Sharififard, H.; Zokaee Ashtiani, F.; Soleimani, M. Adsorption of palladium and platinum from aqueous solutions by chitosan and activated carbon coated with chitosan. *Asia-Pac.J. Chem. Eng.* **2013**, *8*, 384–395. [CrossRef]
286. Gonçalves, J.O.; Santos, J.P.; Rios, E.C.; Crispim, M.M.; Dotto, G.L.; Pinto, L.A.A. Development of chitosan based hybrid hydrogels for dyes removal from aqueous binary system. *J. Mol. Liq.* **2017**, *225*, 265–270. [CrossRef]
287. Ge, H.; Fan, X. Adsorption of Pb^{2+} and Cd^{2+} onto a novel activated carbon-chitosan complex. *Chem. Eng. Technol.* **2011**, *34*, 1745–1752. [CrossRef]
288. Li, J.; Jiang, B.; Liu, Y.; Qiu, C.; Hu, J.; Qian, G.; Guo, W.; Ngo, H.H. Preparation and adsorption properties of magnetic chitosan composite adsorbent for Cu^{2+} removal. *J. Clean. Prod.* **2017**, *158*, 51–58. [CrossRef]
289. Chang, Y.-H.; Hsieh, K.-H.; Chang, F.-C. Removal of Hg^{2+} from aqueous solution using a novel composite carbon adsorbent. *J. Appl. Polym. Sci.* **2009**, *112*, 2445–2454. [CrossRef]
290. Villarante, N.R.; Bautista, A.P.R.; Sumalapao, D.E.P. Batch adsorption study and kinetic profile of Cr(VI) using lumbang (Aleurites moluccana)-derived activated carbon-chitosan composite crosslinked with epichlorohydrin. *Orient. J. Chem.* **2017**, *33*, 1111–1119. [CrossRef]

291. Huang, R.; Yang, B.; Liu, Q.; Liu, Y. Simultaneous adsorption of aniline and Cr(VI) ion by activated carbon/chitosan composite. *J. Appl. Polym. Sci.* **2014**, *131*. [CrossRef]
292. Arumugam, T.K.; Krishnamoorthy, P.; Rajagopalan, N.R.; Nanthini, S.; Vasudevan, D. Removal of malachite green from aqueous solutions using a modified chitosan composite. *Int. J. Biol. Macromol.* **2019**, *128*, 655–664. [CrossRef] [PubMed]
293. Nowruzi, R.; Heydari, M.; Javanbakht, V. Synthesis of a chitosan/polyvinyl alcohol/activate carbon biocomposite for removal of hexavalent chromium from aqueous solution. *Int. J. Biol. Macromol.* **2020**, *147*, 209–216. [CrossRef] [PubMed]
294. Salehi, E.; Farahani, A. Macroporous chitosan/polyvinyl alcohol composite adsorbents based on activated carbon substrate. *J. Porous Mater.* **2017**, *24*, 1197–1207. [CrossRef]
295. Salehi, E.; Soroush, F.; Momeni, M.; Barati, A.; Khakpour, A. Chitosan/polyethylene glycol impregnated activated carbons: Synthesis, characterization and adsorption performance. *Front. Chem. Sci. Eng.* **2017**, *11*, 575–585. [CrossRef]
296. Liu, L.F.; Zhang, P.H.; Yang, F.L. Adsorptive removal of 2,4-DCP from water by fresh or regenerated chitosan/ACF/TiO_2 membrane. *Sep. Purif. Technol.* **2010**, *70*, 354–361. [CrossRef]
297. Zhang, S.; Li, X.-Y.; Chen, J.P. Preparation and evaluation of a magnetite-doped activated carbon fiber for enhanced arsenic removal. *Carbon* **2010**, *48*, 60–67. [CrossRef]
298. Karamipour, A.; Khadiv Parsi, P.; Zahedi, P.; Moosavian, S.M.A. Using Fe_3O_4-coated nanofibers based on cellulose acetate/chitosan for adsorption of Cr(VI), Ni(II) and phenol from aqueous solutions. *Int. J. Biol. Macromol.* **2020**, *154*, 1132–1139. [CrossRef]
299. Danalıoğlu, S.T.; Bayazit, Ş.S.; Kerkez Kuyumcu, Ö.; Salam, M.A. Efficient removal of antibiotics by a novel magnetic adsorbent: Magnetic activated carbon/chitosan (MACC) nanocomposite. *J. Mol. Liq.* **2017**, *240*, 589–596. [CrossRef]
300. Çavuşoğlu, F.C.; Akan, S.; Arı, E.A.; Çetinkaya, E.; Çolak, E.; Daştan, G.N.; Deniz, S.; Erdem, D.; Köksal, M.; Korkmaz, S.; et al. Preparation of magnetic activated carbon-chitosan nanocomposite for crystal violet adsorption. *Kor. J. Chem. Eng.* **2019**, *36*, 1915–1921. [CrossRef]
301. Malakootian, M.; Nasiri, A.; Mahdizadeh, H. Preparation of $CoFe_2O_4$/activated carbon@chitosan as a new magnetic nanobiocomposite for adsorption of ciprofloxacin in aqueous solutions. *Water Sci. Technol.* **2018**, *78*, 2158–2170. [CrossRef]
302. Sanlier, S.H.; Ak, G.; Yilmaz, H.; Ozbakir, G.; Cagliyan, O. Removal of textile dye, direct red 23, with glutaraldehyde cross-linked magnetic chitosan beads. *Prep. Biochem. Biotechnol.* **2013**, *43*, 163–176. [CrossRef] [PubMed]
303. Soni, U.; Bajpai, J.; Singh, S.K.; Bajpai, A.K. Evaluation of chitosan-carbon based biocomposite for efficient removal of phenols from aqueous solutions. *J. Water Process Eng.* **2017**, *16*, 56–63. [CrossRef]
304. Mashile, G.P.; Mpupa, A.; Nqombolo, A.; Dimpe, K.M.; Nomngongo, P.N. Recyclable magnetic waste tyre activated carbon-chitosan composite as an effective adsorbent rapid and simultaneous removal of methylparaben and propylparaben from aqueous solution and wastewater. *J. Water Process Eng.* **2020**, *33*. [CrossRef]
305. Jawad, A.H.; Abdulhameed, A.S.; Mastuli, M.S. Mesoporous crosslinked chitosan-activated charcoal composite for the removal of thionine cationic dye: Comprehensive adsorption and mechanism study. *J. Polym. Environ.* **2020**, *28*, 1095–1105. [CrossRef]
306. Gonçalves, J.O.; da Silva, K.A.; Rios, E.C.; Crispim, M.M.; Dotto, G.L.; de Almeida Pinto, L.A. Single and binary adsorption of food dyes on chitosan/activated carbon hydrogels. *Chem. Eng. Technol.* **2019**, *42*, 454–464. [CrossRef]
307. Saloglu, D.; Ozcan, N. Activated carbon embedded chitosan/polyvinyl alcohol biocomposites for adsorption of nonsteroidal anti-inflammatory drug-naproxen from wastewater. *Desalin. Water Treat.* **2018**, *107*, 72–84. [CrossRef]
308. Amouzgar, P.; Vakili, M.; Chan, E.-S.; Salamatinia, B. Effects of beading parameters for development of chitosan-nano-activated carbon biocomposite for acetaminophen elimination from aqueous sources. *Environ. Eng. Sci.* **2017**, *34*, 805–815. [CrossRef]
309. Liu, Q.; Zhang, L.; Hu, P.; Huang, R. Removal of aniline from aqueous solutions by activated carbon coated by chitosan. *J. Water Reuse Desalin.* **2015**, *5*, 610–618. [CrossRef]
310. Ji, G.; Bao, W.; Gao, G.; An, B.; Zou, H.; Gan, S. Removal of Cu (II) from aqueous solution using a novel crosslinked alumina-chitosan hybrid adsorbent. *Chin. J. Chem. Eng.* **2012**, *20*, 641–648. [CrossRef]
311. Jagtap, S.; Yenkie, M.K.N.; Labhsetwar, N.; Rayalu, S. Defluoridation of drinking water using chitosan based mesoporous alumina. *Microporous Mesoporous Mater.* **2011**, *142*, 454–463. [CrossRef]
312. Wang, J.; Zhou, Q.; Song, D.; Qi, B.; Zhang, Y.; Shao, Y.; Shao, Z. Chitosan–silica composite aerogels: Preparation, characterization and Congo red adsorption. *J. Sol-Gel Sci. Technol.* **2015**, *76*, 501–509. [CrossRef]
313. Nuryono, N.; Miswanda, D.; Sakti, S.C.W.; Rusdiarso, B.; Krisbiantoro, P.A.; Utami, N.; Otomo, R.; Kamiya, Y. Chitosan-functionalized natural magnetic particle@silica modified with (3-chloropropyl)trimethoxysilane as a highly stable magnetic adsorbent for gold(III) ion. *Mater. Chem. Phys.* **2020**, *255*, 123507. [CrossRef]
314. Walcarius, A.; Mercier, L. Mesoporous organosilica adsorbents: Nanoengineered materials for removal of organic and inorganic pollutants. *J. Mater. Chem.* **2010**, *20*, 4478–4511. [CrossRef]
315. Shahat, A.; Hassan, H.M.A.; Azzazy, H.M.E.; El-Sharkawy, E.A.; Abdou, H.M.; Awual, M.R. Novel hierarchical composite adsorbent for selective lead(II) ions capturing from wastewater samples. *Chem. Eng. J.* **2018**, *332*, 377–386. [CrossRef]
316. Dos Santos, S.M.L.; Nogueira, K.A.B.; De Souza Gama, M.; Lima, J.D.F.; Da Silva Júnior, I.J.; De Azevedo, D.C.S. Synthesis and characterization of ordered mesoporous silica (SBA-15 and SBA-16) for adsorption of biomolecules. *Microporous Mesoporous Mater.* **2013**, *180*, 284–292. [CrossRef]
317. Wang, J.; Mao, M.; Atif, S.; Chen, Y. Adsorption behavior and mechanism of aqueous Cr(III) and Cr(III)-EDTA chelates on DTPA-chitosan modified $Fe_3O_4@SiO_2$. *React. Funct. Polym.* **2020**, *156*, 104720. [CrossRef]

318. Aden, M.; Ubol, R.N.; Knorr, M.; Husson, J.; Euvrard, M. Efficent removal of nickel(II) salts from aqueous solution using carboxymethylchitosan-coated silica particles as adsorbent. *Carbohydr. Polym.* **2017**, *173*, 372–382. [CrossRef]
319. Malwal, D.; Gopinath, P. Silica stabilized magnetic-chitosan beads for removal of arsenic from water. *Colloids Interface Sci. Commun.* **2017**, *19*, 14–19. [CrossRef]
320. Darvishi Cheshmeh Soltani, R.; Khataee, A.R.; Safari, M.; Joo, S.W. Preparation of bio-silica/chitosan nanocomposite for adsorption of a textile dye in aqueous solutions. *Int. Biodeterior. Biodegrad.* **2013**, *85*, 383–391. [CrossRef]
321. Ebisike, K.; Okoronkwo, A.E.; Alaneme, K.K. Adsorption of Cd(II) on chitosan–silica hybrid aerogel from aqueous solution. *Environ. Technol. Innov.* **2019**, *14*, 100337. [CrossRef]
322. Zhao, H.; Xu, J.; Lan, W.; Wang, T.; Luo, G. Microfluidic production of porous chitosan/silica hybrid microspheres and its Cu(II) adsorption performance. *Chem. Eng. J.* **2013**, *229*, 82–89. [CrossRef]
323. Xu, X.; Dong, P.; Feng, Y.; Li, F.; Yu, H. A simple strategy for preparation of spherical silica-supported porous chitosan matrix based on sol-gel reaction and simple treatment with ammonia solution. *Anal. Methods* **2010**, *2*, 546–551. [CrossRef]
324. Wong, J.S.H.; Hicks, R.E.; Probstein, R.F. EDTA-enhanced electroremediation of metal-contaminated soils. *J. Hazard. Mater.* **1997**, *55*, 61–79. [CrossRef]
325. Repo, E.; Warchoł, J.K.; Bhatnagar, A.; Sillanpää, M. Heavy metals adsorption by novel EDTA-modified chitosan-silica hybrid materials. *J. Colloid Interface Sci.* **2011**, *358*, 261–267. [CrossRef]
326. Blachnio, M.; Budnyak, T.M.; Derylo-Marczewska, A.; Marczewski, A.W.; Tertykh, V.A. Chitosan-silica hybrid composites for removal of sulfonated azo dyes from aqueous solutions. *Langmuir* **2018**, *34*, 2258–2273. [CrossRef]
327. Li, Y.; Zhou, Y.; Nie, W.; Song, L.; Chen, P. Highly efficient methylene blue dyes removal from aqueous systems by chitosan coated magnetic mesoporous silica nanoparticles. *J. Porous Mater.* **2015**, *22*, 1383–1392. [CrossRef]
328. Ting, C.-C.; Wu, H.-Y.; Vetrivel, S.; Saikia, D.; Pan, Y.-C.; Fey, G.T.K.; Kao, H.-M. A one-pot route to synthesize highly ordered mesoporous carbons and silicas through organic-inorganic self-assembly of triblock copolymer, sucrose and silica. *Microporous Mesoporous Mater.* **2010**, *128*, 1–11. [CrossRef]
329. Chen, D.; Hu, B.; Huang, C. Chitosan modified ordered mesoporous silica as micro-column packing materials for on-line flow injection-inductively coupled plasma optical emission spectrometry determination of trace heavy metals in environmental water samples. *Talanta* **2009**, *78*, 491–497. [CrossRef]
330. He, H.; Meng, X.; Yue, Q.; Yin, W.; Gao, Y.; Fang, P.; Shen, L. Thiol-ene click chemistry synthesis of a novel magnetic mesoporous silica/chitosan composite for selective Hg(II) capture and high catalytic activity of spent Hg(II) adsorbent. *Chem. Eng. J.* **2021**, *405*. [CrossRef]
331. Sabourian, V.; Ebrahimi, A.; Naseri, F.; Irani, M.; Rahimi, A. Fabrication of chitosan/silica nanofibrous adsorbent functionalized with amine groups for the removal of Ni(II), Cu(II) and Pb(II) from aqueous solutions: Batch and column studies. *RSC Adv.* **2016**, *6*, 40354–40365. [CrossRef]
332. Gao, Q.; Zhu, H.; Luo, W.-J.; Wang, S.; Zhou, C.-G. Preparation, characterization, and adsorption evaluation of chitosan-functionalized mesoporous composites. *Microporous Mesoporous Mater.* **2014**, *193*, 15–26. [CrossRef]
333. Mousavi, S.J.; Parvini, M.; Ghorbani, M. Adsorption of heavy metals (Cu^{2+} and Zn^{2+}) on novel bifunctional ordered mesoporous silica: Optimization by response surface methodology. *J. Taiwan Inst. Chem. Eng.* **2018**, *84*, 123–141. [CrossRef]
334. Du, Q.; Sun, J.; Li, Y.; Yang, X.; Wang, X.; Wang, Z.; Xia, L. Highly enhanced adsorption of congo red onto graphene oxide/chitosan fibers by wet-chemical etching off silica nanoparticles. *Chem. Eng. J.* **2014**, *245*, 99–106. [CrossRef]
335. Zhang, L.; Luo, H.; Liu, P.; Fang, W.; Geng, J. A novel modified graphene oxide/chitosan composite used as an adsorbent for Cr(VI) in aqueous solutions. *Int. J. Biol. Macromol.* **2016**, *87*, 586–596. [CrossRef]
336. Shahzad, A.; Miran, W.; Rasool, K.; Nawaz, M.; Jang, J.; Lim, S.-R.; Lee, D.S. Heavy metals removal by EDTA-functionalized chitosan graphene oxide nanocomposites. *RSC Adv.* **2017**, *7*, 9764–9771. [CrossRef]
337. Hadi Najafabadi, H.; Irani, M.; Roshanfekr Rad, L.; Heydari Haratameh, A.; Haririan, I. Removal of Cu^{2+}, Pb^{2+} and Cr^{6+} from aqueous solutions using a chitosan/graphene oxide composite nanofibrous adsorbent. *RSC Adv.* **2015**, *5*, 16532–16539. [CrossRef]
338. Travlou, N.A.; Kyzas, G.Z.; Lazaridis, N.K.; Deliyanni, E.A. Graphite oxide/chitosan composite for reactive dye removal. *Chem. Eng. J.* **2013**, *217*, 256–265. [CrossRef]
339. Li, X.; Zhou, H.; Wu, W.; Wei, S.; Xu, Y.; Kuang, Y. Studies of heavy metal ion adsorption on chitosan/sulfydryl-functionalized graphene oxide composites. *J. Colloid Interface Sci.* **2015**, *448*, 389–397. [CrossRef]
340. Hosseinzadeh, H.; Ramin, S. Effective removal of copper from aqueous solutions by modified magnetic chitosan/graphene oxide nanocomposites. *Int. J. Biol. Macromol.* **2018**, *113*, 859–868. [CrossRef]
341. Sherlala, A.I.A.; Raman, A.A.A.; Bello, M.M.; Buthiyappan, A. Adsorption of arsenic using chitosan magnetic graphene oxide nanocomposite. *J. Environ. Manag.* **2019**, *246*, 547–556. [CrossRef]
342. Samuel, M.S.; Shah, S.S.; Subramaniyan, V.; Qureshi, T.; Bhattacharya, J.; Pradeep Singh, N.D. Preparation of graphene oxide/chitosan/ferrite nanocomposite for chromium(VI) removal from aqueous solution. *Int. J. Biol. Macromol.* **2018**, *119*, 540–547. [CrossRef] [PubMed]
343. Sun, W.; Li, L.; Luo, C.; Fan, L. Synthesis of magnetic graphene nanocomposites decorated with ionic liquids for fast lead ion removal. *Int. J. Biol. Macromol.* **2016**, *85*, 246–251. [CrossRef] [PubMed]
344. Nasiri, R.; Arsalani, N.; Panahian, Y. One-pot synthesis of novel magnetic three-dimensional graphene/chitosan/nickel ferrite nanocomposite for lead ions removal from aqueous solution: RSM modelling design. *J. Clean. Prod.* **2018**, *201*, 507–515. [CrossRef]

345. Subedi, N.; Lähde, A.; Abu-Danso, E.; Iqbal, J.; Bhatnagar, A. A comparative study of magnetic chitosan (Chi@Fe_3O_4 and graphene oxide modified magnetic chitosan (Chi@Fe_3O_4/GO) nanocomposites for efficient removal of Cr(VI) from water. *Int. J. Biol. Macromol.* **2019**, *137*, 948–959. [CrossRef]
346. Yang, A.; Yang, P.; Huang, C.P. Preparation of graphene oxide–chitosan composite and adsorption performance for uranium. *J. Radioanal. Nucl. Chem.* **2017**, *313*, 371–378. [CrossRef]
347. Li, C.; Yan, Y.; Zhang, Q.; Zhang, Z.; Huang, L.; Zhang, J.; Xiong, Y.; Tan, S. Adsorption of Cd^{2+} and Ni^{2+} from aqueous single-metal solutions on graphene oxide-chitosan-poly(vinyl alcohol) hydrogels. *Langmuir* **2019**, *35*, 4481–4490. [CrossRef]
348. Cheng, J.; Gao, M.; Yang, L.; Zhang, L.; Zhu, B. Coral-inspired "nanotentaclization" porous composite gel for efficient removal of lead(II) from aqueous solution. *Mater. Des.* **2020**, *195*. [CrossRef]
349. Dou, J.; Gan, D.; Huang, Q.; Liu, M.; Chen, J.; Deng, F.; Zhu, X.; Wen, Y.; Zhang, X.; Wei, Y. Functionalization of carbon nanotubes with chitosan based on MALI multicomponent reaction for Cu^{2+} removal. *Int. J. Biol. Macromol.* **2019**, *136*, 476–485. [CrossRef]
350. Huang, Y.; Lee, X.; Macazo, F.C.; Grattieri, M.; Cai, R.; Minteer, S.D. Fast and efficient removal of chromium(VI) anionic species by a reusable chitosan-modified multi-walled carbon nanotube composite. *Chem. Eng. J.* **2018**, *339*, 259–267. [CrossRef]
351. Ouyang, J.; Wang, Y.; Li, T.; Zhou, L.; Liu, Z. Immobilization of carboxyl-modified multiwalled carbon nanotubes in chitosan-based composite membranes for U(VI) sorption. *J. Radioanal. Nucl. Chem.* **2018**, *317*, 1419–1428. [CrossRef]
352. Marques Neto, J.D.O.; Bellato, C.R.; Silva, D.D.C. Iron oxide/carbon nanotubes/chitosan magnetic composite film for chromium species removal. *Chemosphere* **2019**, *218*, 391–401. [CrossRef] [PubMed]
353. Xiao, F.; Cheng, J.; Cao, W.; Yang, C.; Chen, J.; Luo, Z. Removal of heavy metals from aqueous solution using chitosan-combined magnetic biochars. *J. Colloid Interface Sci.* **2019**, *540*, 579–584. [CrossRef] [PubMed]
354. Arabyarmohammadi, H.; Darban, A.K.; Abdollahy, M.; Yong, R.; Ayati, B.; Zirakjou, A.; van der Zee, S.E.A.T.M. Utilization of a novel chitosan/clay/biochar nanobiocomposite for immobilization of heavy metals in acid soil environment. *J. Polym. Environ.* **2018**, *26*, 2107–2119. [CrossRef]
355. Shariful, M.I.; Sepehr, T.; Mehrali, M.; Ang, B.C.; Amalina, M.A. Adsorption capability of heavy metals by chitosan/poly(ethylene oxide)/activated carbon electrospun nanofibrous membrane. *J. Appl. Polym. Sci.* **2018**, *135*, 45851. [CrossRef]
356. Li, K.; Wang, Y.; Huang, M.; Yan, H.; Yang, H.; Xiao, S.; Li, A. Preparation of chitosan-graft-polyacrylamide magnetic composite microspheres for enhanced selective removal of mercury ions from water. *J. Colloid Interface Sci.* **2015**, *455*, 261–270. [CrossRef] [PubMed]
357. Pourebrahim, F.; Ghaedi, M.; Dashtian, K.; Heidari, F.; Kheirandish, S. Simultaneous removing of Pb^{2+} ions and alizarin red S dye after their complexation by ultrasonic waves coupled adsorption process: Spectrophotometry detection and optimization study. *Ultrason. Sonochem.* **2017**, *35*, 51–60. [CrossRef]
358. Mousavi, S.J.; Parvini, M.; Ghorbani, M. Experimental design data for the zinc ions adsorption based on mesoporous modified chitosan using central composite design method. *Carbohydr. Polym.* **2018**, *188*, 197–212. [CrossRef]
359. Liu, J.; Chen, Y.; Han, T.; Cheng, M.; Zhang, W.; Long, J.; Fu, X. A biomimetic SiO_2@chitosan composite as highly-efficient adsorbent for removing heavy metal ions in drinking water. *Chemosphere* **2019**, *214*, 738–742. [CrossRef]
360. Zarei, F.; Marjani, A.; Soltani, R. Novel and green nanocomposite-based adsorbents from functionalised mesoporous KCC-1 and chitosan-oleic acid for adsorption of Pb(II). *Eur. Polym. J.* **2019**, *119*, 400–409. [CrossRef]
361. Jin, X.; Li, K.; Ning, P.; Bao, S.; Tang, L. Removal of Cu(II) Ions from aqueous solution by magnetic chitosan-tripolyphosphate modified silica-coated adsorbent: Characterization and mechanisms. *Water. Air. Soil Pollut.* **2017**, *228*, 302. [CrossRef]
362. Srinivasan, S.; Chelliah, P.; Srinivasan, V.; Stantley, A.B.; Subramani, K. Chitosan and reinforced chitosan films for the removal of Cr(VI) heavy metal from synthetic aqueous solution. *Orient. J. Chem.* **2016**, *32*, 671–680. [CrossRef]
363. Plohl, O.; Ajdnik, U.; Gyergyek, S.; Ban, I.; Vesel, A.; Glaser, T.K.; Zemljič, L.F. Superior stability and high biosorbent efficiency of carboxymethylchitosan covalently linked to silica-coated core-shell magnetic nanoparticles for application in copper removal. *J. Environ. Chem. Eng.* **2019**, *7*, 102913. [CrossRef]
364. Shariatinia, Z.; Esmaeilzadeh, A. Hybrid silica aerogel nanocomposite adsorbents designed for Cd(II) removal from aqueous solution. *Water Environ. Res.* **2019**, *91*, 1624–1637. [CrossRef] [PubMed]
365. Lai, K.C.; Hiew, B.Y.Z.; Lee, L.Y.; Gan, S.; Thangalazhy-Gopakumar, S.; Chiu, W.S.; Khiew, P.S. Ice-templated graphene oxide/chitosan aerogel as an effective adsorbent for sequestration of metanil yellow dye. *Bioresour. Technol.* **2019**, *274*, 134–144. [CrossRef] [PubMed]
366. Blackburn, R.S. Natural polysaccharides and their interactions with dye molecules: Applications in effluent treatment. *Environ. Sci. Technol.* **2004**, *38*, 4905–4909. [CrossRef]
367. Nekouei Marnani, N.; Shahbazi, A. A novel environmental-friendly nanobiocomposite synthesis by EDTA and chitosan functionalized magnetic graphene oxide for high removal of Rhodamine B: Adsorption mechanism and separation property. *Chemosphere* **2019**, *218*, 715–725. [CrossRef]
368. Luo, T.; Zhang, L. Comment on "A novel environmental-friendly nanobiocomposite synthesis by EDTA and Chitosan functionalized magnetic graphene oxide for high removal of rhodamine B: Adsorption mechanism and separation property" [Chemosphere, 218 (2019) 715–725]. *Chemosphere* **2019**, *233*, 983–984. [CrossRef]
369. Li, Z.; Wang, X.; Zhang, X.; Yang, Y.; Duan, J. A high-efficiency and plane-enhanced chitosan film for cefotaxime adsorption compared with chitosan particles in water. *Chem. Eng. J.* **2020**, 127494. [CrossRef]

370. Liu, Q.-S.; Zheng, T.; Wang, P.; Jiang, J.-P.; Li, N. Adsorption isotherm, kinetic and mechanism studies of some substituted phenols on activated carbon fibers. *Chem. Eng. J.* **2010**, *157*, 348–356. [CrossRef]
371. Wang, Y.; Liu, X.; Wang, H.; Xia, G.; Huang, W.; Song, R. Microporous spongy chitosan monoliths doped with graphene oxide as highly effective adsorbent for methyl orange and copper nitrate ($Cu(NO_3)_2$) ions. *J. Colloid Interface Sci.* **2014**, *416*, 243–251. [CrossRef]
372. Sheshmani, S.; Ashori, A.; Hasanzadeh, S. Removal of Acid Orange 7 from aqueous solution using magnetic graphene/chitosan: A promising nano-adsorbent. *Int. J. Biol. Macromol.* **2014**, *68*, 218–224. [CrossRef] [PubMed]
373. Jiang, Y.; Gong, J.-L.; Zeng, G.-M.; Ou, X.-M.; Chang, Y.-N.; Deng, C.-H.; Zhang, J.; Liu, H.-Y.; Huang, S.-Y. Magnetic chitosan-graphene oxide composite for anti-microbial and dye removal applications. *Int. J. Biol. Macromol.* **2016**, *82*, 702–710. [CrossRef] [PubMed]
374. Omidi, S.; Kakanejadifard, A. Eco-friendly synthesis of graphene-chitosan composite hydrogel as efficient adsorbent for Congo red. *RSC Adv.* **2018**, *8*, 12179–12189. [CrossRef]
375. Zhu, H.; Fu, Y.; Jiang, R.; Yao, J.; Liu, L.; Chen, Y.; Xiao, L.; Zeng, G. Preparation, characterization and adsorption properties of chitosan modified magnetic graphitized multi-walled carbon nanotubes for highly effective removal of a carcinogenic dye from aqueous solution. *Appl. Surf. Sci.* **2013**, *285*, 865–873. [CrossRef]
376. Wang, S.; Zhai, Y.-Y.; Gao, Q.; Luo, W.-J.; Xia, H.; Zhou, C.-G. Highly efficient removal of acid red 18 from aqueous solution by magnetically retrievable chitosan/carbon nanotube: Batch study, isotherms, kinetics, and thermodynamics. *J. Chem. Eng. Data* **2014**, *59*, 39–51. [CrossRef]
377. Abbasi, M. Synthesis and characterization of magnetic nanocomposite of chitosan/SiO_2/carbon nanotubes and its application for dyes removal. *J. Clean. Prod.* **2017**, *145*, 105–113. [CrossRef]
378. Abbasi, M.; Habibi, M.M. Optimization and characterization of Direct Blue 71 removal using nanocomposite of Chitosan-MWCNTs: Central composite design modeling. *J. Taiwan Inst. Chem. Eng.* **2016**, *62*, 112–121. [CrossRef]
379. Idohou, E.A.; Fatombi, J.K.; Osseni, S.A.; Agani, I.; Neumeyer, D.; Verelst, M.; Mauricot, R.; Aminou, T. Preparation of activated carbon/chitosan/Carica papaya seeds composite for efficient adsorption of cationic dye from aqueous solution. *Surf. Interf.* **2020**, *21*, 100741. [CrossRef]
380. Vakili, M.; Zwain, H.M.; Mojiri, A.; Wang, W.; Gholami, F.; Gholami, Z.; Giwa, A.S.; Wang, B.; Cagnetta, G.; Salamatinia, B. Effective adsorption of reactive black 5 onto hybrid hexadecylamine impregnated chitosan-powdered activated carbon beads. *Water* **2020**, *12*, 2242. [CrossRef]
381. Luo, X.; Liu, L.; Wang, L.; Liu, X.; Cai, Y. Facile synthesis and low concentration tylosin adsorption performance of chitosan/cellulose nanocomposite microspheres. *Carbohydr. Polym.* **2019**, *206*, 633–640. [CrossRef]
382. Yan, H.; Li, H.; Yang, H.; Li, A.; Cheng, R. Removal of various cationic dyes from aqueous solutions using a kind of fully biodegradable magnetic composite microsphere. *Chem. Eng. J.* **2013**, *223*, 402–411. [CrossRef]
383. Zhang, B.; Huyan, Y.; Wang, J.; Chen, X.; Zhang, H.; Zhang, Q. Fe_3O_4@SiO_2@CCS porous magnetic microspheres as adsorbent for removal of organic dyes in aqueous phase. *J. Alloys Compd.* **2018**, *735*, 1986–1996. [CrossRef]
384. Hassan, H.; Salama, A.; El-ziaty, A.K.; El-Sakhawy, M. New chitosan/silica/zinc oxide nanocomposite as adsorbent for dye removal. *Int. J. Biol. Macromol.* **2019**, *131*, 520–526. [CrossRef] [PubMed]
385. Mahmoodi, N.M.; Mokhtari-Shourijeh, Z.; Abdi, J. Preparation of mesoporous polyvinyl alcohol/chitosan/silica composite nanofiber and dye removal from wastewater. *Environ. Prog. Sustain. Energy* **2019**, *38*, S100–S109. [CrossRef]
386. Afzal, M.Z.; Sun, X.-F.; Liu, J.; Song, C.; Wang, S.-G.; Javed, A. Enhancement of ciprofloxacin sorption on chitosan/biochar hydrogel beads. *Sci. Total Environ.* **2018**, *639*, 560–569. [CrossRef] [PubMed]

molecules

MDPI

Review

Extremophilic Microorganisms for the Treatment of Toxic Pollutants in the Environment

Sun-Wook Jeong and Yong Jun Choi *

School of Environmental Engineering, University of Seoul, Seoul 02504, Korea; jeongsunwook@gmail.com
* Correspondence: yongjun2165@uos.ac.kr; Tel.: +82-02-6490-2873; Fax: +82-02-6490-2859

Academic Editors: Chiara Bisio and Monica Pica
Received: 17 September 2020; Accepted: 23 October 2020; Published: 23 October 2020

check for updates

Abstract: As concerns about the substantial effect of various hazardous toxic pollutants on the environment and public health are increasing, the development of effective and sustainable treatment methods is urgently needed. In particular, the remediation of toxic components such as radioactive waste, toxic heavy metals, and other harmful substances under extreme conditions is quite difficult due to their restricted accessibility. Thus, novel treatment methods for the removal of toxic pollutants using extremophilic microorganisms that can thrive under extreme conditions have been investigated during the past several decades. In this review, recent trends in bioremediation using extremophilic microorganisms and related approaches to develop them are reviewed, with relevant examples and perspectives.

Keywords: bioremediation; toxic pollutants; extreme conditions; extremophilic microorganism

1. Introduction

Due to the rapid industrial growth, the environment and public health are threatened by the huge amount of toxic pollutants that have accumulated in the environment. Therefore, maintaining and protecting the environment from toxic pollutants has become a great challenge for mankind over the past few decades. Recently, various strategies have been intensively exploited to protect the environment by preventing the dispersion of toxic pollutants into it. For example, physicochemical methods such as electrochemical treatments, excavation, ion exchange, precipitation, reverse osmosis, evaporation, and sorption have been developed for the removal of toxic substances [1–4]. However, many of these techniques are not yet commonly applied to the actual treatment of contamination due to critical drawbacks such as high cost and secondary contamination possibly associated with them [5–7]. As an alternative, microbial bioremediation has attracted much attention as a promising technology that can overcome the shortcomings of the currently used physicochemical methods (Figure 1) [8–10]. Specifically, extremophilic microorganisms offer the most suitable approach for the treatment of toxic pollutants [11–14] because not only can they detoxify toxic pollutants through microbial cellular metabolism but also they can withstand extremely harsh conditions [11,13–15]. Herein, we focus on recent trends in bioremediation processes for the treatment of toxic pollutants such as inorganic heavy metals, harmful organic substances, and radioactive elements using extremophilic microorganisms and on the perspectives of this approach in public health.

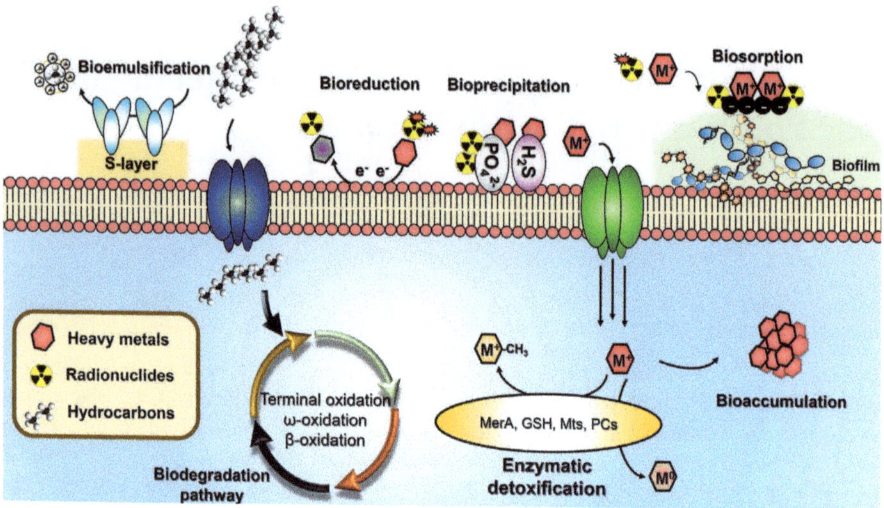

Figure 1. Current microbial bioremediation strategies for the removal of diverse toxic pollutants. Biosorption, a metabolically independent process based on ionic interactions between the extracellular surface of biomass and metal ions; bioaccumulation, a metabolically active process in which microorganisms use proteins to absorb metal ions inside their intracellular space; bioprecipitation, a process of immobilizing soluble metal ions through redox reactions, enzymes, and metabolites on the extracellular surface of microorganisms; bioreduction, a process of transformation of toxic metals/metalloids to non-toxic elements through a biological reduction and oxidation process; bioemulsification, a biological process of using proteins or metabolites to form emulsions in two immiscible liquid phases.

2. Survival Strategies of Extremophilic Microorganisms under Extreme Conditions

Extreme environments are defined as habitats that make the prospect of survival difficult for most organisms on earth. These are mostly natural conditions such as extreme temperatures, salinity, pH, and desiccation observed in environments such as deep sea, volcanoes, and deserts. However, these extreme conditions can also appear in polluted areas containing harmful organic substances [16], heavy metals [17], and/or radioactive waste [18]. Under extremely polluted conditions, the clean-up process of pollutants by using physicochemical methods is not always successful due to limited accessibility to the pollutants and secondary contamination. Thus, there is a need to combine microbial biotechnology and chemistry to advance the remediation processes. Over the past century, extremophilic microorganisms have adapted and evolved in various ways to thrive under extreme conditions through unique biological mechanisms. During the process of adaptation, extremophilic microorganisms have evolved not only to convert unstable toxic pollutants into sufficiently stable beneficial resources for their cellular metabolism but also to become highly tolerant to toxic matter. Thus, many studies have been attempted to develop sustainable bioremediation processes using the survival strategies of extremophilic microorganisms. Here, we briefly describe the adaptation and survival mechanisms that can be used for bioremediation.

2.1. *Acidophilic and Alkaliphilic Microorganisms*

Acidophilic microorganisms can survive under extremely low pH (less than pH 3) conditions, maintaining pH homeostasis by controlling proton permeation [19]. For example, microorganisms from the genera *Thermoplasma*, *Ferroplasma*, and *Sulfolobus* can regulate proton permeation under extremely low pH conditions due to a highly impermeable cell membrane mainly composed of tetraether lipids having a diverse array of polar head groups and a bulky isoprenoid core [20–23].

The modulation of the influx of protons through the proton pump system is important to survive at low pH, and putative proton pump proteins such as H^+-ATPase, symporters, and antiporters from *Ferroplasma* type II and *Leptospirillium* group II are involved in maintaining pH homeostasis [21,24,25]. Moreover, F_0F_1-type adenosine triphosphate synthase in *Bacillus acidocaldarus*, *Thermoplasma acidophilum*, and *Leptospirillium ferriphilum* is known to play a critical role in regulating proton permeation [25]. In addition to these mechanisms, several other auxiliary mechanisms, involving for example, chaperone proteins and cytoplasmic buffering capacity contribute to survival strategies under extremely low pH conditions by protecting intracellular molecules such as DNA, RNA, and proteins [25].

Contrary to acidophilic microorganisms, alkaliphilic microorganisms can resistant high pH. To date, three key biological mechanisms have been identified as survival strategies in these microorganisms. First, under extremely high pH conditions, some alkaliphilic *Bacillus* spp. can increase the generation of proton motive force through synthesizing a secondary acidic cell membrane consisting primarily of peptidoglycan, teichuronic acid, and teichuronopeptide [26,27]. Increasing the proton motive force contributes to not only energy generation but also pH balance [28–30]. Second, sodium motive force can also promote pH balance under extremely high pH conditions [31,32]. Under high Na^+ ion conditions, Na^+/H^+ antiporters extrude Na^+ ions and absorb a greater amount of extracellular H^+ ions than that of extruded Na^+ ions, thereby activating a bioenergetic process and regulating the internal pH [33]. Finally, the production of organic acids that can be used for pH calibration is known to be an important biological process in maintaining pH balance [34,35].

2.2. *Halophilic Microorganisms*

Halophilic microorganisms can thrive in a high-salt environment which hinders organisms' survival due to osmolar imbalance and metabolic problems [36,37]. Previous studies on halophilic microorganisms reported two fundamental adaptation strategies to survive under extremely high salt conditions. The first is to use a "salt-in" strategy that refers to the accumulation of inorganic osmoprotectants such as KCl inside the cell to maintain the osmotic balance both inside and outside the cell [37]. It has been demonstrated that *Halobacterium salinarum* can accumulate 3.97 M and 4.57 M of K^+ and Cl^- ions, respectively, inside the cell using the ATP-dependent K^+ transport system (the KdpFABC complex and cationic amino acid transporter-3 (Cat3) and Na^+ efflux antiporters (NhaC) to balance the osmotic gradient under high-salt conditions [38–41]. Moreover, halophilic microorganisms have evolved an abundance of negatively charged aspartate and glutamate residues on protein surfaces that can interact with water molecules to form a water cage that prevents protein precipitation and dehydration [41–44].

As another adaptation strategy, some halophilic and halotolerant bacteria use the 'compatible solutes adaptation' strategy to maintain osmotic balance by using compatible organic solutes such as polyols, glucosylglycerol, sucrose, trehalose, ectoine, and betaine [45,46]. For example, the halophilic bacterium *Spiribacter salinus* M19-40 produces enhanced levels of compatible solutes such as ectoine and trehalose when they are exposed to a high NaCl concentration [45]. These organic solutes have a critical role in reducing the thermodynamic activity of water to compensate for the external osmotic pressure [47].

2.3. *Psychrophilic and Thermophilic Microorganisms*

Psychrophilic microorganisms usually have a preferred temperature range of 1–4 °C. Unlike mesophilic microorganisms, whose preferred temperature range is 30–37 °C, psychrophilic microorganisms can fully maintain cellular metabolism even at temperatures below 0 °C. To adapt to these harsh conditions, they have evolved several physiological adaptation mechanisms, including membrane fluidity control, molecular chaperones' action, and antifreeze molecules' synthesis [48,49]. For example, they can modulate membrane fluidity by altering its lipid composition, increasing the amount of polyunsaturated fatty acids and polar/non-polar carotenoids and decreasing the size of the lipid head groups [19,49]. A variety of temperature-induced enzymes such as cold-shock proteins

(Csps) and heat-shock proteins (Hsps) are also involved in cold-shock resistance by regulating signaling cascades that protect damaged proteins and cofactors [50]. Moreover, various antifreeze proteins and polysaccharides such as trehalose, mannitol, and exopolysaccharides, which are constituents of biofilm, can act as cryoprotectants [51].

Thermophilic microorganisms with a preferred temperature above 60 °C activate similar survival mechanisms to psychrophilic microorganisms. For example, *B. acidocalidus*, a thermophilic spore-forming bacterium, modulates membrane lipid fluidity by increasing hopanoids (a subclass of triterpenoids) to resist high temperatures [52]. The thermophilic archaeon *Metahnocaldococcus jannaschii* can resist high temperatures by regulating membrane lipid composition. When these microorganisms were exposed to high temperature, the diether lipids decreased from 80% to 20%, while the caldarchaeol-based and cyclic archaeol-based lipids increased from 10% to 40% [53,54]. In addition, thermophilic microorganisms have evolved various biomolecules to induce thermal stability, e.g., by increasing the guanine/cytosine content of DNA or developing a positive supercoiled DNA structure [55]. Moreover, they not only possess very rich ribosomal proteins but also have a well-developed heat-shock response to allow normal protein synthesis even at high temperatures [56,57].

2.4. Radiophilic Microorganisms

Radiophilic (radio-tolerant) microorganisms can thrive in environments with high levels of radiation, including ultraviolet light and gamma rays. Previous studies on how they can adapt and survive under high-dose radiation and oxidative stress conditions have revealed that they possess robust DNA repair systems and antioxidation mechanisms to withstand intensive irradiation stress [58–63]. For example, RecA proteins from *Deinococcus radiodurans* R1, which is a representative radiophilic microorganism, plays a crucial role in repairing damaged DNA under gamma ray irradiation [63,64]. When it is exposed to a high dose of irradiation, the expression levels of several novel proteins (PprA, PprM, PprI, and DdrABCDO) and of DNA damage response regulons are dramatically increased and contribute to DNA repair and damaged genome reconstruction [65–68].

Radiophilic microorganisms also have efficient antioxidant enzymes, such as catalase (CAT), superoxide dismutase (SOD), and peroxidase, which are responsible for the scavenging of reactive oxygen species (ROS) [63,69]. For example, CATs and SODs from *D. radiodurans* exhibit a 30-fold higher ROS scavenging activity than radiation-sensitive bacteria such as *Escherichia coli* and *Saccharomyces cerevisiae* [63]. Moreover, non-enzymatic factors such as relatively high intracellular manganese concentrations, polyphosphate granules, carotenoids, and pyrroloquinoline quinone are also involved in the efficient scavenging of various ROSs as well as in the protection against protein damage [70–73]. Other non-enzymatic factors protecting biomolecules from ionizing radiation are a high intracellular Mn/Fe concentration ratio, orthophosphates, large amounts of free amino acids, and small peptides that have been found in the polyextremophilic microorganism *H. salinarum* [74].

3. Bioremediation Using Extremophiles

3.1. Treatment of Heavy Metal Pollutants

Concerns about the toxicity of heavy metals have been drastically increasing because even a tiny amount can be dangerous for public health and the environment. Moreover, currently used chemical treatments of toxic heavy metals under extreme conditions is often hampered by their poor accessibility. Thus, the development of sustainable bioremediation methods using extremophilic microorganisms for the treatment of heavy metals has been investigated during the past several decades (Table 1). In the case of extremely acidic conditions, acidophilic microorganisms that can thrive under low pH conditions have been used as host strains for the detoxification of heavy metals through biomining processes such as bioleaching and bio-oxidation [75–78]. There have been several reports on the development of bioremediation processes using *Acidothiobacillus* strains, which are the most common acidophilic and chemolithotrophic microorganisms. For example, industrial-scale bioleaching has been performed using *Acidothiobacillus ferrooxidans* [79–81]. Romero-González et al. [82] reported the bioremediation of 100 mg/L of U(IV) ex situ from polluted mine water using *At. ferrooxidans* NCIMB 8455, while Jameson et al. [83] demonstrated the utility of *At. ferrooxidans* and *Acidothiobacillus ferrivorans* strains for hydrogen sulfide (H_2S)-assisted copper precipitation (>99%) under acidic conditions (pH 2.5–2.6). In other studies, the efficient reduction of vanadium ions [vanadate; V(V)] to V(IV) and the biosorption of cadmium cations were successfully achieved by *Acidocella aromatica* PFBC and *Acidiphilium symbioticum* H8, respectively, under highly acidic conditions [84,85].

More efficient decontamination of toxic heavy metals can be obtained using a microbial consortium, a major advantage of which is to synergize different enzymatic systems and metabolic pathways of individual microorganisms. Recently, the bioaugmentation of heavy metals using an acid mine drainage (AMD)-isolated acidophilic microorganism consortium was performed on polluted port sediment. The extraction of more than 90% Cu^{2+}, Cd^{2+}, Hg^{2+}, and Zn^{2+} was successfully achieved using an acidophilic microbial consortium consisting of *Acidothiobacillus thiooxidans, At. ferrooxidans, Acidiphilium cryptum*, and *Leptospirillum ferrooxidans* [86]. Another study also reported the in situ bioremediation of AMD soil defined as highly acidic (pH 3.21), sulfate (6285 mg/L), and heavy metals. The introduction of an enriched microbial consortium composed of acidophilic microorganisms and metal-resistant strains of *Chloroflexi* (29%), *Acidobacteria* (21%), *Proteobacteria* (16%), and *Firmicutes* (2%) into AMD soil enabled 97% reduction of dissolved sulfate and increased the pH to 7.5 [87].

Halophilic microorganisms offer great advantages in the treatment of toxic pollutants in high-salt environments. For example, bioremediation using marine bacteria is a promising solution for the decontamination of seawater from toxic heavy metals, as these bacteria can survive at high salt concentrations. There have been a few reports on the removal of toxic heavy metals using several marine bacteria. For instance, *Vibrio harveyi* showed a good capability to accumulate cadmium cations inside the cell with a high adsorption capacity (up to 23.3 mg Cd^{2+}/g of dry cells) [88]. Another marine bacterium, *Enterobacter cloaceae*, can chelate Cd, Cu, and Co by up to 65%, 20%, and 8%, respectively, from mixed-salts solutions [89]. In addition to marine bacteria, some thermophilic microorganisms such as *Geobacillus thermantarcticus* and *Anoxybacillus amylolyticus* have considerable biosorption capacity for heavy metals, which suggests their applicability for the removal of heavy metals in polluted environments [90].

As the development of biotechnology progresses, more advanced bioremediation methods that are superior to traditional methods have been reported. Unlike conventional bioremediation methods whose principle is based on the microorganism itself, new methods present improved efficiency and specificity thanks to the use of biomolecular engineering approaches. For instance, S-layer proteins, which have high stability and activity toward various heavy metals, are produced by lactic acid bacteria and are promising biomolecules for toxic heavy metal decontamination under very low pH (pH 2) conditions [91]. The S-layer proteins from *Lactobacillus plantarum* YW11 showed 99.9% Pb adsorption capacity [92]; scanning electron microscopy–energy dispersive X-ray analysis demonstrated that the Pb^{2+} ions were efficiently adsorbed and accumulated on the cell surface of *L. plantarum* YW11 in a process mediated via S-layer proteins. The interaction of S-layer proteins from two *Lactobacillus kefiri* strains (CIDCA 8348 and JCM 5818) has also been investigated for the adsorption of various metal ions such as Cd^{2+}, Zn^{2+}, Pb^{2+}, and Ni^{2+} [93].

Table 1. Extremophilic microorganisms used in the removal of heavy metals.

Heavy Metal	Method/Mechanism	Extremophile	Resistance [1]	Removal Efficiency (Initial Concentration) [2]	Reference
As(III)	Bioleaching [3]	*Acidothiobacillus ferrooxidans* BY-3	Low pH (pH < 1.8)	35.9% (ND)	[80]
U(VI)	Bioleaching	*At. ferrooxidans*	Low pH (pH 1.5–4.5)	50% (100 mg/L)	[82]
Cu(II)	Bioprecipitation	*Acidothiobacillus ferrivorans*	Low pH (pH 2.5)	>99% (50 mM)	[83]
V(V)	Bioreduction	*Acidocella aromatica*	Low pH (pH 2.5)	70% (1 mM)	[84]
Cd(II)	Biosorption	*Acidiphilium symbioticum* H8	ND	248.62 mg Cd(II)/g biomass (250 mg/L)	[85]
	Bioaccumulation	*Vibrio harveyi*	60 mg/L MIC	84% (30–60 mg/L)	[88]
	Biosorption	*Enterobacter cloaceae*	ND	65% (100 mg/L)	[89]
	Biosorption	*Geobacillus thermantarcticus, Anoxybacillus amylolyticus*	High temperature (80 °C)	85.4%, 74.1% (50 mg/L)	[90]
Cr(VI)	Bioreduction	*Pyrobaculum islandicum*	High temperature (100 °C)	100% (600 μM)	[94]

[1] Either the experimental conditions or the tolerance of the species. ND, not determined; MIC, minimum inhibitory concentration. [2] Initial concentration of contaminant in the test. [3] Bioleaching, a metal solubilization process mediated by sulfur-/iron-oxidizing bacteria.

3.2. Biodegradation of Organic Pollutants

A variety of microorganisms can transform toxic organic pollutants into non-toxic substances such as petroleum hydrocarbons, aromatic petrochemicals, and various halogenated compounds (Table 2). Such complete transformation requires not only strong resistance to toxic organic pollutant exposure but also the ability to utilize toxic organic contaminants for their cellular metabolism. Therefore, extremophilic microorganisms that have adapted to harsh environments such as extreme temperatures and high salt concentrations over a long time period can potentially be widely used for the treatment of organic toxic pollutants under the corresponding condition. For example, the decontamination of polycyclic aromatic hydrocarbons and long-chain alkanes (C_{10} to C_{32}) using thermophilic *Bacillus*, *Thermus*, and *Geobacillus* strains isolated from oil-contaminated areas has been reported [95–99]; a *Geobacillus* SH-1 strain isolated from a deep oil well was also able to degrade saturated alkanes ranging from C_{12} to C_{33} and naphthalene. In another study, C_{12}–C_{21} *n*-alkanes were completely decomposed within 8 days, and 100 ppm of naphthalene was almost degraded within 72 h [100]. Furthermore, bioaugmentation through introduction of various extremophilic microorganisms including *Geobacillus thermoparaffinivorans* IR2, *Geobacillus stearothermophilus* IR4, and *Bacillus licheniformis* increased the decontamination of long alkyl (C_{32} and C_{40}) substances [101].

Table 2. Extremophilic microorganisms used in the removal of hydrocarbons.

Hydrocarbons	Extremophile	Resistance	Removal Efficiency (Initial Concentration)	Reference
acenaphthene, fluoranthene, pyrene, benzo[e]pyrene	*Bacillus* spp., *Thermus* sp.	High temperature (60–70 °C)	35–77% (30–60 mg/L)	[95]
Pentadecane, octadecane, octacosane	*Geobacillus* sp. SH-1	High temperature (70 °C)	>70% (100 mg/L)	[100]
Rotricontane, tetracotane	*Geobacillus thermoparaffinivorans, Geobacillus stearothermophillus, Bacillus licheniformis*	High temperature (50 °C)	>87% (1 g/L)	[101]
Mixed oil	*Pseudoalteromonas* sp. P29	Low temperature (5 °C)	90% (2 g/L)	[102]
Diesel, jet fuel, crude oil	*Oleispira antarctica* RB-8T	Low temperature (4–15 °C)	53.7–79.4% (1 g/L)	[103]
Biphenyl, phenanthrene, anthracene, naphthalene	*Marinobacter sedimentalis, Marinobacter falvimaris, Marinobacter nanhaiticus*	High salinity (5M NaCl)	70–90% (0.2–3 g/L)	[104,105]

In addition to thermophilic microorganisms, psychrotrophic and halophilic microorganisms have shown excellent performance in the treatment of organic hydrocarbon pollutants. Low-temperature-adapted *Pseudoalteromonas* sp. P29 and *Oleispira antarctica* RB-8^{T} exhibited high efficiencies in the degradation of hydrocarbon mixtures composed of diesel, military jet fuel, and crude oil [102,103], while the halotolerant microorganisms *Marinobacter sedimentalis*, *Marinobacter falvimaris*, and *Marinobacter nanhaiticus* D15-8W were able to transform biphenyl, phenanthrene, anthracene, and naphthalene into useful carbon sources in hypersaline environments (e.g., salt lakes, salt marshes, and highly saline soils) [104,105]. In particular, extracellular polymeric substances (EPSs), which are cellular components of halophilic microorganisms, play a critical role in the remediation of organic pollutants from hypersaline environments. Exopolysaccharides secreted by halophiles can act as biosurfactants that contribute toward aggregating oils and emulsifying hydrocarbons, as well as offer cellular resistance toward toxic heavy metals. Halophilic microorganism *Halobacillus* sp. EG1HP4QL develops the ability to utilize crude oil as the sole carbon source within 12 days and to degrade paraffin (34.5%), naphthalene (49.6%), mono- and bicyclic aromatic hydrocarbons (51.2%), polycyclic aromatic hydrocarbon (43.5%), and alcohol–benzene resins (25.5%) [106]. EPS-producing *Halomonas* strain TG39 was also used for bioremediation of a hydrocarbon-contaminated Deepwater Horizon spill site [107]; the extracted EPS was effective not only in increasing the solubilization of aromatic hydrocarbons but also in enhancing the degradation rate of phenanthrene. Hence, bioremediation using extremophilic microorganisms is a promising method for the treatment of organic contaminant-polluted areas under extreme conditions because the organic pollutants can be metabolized by the microorganisms.

3.3. Microbial Treatment of Radioactive Waste

Recent advances in synthetic chemistry and separation methods have led to the design of various adsorbent systems including surface-modified nanomaterials and/or hybrid composites for the treatment of radionuclides in soil or aqueous media. For example, surface-modified iron oxide (Fe_3O_4) nanoparticles have been applied to selectively adsorb toxic heavy metals such as Cr(III), Co(II), Ni(II), Cd(II), Pb(II), and As^{3+} from aqueous media [108]. Furthermore, engineered Au nanomaterials have been developed that are excellent adsorbents for the desalination of non-radioactive and radioactive iodine anions [109–111]. However, there are still several problems in the practical application of these methods. First, a large volume of secondary radioelement-contaminated solid adsorbents is generated during the desalination procedure, and so the removal of unsettled adsorbents after the treatment requires an additional expensive step. Second, small- (nano- or micro-) sized adsorbents tend to lose their stability and properties under particularly harsh conditions such as high salt concentration and high radiation. Therefore, employing extremophilic microorganisms that can be used as a live cleaning agent offer a useful alternative for the treatment of radioactive waste (Table 3).

The microbial treatment of radioactive waste can be accomplished through the interactions between microorganisms and radioisotopes, such as biomineralization, biotransformation, and biosorption [112–115]. Among these, mineralization of the target element inside bacterial cells has been proposed as the main strategy for the removal of radionuclides from a contaminated area [116,117]. As an example, *Shewanella* and *Geobacter* strains can reduce some alpha nuclides such as U(VI), Pu(IV), Am(V), and Th(IV) to make them harmless [15,114,116,118,119]. Anderson et al. reported the removal of uranium from aqueous media by using acetate-stimulating *Geobacter* species, while enhanced removal efficiency was demonstrated by supplementation with glucose, ethanol, and acetate as an electron donor [120]. Since the 1990s, a variety of extremophilic microorganisms that can thrive under high levels of ionizing radiation conditions (>15 kGy) have been identified [121–123]. Among these, *D. radiodurans,* which is one of the most radio-resistant microorganisms, has received much attention as a biological material for on-site treatment of radionuclide-contaminated environments [124,125] (Table 3). Moreover, a variety of studies investigating the development of the bioremediation processes using *D. radiodurans* for the removal of radionuclides pollutants have been reported [123,126–129]. A genetically engineered *D. radiodurans* strain expressing a non-specific acid phosphatase from *Salmonella enterica* serovar Typhi [127–129] or bacterial Ni/Co transporter (NiCoT) [130] can precipitate the oxidized form of uranium pollutants and radioactive cobalt (^{60}Co), respectively.

In recent years, the combination of extremophilic microorganisms with nanotechnology has emerged as a central strategy in efforts to treat polluted environments. A few case studies including the biosynthesis of various nanomaterials using extremophilic microorganisms have been reported [131–135]. With the advent of nano-biotechnology, the combination of extremophilic microorganisms with nanomaterials (nano-adsorbents and reductants) will be a promising technology for useful bioremediation applications. For example, a highly efficient and stable method for the removal of radioactive iodine (^{125}I) using *D. radiodurans* with biogenic Au nanoparticles has been reported [131], in which more than 3.7 MBq of ^{125}I was efficiently removed (>99%) within 30 min. More recently, the thermo-acidophilic archeon *S. tokodaii* 7^T (NBRC 100140) capable of synthesizing biogenic Pd(0) nanoparticles (mean diameter: 8.7 nm) showed four-fold increased Cr(IV) reduction with 2.0 mg Cr(VI)/L/h/Pd(0) compared to a commercial Pd/C catalyst [(0.5 mg Cr(VI)/L/h/Pd(0)] [136]. Another study also demonstrated efficient Cr(IV) reduction using Pd(0) nanoparticles synthesized by the acidophilic Fe^{3+}-reducing bacteria *Ac. aromatica* PFBCT and *Ap. cryptum* SJH via a one-step microbiological reaction [137].

Table 3. Extremophilic microorganisms used in radioactive waste bioremediation.

Radionuclide	Extremophile	Resistance	Removal Efficiency	Reference
U(VI), Cr(VI), Tc(VII)	*Deinococcus geothermalis*	Radiation (12 kGy), high temperature (55 °C)	>90%	[123]
	D. radiodurans	ND	95–100%	[126]
U(VI)	*D. radiodurans* expressing PhoN	Radiation (6 kGy)	>90%	[127]
Co-60	*D. radiodurans* expressing NiCoT	Radiation (6.4 kGy)	>60%	[130]
I-125	*D. radiodurans*	Radiation (8 kGy)	>99%	[131]

4. The Future Direction

Pollution, which has emerged as a side effect of the rapid growth of industrialization and urbanization, is a worldwide threat to the environment and public health. Thus, the development of highly efficient and stable methods for cleaning up polluted environments has become a major challenge. Although a variety of conventional methods to remove toxic pollutants have been developed over the past several decades, there are still many hurdles that need to be overcome to realize practical applications [138]. Hence, extremophilic microorganisms, which can thrive under harsh conditions, have been receiving particular interest as bioagents for the removal of toxic pollutants.

Although conventional microbial bioremediation processes have succeeded in the removal of various toxic pollutants, current methods still require much effort to overcome their limitations in terms of cost-effectiveness, removal efficiency, and practicality. *E. coli* and *Bacillus* spp. are commonly considered host strains for microbial bioremediation processes, being well known due to their broad use with well-established genetic engineering tools [139,140]. However, despite intensive genetic engineering, the practical use of these microorganisms for on-site remediation is extremely limited, owing to their relatively weak resistance to harsh conditions and low removal efficiency. Thus, to overcome these limitations, subsequent strategies based on the combination of extremophilic microorganisms with advanced biotechnology from fields such as systems metabolic engineering, synthetic biology, and nanotechnology have enhanced the performance of bioremediation through reprogramming the nature of wild-type microorganisms [141,142]. Several approaches based on biotechnology and nanotechnology are **(1)** screening and identification of microorganisms that have a strong tolerance for harsh conditions, **(2)** making microorganisms capable of degrading a variety of environmental toxic pollutants, **(3)** increasing the removal capacity and specificity of microorganisms toward target pollutants, and **(4)** expanding the removal spectrum of microorganisms using biogenic nanoparticles. Moreover, a variety of advanced tools in bioengineering, such as in silico flux analysis, biostatistics, and multi-omics analysis, will allow us to access the possibly infinite potential of extremophilic microorganisms for the treatment of environmental toxic pollutants.

5. Conclusions

When considering all the aspects presented in this review, extremophilic microorganisms appear as attractive bioagents for the clean-up of toxic pollutants contaminating the environment, due to their unique characteristics such as toughness, adaptability, and strong resistance to extreme conditions. Although many challenges still need to be addressed, the adoption of extremophilic microorganisms for the development of bioremediation processes is an environmental imperative for us to meet the needs of global public health. Indeed, combining extremophilic microorganisms with biotechnology and nanotechnology will open new avenues toward developing highly efficient and eco-friendly methods for the treatment of toxic pollutants (Figure 2).

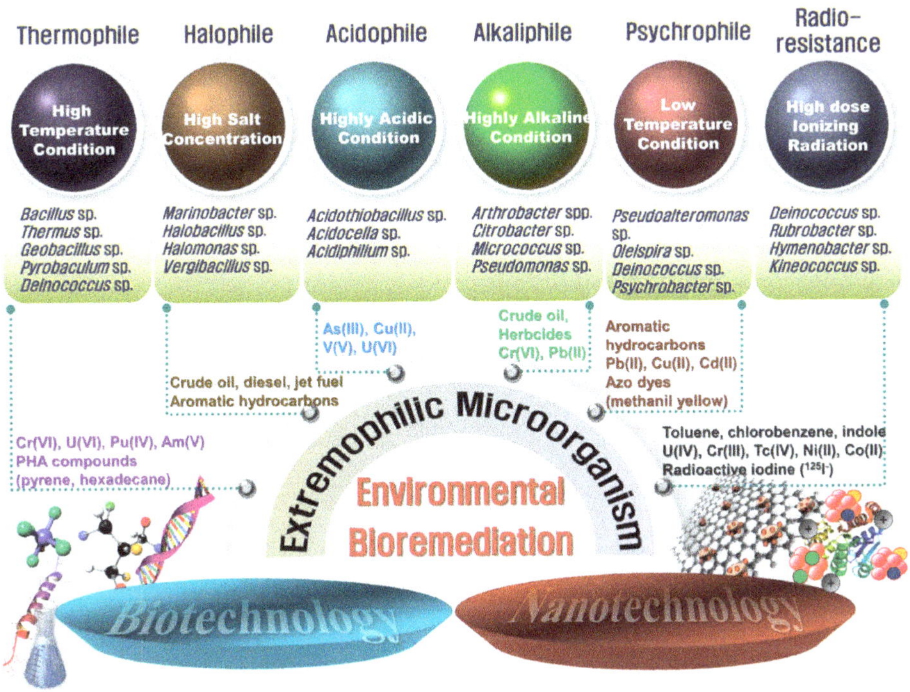

Figure 2. A schematic diagram of advanced bioremediation using extremophilic microorganisms combined with biotechnology and nanotechnology. Representative candidates that can be used as a host strain for the treatment of pollutants in the environment are shown.

Author Contributions: S.-W.J. and Y.J.C. conceived and designed the review concept. S.-W.J. contributed to data curation and analysis. S.-W.J. and Y.J.C. wrote the manuscript. Y.J.C. reviewed and edited the manuscript. All authors have read and agreed to the published version of the manuscript.

Funding: This research was supported by Basic Science Research Program through the National Research Foundation of Korea (NRF) funded by the Ministry of Education (2019R1A6A3A01092533) and the Korea government (MSIT) (2020R1A2C4001737). This work was also supported by the C1 Gas Refinery Program (NRF-2017M3D3A1A01037019).

Conflicts of Interest: The authors declare no conflict of interest.

References

1. Muddemann, T.; Haupt, D.; Sievers, M.; Kunz, U. Electrochemical reactors for wastewater treatment. *ChemBioEng Rev.* **2019**, *6*, 142–156. [CrossRef]

2. Ouyang, W.; Chen, T.; Shi, Y.; Tong, L.; Chen, Y.; Wang, W.; Yang, J.; Xue, J. Physico-chemical process. *Water Environ. Res.* **2019**, *91*, 1350–1377. [CrossRef] [PubMed]
3. Xiang, Q.; Nomura, Y.; Fukahori, S.; Mizuno, T.; Tanaka, H.; Fujisawa, T. Innovative treatment of organic contaminants in reverse osmosis concentrate from water reuse: A mini review. *Curr. Pollut. Rep.* **2019**, *5*, 294–307. [CrossRef]
4. Gebreeyessus, G.D. Status of hybrid membrane-ion-exchange systems for desalination: A comprehensive review. *Appl. Water. Sci.* **2019**, *9*, 135. [CrossRef]
5. Abdelfattah, A.; Hossain, M.I.; Cheng, L. High-strength wastewater treatment using microbial biofilm reactor: A critical review. *World J. Microb. Biot.* **2020**, *36*, 75. [CrossRef] [PubMed]
6. Zhang, C.; Wu, D.; Ren, H. Bioremediation of oil contaminated soil using agricultural wastes via microbial consortium. *Sci. Rep.* **2020**, *10*, 9188. [CrossRef] [PubMed]
7. Dixit, R.W.; Malaviya, D.; Pandiyan, K.; Singh, U.B.; Sahu, A.; Shukia, R.; Singh, B.P.; Rai, J.P.; Sharma, P.K.; Lade, H.; et al. Bioremediation of heavy metals from soil and aquatic environment: An overview of principles and criteria of fundamental processes. *Sustainability* **2015**, *7*, 2189–2212. [CrossRef]
8. Mishra, G.K. Microbes in heavy metal bioremediation: A review on current trends and patents. *Recent Pat. Biotechnol.* **2017**, *11*, 188–196. [CrossRef]
9. Diep, P.; Mahadevan, R.; Yakunin, A.F. Heavy metal removal by bioaccumulation using genetically engineered microorganisms. *Front. Bioeng. Biotechnol.* **2018**, *6*, 157. [CrossRef]
10. Igiri, B.E.; Okoduwa, S.I.R.; Idoko, G.O.; Akabuogu, E.P.; Adeyi, A.O.; Ejiogu, I.K. Toxicity and bioremediation of heavy metals contaminated ecosystem from Tannery wastewater: A review. *J. Toxicol.* **2018**, *2018*, e2568038. [CrossRef]
11. Singh, S.; Kang, S.H.; Mulchandani, A.; Chen, W. Bioremediation: Environmental clean-up through pathway engineering. *Curr. Opin. Biotechnol.* **2008**, *19*, 437–444. [CrossRef] [PubMed]
12. Kiadehi, M.S.H.; Amoozegar, M.A.; Asad, S.; Siroosi, M. Exploring the potential of halophilic archaea for the decolorization of azo dyes. *Water Sci. Technol.* **2018**, *77*, 1602–1611. [CrossRef] [PubMed]
13. Azubuike, C.C.; Chikere, C.B.; Okpokwasili, G.C. Bioremediation techniques-classification based on site of application: Principles, advantages, limitations and prospects. *World J. Microbiol. Biotechnol.* **2016**, *32*, 180. [CrossRef] [PubMed]
14. Waigi, M.G.; Sun, K.; Gao, Y.Z. Sphingomonads in microbe-assisted phytoremediation: Tackling soil pollution. *Trends Biotechnol.* **2017**, *35*, 883–899. [CrossRef] [PubMed]
15. Tkacv, R.; Matrosova, V.Y.; Grichenko, O.E.; Gostinčar, C.; Volpe, R.P.; Klimenkova, P.; Gaidamakova, E.K.; Zhou, C.E.; Stewart, B.J.; Lyman, M.G.; et al. Prospects for fungal bioremediation of acidic radioactive waste sites: Characterization and genome sequence of *Rhodotorula taiwanensis* MD1149. *Front. Microbiol.* **2018**, *8*, 2528. [CrossRef]
16. Ali, N.; Dashti, N.; Khanafer, M.; Al-Awadhi, H.; Radwan, S. Bioremediation of soils saturated with spilled crude oil. *Sci. Rep.* **2020**, *10*, 1–9. [CrossRef]
17. Rambabu, K.; Banat, F.; Pham, Q.M.; Ho, S.-H.; Ren, N.-Q.; Show, P.L. Biological remediation of acid mine drainage: Review of past trends and current outlook. *Environ. Sci. Technol.* **2020**, *2*, 100024. [CrossRef]
18. Cleary, A.; Lloyd, J.R.; Newsome, L.; Shaw, S.; Boothman, C.; Boshoff, G.; Atherton, N.; Morris, K. Bioremediation of strontium and technetium contaminated groundwater using glycerol phosphate. *Chem. Geol.* **2019**, *509*, 213–222. [CrossRef]
19. Siliakus, M.F.; Van der Oost, J.; Kengen, S.W.M. Adaptations of archaeal and bacterial membranes to variations in temperature, pH and pressure. *Extremophiles* **2017**, *21*, 651–670. [CrossRef]
20. Rastädter, K.; Wurm, D.J.; Spadiut, O.; Quehenberger, J. The cell membrane of *Sulfolobus* spp.–homeoviscous adaptation and biotechnological applications. *Int. J. Mol. Sci.* **2020**, *21*, 3935. [CrossRef]
21. Vergara, E.; Neira, G.; González, C.; Cortez, D.; Dopson, M.; Holmes, D.S. Evolution of predicted acid resistance mechanisms in the extremely acidophilic *Leptospirillum* genus. *Genes* **2019**, *11*, 389. [CrossRef] [PubMed]
22. Guan, N.; Liu, L. Microbial response to acid stress: Mechanisms and applications. *Appl. Microbiol. Biotechnol.* **2020**, *104*, 51–65. [CrossRef] [PubMed]
23. Salvador-Castell, M.; Tourte, M.; Ogar, P.M. In search for the membrane regulators of archaea. *Int. J. Mol. Sci.* **2019**, *20*, 4434. [CrossRef] [PubMed]

24. Golyshina, O.V.; Tran, H.; Reva, O.N.; Lemak, S.; Yakunin, A.F.; Goesmann, A.; Nechitaylo, T.Y.; LaCono, V.; Smedile, F.; Slesarev, A.; et al. Metabolic and evolutionary patterns in the extreme acidophilic archaeon *Ferroplasma acidiphilum* Y^T. *Sci. Rep.* **2017**, *7*, 3682. [CrossRef] [PubMed]
25. Baker-Austin, C.; Dopson, M. Life in acid: pH homeostasis in acidophiles. *Trends Microbiol.* **2007**, *15*, 165–171. [CrossRef]
26. Aono, R.; Ito, M.; Machida, T. Contribution of the cell wall component teichuronopeptide to pH homeostasis and alkaliphily in the alkaliphile *Bacillus lentus* C-125. *J. Bacteriol.* **1999**, *181*, 6600–6606. [CrossRef]
27. Aono, R. Assignment of facultatively alkaliphilic *Bacillus* sp. strain C-125 to *Bacillus lentus* group 3. *Int. J. Syst. Bacteriol.* **1995**, *45*, 582–585. [CrossRef]
28. Calamita, H.G.; Ehringer, W.D.; Koch, A.L.; Doyle, R.J. Evidence that the cell wall of *Bacillus subtilis* is protonated during respiration. *Proc. Natl. Acad. Sci. USA* **2001**, *98*, 15260–15263. [CrossRef]
29. Padan, E.; Bibi, E.; Ito, M.; Krulwich, T.A. Alkaline pH homeostasis in bacteria: New insights. *Biochim. Biophys. Acta Biomembr.* **2005**, *2*, 67–88. [CrossRef]
30. Kitada, M.; Kosono, S.; Kudo, T. The Na^+/H^+ antiporter of alkaliphilic *Bacillus* sp. *Extremophiles* **2000**, *4*, 253–258. [CrossRef]
31. Fang, H.; Qin, X.-Y.; Zhang, K.-D.; Nie, Y.; Wu, X.-L. Role of the Group 2 Mrp sodium/proton antiporter in rapid response to high alkaline shock in the alkaline- and salt-tolerant *Dietzia* sp. DQ12-45-1b. *Appl. Microbiol. Biotechnol.* **2018**, *102*, 3765–3777. [CrossRef] [PubMed]
32. Matsuno, T.; Goto, T.; Ogami, S.; Morimoto, H.; Yamazaki, K.; Inoue, N.; Matsuyama, H.; Yoshimune, K.; Yumoto, I. Formation of proton motive force under low-aeration alkaline conditions in alkaliphilic bacteria. *Front. Microbiol.* **2018**, *9*, 2331. [CrossRef] [PubMed]
33. Stancik, L.M.; Stancik, D.M.; Schmidt, B.; Barnhart, D.M.; Yoncheva, Y.N.; Slonczewski, J.L. pH-Dependent expression of periplasmic proteins and amino acid catabolism in *Escherichia coli*. *J. Bacteriol.* **2002**, *184*, 4246–4258. [CrossRef] [PubMed]
34. Slonczewski, J.L.; Fujisawa, M.; Dopson, M.; Krulwich, T.A. Cytoplasmic pH measurement and homeostasis in bacteria and archaea. *Adv. Microb. Physiol.* **2009**, *55*, 1–79. [CrossRef] [PubMed]
35. Wernick, D.G.; Pontrelli, S.P.; Pollock, A.W.; Liao, J.C. Sustainable biorefining in wastewater by engineered extreme alkaliphile *Bacillus marmarensis*. *Sci. Rep.* **2016**, *6*, 20224. [CrossRef]
36. Setati, M.E. Diversity and industrial potential of hydrolase-producing halophilic/halotolerant eubacteria. *Afr. J. Biotechnol.* **2010**, *9*, 1555–1560. [CrossRef]
37. Gunde-Cimerman, N.; Plemenitaš, A.; Oren, A. Strategies of adaptation of microorganisms of the three domains of life to high salt concentrations. *FEMS Microbiol. Rev.* **2018**, *42*, 353–375. [CrossRef]
38. Starhl, H.; Greie, J.-C. The extremely halophilic archaeon *Halobacterium salinarum* R1 responds to potassium limitation by expression of the K^+-transporting KdpFABC P-type ATPase and by a decrease in intracellular K^+. *Extremophiles* **2008**, *12*, 741–752. [CrossRef]
39. Engel, M.B.; Catchpole, H.R. A microprobe analysis of inorganic elements in *Halobacterium salinarum*. *Cell Biol. Int.* **2005**, *29*, 616–622. [CrossRef]
40. Coker, J.A.; DasSarma, P.; Kumar, J.; Müller, J.A.; DasSarma, S. Transcriptional profiling of model Archeon *Halobacterium* sp. NRC-1: Responses to changes in salinity and temperature. *Saline Syst.* **2007**, *3*, 6. [CrossRef]
41. Corral, P.; Amoozegar, M.A.; Ventosa, A. Halophiles and their biomolecules: Recent advances and future applications in biomedicine. *Mar. Drugs* **2020**, *18*, 33. [CrossRef]
42. Reed, C.J.; Bushnell, S.; Evilia, C. Circular dichroism and fluorescence spectroscopy of cysteine-tRNA synthetase from *Halobacterium salinarum* ssp. NRC-1 demonstrates that group I cations are particularly effective in providing structure and stability to this halophilic protein. *PLoS ONE* **2014**, *9*, e89452. [CrossRef] [PubMed]
43. Brininger, C.; Spradlin, S.; Cobani, L.; Evilia, C. The more adaptive to change, the more likely you are to survive: Protein adaptation in extremophiles. *Semin. Cell Dev. Biol.* **2018**, *84*, 158–169. [CrossRef]
44. Elcock, A.H.; McCammon, J.A. Electrostatic contributions to the stability of halophilic proteins. *J. Mol. Biol.* **1998**, *280*, 731–748. [CrossRef]
45. León, M.J.; Hoffmann, T.; Sánchez-Porro, C.; Heider, J.; Ventosa, A.; Bremer, E. Compatible solute synthesis and imported by the moderate halophile *Spiribacter salinus*: Physiology and genomics. *Front. Microbiol.* **2018**, *9*, 108. [CrossRef] [PubMed]

46. Le Borgne, S.; Paniagua, D.; Vazques-Duhalt, R. Biodegradation of organic pollutants by halophilic bacteria and archaea. *J. Mol. Microbiol. Biotechnol.* **2008**, *15*, 74–92. [CrossRef] [PubMed]
47. Vandrich, J.; Pfeiffer, F.; Alfaro-Espinoza, G.; Kunte, H.J. Contribution of mechanosensitive channels to osmoadaptation and ectoine excretion in *Halomonas elongata*. *Extremophile* **2020**, *24*, 421–432. [CrossRef]
48. Collins, T.; Margesin, R. Psychrophilic lifestyles: Mechanisms of adaptation and biotechnological tools. *Appl. Microbiol. Biotechnol.* **2019**, *103*, 2857–2871. [CrossRef]
49. De Maayer, P.; Anderson, D.; Cary, C.; Cowan, D.A. Some like it cold: Understanding the survival strategies of psychrophiles. *EMBO Rep.* **2014**, *15*, 508–517. [CrossRef]
50. Yoshimune, K.; Galkin, A.; Kulakova, L.; Yoshimura, T.; Esaki, N. Cold-active DnaK of an Antarctic psychrotroph *Shewanella* sp. Ac10 supporting the growth of dnaK-null mutant of *Escherichia coli* at cold temperatures. *Extremophiles* **2005**, *9*, 145–150. [CrossRef]
51. Białkowska, A.; Majewska, E.; Olczak, A.; Twarda-Clapa, A. Ice binding proteins: Diverse biological roles and applications in different types of industry. *Biomolecules* **2020**, *10*, 274. [CrossRef]
52. Ranawat, P.; Rawat, S. Stress response physiology of thermophiles. *Arch. Microbiol.* **2017**, *199*, 391–414. [CrossRef]
53. Sprott, G.D.; Meloche, M.; Richards, J.C. Proportions of diether, macrocyclic diether, and tetraether lipids in *Methanococcus jannaschii* grown at different temperatures. *J. Bacteriol.* **1991**, *173*, 3907–3910. [CrossRef]
54. Mansilla, M.C.; Cybulski, L.E.; Albanesi, D.; de Mendoza, D. Control of membrane lipid fluidity by molecular thermosensors. *J. Bacteriol.* **2004**, *186*, 6681–6688. [CrossRef]
55. Valenti, A.; Perugino, G.; Rossi, M.; Ciaramella, M. Positive supercoiling in thermophiles and mesophiles: Of the good and evil. *Biochem. Soc. Trans.* **2011**, *39*, 58–63. [CrossRef] [PubMed]
56. Sieck, G. Life at the extreme: Physiological adaptation. *Physiology* **2015**, *30*, 84–85. [CrossRef] [PubMed]
57. Wang, Q.; Cen, Z.; Zhao, J. The survival mechanisms of thermophiles at high temperatures: An angle of omics. *Physiology* **2015**, *30*, 97–106. [CrossRef]
58. Ferreira, A.C.; Nobre, M.F.; Rainey, F.A.; Silva, M.T.; Wait, R.; Burghardt, J.; Chung, A.P.; Da Costa, M.S. *Deinococcus geothermalis* sp. nov. and *Deinococcus murrayi* sp. nov., two extremely radiation-resistant and slightly thermophilic species from hot springs. *Int. J. Syst. Evol.* **1997**, *47*, 939–947. [CrossRef] [PubMed]
59. Liu, Z.; Kim, M.C.; Wang, L.; Zhu, G.; Zhang, Y.; Huang, Y.; Wei, Z.; Danzeng, W.; Peng, F. *Deinococcus taklimakanensis* sp. nov. isolated from desert soil. *Int. J. Syst. Evol.* **2017**, *67*, 4311–4316. [CrossRef]
60. Srinivasan, S.; Lim, S.Y.; Lim, J.-H.; Jung, H.-Y.; Kim, M.K. *Deinococcus rubrus* sp. nov., a bacterium isolated from Antarctic coastal sea water. *J. Microbiol. Biotechnol.* **2017**, *27*, 535–541. [CrossRef]
61. Park, M.R.; Song, J.H.; Nam, G.G.; Joung, Y.C.; Zhao, L.; Kim, M.-K.; Cho, J.C. *Deinococcus lacus* sp. nov., a gamma radiation-resistant bacterium isolated from an artificial freshwater pond. *Int. J. Syst. Evol.* **2018**, *68*, 1372–1377. [CrossRef] [PubMed]
62. Tanner, K.; Molina-Menor, E.; Latorre-Pérez, A.; Vidal-Verdú, À.; Vilanova, C.; Peretó, J.; Porcar, M. Extremophilic microbial communities on photovoltaic panel surfaces: A two-year study. *Microb. Biotechnol.* **2020**, *13*, 1819–1830. [CrossRef]
63. Slade, D.; Radman, M. Oxidative stress resistance in *Deinococcus radiodurans*. *Microbiol. Mol. Biol. Rev.* **2011**, *75*, 133–191. [CrossRef] [PubMed]
64. Radjpurohit, Y.S.; Bihani, S.C.; Waldor, M.K.; Misra, H.S. Phosphorylation of *Deinococcus radiodurans* RecA regulates its activity and may contribute to radioresistance. *J. Biol. Chem.* **2016**, *291*, 16672–16685. [CrossRef] [PubMed]
65. Tanaka, M.; Earl, A.M.; Howell, H.A.; Park, M.J.; Eisen, J.A.; Peterson, S.N.; Battista, J.R. Analysis of *Deinococcus radiodurans*'s transcriptional response to ionizing radiation and desiccation reveals novel proteins that contribute to extreme radioresistance. *Genetics* **2004**, *168*, 21–33. [CrossRef] [PubMed]
66. Lim, S.Y.; Jung, J.H.; Blanchard, L.; de Groot, A. Conservative and diversity of radiation and oxidative stress resistance mechanisms in *Deinococcus radiodurans*. *FEMS Microbiol. Rev.* **2019**, *43*, 19–52. [CrossRef] [PubMed]
67. Jin, M.; Xiao, A.; Zhang, Z.; Huang, H.; Jiang, L. The diversity and commonalities of the radiation-resistance mechanisms of *Deinococcus* and its up-to-date applications. *AMB Express.* **2019**, *9*, 138. [CrossRef] [PubMed]
68. Floc'h, K.; Lacroix, F.; Servant, P.; Wong, Y.-S.; Kelman, J.-P.; Bourgeois, D.; Timmins, J. Cell morphology and nucleoid dynamic in dividing *Deinococcus radiodurans*. *Nat. Commun.* **2019**, *10*, 3815. [CrossRef]

69. Jeong, S.-W.; Jung, J.H.; Kim, M.K.; Seo, H.S.; Lim, H.-M.; Lim, S.Y. The three catalases in *Deinococcus radiodurans*: Only two show catalase activity. *Biochem. Biophys. Res. Commun.* **2016**, *469*, 443–448. [CrossRef]
70. Maqbool, I.; Sudharsan, M.; Kanimozhi, G.; Alrashood, S.T.; Khan, H.A.; Prasad, N.R. Crude cell-free extract from *Deinococcus radiodurans* exhibit anticancer activity by inducing apoptosis in triple-negative breast cancer cells. *Front. Cell Dev. Biol.* **2020**, *8*, 707. [CrossRef]
71. Choi, J.Y.; Lee, K.J.; Lee, P.C. Characterization of carotenoid biosynthesis in newly isolated *Deinococcus* sp. AJ005 and investigation of the effects of environmental conditions on cell growth and carotenoid biosynthesis. *Mar. Drugs* **2019**, *17*, 705. [CrossRef] [PubMed]
72. Daly, M.J.; Gaidamakova, E.K.; Matrosova, V.Y.; Kiang, J.G.; Fukumoto, R.; Lee, D.-Y.; Wehr, N.B.; Viteri, G.A.; Berlett, B.S.; Levine, R.L. Small-molecule antioxidant proteome-shields in *Deinococcus radiodurans*. *PLoS ONE* **2010**, *5*, e12570. [CrossRef]
73. Santos, S.P.; Yang, Y.; Rosa, M.T.G.; Rodrigues, M.A.A.; De La Tour, C.B.; Sommer, S.; Teixeira, M.; Carrondo, M.A.; Cloetens, P.; Abreu, I.A.; et al. The interplay between Mn and Fe in *Deinococcus radiodurans* triggers cellular protection during paraquat-induced oxidative stress. *Sci. Rep.* **2019**, *9*, 17217. [CrossRef] [PubMed]
74. Robinson, C.K.; Webb, K.; Kaur, A.; Jaruga, P.; Dizdaroglu, M.; Baliga, N.S.; Place, A.; DiRuggiero, J. A major role for nonenzymatic antioxidant processes in the radioresistance of *Halobacterium salinarum*. *J. Bacteriol.* **2011**, *193*, 1653–1662. [CrossRef]
75. Gumulya, Y.; Boxall, N.J.; Khaleque, H.N.; Santala, V.; Carlson, R.P.; Kaksonen, A.H. In a quest for engineering acidophiles for biomining applications: Challenges and opportunities. *Genes* **2018**, *9*, 116. [CrossRef]
76. Navarro, C.A.; Von Bernath, D.; Jerez, C.A. Heavy metal resistance strategies of acidophilic bacteria and their acquisition: Importance for biomining and bioremediation. *Biol. Res.* **2013**, *46*, 363–371. [CrossRef]
77. Saavedra, A.; Aguirre, P.; Gentina, J.C. Biooxidation of iron by *Acidithiobacillus ferroxidans* in the presence of D-galactose: Understanding its influence on the production of EPS and cell tolerance to high concentration of iron. *Front. Microbiol.* **2020**, *11*, 759. [CrossRef]
78. Jafari, M.; Abdollahi, H.; Shafaei, S.Z.; Gharabaghi, M.; Jafari, H.; Akcil, A. Acidophilic bioleaching: A review on the process and effect of organic-inorganic reagents and materials on its efficiency. *Min. Proc. Ext. Met. Rev.* **2019**, *2*, 87–107. [CrossRef]
79. Brierley, J.A. A perspective on developments in biohydrometallurgy. *Hydrometallurgy* **2008**, *94*, 2–7. [CrossRef]
80. Chen, P.; Yan, L.; Leng, F.; Nan, W.; Yue, X.; Zheng, Y.; Feng, N.; Li, H. Bioleaching of realgar by *Acidithiobacillus ferrooxidans* using ferrous iron and elemental sulfur as the sole and mixed energy sources. *Bioresour. Technol.* **2011**, *102*, 3260–3267. [CrossRef]
81. Zhang, S.; Yan, L.; Xing, W.; Chen, P.; Zhang, Y.; Wang, W. *Acidothiobacillus ferrooxidans* and its potential application. *Extremophiles* **2018**, *22*, 563–579. [CrossRef] [PubMed]
82. Romero-González, M.; Nwaobi, B.C.; Hufton, J.M.; Gilmour, D.J. Ex-situ bioremediation of U(VI) from contaminated mine water using *Acidothiobacillus ferrooxidans* strains. *Front. Environ. Sci.* **2016**, *4*. [CrossRef]
83. Jameson, E.; Rowe, O.F.; Hallberg, K.B.; Johnson, D.B. Sulfidogenesis and selective precipitation of metals at low pH mediated by *Acidithiobacillus* spp. and acidophilic sulfate-reducing bacteria. *Hydrometallurgy* **2010**, *104*, 488–493. [CrossRef]
84. Okibe, N.; Maki, M.; Nakayama, D.; Sasaki, K. Microbial recovery of vanadium by the acidophilic bacterium, *Acidocella aromatica*. *Biotechnol. Lett.* **2016**, *38*, 1475–1481. [CrossRef]
85. Chakravarty, R.; Banerjee, P.C. Mechanism of cadmium binding on the cell wall of an acidophilic bacterium. *Bioresour. Technol.* **2012**, *108*, 176–183. [CrossRef]
86. Beolchini, F.; Dell'Anno, A.; De Propris, L.; Ubaldini, S.; Cerrone, F.; Danovaro, R. Auto- and heterotrophic acidophilic bacteria enhance the bioremediation efficiency of sediments contaminated by heavy metals. *Chemosphere* **2009**, *74*, 1321–1326. [CrossRef]
87. Gupta, A.; Sar, P. Characterization and application of an anaerobic, iron and sulfate reducing bacterial culture in enhanced bioremediation of acid mine drainage impacted soil. *J. Environ. Sci. Health C* **2020**, *4*, 464–482. [CrossRef]
88. Abd-Elnaby, H.; Abou-Elela, G.M.; El-Sersy, N.A. Cadmium resisting bacteria in Alexandria Eastern Harbor (Egypt) and optimization of cadmium bioaccumulation by *Vibrio harveyi*. *Afr. J. Biotechnol.* **2011**, *10*, 3412–3423. [CrossRef]

89. Iyer, A.; Mody, K.; Jha, B. Biosorption of heavy metals by a marine bacterium. *Mar. Pollut. Bull.* **2005**, *50*, 340–343. [CrossRef] [PubMed]
90. zdemir, S.; Kilinc, E.; Poli, A.; Nicolaus, B. Biosorption of heavy metals (Cd^{2+}, Cu^{2+}, Co^{2+}, and Mn^{2+}) by thermophilic bacteria, *Geobacillus thermantarcticus* and *Anoxybacillus amylolyticus*: Equilibrium and kinetic studies. *Bioremediat. J.* **2013**, *17*, 86–96. [CrossRef]
91. Eslami, N.; Kermanshahi, R.K.; Erfan, M. Studying the stability of S-layer protein of *Lactobacillus acdiophilius* ATCC 4356 in simulated gastrointestinal fluids using SDS-PAGE and circular dichroism. *Iran J. Pharm. Res.* **2013**, *12*, 47–56. [PubMed]
92. Liu, S.; Zheng, Y.; Ma, Y.; Sarwar, A.; Zhao, X.; Luo, T.; Yang, Z. Evaluation and proteomic analysis of lead adsorption by lactic acid bacteria. *Int. J. Mol. Sci.* **2019**, *20*, 5540. [CrossRef] [PubMed]
93. Gerbino, E.; Mobili, P.; Tymczyszyn, E.; Fausto, R.; Gómez-Zavaglia, A. FTIR spectroscopy structural analysis of the interaction between *Lactobacillus kefir* S-layers and metal ions. *J. Mol. Struct.* **2011**, *1–3*, 186–192. [CrossRef]
94. Kashefi, K.; Lovely, D.R. Reduction of Fe(III), Mn(IV), and toxic metals at 100 °C by *Pyrobacculum islandicum*. *Appl. Environ. Microbiol.* **2000**, *66*, 1050–1056. [CrossRef]
95. Feitkenhauer, H.; Muller, R.; Markl, H. f. Degradation of polycyclic aromatic hydrocarbons and long chain alkanes at 60-70 degrees C by *Thermus* and *Bacillus* spp. *Biodegradation* **2003**, *14*, 367–372. [CrossRef]
96. Nazina, T.N.; Tourova, T.P.; Poltaraus, A.B.; Novikova, E.V.; Grigoryan, A.A.; Ivanova, A.E.; Lysenko, A.M.; Petrunyaka, V.V.; Osipov, G.A.; Belyaev, S.S.; et al. Taxonomic study of aerobic thermophilic bacilli: Descriptions of Geobacillus subterraneus gen. nov., sp. nov., and Geobacillus uzenensis sp. nov. from petroleum reservoirs and transfer of Bacillus stearothermophilus, Bacillus thermocatenulatus, Bacillus thermoleovorans, Bacillus kaustophilus, Bacillus thermodenitrificans to Geobacillus as the new combinations G. stearothermophilus, G. th. *Int. J. Syst. Evol. Microbiol.* **2001**, *51*, 433–446. [CrossRef]
97. Feng, L.; Wang, W.; Cheng, J.; Ren, Y.; Zhao, G.; Gao, C.; Tang, Y.; Liu, X.; Han, W.; Peng, X.; et al. Genome and proteome of long-chain alkane degrading *Geobacillus thermodenitrificans* NG80-2 isolated from a deep-subsurface oil reservoir. *Proc. Natl. Acad. Sci. USA* **2007**, *104*, 5602–5607. [CrossRef]
98. Sood, N.; Lal, B. (2008). Isolation and characterization of a potential paraffin-wax degrading thermophilic bacterial strain *Geobacillus kaustophilus* TERI NSM for application in oil wells with paraffin deposition problems. *Chemosphere* **2008**, *70*, 1445–1451. [CrossRef]
99. Sun, Y.; Ning, Z.; Yang, F.; Li, X. Characteristics of newly isolated *Geobacillus* sp. ZY-10 degrading hydrocarbons in crude oil. *Pol. J. Microbiol.* **2015**, *64*, 253–263. [CrossRef]
100. Zhang, J.; Zhang, X.; Liu, J.; Li, R.; Shen, B. Isolation of a thermophilic bacterium, *Geobacillus* sp. SH-1, capable of degrading aliphatic hydrocarbons and naphthalene simultaneously, and identification of its naphthalene degrading pathway. *Bioresour. Technol.* **2012**, *124*, 83–89. [CrossRef]
101. Elumalai, P.; Parthipan, P.; Karthikeyan, O.P.; Rajasekar, A. Enzyme-mediated biodegradation of long-chain n-alkanes (C_{32} and C_{40}) by thermophilic bacteria. *3 Biotech.* **2017**, *7*, 116. [CrossRef] [PubMed]
102. Lin, X.; Yang, B.; Shen, J.; Du, N. Biodegradation of crude oil by an Arctic psychrotrophic bacterium *Pseudoalteromomas* sp. P29. *Curr. Microbiol.* **2009**, *59*, 341–345. [CrossRef]
103. Gentile, G.; Bonsignore, M.; Santisi, S.; Catalfamo, M.; Giuliano, L.; Genovese, L.; Yakimov, M.M.; Denaro, R.; Genovese, M.; Cappello, S. Biodegradation potentiality of psychrophilic bacterial strain *Oleispira antarctica* RB-8(T). *Mar. Pollut. Bull.* **2016**, *105*, 125–130. [CrossRef] [PubMed]
104. Gao, W.; Cui, Z.; Li, Q.; Xu, G.; Jia, X.; Zheng, L. *Marinobacter nanhaiticus* sp. nov., polycyclic aromatic hydrocarbon-degrading bacterium isolated from the sediment of the South China Sea. *Antonie Van Leeuwenhoek* **2013**, *103*, 485–491. [CrossRef] [PubMed]
105. Al-Mailem, D.M.; Eliyas, M.; Radwan, S.S. Oil-bioremediation potential of two hydrocarbonoclastic, diazotrophic *Marinobacter* strains from hypersaline areas along the Arabian Gulf coasts. *Extremophiles* **2013**, *17*, 463–470. [CrossRef] [PubMed]
106. Ibrahim, I.M.; Konova, S.V.; Sigida, E.N.; Lyubun, E.V.; Muratova, A.Y.; Fedonenko, Y.P.; Elbanna, K. Bioremediation potential of a halophilic *Halobacillus* sp. strain EG1HP4QL: Exopolysaccharide production, crude oil degradation, and heavy metal tolerance. *Extremophile* **2020**, *24*, 157–166. [CrossRef]
107. Gutierrez, T.; Berry, D.; Yang, T.; Mishamandani, S.; McKay, L.; Teske, A.; Aitken, M.D. Role of exopolysaccharide (EPS) in the fate of the oil released during the Deepwater Horizon oil spill. *PLoS ONE* **2013**, *8*, e67717. [CrossRef]

108. Sadegh, H.; Ali, G.A.M.; Gupta, V.K.; Makhlouf, A.S.H.M.; Shahryari-ghoshekandi, R.; Nadagouda, M.N.; Sillanpää, M.; Megiel, E. The role of nanomaterials as effective adsorbents and their applications in wastewater treatment. *J. Nanostruct. Chem.* **2017**, *7*, 1–14. [CrossRef]
109. Kim, Y.-H.; Jeon, J.H.; Hong, S.H.; Rhim, W.-K.; Lee, Y.-S.; Youn, H.W.; Chung, J.-K.; Lee, M.C.; Lee, D.S.; Kang, K.W.; et al. Tumor targeting and imaging using cyclic RGD-PEGlated gold nanoparticle probes with directly conjugated iodine-125. *Small* **2011**, *7*, 2052–2060. [CrossRef]
110. Choi, M.H.; Shim, H.E.; Yun, S.J.; Park, S.H.; Choi, D.S.; Jang, B.-S.; Choi, Y.J.; Jeon, J.H. Gold-nanoparticle-immobilized desalting column for highly efficient and specific removal of radioactive iodine in aqueous media. *Acs Appl. Mater. Interspace* **2016**, *8*, 29227–29231. [CrossRef]
111. Mushtaq, S.; Yun, S.-J.; Yang, J.E.; Jeong, S.-W.; Shim, H.E.; Choi, M.H.; Park, S.H.; Choi, Y.J.; Jeon, J.H. Efficient and selective removal of radioactive iodine anions using engineered nanocomposite membranes. *Environ. Sci. Nano* **2017**, *4*, 2157–2163. [CrossRef]
112. Lloyd, J.R. Microbial reduction of metals and radionuclides. *FEMS Microbiol. Rev.* **2003**, *27*, 411–425. [CrossRef]
113. Kumar, R.; Singh, S.; Singh, O.V. Bioremediation of radionuclides: Emerging technologies. *OMICS* **2007**, *11*, 295–304. [CrossRef]
114. Prakash, D.; Gabani, P.; Chandel, A.K.; Ronen, Z.; Singh, O.V. Bioremediation: A genuine technology to remediate radionuclides from the environment. *Microb. Biotechnol.* **2013**, *6*, 349–360. [CrossRef]
115. Shukla, A.; Parmar, P.; Sarar, M. Radiation, radionuclides and bacteria: An in-perspective review. *J. Environ. Radioact.* **2017**, *180*, 27–35. [CrossRef] [PubMed]
116. Wildung, R.E.; Gorby, Y.A.; Krupka, K.M.; Hess, N.J.; Li, S.W.; Plymale, A.E.; McKinley, J.P.; Fredrickson, J.K. Effect of electron donor and solution chemistry on products of dissimilatory reduction of technetium by *Shewanella putrefaciens*. *Appl. Environ. Microbiol.* **2000**, *66*, 2451–2460. [CrossRef]
117. Istok, J.D.; Senko, J.M.; Krumholz, L.R.; Watson, D.; Bogle, M.A.; Peacock, A.; Chang, Y.-J.; White, D.C. In situ bioreduction of technetium and uranium in a nitrate-contaminated aquifer. *Environ. Sci. Technol.* **2004**, *38*, 468–475. [CrossRef]
118. Panak, P.J.; Nitsche, H. Interaction of aerobic soil bacteria with plutonium(VI). *Radiochim. Acta.* **2001**, *89*, 499–504. [CrossRef]
119. Kim, S.-J.; Koh, D.-C.; Park, S.-J.; Cha, I.-T.; Park, J.-W.; Na, J.-H.; Roh, Y.; Ko, K.-S.; Kim, K.J.; Rhee, S.-K. Molecular analysis of spatial variation of iron-reducing bacteria in riverine alluvial aquifers of the Mankyeong River. *J. Microbiol.* **2012**, *50*, 207–217. [CrossRef]
120. Anderson, R.T.; Vrionis, H.A.; Ortiz-Bernard, I.; Resch, C.T.; Long, P.E.; Dayvault, R.; Karp, K.; Metzler, D.R.; Peacock, A.; White, D.C.; et al. Stimulating the in situ activity of *Geobacter* species to remove uranium from the groundwater of a uranium-contaminated aquifer. *Appl. Environ. Microbiol.* **2003**, *69*, 5884–5891. [CrossRef]
121. Ferreira, A.C.; Nobre, M.F.; Moore, E.; Rainey, F.A.; Battista, J.R.; Da Costa, M.S. Characterization and radiation resistance of new isolates of *Rubrobacter radiotolerans* and *Rubrobacter xylanophilus*. *Extremophiles* **1999**, *3*, 235–238. [CrossRef]
122. Billi, D.; Friedmann, E.I.; Hofer, K.G.; Caiola, M.G.; Ocampo-Friedmann, R. Ionizing-radiation resistance in the desiccation-tolerant cyanobacterium *Chroococcidiopsis*. *Appl. Environ. Microbiol.* **2000**, *66*, 1489–1492. [CrossRef] [PubMed]
123. Brim, H.; Mcfarlan, S.C.; Fredrickson, J.K.; Minton, K.W.; Zhai, M.; Wackett, L.P.; Daly, M.J. Engineering *Deinococcus radiodurans* for metal remediation in radioactive mixed waste environments. *Nat. Biotechnol.* **2000**, *18*, 85–90. [CrossRef] [PubMed]
124. Daly, M.J. Engineering radiation-resistant bacteria for environmental biotechnology. *Curr. Opin. Biotechnol.* **2000**, *11*, 280–285. [CrossRef]
125. Jeong, S.-W.; Choi, Y.J. Research perspective of an extremophilic bacterium, *Deinococcus radiodurans* on bioremediation of radioactive wastes. *Appl. Chem. Eng.* **2017**, *28*, 133–140. [CrossRef]
126. Fredrickson, J.K.; Kostandarithes, H.M.; Li, S.W.; Plymale, A.E.; Daly, M.J. Reduction of Fe(III), Cr(VI), U(VI), and Tc(VII) by *Deinococcus radiodurans* R1. *Appl. Environ. Microbiol.* **2000**, *66*, 2006–2011. [CrossRef]
127. Appukuttan, D.; Rao, A.S.; Apte, S.K. Engineering of *Deinococcus radiodurans* R1 for bioprecipitation of uranium from dilute nuclear waste. *Appl. Environ. Microbiol.* **2006**, *72*, 7873–7878. [CrossRef]

128. Misra, C.S.; Appukuttan, D.; Kantamreddi, V.S.S.; Rao, A.S.; Apte, S.K. Recombinant, *D. radiodurans* cells for bioremediation of heavy metals from acidic/neutral aqueous wastes. *Bioeng. Bugs* **2012**, 3, 44–48. [CrossRef] [PubMed]
129. Appukuttan, D.; Seetharam, C.; Padma, N.; Rao, A.S.; Apte, S.K. PhoN-expressing, lyophilized, recombinant *Deinococcus radiodurans* cells for uranium bioprecipitation. *J. Biotechnol.* **2011**, *154*, 285–290. [CrossRef]
130. Gogada, R.; Singh, S.S.; Lunavat, S.K.; Pamarthi, M.M.; Rodrigue, A.; Vadivelu, B.; Phanithi, P.-B.; Gopala, V.; Apte, S.K. Engineered *Deinococcus radiodurans* R1 with NiCoT genes for bioremoval of trace cobalt from spent decontamination solutions of nuclear power reactors. *Appl. Microbiol. Biotechnol.* **2015**, *99*, 9203–9213. [CrossRef]
131. Choi, M.H.; Jeong, S.-W.; Shim, H.E.; Yun, S.J.; Mushtaq, S.; Choi, D.S.; Jang, B.-S.; Jung, E.Y.; Choi, Y.J.; Jeon, J.H. Efficient bioremediation of radioactive iodine using biogenic gold nanomaterial-containing radiation-resistant bacterium, *Deinococcus radiodurans* R1. *Chem. Comm.* **2017**, *53*, 3937–3940. [CrossRef] [PubMed]
132. Kulkarni, R.R.; Shaiwale, N.S.; Deobagkar, D.N.; Deobagkar, D.D. Synthesis and extracellular accumulation of silver nanoparticles by employing radiation-resistant *Deinococcus radiodurans*, their characterization, and determination of bioactivity. *Int. J. Nanomed.* **2015**, *10*, 963–974. [CrossRef]
133. Beeler, E.; Singh, O.V. Extremophiles as sources of inorganic bio-nanoparticles. *World, J. Microbiol. Biotechnol.* **2016**, *32*, 156. [CrossRef]
134. Li, J.; Li, Q.; Ma, X.; Tian, B.; Yu, J.; Dai, S.; Weng, Y.; Hua, Y. Biosynthesis of gold nanoparticles by the extreme bacterium *Deinococcus radiodurans* and an evaluation of their antibacterial properties. *Int. J. Nanomed.* **2016**, *11*, 5931–5944. [CrossRef] [PubMed]
135. Chen, A.; Contreras, L.M.; Keitz, B.K. Imposed environmental stresses facilitate cell-free nanoparticle formation by *Deinococcus radiodurans*. *Appl. Environ. Microbiol.* **2017**, *83*, e00798-17. [CrossRef]
136. Kitjanukit, S.; Sasaki, K.; Okibe, N. Production of highly catalytic, archaeal Pd(0) bionanoparticles using *Sulfolobus tokodaii*. *Extremophiles* **2019**, *20*, 549–556. [CrossRef] [PubMed]
137. Okibe, N.; Nakayama, D.; Matsumoto, T. Palladium bionanoparticles production from acidic Pd(II) solutions and spent catalyst leachate using acidophilic Fe(II)-reducing bacteria. *Extremophiles* **2017**, *21*, 1091–1100. [CrossRef]
138. Crini, G.; Lichtfouse, E. Advantages and disadvantages of techniques used for wastewater treatment. *Environ. Chem. Lett.* **2019**, *17*, 145–155. [CrossRef]
139. Ren, J.; Lee, J.G.; Na, D.K. Recent advances in genetic engineering tools based on synthetic biology. *J. Microbiol.* **2020**, *58*, 1–10. [CrossRef]
140. Wang, Y.; Wang, D.; Wang, X.; Tao, H.; Feng, E.; Zhu, L.; Pan, C.; Wang, B.; Liu, C.; Liu, X.; et al. Highly efficient genome engineering in *Bacillus anthracis* and *Bacillus cereus* using CRISPR/Cas9 system. *Front. Microbiol.* **2019**, *10*, 1932. [CrossRef]
141. Marques, C.R. Extremophilic microfactories: Application in metal and radionuclide bioremediation. *Front. Microbiol.* **2018**, *9*, 1191. [CrossRef] [PubMed]
142. Giovanella, P.; Vieira, G.A.L.; Ramos Otero, I.V.; Pellizzer, E.P.; de Jesus Fontes, B.; Sette, L.D. Metal and organic pollutants bioremediation by extremophile microorganisms. *J. Hazard. Mater.* **2020**, *382*, 121024. [CrossRef] [PubMed]

molecules

MDPI

Article

Highly Efficient Methylene Blue Dye Removal by Nickel Molybdate Nanosorbent

Souad Rakass [1,*], Hicham Oudghiri Hassani [2], Ahmed Mohmoud [3,4], Fethi Kooli [5], Mostafa Abboudi [4], Eman Assirey [4] and Fahd Al Wadaani [4]

1 Laboratory of Applied Organic Chemistry (LCOA), Chemistry Department, Faculty of Sciences and Techniques, Sidi Mohamed Ben Abdellah University, Imouzzer Road, P.O. Box 2202, 30000 Fez, Morocco
2 Engineering Laboratory of Organometallic, Molecular Materials and Environment (LIMOME), Faculty of Sciences, Chemistry Department, Sidi Mohamed Ben Abdellah University, P.O. Box 1796 (Atlas), 30000 Fez, Morocco; oudghiri_hassani_hicham@yahoo.com
3 Petroleum Technology, Operated Offshore Oil Field Development, Qatar Petroleum, P.O. Box 3212, Doha, Qatar; caadil77@yahoo.co.uk
4 Chemistry Department, College of Science, Taibah University, Al-Madinah Al-Munawwarah 30002, Saudi Arabia; abboudi14@hotmail.com (M.A.); eman_assirey@hotmail.com (E.A.); fwadaani@taibahu.edu.sa (F.A.W.)
5 Department of Chemistry, Faculty of Science, Islamic University of Madinah, Al-Madinah Al-Munawwarah 42351, Saudi Arabia; fethi_kooli@yahoo.com
* Correspondence: rakass_souad@yahoo.fr; Tel.: +212-6449950091

Citation: Rakass, S.; Oudghiri Hassani, H.; Mohmoud, A.; Kooli, F.; Abboudi, M.; Assirey, E.; Al Wadaani, F. Highly Efficient Methylene Blue Dye Removal by Nickel Molybdate Nanosorbent. *Molecules* **2021**, *26*, 1378. https://doi.org/10.3390/molecules26051378

Academic Editors: Chiara Bisio and Monica Pica

Received: 31 December 2020
Accepted: 27 February 2021
Published: 4 March 2021

Abstract: Removing methylene blue (MB) dye from aqueous solutions was examined by the use of nickel molybdate (α-$NiMoO_4$) as an adsorbent produced by an uncomplicated, rapid, and cost-effective method. Different results were produced by varying different parameters such as the pH, the adsorbent dose, the temperature, the contact time, and the initial dye concentration. Adsorbent dose and pH had a major removal effect on MB. Interestingly, a lower amount of adsorbent dose caused greater MB removal. The amount of removal gained was efficient and reached a 99% level with an initial methylene blue solution concentration of $\leq$160 ppm at pH 11. The kinetic studies indicated that the pseudo-second-order kinetic model relates very well with that of the obtained experimental results. The thermodynamic studies showed that removing the MB dye was favorable, spontaneous, and endothermic. Impressively, the highest quantity of removal amount of MB dye was 16,863 mg/g, as shown by the Langmuir model. The thermal regeneration tests revealed that the efficiency of removing MB (11,608 mg/g) was retained following three continuous rounds of recycled adsorbents. Adsorption of MB onto α-$NiMoO_4$ nanoparticles and its regeneration were confirmed by Fourier transform infrared spectroscopy (FTIR) analysis and scanning electron microscopy (SEM) analysis. The results indicated that α-$NiMoO_4$ nanosorbent is an outstanding and strong candidate that can be used for removing the maximum capacity of MB dye in wastewater.

Keywords: nanosorbent; regeneration; α-$NiMoO_4$; methylene blue; removal

1. Introduction

Dyes have recently been extensively utilized in several industrial and manufacturing applications, e.g., printing, textile, paper, carpet, and cosmetics. Dyes are considered toxic, hazardous pollutants and require removal before their discharge into the environment [1–7].

Numerous methods have been developed for dye removal from wastewater and industrial waste matter, including adsorption, coagulation, photodegradation, flocculation, membrane separation, ion exchange, biological treatment, chemical oxidation, and extraction [8–15].

Adsorption application is extensively employed due to its ease of process and guarantee of superb minimal cost from among those above-mentioned methods [1,16–21]. Several natural adsorbents have been successful in the elimination of color from aqueous waste matter [22–25]. A universally used example is activated carbon owing to the presence of its large surface area [26,27]. There are still some difficulties that limit its use, such as its high cost of production, low-quality mechanical properties, regeneration issues, and phase separation strain [28]. The challenge faced by the researchers is to develop novel adsorbents with boundless adsorption capabilities that are capable of being regenerated for the recovery of reusable compounds.

In recent years, researchers have shown a great interest in binary metal oxides due to their potential performances for different materials [29]. In recent years, scientists have paid extensively studied the group of metal molybdates which has given the most favorable examples of mixed metal oxides [30–33]. Nickel molybdate ($NiMoO_4$) has various applications in catalysis such as hydrodesulfurization and hydrodenitrogenation reactions [34,35], oxidative dehydrogenation of light alkanes [36–40], partial oxidation of hydrocarbons [41], and microwave applications [42]. It is also used in humidity sensors [43], supercapacitors [44,45], optical fibers, and military devices [46]. Nickel molybdate has attractive structures and electrochemical and magnetic properties [47,48], and it can be found in two crystalline forms, α-$NiMoO_4$ and β-$NiMoO_4$.

Various methods of $NiMoO_4$ synthesis have been presented in the literature, including sonochemical [49,50], hydrothermal [46,51,52], precipitation [53,54], sol–gel [53], mechanochemical synthesis [55], solid state at high temperature [56,57], and microwave-assisted methods [58].

Recently, molybdate compounds have attracted great interest for their utilization in environmental applications such as the photocatalytic oxidation of dyes [59–62], the oxidation of methylene blue (MB) dye [63], and the sorption of water-soluble dyes [30,61].

In particular, α-$NiMoO_4$ synthesized by the microwave-assisted method has shown good photocatalytic activity for methylene blue photodegradation [58]. In addition, $NiMoO_4$ nanostructures synthesized by the coprecipitation method were efficiently used as a catalyst for methyl orange photooxidation under UV irradiation [64]. Furthermore, hydrothermally synthesized β-$NiMoO_4$ was recently used as a sono-photocatalyst for the degradation of methylene blue (MB) under diffused sunlight [65]. However, α-$NiMoO_4$ has not yet been explored for the removal of dyes by adsorption.

In the present work, nickel molybdate nanoparticles, synthesized using a facile and easy method without the use of any solvents, were evaluated for removing methylene blue dye (MB) as adsorbents. MB dye was utilized as an ideal dye owing to its extensive manufacturing uses as a food coloring agent and for cotton, wool, silk, and leather, to name a few examples [60]. The influence of diverse parameters, namely solution pH, initial concentration, adsorbent dose, and contact time, on the removal of methylene blue by synthesized α-$NiMoO_4$ nanosorbents was examined. The kinetics and adsorption isotherms were evaluated. In addition, after the nanosorbent had been regenerated by calcinating at high temperature, the removal efficiency was likewise investigated.

2. Results and Discussion

2.1. *Removal of MB*

2.1.1. pH Point of Zero Charge (pH_{pzc})

The pH point of zero charge (pH_{pzc}) can provide information regarding the surface charge of a material. The results found for this parameter are given in Figure 1. In fact, the pH_{pzc} of nickel molybdate was determined from graphs where the initial pH is equal to the final pH (intersection of curves). As shown in Figure 1, the nickel molybdate presented a surface charged negatively (pH_{pzc} = 8.96). The surface of nickel molybdate became negatively charged for a pH of solutions >8.96 and acquired a positive charge when the pH was lower than 8.96. According to the literature, the cation uptake is favorable at a pH > pH_{pzc}, whereas the uptake of anions is encouraged at a pH < pH_{pzc} of sorbent [66]. The obtained pH point of zero charge value is close to those obtained for some metal oxides such as CuO and NiO, which are in the range of 9–10 [67].

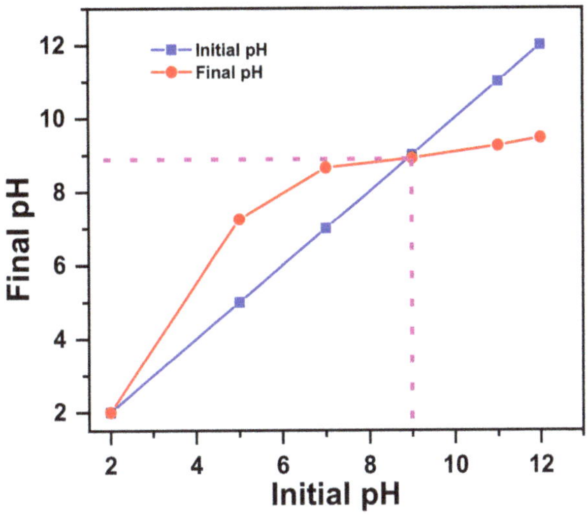

Figure 1. pH_{pzc} for nickel molybdate.

2.1.2. Effect of pH

pH is essential in terms of controlling the removal of dyes. It has no effect on altering the separation of the adsorbent site; nevertheless, it modifies the structure and the chemistry of the dye [68]. Moreover, the charge and surface potential of the oxide are primarily pH-dependent. It is possible, in some cases, to adjust the pH conditions of the slurry in such a way that all the particles exhibit the same charge polarity [68]. Hence, the effect of pH on the removal of MB by nickel molybdate (α-$NiMoO_4$) nanosorbent was assessed by varying pH values between 3 and 11 at a controlled temperature of 20 °C with the initial concentration of 100 ppm. Figure 2 demonstrates that methylene blue removal depended on the effect of pH. By increasing the pH from 3 to 7, the removal percentage did not substantially change and remained at about 29%. The further increase of pH to 9 slightly increased the removal percentage to 35%. The increase of pH to 11 led to the highly efficient removal of MB, as the removal percentage reached 93%. Moreover, there was an increase in the quantity of dye eliminated for every unit mass of the adsorbent at its equilibrium (qe) from 25 to 93 mg/g.

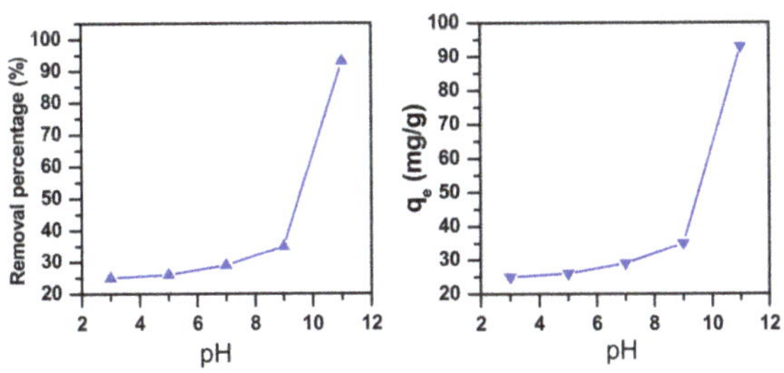

Figure 2. Effect of pH on dye removal performance of α-$NiMoO_4$ in a 100 ppm methylene blue solution (m_{ads} = 0.1 g, T = 20 °C, t = 30 min).

Strong electrostatic interactions took place between the charges of the MB dyes and those of α-$NiMoO_4$ adsorbent, as demonstrated by the increase in removal percentage obtained with the increase of the pH values. Furthermore, in the solution at pH 11, the hydroxyl group (OH^-) favors the direction of the positively charged MB as the pKa equals

3.8 [69]. Nevertheless, the lower removal performance at acidic values may well be related to the extra proton ions within the solution which are in concurrency with those of the basic dye cations on the removal sites of α-$NiMoO_4$. Comparable outcomes were reported by Kooli et al. in the examination of waste bricks utilized as favorable agents for the removal of basic blue 41 from liquid solutions [70].

On the other hand, these results can be explained by the point of zero charge value measured for nickel molybdate (pH_{pzc} = 8.96). The literature reports that at lower pH ($pH < pH_{pzc}$), the surface charge may become positive, thus allowing H^+ ions to compete effectively with dye cations and causing a decrease in the amount of dye adsorbed [71]. At higher pH ($pH > pH_{pzc}$), the nickel molybdate may become negatively charged, which enhances the positively charged cationic dye through electrostatic forces of attraction.

Thus, pH 11 was found to be the best value for the removal of MB when employing α-$NiMoO_4$ nanosorbent.

2.1.3. Effect of Adsorbent Dose

One crucial parameter in adsorption processes is the adsorbent dose [72]. The MB dye removal using α-$NiMoO_4$ was explored with varying adsorbent doses of 0.001 to 0.5 g/L with an initial dye concentration of 160 ppm. As can be seen clearly in Figure 3, the percentage (%) and the amount (mg/g) of MB removed were reduced as the nanosorbent dose increased from 0.001 to 0.5 g/L. The decrease in the removal effectiveness might be due to the performance of particle interaction (e.g., aggregation) as a result of the high dosage of the adsorbent. Such aggregation could increase diffusional path duration. Furthermore, the adsorption sites remain unsaturated throughout the course of sorption under these conditions. In fact, all of these factors can lead to a decrease in available particle size [73,74].

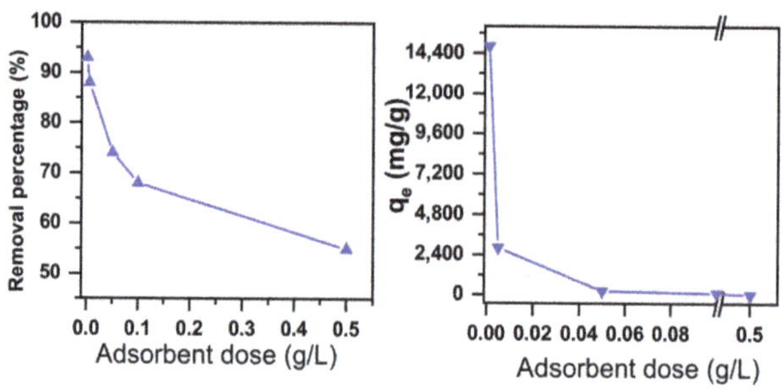

Figure 3. Effect of adsorbent dose on the dye removal performance of α-$NiMoO_4$ in a 160 ppm methylene blue solution (t = 30 min, T = 20 °C).

2.1.4. Effects of Initial Concentration and Contact Time

Figure 4 presents the effects of initial MB dye concentration and contact time on dye removal. The removal of MB was enhanced with the increase in contact time, reaching the highest value of 98% at around 30 min for initial methylene blue concentrations of 100 ppm. For MB concentrations of 120, 140, and 160 ppm, a maximum value of 99% removal was reached at around 60 min. However, the removal percentage of MB decreased from 99% to 76% as the Ci value was increased from 160 to 200 ppm. The removal amount remarkably increased from 9993 mg/g to 15,900 mg/g when the initial dye concentration increased from 100 to 160 ppm and remained stable when the concentration was increased to 200 ppm. This showed that the concentration gradient is an essential factor that drives the overcoming of the mass transfer resistances within the solid and liquid phases. The ratio of the solution connected with the α-$NiMoO_4$ surface was higher at lower MB concentrations, which triggered a rise in removal efficiency. In contrast, at higher MB dye concentrations, the decrease in the adsorption percentage was affected by the saturation of active sites on the α-$NiMoO_4$ surface [75].

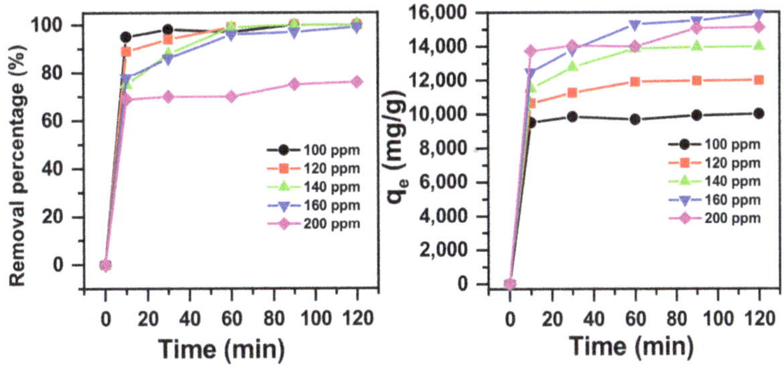

Figure 4. Impact of the initial dye concentration and contact time on the methylene blue (MB) dye removal performance of α-$NiMoO_4$ ($m_{adsorbent}$ = 0.001 g, T = 20°C, pH = 11).

2.1.5. Temperature Effect

The temperature is an essential factor that greatly affects the removal of dyes [76]. The procedure for removing the methylene blue dye was examined from 20 to 70 °C, as shown in Figure 5. The outcome of temperature experiments shows that the removal percentage increased from 70% to 100% and the removal capacity increased from 14,047 to 19,990 mg/g at an initial MB dye concentration of 200 ppm. The efficiency progression of MB removal with a rise in temperature was due to the intensity of attractive forces between removal sites and the MB, which shows an endothermic process [70]. In addition, increasing the temperature improves the removal motion of the adsorbent sites and the dye molecule motion [76].

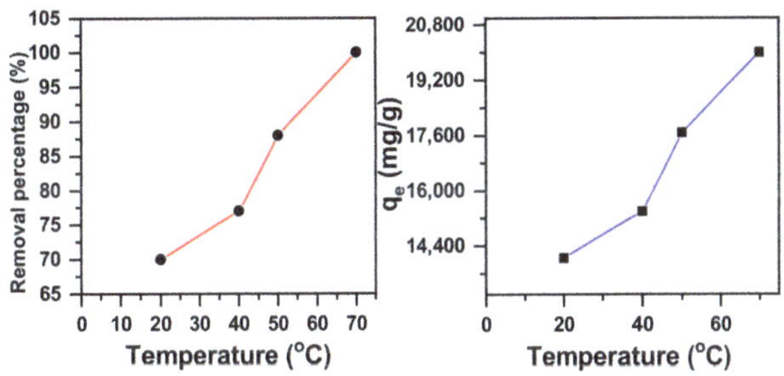

Figure 5. The effect of temperature on the dye removal capacity of $NiMoO_4$ in a 200 ppm methylene blue solution (t = 30 min, pH = 11).

Thermodynamic factors are also important factors in adsorption processes [77,78]. The probability and the mechanism of adsorption can be predicted by the thermodynamic factors [77]. The following equations are used to determine the thermodynamic parameters:

$$\Delta G^o = -RTLnK_d \tag{1}$$

$$K_d = \frac{C_a}{C_e} \tag{2}$$

$$LnK_d = \frac{\Delta S^o}{R} - \frac{H^o}{RT} \tag{3}$$

where R is the gas constant (J mol^{-1} K^{-1}), ΔG° is the free energy, K_d is the distribution constant, T is absolute temperature (K), C_a is the quantity of dye adsorbed by the adsorbent at equilibrium (mol/L), C_e is the equilibrium concentration, ΔH° is the standard enthalpy, and ΔS° is the standard entropy. ΔS° and ΔH° values were obtained from the intercept and slope of the ln K_d versus 1/T plot (Figure 6). ΔG° values were obtained from Equation (1) and are shown in Table 1. The adsorption is favorable and spontaneous, and this is revealed by the negative value obtained for ΔG°. In fact, Gibbs free energy change ($\Delta G°$) values can discern whether a process is spontaneous or not, and negative values of $\Delta G°$ imply a spontaneous process. The enthalpy change ($\Delta H°$) provides information about the exothermic or endothermic nature of the process and differentiates between physical and chemical adsorption processes. Therefore, the positive value of ΔH° (35.12 kJ mol^{-1}) shows that methylene blue removal followed an exothermic process. In addition, the (ΔH°) value was found to be less than 40 kJ/ mol, which indicates that the adsorption of MB by nickel molybdate is physisorption [79]. The present results are similar to the results reported by Xia [80] for adsorption of congo red from aqueous solution by CTAB–hectorite and ODA–hectorite composites. The enhanced anarchy and uncertainty in the solid solution interface of methylene blue and α-$NiMoO_4$ are shown by the positive values of ΔS°. The adsorbate molecules move the adsorbed water molecules; consequently, translational energy is gained rather than lost, which indicates that this approach takes place randomly [81].

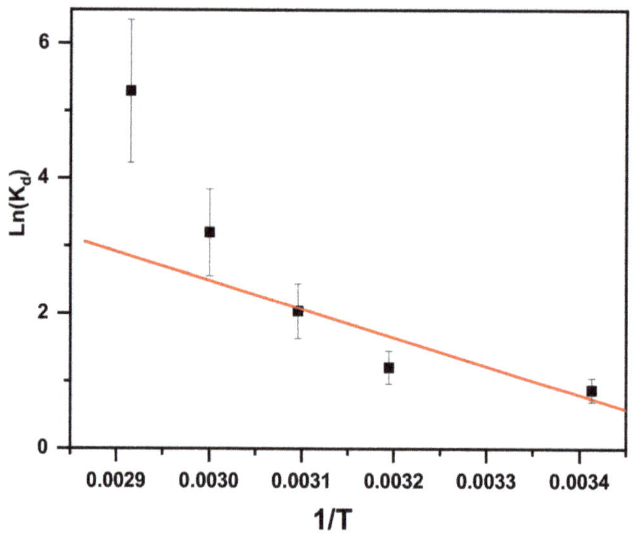

Figure 6. Van't Hoff plot presenting the impact of temperature on methylene blue dye removal utilizing α-$NiMoO_4$.

Table 1. Thermodynamic parameters for the removal of MB dye utilizing α-$NiMoO_4$.

<table>
<tr><th>Adsorbent</th><th>Adsorbate</th><th>Temperature (K)</th><th>K_d</th><th>ΔH° (kJ mol^{-1})</th><th>ΔS° (kJ mol^{-1} K)</th><th colspan="5">ΔG° (kJ mol^{-1})</th></tr>
<tr><td rowspan="5">α-$NiMoO_4$</td><td rowspan="5">MB</td><td>293</td><td>2.401</td><td rowspan="5">35.12</td><td rowspan="5">0.126</td><td rowspan="3">293K</td><td rowspan="3">313K</td><td rowspan="3">323K</td><td rowspan="3">333K</td><td rowspan="3">343K</td></tr>
<tr><td>313</td><td>3.348</td></tr>
<tr><td>323</td><td>7.696</td></tr>
<tr><td>333</td><td>23.390</td><td rowspan="2">−2.134</td><td rowspan="2">−3.144</td><td rowspan="2">−5.480</td><td rowspan="2">−8.727</td><td rowspan="2">−14.822</td></tr>
<tr><td>343</td><td>180.818</td></tr>
</table>

2.2. Kinetic Study

Kinetic study of removal of methylene blue dye has been conducted as it provides an indication regarding the adsorption system [82].

The data found from the kinetics of removing MB dye using α-$NiMoO_4$ nanosorbent were examined by pseudo-first-order, pseudo-second-order, and intraparticle diffusion models. Equations of the studied models are shown in Table 2.

Table 2. Kinetic model equations.

Model	Equation	Parameters
Pseudo-first-order (PFD) [83]	$Ln(q_e - q_t) = Lnq_e + K_1t$ (4)	K_1: the rate constant of pseudo-first-order adsorption (1/min) q_e: the removal capacity at equilibrium (mg/g) q_t: the removal capacity at time t (mg/g)
Pseudo-second-order (PSD) [83]	$\frac{t}{q_t} = \frac{1}{K_2q_e^2} + \frac{t}{q_e}$ (5)	K_2: the pseudo-second-order rate constant ($g \cdot mg^{-1} \cdot min^{-1}$) q_e: the removal capacity at equilibrium (mg/g) q_t: the removal capacity at time t (mg/g)
Intraparticle diffusion (IPD) [77]	$q_t = K_It^{0.5} + 1$ (6)	q_t: the removal capacity (mg/g) at time t t: the contact time (min) I (mg/g) and K_I ($mg/(g \cdot min^{0.5})$): the intraparticle diffusion constants

Three model parameters, namely pseudo-first-order, pseudo-second-order, and intraparticle diffusion, are presented in Table 3 and displayed in Figure 7, Figure 8, Figure 9 respectively. Regression correlation coefficients (R^2) of the three models vary. Intraparticle diffusion is 0.934 to 0.981, pseudo-first-order ranges from 0.952 to 0.988, and pseudo-second-order is 0.999 to 1.000, varying with their concentrations used. The R^2 for pseudo-second-order is equal to or near 1, and hence this model fits very well.

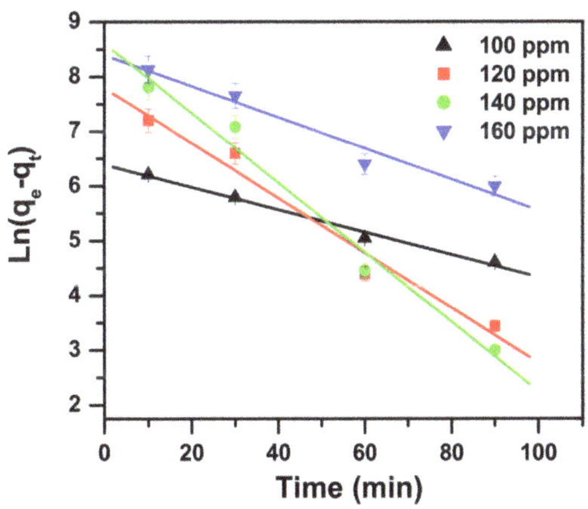

Figure 7. Pseudo-first-order model plot showing the impact of contact time and initial dye concentration on methylene blue removal utilizing α-$NiMoO_4$.

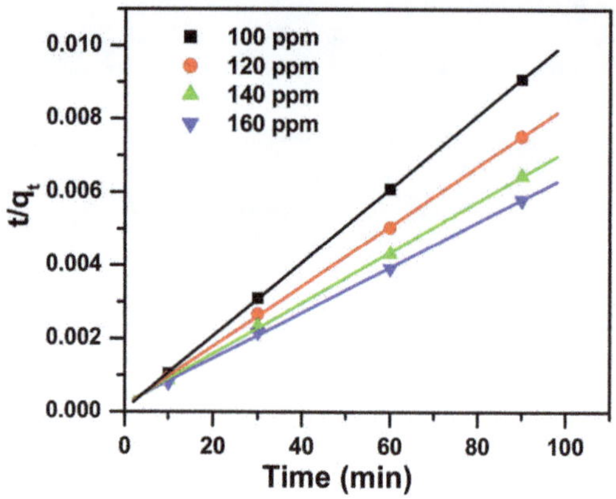

Figure 8. Pseudo-second-order model plot showing the impact of contact time and initial dye concentration on methylene blue dye removal utilizing α-$NiMoO_4$.

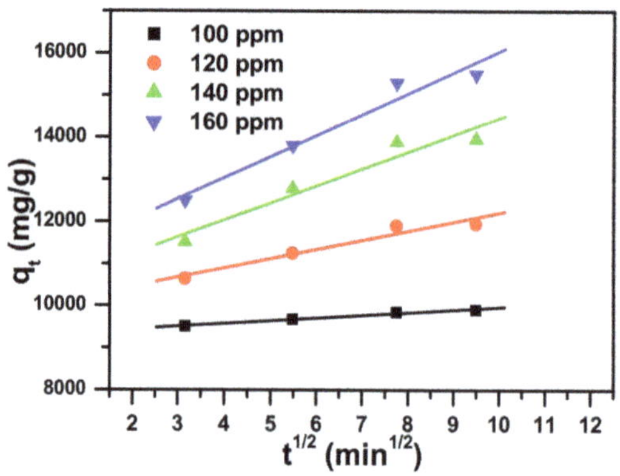

Figure 9. Intraparticle diffusion model plot showing the impact of contact time and initial dye concentration on methylene blue dye removal utilizing α-$NiMoO_4$.

Table 3. Kinetic parameters for removal of methylene blue utilizing α-$NiMoO_4$.

C_i mg/L	Pseudo-First-Order				Pseudo-Second-Order			Intraparticle Diffusion Model		
	q_{exp} (mg/g)	q_e (mg/g)	k_1 (1/min)	R_1^2	q_e (mg/g)	k_2 (g/mg min)	R_2^2	I (mg/g)	k_i (mg/g $min^{0.5}$)	R_3^2
100	10,000	584	0.020	0.988	9964	0.00015	1.000	9311	65	0.981
120	11,981	2101	0.048	0.966	12,200	0.00004	1.000	10,027	218	0.946
140	13,970	4770	0.062	0.982	14,436	0.00002	1.000	10,426	403	0.934
160	15,900	4218	0.028	0.952	16,136	0.00002	0.999	11,036	501	0.959

2.3. Adsorption Isotherms

When planning adsorption methods, adsorption isotherms, which are known to be essential, are taken into consideration due to their perfect explanation [84]. Four adsorption models have been examined, namely Dubinin–Radushkevich, Temkin, Freundlich, and Langmuir models. Equations of the four examined models are presented in Table 4.

Table 4. Adsorption isotherm models for the removal of methylene blue dye utilizing α-$NiMoO_4$.

Model	Equation		Parameters
Freundlich [78]	$Lnq_e = Lnq_F + \frac{1}{n}LnC_e$	(7)	C_e: concentration of MB at equilibrium (ppm) n: the heterogeneity factor (g/L) q_F: the Freundlich constant ($mg^{(1-1/n)}\ L^{1/n}\ g^{-1}$) q_e: the methylene blue dye quantity adsorbed by α-NiMoO4 at equilibrium (mg/g)
Langmuir [85]	$\frac{C_e}{q_e} = \frac{1}{q_m K_L} + \frac{C_e}{q_m}$	(8)	q_e: the methylene blue dye quantity adsorbed by α-$NiMoO_4$ at equilibrium (mg/g) C_e: concentration of MB at equilibrium (ppm) q_m: the maximum quantity of methylene blue dye removed by α-$NiMoO_4$ (mg/g) K_L: Langmuir constant of adsorption (L/mg)
	$R_L = \frac{1}{1 + K_L C_i}$	(9)	C_i: the initial concentration of methylene blue K_L: the Langmuir constant R_L: values indicate that the removal of methylene blue dye could be linear (R_L = 1), irreversible (R_L = 0), favorable (0 < R_L< 1), or unfavorable (R_L > 1)
Dubinin–Radushkevich (D-R) [84]	$Lnq_e = Lnq_m - K\varepsilon^2$	(10)	K: constant for the sorption energy (mol^2/kJ^2) ε: the Polanyi potential T: the temperature (K) R: the universal gas constant (8.314 J.mol^{-1} K^{-1}) q_m: the theoretical saturation capacity C_e: the equilibrium concentration of the methylene blue dye left in the solution (ppm)
	$\varepsilon = RTLn\left(1 + \frac{1}{C_e}\right)$	(11)	
Temkin [86]	$q_e = B_T LnA_T + B_T LnC_e$	(12)	b_T: the Temkin constant related to heat of sorption (J/mol) $B_T = R_T/b_T$ R: the gas constant (8.314 J/mol K) A_T: the Temkin isotherm constant (L/g) T: the absolute temperature (K)

The models employed to match the investigational data were Freundlich, Langmuir, Temkin, and D–R isotherm. Standards of regression correlation coefficients (R^2) and the model parameters are displayed in Figure 10 and contained in Table 5. Langmuir equation demonstrated the highest value of R^2 (0.999), and D–R model revealed the lowest value of R^2 (0.782), while intermediary values were attained for Temkin and Freundlich (0.960 and 0.948, respectively). The Langmuir model fits the experimental results well; the methylene blue removal occurred on a homogeneous surface, establishing a monolayer on the α-$NiMoO_4$ adsorbent, with a high adsorption capacity of 16,863 mg/g. Methylene blue dye removal by α-$NiMoO_4$ is favorable and is revealed by the RL separation factor ranging from 0.0004 to 0.0006.

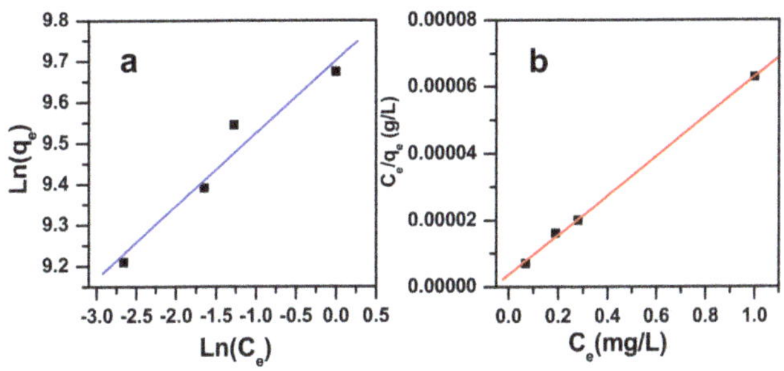

Figure 10. Freundlich (**a**) and Langmuir (**b**) isotherm model plots presenting the results of initial dye concentration and the removal of methylene blue dye utilizing α-$NiMoO_4$.

Table 5. Isotherm parameters for the removal of MB dye utilizing α-$NiMoO_4$.

Langmuir			Freundlich				Temkin			Dubinin–Radushkevich		
q_m (mg/g)	K_L (L/mg)	R^2	Range R_L	q_F($mg^{(1-1/n)}$ $L^{1/n}g^{-1}$)	1/n	R^2	A_T (L/g)	B_T	R^2	q_m (mg/g)	R^2	E (Kj/mol)
16,863	16	0.999	0.0004–0.0006	16,318	6	0.948	1220	2269	0.960	14,429	0.782	43

Table 6 presents previous reports of the maximum amount of methylene blue dye removed. When compared with many nanosorbents, Nickel-based nanosorbents NiO (Q_{max} = 10,585 mg/g) and α-$NiMoO_4$ (Q_{max} = 16,863.00 mg/g) show a considerably higher rate of adsorption for MB. The adsorption capacity of $Fe_2(MoO_4)_3$ (Q_{max} = 6173.00 mg/g) is lower than that obtained by α-$NiMoO_4$, which can be related to the difference in their specific surface area (8.03 versus 29.86 m^2/g). Thus, α-$NiMoO_4$ has the advantage of being able to be synthesized at a rather low temperature via a relatively cost-effective, very simple procedure for use in potential novel, more efficient decontamination processes aimed at the removal of methylene pollutants.

Table 6. Previous reports of the maximum amount of methylene blue dye removed (q_m).

Nanosorbent	Q_{max} (mg/g)	Reference
$ZnMoO_4$ nanoparticles	217.86	[30]
$Fe_2(MoO_4)_3$ nanoparticles	6173.00	[87]
α-MoO_3 nanoparticles	152.00	[88]
Chemically reduced graphene oxide	1519.60	[89]
Magnetic β-cyclodextrin–chitosan nanoparticles	2783.30	[90]
ZnO	7918.02–9197.70	[91]
Fe_2O_3	1124.70	[92]
CoO	5501.93	[92]
NiO	10,585.00	[92]
Nickel molybdate (α-$NiMoO_4$)	16,863.00	This work

2.4. *Regeneration and Characterization of the α-$NiMoO_4$ Nanosorbent*

2.4.1. Regeneration Performance

The repeatability and regeneration of nanosorbents are quite important for their practical applications. Regeneration techniques suggested in the literature include chemical extraction, thermal treatment, supercritical regeneration, microwave irradiation, bio-regeneration, etc. [25,88,93–95]. The thermal regeneration, which has been applied for molybdenum oxide nanosorbent, was described in our previously published work [88]. In this investigation, the thermal treatment technique was examined for purposes of regeneration testing, as the structure of the α-$NiMoO_4$ removal agent was steady. The adsorbed MB was completely oxidized and decomposed during the calcination process.

The results showed that α-$NiMoO_4$ could be re-stimulated through thermal treatment. Figure 11 indicates the reused performance of α-$NiMoO_4$ in the removal of MB in three cycles. As a matter of fact, the data show a decrease in dye removal from 99% to 73% with a decrease in removal capacity from 15,900 to 11,608mg/g. The maximum adsorption capacity obtained after four cycles of use (11,608 mg/g) was higher than that obtained by several nanosorbents [30,87–92]. The high level of removal efficiency showed that the adsorbent regeneration by way of calcination under atmospheric air at a temperature of 400 °C was extremely efficient and indicative of outstanding recycling capability.

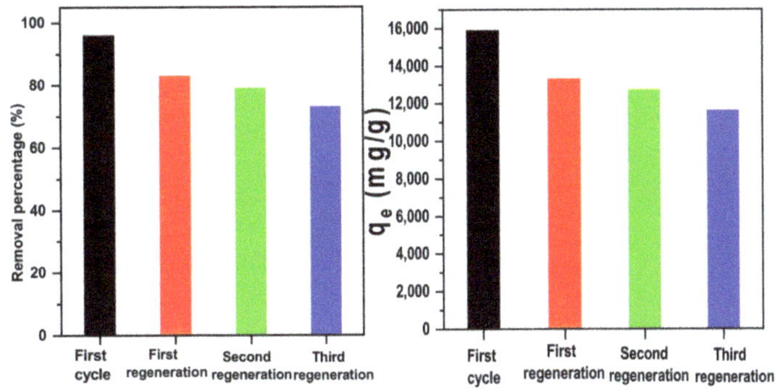

Figure 11. Recycled performance of α-$NiMoO_4$ in the removal of MB dye.

2.4.2. Fourier Transform Infrared Spectroscopy

To completely recognize the method by which α-$NiMoO_4$ nanosorbent removes MB dye, the components subjected to MB dye were examined by FT-IR spectroscopy. Figure 12 shows the FTIR spectra for the α-$NiMoO_4$ sample before and after the removal of methylene blue dye. As observed, the characteristics of flexing and stretching vibrations of the metal–oxygen bonds at 966 and 930 cm^{-1} and the broad, centered bonds at 650 cm^{-1} correspond to nickel molybdate [96]. The FTIR spectrum of the pure methylene blue displayed bands between 1700 and 1000 cm^{-1} [97]. The FTIR spectrum of $NiMoO_4$ after adsorption of methylene blue ($NiMoO_4$-MB) displayed further bands located at 1600 cm^{-1}, related to the C=C stretching of methylene blue, because of the presence of the methylene blue attached to the active sites of $NiMoO_4$ [98]. The FTIR spectrum of the regenerated $NiMoO_4$ ($NiMoO_4$-MB-Reg) after thermal treatment and the FTIR spectrum of fresh $NiMoO_4$ were alike, indicating thorough combustion of the attached methylene blue on the surface, and the resulting spectrum showed the cleanness and performance of the regenerated adsorbent.

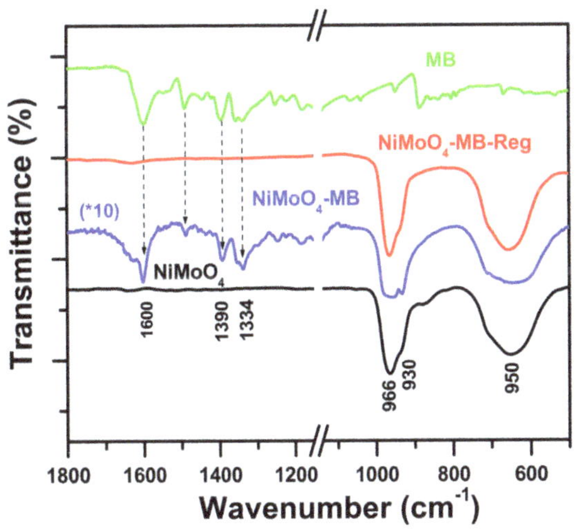

Figure 12. Fourier transform infrared spectra of $NiMoO_4$, $NiMoO_4$-MB, $NiMoO_4$-MB-Reg, and MB.

2.5. *Removal Mechanism of MB*

The removal of MB by α-$NiMoO_4$ nanoparticles was discovered to be due to the adsorption mechanism. Moreover, FTIR spectroscopy revealed that the removed methylene blue cations were triggered by the adsorption method without any intermediate compounds produced due to the absence of MB decomposition. Additionally, by using α-$NiMoO_4$ nanoparticles, the effectiveness of MB dye removal increased with the increase in pH up to pH 11, and this may be credited to its basic media. A reasonable mechanism can be proposed (Figure 13) on the basis of these findings. Furthermore, the positive charge of the MB dye is sustained in the first step at pH 11 since the pKa is equal to 3.8 [69]. Additionally, α-$NiMoO_4$ reacts with the hydroxyl groups (OH^-) in the solution to generate the ion nickel molybdate ($NiMoO_5^{2-}$) with no intermediate compounds present [99]. Hence, the adsorption is directed by the strong electrostatic interactions between the negatively charged surface of nickel molybdate ($NiMoO_5^{2-}$) and the positive charge of methylene blue cations [88].

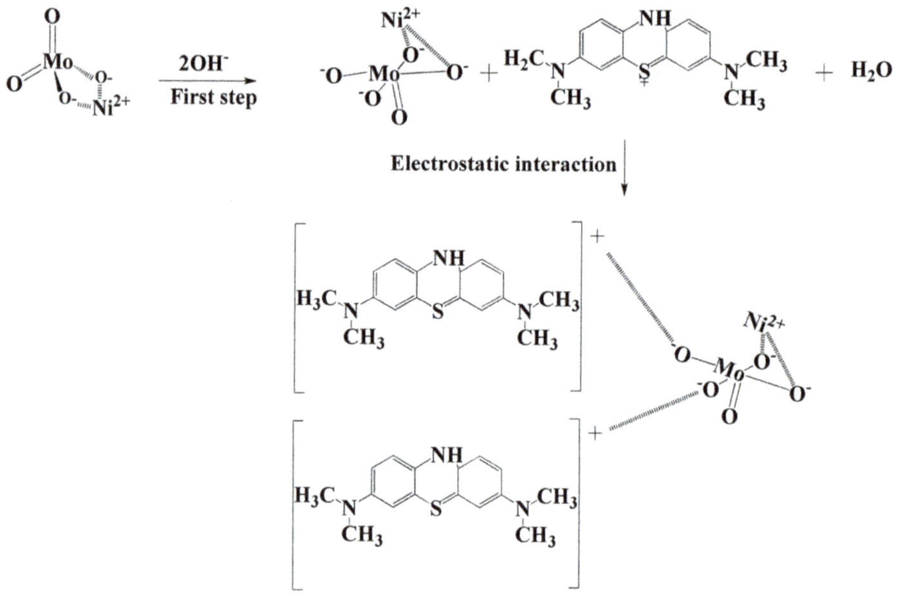

Figure 13. Schematic mechanism of the methylene blue dye removal by α-$NiMoO_4$ nanosorbent.

It is important to adhere to the progression of the α-$NiMoO_4$ morphology at different phases of the adsorption study. The SEM micrograph in Figure 14A gives an idea about how the particles form aggregates, showing a good porosity that can allow the improved adsorption of the dye. Nevertheless, the micrographs in Figure 14B,D,F show much less porous powder after the adsorption experiments; the methylene blue molecules filled the pores present in the starting samples. Figure 14C,E,G shows that the morphology of the sample did not change after the regeneration or the first and second reuses. In all cases, the particles were less agglomerated, displaying exceptionally porous powder. The morphology of α-$NiMoO_4$ was not significantly altered, even after the second and third reuses (Figure 14E,G).

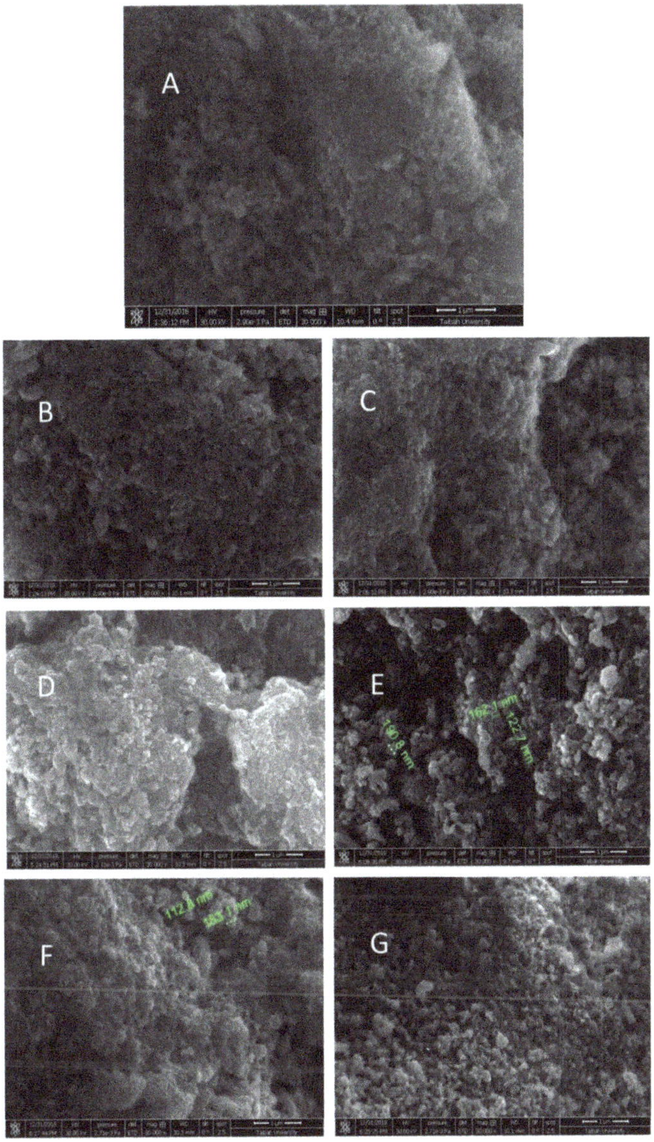

Figure 14. SEM micrographs: (**A**) the starting material, pure nickel molybdate (α-$NiMoO_4$); (**B**) the material after the MB dye had been removed; (**C**) the regenerated α-$NiMoO_4$; (**D**) the material after the second regeneration and/or removal cycle of the methylene blue dye; (**E**) the morphology of α-$NiMoO_4$ after the second regeneration process; (**F**) the material after the third regeneration/removal cycle of methylene blue dye; (**G**) the morphology of α-$NiMoO_4$ after the third regeneration.

3. Materials and Methods

3.1. Nickel Molybdate Nanosorbent Preparation

All synthetic compounds aside from the methylene blue (provided by Panreac, Barcelona, Spain) were purchased from Sigma-Aldrich (St. Louis, MO, USA) and utilized as received with no alterations.

Nickel molybdate ($NiMoO_4$) was formed by thermal breakdown of a nickel molybdenum complex obtained from the reaction of oxalic acid dihydrate $H_2C_2O_4{\cdot}2H_2O$, nickel nitrate $Ni(NO_3)_2{\cdot}6H_2O$, and ammonium molybdate $(NH_4)_6Mo_7O_{24}{\cdot}4H_2O$ in its solid form, as defined previously in the literature [32]. Nickel nitrate, oxalic acid dihydrate, and ammonium molybdate were blended jointly together in a molar proportion of 1/10/0.143. The mixture was powdered homogeneously and placed on a hot plate at a temperature of 160 °C for heating. The obtained nickel molybdenum complex was then decomposed under the control with static air at 500 °C for 2 h inside a cylindrical furnace, which was open at the two ends.

3.2. *Adsorption Experiments*

Experimental adsorption batches were set up for the removal of the MB dye [75]. The elimination of MB dye by $NiMoO_4$ was undertaken by the constant stirring of a precise quantity of adsorbent into a 100 mL MB dye solution with known concentrations and varied temperatures (such as T = 20, 50, and 70 °C) with various contact times (such as 10, 30, 60, 90, and 120 min). Toward the end of prearranged time intervals, before examination with a UV-Visible spectrometer, 0.22 µm syringe filters (Whatman) were used for filtration. Using 0.01 N NaOH or 0.01 N HCl, the pH of the methylene blue solution was easily adjusted. The removed quantity and percentage (%) of methylene blue dye at its equilibrium q_e (mg/g) were calculated by the following equations:

$$\text{Removal }\% = \frac{C_0 - C_e}{C_0} \times 100 \quad (13)$$

$$q_e = \frac{(C_0 - C_e)}{M} \times V \quad (14)$$

where M stands for the mass of α-$NiMoO_4$ (g) added; V is the quantity of solution used (L); and C_e and C_0 are equilibrium and initial concentrations of MB (ppm), respectively. The results were tested three consecutive times, and the percentage uncertainty was found to be around 3%.

3.3. *Adsorbent Regeneration Method*

A solution of 160 ppm was used for regeneration experiments, and the removal equilibrium period was extended by 1 h. The fresh α-$NiMoO_4$ used was filtered and dried under a constant temperature of 100 °C and then calcined at 400 °C for 1 h under atmospheric air. The calcined α-$NiMoO_4$ was examined to assess the recycling objectives under conditions similar to those for the freshly used α-$NiMoO_4$. Once the first recycling test worked well, the restoration process was replicated for three consecutive cycles under consistent conditions.

3.4. *pH Point of Zero Charge (pH_{pzc}) Measure*

The pH point of zero charge (pHpzc) of the material was measured by the electrochemical method reported by Altenor et al. [100]. First, 50 mL of a 0.01 M NaCl solution was placed in a 100 mL beaker. Then, the pH was adjusted to successive initial values between 2 and 12 by using either NaOH or HCl (0.1 M), and 0.001 g of nickel molybdate was added to the solutions. After a contact time of 24 h, the final pH was measured and plotted against the initial pH.

3.5. *Characterization*

Analysis of XRD patterns (X-ray diffractometer 6000, Shimadzu, Tokyo, Japan, installed with $\lambda_{Cu\text{-}K\alpha}$ = 1.5406 Å and Ni filter) was conducted to characterize the phase composition of the synthesized α-$NiMoO_4$ nanosorbent, as presented in Figure 15.

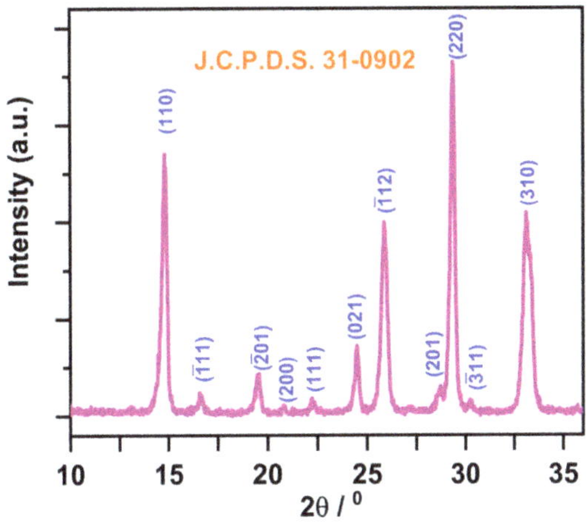

Figure 15. X-ray diffraction pattern of the synthesized $NiMoO_4$ nanoparticle powder. The Joint Committee on Powder Diffraction Standards (J.C.P.D.S) index file number is 31-0902.

The specific surface area characterization was completed by using nitrogen isotherm adsorption in the same way as stated in our previous research work [32]. The recorded specific surface area was 29.86 m^2/g.

FTIR spectroscopy in the range of 400 to 4000 cm^{-1} (IR Affinity-1S Shimadzu apparatus, Tokyo, Japan) using KBr pellets confirmed the existence of methylene blue dye on α-$NiMoO_4$ nanoparticles following the experimental adsorption and regeneration studies.

SEM analysis (Quanta Feg 250, Thermo Fisher Scientific, Hillsboro, OR, USA) was conducted. UV-Visible spectrophotometer (Thermo Scientific Genesys 10S, Madison, WI, USA) was used to determine the concentration at equilibrium.

4. Conclusions

α-$NiMoO_4$ nanosorbent was synthesized and investigated as a material for the removal of MB dye from aqueous solutions. Removal of MB was extremely reliant on the pH, and this resulted in the achievement of 99% removal efficiency for initial dye concentrations between 100 and 160 ppm at pH 11. The kinetic findings suggested that the removal of methylene blue followed the pseudo-second-order model, and the equilibrium adsorption results were best fitted with the Langmuir isotherm model. The highest removal amount achieved was 16,863 mg/g, as determined by the Langmuir model. Calcination at 400 °C was effectively sufficient to regenerate the adsorbent for further reuse. Even after three cycles of reusing the adsorbent, the MB removal efficiency of $NiMoO_4$ was still high. The data proved that α-$NiMoO_4$ could be an effective nanosorbent offering outstanding performance in removing MB dye even after being recycled.

Author Contributions: Conceptualization, S.R., H.O.H., and A.M.; Methodology, S.R., H.O.H., and A.M.; Validation, S.R., H.O.H., F.K., M.A., E.A., F.A.W. and A.M.; Formal Analysis, H.O.H., S.R., A.M., M.A., and F.K.; Investigation, S.R., H.O.H., A.M., F.K., M.A. and E.A.; Resources, H.O.H., M.A., F.K., F.A.W., E.A. and S.R.; Data Curation, H.O.H., S.R., A.M., M.A., F.K. and E.A.; Writing-Original Draft Preparation, S.R., H.O.H. and A.M.; Writing-Review & Editing, F.K., M.A, S.R., A.M. and H.O.H.; Visualization, S.R., H.O.H., F.K., A.M., M.A., E.A. and F.A.W.; Supervision, S.R., and H.O.H.; Project Administration, S.R. and H.O.H.; Funding Acquisition, E.A., F.A.W. and M.A. All authors have read and agreed to the published version of the manuscript.

Funding: This research received no external funding.

Data Availability Statement: The data that support the findings of this study are available from the corresponding author upon reasonable request.

Conflicts of Interest: The authors declare no conflict of interest.

Sample Availability: Samples of the compounds NiMoO4 are available from the corresponding author.

References

1. Khin, M.M.; Nair, A.S.; Babu, V.J.; Murugana, R.; Ramakrishna, S. A review onnanomaterials for environmental remediation. *Energy Environ. Sci.* **2012**, *5*, 8075–8109. [CrossRef]
2. Colmenares, J.C.; Luque, R. Heterogeneous photocatalytic nanomaterials: Prospects and challenges in selective transformations of biomass-derived compounds. *Chem. Soc. Rev.* **2014**, *43*, 765–778. [CrossRef]
3. Madrakian, T.; Afkhami, A.; Ahmadi, M.; Bagheri, H. Removal of some cationic dyes from aqueous solutions using magnetic modified multi-walled carbon nanotubes. *J. Hazard. Mater.* **2011**, *196*, 109–114. [CrossRef]
4. Yang, N.; Zhu, S.; Zhang, D.; Xu, S. Synthesis and properties of magnetic Fe_3O_4-activated carbon nanocomposite particles for dye removal. *Mater. Lett.* **2008**, *62*, 645–647. [CrossRef]
5. Sucharita, A. Textile Dyes: Its Impact on Environment and its Treatment. *J. Bioremediat. Biodegrad.* **2014**, *5*, 1.
6. Solis, M.; Solis, A.; Perez, H.I.; Manjarrez, N.; Flores, M. Microbial decolouration of azo dyes: A review. *Process Biochem.* **2012**, *47*, 1723–1748. [CrossRef]
7. Elemen, S.; Kumbasar, E.P.A.; Yapar, S. Modeling the adsorption of textile dye on organoclay using an artificial neural network. *Dye. Pigment.* **2012**, *95*, 102–111. [CrossRef]
8. Greluk, M.; Hubicki, Z. Effect of basicity of anion exchangers and number and positions of sulfonic groups of acid dyes on dyes adsorption on macroporous anion exchangers with styrenic polymer matrix. *Chem. Eng. J.* **2013**, *215–216*, 731–739. [CrossRef]
9. Turgay, O.; Ersoz, G.; Atalay, S.; Forss, J.; Welander, U. The treatment of azo dyes found in textile industry wastewater by anaerobic biological method and chemical oxidation. *Sep. Purif. Technol.* **2011**, *79*, 26–33. [CrossRef]
10. Verma, A.K.; Dash, R.R.; Bhunia, P. A review on chemical coagulation/flocculation technologies for removal of colour from textile wastewaters. *J. Environ. Manag.* **2012**, *93*, 154–168. [CrossRef]
11. Kanagaraj, J.; Senthilvelan, T.; Panda, R.C. Degradation of azo dyes by laccase: Biological method to reduce pollution load in dye wastewater. *Clean Technol. Environ. Policy* **2015**, *17*, 1443–1456. [CrossRef]
12. Cornelia, P.; Oana, P.; Robert, I.; Simona, G.M. Effective removal of methylene blue from aqueous solution using a new magnetic iron oxide nanosorbent prepared by combustion synthesis. *Clean Technol. Environ. Policy* **2016**, *18*, 705–715.
13. Vanhulle, S.; Trovaslet, M.; Enaud, E.; Lucas, M.; Taghavi, S.; van der Lelie, D.; van Aken, B.; Foret, M.; Onderwater, R.C.A.; Wesenberg, D.; et al. Decolorization, cytotoxicity and genotoxicity reduction during a combined ozonation/fungal treatment of dye-contaminated wastewater. *Environ. Sci. Technol.* **2008**, *42*, 584–589. [CrossRef]
14. Forgacs, E.; Cserhati, T.; Oros, G. Removal of synthetic dyes from wastewaters: A review. *Environ. Int.* **2004**, *30*, 953–971. [CrossRef]
15. Filice, S.; Angelol, D.D.; Libertinol, S.; Kosma, V.; Nicotera, I.; Privitera, V.; Scalese, S. Graphene oxide and titania hybrid Nation membranes for efficient removal of methyl orange dye from water. *Carbon* **2015**, *82*, 489–499. [CrossRef]
16. Kang, S.; Zhao, Y.; Wang, W.; Zhang, T.; Chen, T.; Yi, H.; Rao, F.; Song, S. Removal of methylene blue from water with montmorillonite nanosheets/chitosan hydrogels as adsorbent. *Appl. Surf. Sci.* **2018**, *448*, 203–211. [CrossRef]
17. Miyah, Y.; Lahrichi, A.; Idrissi, M.; Khalil, A.; Zerrouq, F. Adsorption of methylene blue dye from aqueous solutions onto walnut shells powder: Equilibrium and kinetic studies. *Surf. Interface* **2018**, *11*, 74–81. [CrossRef]
18. Ozdemir, U.; Ozbay, I.; Ozbay, B.; Veli, S. Application of economical models for dye removal from aqueous solutions: Cash flow, cost–benefit and alternative selection methods. *Clean Technol. Environ. Policy* **2014**, *16*, 423–429. [CrossRef]
19. Chen, Y.H. Synthesis, characterization and dye adsorption of ilmenite nanoparticles. *J. Non-Cryst. Solids* **2011**, *357*, 136–139. [CrossRef]
20. Sadhukhan, B.; Mondal, N.K.; Chattoraj, S. Biosorptive removal of cationic dye from aqueous system: A response surface methodological approach. *Clean Technol. Environ. Policy* **2014**, *16*, 1015–1025. [CrossRef]
21. George, Z.; Kyzas, J.F.; Kostas, A.M. The Change from Past to Future for Adsorbent Materials in Treatment of Dyeing Wastewaters. *Materials* **2013**, *6*, 5131–5158.
22. Mouni, L.; Belkhiri, L.; Bollinger, J.C.; Bouzaza, A.; Assadi, A.; Tirri, A.; Dahmoune, F.; Madani, K.; Remini, H. Removal of methylene blue from aqueous solutions by adsorption on Kaolin: Kinetic and equilibrium studies. *Appl. Clay Sci.* **2018**, *153*, 38–45. [CrossRef]
23. Bentahar, S.; Dbik, A.; El Khomri, M.; El Messaoudi, N.; Lacherai, A. Removal of a cationic dye from aqueous solution by natural clay. *Groundw. Sustain. Dev.* **2018**, *6*, 255–262. [CrossRef]
24. Low, S.K.; Tan, M.C. Dye adsorption characteristic of ultrasound pre-treated pomelo peel. *J. Environ. Chem. Eng.* **2018**, *6*, 3502–3509. [CrossRef]
25. Rakass, S.; Mohmoud, A.; Oudghiri-Hassani, H.; Abboudi, M.; Kooli, F.; Al Wadaani, F. Modified Nigella Sativa Seeds as a Novel Efficient Natural Adsorbent for Removal of Methylene Blue Dye. *Molecules* **2018**, *23*, 1950. [CrossRef] [PubMed]
26. Ghaedi, M.; Tavallali, H.; Sharifi, M.; NasiriKokhdan, S.; Asghari, A. Preparation of low cost activated carbon from Myrtuscommunis and pomegranate and their efficient application for removal of Congo red from aqueous solution. *Spectrochim. Acta Part A* **2012**, *86*, 107–114. [CrossRef] [PubMed]

27. Taghizadeh, F.; Ghaedi, M.; Kamali, K.; Sharifpour, E.; Sahraie, R.; Purkait, M.K. Comparison of nickel and/or zinc selenide nanoparticle loaded on activated carbon as efficient adsorbents for kinetic and equilibrium study of removal of Arsenazo (III) dye. *Powder Technol.* **2013**, *245*, 217–226. [CrossRef]
28. Singh, K.P.; Gupta, S.; Singh, A.K.; Sinha, S. Optimizing adsorption of crystal violet dye from water by magnetic nanocomposite using response surface modeling approach. *J. Hazard. Mater.* **2011**, *186*, 1462–1473. [CrossRef]
29. Yufei, Z.; Laiquan, L.; Haiquan, S.; Wei, H.; Xiaochen, D. Binary metal oxide: Advanced energy storage materials in supercapacitors. *J. Mater. Chem. A* **2015**, *3*, 43–59.
30. Oudghiri-Hassani, H.; Rakass, S.; Abboudi, M.; Mohmoud, A.; Al Wadaani, F. Preparation and Characterization of α-Zinc Molybdate Catalyst: Efficient Sorbent for Methylene Blue and Reduction of 3-Nitrophenol. *Molecules* **2018**, *23*, 1462. [CrossRef]
31. Oudghiri-Hassani, H. Synthesis, characterization and catalytic performance of iron molybdate Fe_2 $(MoO_4)_3$ nanoparticles. *Catal. Commun.* **2015**, *60*, 19–22. [CrossRef]
32. Oudghiri-Hassani, H.; Al Wadaani, F.T. Preparation, Characterization and Catalytic Activity of Nickel Molybdate ($NiMoO_4$) Nanoparticles. *Molecules* **2018**, *23*, 273. [CrossRef]
33. Al-Wadaani, F.; Omer, A.; Abboudi, M.; Oudghiri-Hassani, H.; Rakass, S.; Messali, M.; Benaissa, M. High Catalytic Efficiency of Nanostructured β-$CoMoO_4$ in the Reduction of the Ortho-, Meta- and Para-Nitrophenol Isomers. *Molecules* **2018**, *23*, 364. [CrossRef] [PubMed]
34. Madeley, R.A.; Wanke, S. Variation of the dispersion of active phases in commercial nickel—molybdenum/γ-alumina hydrotreating catalysts during oxidative regeneration. *Appl. Catal.* **1988**, *39*, 295–314. [CrossRef]
35. Gates, B.C.; Katzer, J.R.; Schuit, G.C.A. *Chemistry of Catalytic Processes*; McGraw-Hill: New York, NY, USA, 1979; p. 390.
36. Kaddouri, A.; Anouchinsky, R.; Mazzocchia, C.; Madeira, L.M.; Portela, M.F. Oxidative dehydrogenation of ethane on the α and β phases of $NiMoO_4$. *Catal. Today* **1998**, *40*, 201–206. [CrossRef]
37. Pillay, B.; Mathebula, M.R.; Friedrich, H.B. The oxidative dehydrogenation of *n*-hexane over Ni–Mo–O catalysts. *Appl. Catal. A* **2009**, *361*, 57–64. [CrossRef]
38. Rodriguez, J.A.; Chaturvedi, S.; Hanson, J.C.; Brito, J.L. Reaction of H_2 and H_2S with $CoMoO_4$ and $NiMoO_4$: TPR, XANES, Time-Resolved XRD, and Molecular-Orbital Studies. *J. Phys. Chem.* **1999**, *103*, 770–781. [CrossRef]
39. Sundaram, R.; Nagaraja, K.S. Solid state electrical conductivity and humidity sensing studies on metal molybdate–molybdenum trioxide composites (M = Ni^{2+}, Cu^{2+} and Pb^{2+}). *Sens. Actuators B Chem.* **2004**, *101*, 353–360. [CrossRef]
40. Mi, Y.; Huang, Z.; Hu, F.; Jiang, J.; Li, Y. Controlled synthesis and growth mechanism of alpha nickel molybatemicrohombohedron. *Mater. Lett.* **2010**, *64*, 695–697. [CrossRef]
41. Brito, J.L.; Barbosa, A.L.; Albornoz, A.; Severino, F. Nickel molybdate as precursor of HDS catalysts: Effect of phase composition. *Catal. Lett.* **1994**, *26*, 329–337. [CrossRef]
42. Ryu, J.H.; Koo, S.M.; Yoon, J.W.; Lim, C.S.; Shim, K.B. Synthesis of nanocrystalline $MMoO_4$ (M = Ni, Zn) phosphors via a citrate complex route assisted by microwave irradiation and their photoluminescence. *Mater. Lett.* **2006**, *60*, 1702–1705. [CrossRef]
43. Chen, Y.; Meng, F.; Ma, C.; Yang, Z.; Zhu, C.; Ouyang, Q.; Gao, P.; Li, J.; Sun, C. In situ diffusion growth of Fe_2 $(MoO_4)_3$ nanocrystalson the surface of α-MoO_3 nanorods with significantly enhanced ethanol sensing properties. *J. Mater. Chem.* **2012**, *22*, 12900–12906. [CrossRef]
44. Senthilkumar, B.; VijayaSankar, K.; Selvan, R.K.; Danielle, M.; Manickam, M. Nano α-$NiMoO_4$ as a new electrode for electrochemical supercapacitors. *RSC Adv.* **2013**, *3*, 352–357. [CrossRef]
45. Liu, M.; Kong, L.; Lu, C.; Li, X.; Luo, Y.; Kang, L. Waste paper based activated carbon monolith as electrode materials for high performance electric double-layer capacitors. *RSC Adv.* **2012**, *2*, 1890–1896. [CrossRef]
46. Liu, P.; Deng, Y.; Zhang, Q.; Hu, Z.; Xu, Z.; Liu, Y.; Yao, M.; Ai, Z. Facile synthesis and characterization of high-performance $NiMoO_4 \cdot xH_2O$ nanorods electrode material for supercapacitors. *Ionics* **2015**, *21*, 2797–2804. [CrossRef]
47. Cherian, C.T.; Reddy, M.V.; Haur, S.C.; Chowdari, B.V.R. Interconnected Network of $CoMoO_4$ Submicrometer Particles As High Capacity Anode Material for Lithium Ion Batteries. *ACS Appl. Mater. Interfaces* **2013**, *5*, 918–923. [CrossRef] [PubMed]
48. Ding, Y.; Yu, S.H.; Liu, C.; Zang, Z.A. 3D Architectures of Iron Molybdate: Phase Selective Synthesis, Growth Mechanism, and Magnetic Properties. *Chem. Eur. J.* **2007**, *13*, 746–753. [CrossRef]
49. Saberyan, K.; Soofivand, F.; Kianpour, G.; Salavati-Niasari, M.; Bagheri, S. Synthesis and characterization of $NiMoO_4$ via ultrasonic route by a novel precursor. *J. Mater. Sci. Mater. Electron.* **2016**, *27*, 3765–3772. [CrossRef]
50. Kianpour, G.; Salavati-Niasari, M.; Emadi, H. Sonochemical synthesis and characterization of $NiMoO_4$nanorods. *Ultrason. Sonochem.* **2013**, *20*, 418–424. [CrossRef]
51. Cai, D.; Wang, D.; Liu, B.; Wang, Y.; Liu, Y.; Wang, L.; Li, H.; Li, H.Q.; Wang, T. Comparison of the Electrochemical Performance of $NiMoO_4$ Nanorods and Hierarchical Nanospheres for Supercapacitor Applications. *ACS Appl. Mater. Interfaces* **2013**, *5*, 12905–12910. [CrossRef]
52. Ray, S.K.; Dipesh, D.; Yuwaraj, K.K.; Soo, W.L. Cu-α-$NiMoO_4$photocatalyst for degradation of Methylene blue with pathways and antibacterial performance. *J. Photochem. Photobiol. A* **2017**, *348*, 18–32. [CrossRef]
53. Kaddouri, A.; Tempesti, E.; Mazzocchia, C. Comparative study of β-nickel molybdate phase obtained by conventional precipitation and the sol-gel method. *Mater. Res. Bull.* **2004**, *39*, 695–706. [CrossRef]
54. Mosleh, M. Facile approach to synthesize nanocristalline $NiMoO_4$ in the presence of amino acids as capping agent. *J. Mater. Sci. Mater. Electron.* **2017**, *28*, 6788–6793. [CrossRef]

55. Klissurski, D.; Mancheva, M.; Iordanova, R.; Kunev, B. Mechanochemical Synthesis of Nanocrystalline Nickel Molybdates. *J. Alloys Compd.* **2006**, *422*, 53–57. [CrossRef]
56. Maione, A.; Devilers, M. Solid Solutions of Ni and Co Molybdates in Silica-Dispersed and Bulk Catalysts Prepared by Sol-Gel and Citrate Methods. *J. Solid State Chem.* **2004**, *177*, 2339–2349. [CrossRef]
57. Saravanakumar, B.; Ramachandran, S.P.; Ravi, G.; Ganesh, V.; Sakunthala, A.; Yuvakkumar, R. Morphology dependent electrochemical capacitor performance of $NiMoO_4$ nanoparticles. *Mater. Lett.* **2017**, *209*, 1–4. [CrossRef]
58. Azadeh, T.; Raheleh, P.; Mina, I.; Samaneh, E.; Mohammad, S. A simplified microwave-assisted synthesis of $NiMoO_4$ nanoparticles by using organic driving agent and study of photocatalytic activity. In Proceedings of the 18th International Electronic Conference on Synthetic Organic Chemistry Session Microwave Assisted Synthesis, 1–30 November 2014.
59. Edrissi, M.; Samadanian-Isfahani, S.; Soleymani, M. Preparation of cobalt molybdate nanoparticles; Taguchi optimization and photocatalytic oxidation of Reactive Black 8 dye. *Powder Technol.* **2013**, *249*, 378–385. [CrossRef]
60. Ghoreishiana, S.M.; Raju, G.S.R.; Pavitra, E.; Kwak, C.H.; Han, Y.K.; Huh, Y.S. Controlled synthesis of hierarchical α-nickel molybdate with enhanced solar-light-responsive photocatalytic activity: A comprehensive study on the kinetics and effect of operational factors. *Ceram. Int.* **2019**, *45*, 12041–12052. [CrossRef]
61. Ferreira, E.A.C.; Andrade Neto, N.F.; Bomio, M.R.D.; Motta, F.V. Influence of solution pH on forming silver molybdates obtained by sonochemical method and its application for methylene blue degradation. *Ceram. Int.* **2019**, *45*, 11448–11456. [CrossRef]
62. Rashad, M.M.; Ibrahim, A.A.; Rayan, D.A.; Sanad, M.M.S.; Helmy, I.M. Photo-Fenton-like degradation of Rhodamine B dye from waste water using iron molybdate catalyst under visible light irradiation. *Environ. Nanotechnol. Monit. Manag.* **2017**, *8*, 175–186. [CrossRef]
63. Oudghiri-Hassani, H. Synthesis, Characterization and Application of Chromium molybdate for Oxidation of Methylene Blue Dye. *J. Mater. Environ. Sci.* **2018**, *9*, 1051–1057.
64. Kianpour, G.; Soofivand, F.; Badiei, M.; Salavati-Niasari, M.; Hamadanian, M. Facile synthesis and characterization of nickel molybdatenanorods as an effective photocatalyst by co-precipitation method. *J. Mater. Sci. Mater. Electron.* **2016**, *27*, 10244–10251. [CrossRef]
65. Dhanasekar, M.; Ratha, S.; Rout, C.S.; Bhat, S.V. Efficient sono-photocatalytic degradation of methylene blue using nickel molybdate nanosheets under diffused sunlight. *J. Environ. Chem. Eng.* **2017**, *5*, 2997–3004. [CrossRef]
66. Abia, A.A.; Asuquo, E.D. Lead (II) and nickel (II) adsorption kinetics from aqueous metal solutions using chemically modified and unmodified agricultural adsorbents. *Afr. J. Biotechnol.* **2006**, *5*, 1475–1482.
67. Yogesh Kumar, K.; Archana, S.; Vinuth Raj, T.N.; Prasana, B.P.; Raghu, M.S.; Muralidhara, H.B. Superb Adsorption Capacity of Hydrothermally Synthesized Copper oxide and Nickel oxide nanoflakes towards Anionic and Cationic dyes. *J. Sci. Adv. Mater. Dev.* **2017**, *2*, 183–191. [CrossRef]
68. Alver, E.; Metin, A.U. Anionic dye removal from aqueous solutions using modified zeolite: Adsorption kinetics and isotherm studies. *Chem. Eng. J.* **2012**, *200–202*, 59–67. [CrossRef]
69. Jihyun, R.K.; Santiano, B.; Kim, H.; Kan, E. Heterogeneous Oxidation of Methylene Blue with Surface-Modified Iron-Amended Activated Carbon. *Am. J. Anal. Chem.* **2013**, *4*, 115–122.
70. Kooli, F.; Liu, Y.; Abboudi, M.; Oudghiri Hassani, H.; Rakass, S.; Ibrahim, S.M.; Al-Wadaani, F. Waste bricks applied as removal agent of Basic Blue 41 from aqueous solution: Base treatment and their regeneration efficiency. *Appl. Sci.* **2019**, *9*, 1237. [CrossRef]
71. Santhi, T.; Manonmani, S. Adsorption of methylene blue from aqueous solution onto a waste aquacultural shell powders (prawn waste). *Sustain. Environ. Res.* **2012**, *22*, 45–51.
72. Kooli, F.; Liu, Y.; Hbaieb, K.; Ching, O.Y.; Al-Faze, R. Characterization of organo-kenyaites: Thermal stability and their effects on eosin removal characteristics. *Clay Miner.* **2018**, *53*, 91–104. [CrossRef]
73. Alzaydien, A.S. Adsorption Behavior of Methyl Orange onto Wheat Bran: Role of Surface and pH. *Orient. J. Chem.* **2015**, *31*, 643–651. [CrossRef]
74. Garg, V.K.; Gupta, R.; Yadav, A.B.; Kumar, R. Dye removal from aqueous solution by adsorption on treated sawdust. *Bioresour. Technol.* **2003**, *89*, 121–124. [CrossRef]
75. Mahmoud, D.K.; Salleh, M.A.M.; Karim, W.A.W.A.; Idris, A.; Abidin, Z.Z. Batch adsorption of basic dye using acid treated kenaf fibre char: Equilibrium, kinetic and thermodynamic studies. *Chem. Eng. J.* **2012**, *181–182*, 449–457. [CrossRef]
76. Wawrzkiewicz, M.; Hubicki, Z. Removal of tartrazine from aqueous solutions by strongly basic polystyrene anion exchange resins. *J. Hazard. Mater.* **2009**, *164*, 502–509. [CrossRef] [PubMed]
77. Furusawa, T.; Smith, J.M. Intraparticle mass transport in slurries by dynamic adsorption studies. *AIChE J.* **1974**, *20*, 88–93. [CrossRef]
78. Foo, K.Y.; Hameed, B.H. Insights into the modeling of adsorption isotherm systems. *Chem. Eng. J.* **2010**, *156*, 2–10. [CrossRef]
79. Ma, J.; Jia, Y.; Jing, Y.; Yao, Y.; Sun, J. Kinetics and thermodynamics of methylene blue adsorption by cobalt-hectorite composite. *Dye. Pigment.* **2012**, *93*, 1441–1446. [CrossRef]
80. Chenglong, X.; Yan, J.; Yongzhong, J.; Duyuan, Y.; Jun, M.; Xiaojie, Y. Adsorption properties of congo red from aqueous solution on modified hectorite: Kinetic and thermodynamic studies. *Desalination* **2011**, *265*, 81–87.
81. Vadivelan, V.; Kumar, K.V. Equilibrium, kinetics, mechanism and process design for the sorption of methylene blue onto rice husk. *J. Colloid Interface Sci.* **2005**, *286*, 90–100. [CrossRef]

82. Febrianto, J.; Kosasih, A.N.; Sunarso, J.; Ju, Y.; Indraswati, N.; Ismadji, S. Equilibrium and kinetic studies in adsorption of heavy metals using biosorbent: A summary of recent studies. *J. Hazard. Mater.* **2009**, *162*, 616–645. [CrossRef] [PubMed]
83. Ho, Y.S.; McKay, G. Pseudo-second order model for sorption processes. *Process Biochem.* **1999**, *34*, 451–465. [CrossRef]
84. Shahwan, T.; Erten, H.N. Temperature effects in barium sorption on natural kaolinite and chlorite-illite clays. *J. Radioanal. Nucl. Chem.* **2004**, *260*, 43–48. [CrossRef]
85. Langmuir, I. The adsorption of gases on plane surfaces of glass, mica and platinum. *J. Am. Chem. Soc.* **1918**, *40*, 1361–1403. [CrossRef]
86. Dada, A.O.; Olalekan, A.P.; Olatunya, A.M.; Dada, O. Langmuir, Freundlich, Temkin and Dubinin–Radushkevich Isotherms Studies of Equilibrium Sorption of Zn^{2+} Unto Phosphoric Acid Modified Rice Husk. *J. Appl. Chem.* **2012**, *3*, 38–45.
87. Mohmoud, A.; Rakass, S.; Oudghiri-Hassani, H.; Kooli, F.; Abboudi, M.; Ben Aouan, S. Iron Molybdate $Fe_2 (MoO_4)_3$ Nanoparticles: Efficient Sorbent for Methylene Blue Dye Removal from Aqueous Solutions. *Molecules* **2020**, *25*, 5100. [CrossRef]
88. Rakass, S.; Oudghiri-Hassani, H.; Abboudi, M.; Kooli, F.; Mohmoud, A.; Aljuhani, A.; Al Wadaani, F. Molybdenum Trioxide: Efficient Nanosorbent for Removal of Methylene Blue Dye from Aqueous Solutions. *Molecules* **2018**, *23*, 2295. [CrossRef] [PubMed]
89. Wu, T.; Cai, X.; Tan, S.; Li, H.; Liu, J.; Yang, W. Adsorption characteristics of acrylonitrile, p-toluenesulfonic acid, 1-naphthalenesulfonic acid and methyl blue on graphene in aqueous solutions. *Chem. Eng. J.* **2011**, *173*, 144–149. [CrossRef]
90. Fan, L.; Zhang, Y.; Luo, C.; Lu, F.; Qiu, H.; Sun, M. Synthesis and characterization of magnetic beta-cyclodextrin-chitosan nanoparticles as nano-adsorbents for removal of methyl blue. *Int. J. Biol. Macromol.* **2012**, *50*, 444–450. [CrossRef]
91. Zhang, F.; Lan, J.; Yang, Y.; Wei, T.; Tan, R.; Song, W. Adsorption behavior and mechanism of methyl blue on zinc oxide nanoparticles. *J. Nanopart. Res.* **2013**, *15*, 2034–2043. [CrossRef]
92. Li, L.H.; Xiao, J.; Liu, P.; Yang, G.W. Super adsorption capability from amorphousization of metal oxide nanoparticles for dye removal. *Sci. Rep.* **2015**, *5*, 9028. [CrossRef]
93. Maisa El Gamal, M.; Mousa, H.A.; El-Naas, M.H.; Zacharia, R.; Judd, S. Bio-regeneration of activated carbon: A comprehensive review. *Sep. Purif. Technol.* **2018**, *197*, 345–359. [CrossRef]
94. Yanlong Sun, Y.; Zhang, B.; Zheng, T.; Wang, P. Regeneration of activated carbon saturated with chloramphenicol by microwave and ultraviolet irradiation. *Chem. Eng. J.* **2017**, *320*, 264–270.
95. Erfan Sadatshojaei, E.; Esmaeilzadeh, F.; Fathikaljahi, J.; Barzi, S.E.H.S.; Wood, D.A. Regeneration of the Midrex Reformer Catalysts Using Supercritical Carbon Dioxide. *Chem. Eng. J.* **2018**, *343*, 748–758. [CrossRef]
96. Umapathy, V.; Neeraja, P.; Manikandan, A.; Ramu, P. Synthesis of $NiMoO_4$ nanoparticles by sol–gel method and their structural, morphological, optical, magnetic and photocatlytic properties. *Trans. Nonferrous Met. Soc. China* **2017**, *27*, 1785–1793. [CrossRef]
97. Ahmed, F.; Dewani, R.; Pervez, M.K.; Mahboob, S.J.; Soomro, S.A. Non-destructive FT-IR analysis of mono azo dyes. *Bulg. Chem. Commun.* **2016**, *48*, 71–77.
98. Etman, A.S.; Abdelhamid, H.N.; Yuan, Y.; Wang, L.; Zou, X.; Sun, J. Facile Water-Based Strategy for Synthesizing MoO3–x Nanosheets: Efficient Visible Light Photocatalysts for Dye Degradation. *ACS Omega* **2018**, *3*, 2201–2209. [CrossRef]
99. Aracena, A.; Sannino, A.; Jerez, O. Dissolution kinetics of molybdite in KOH media at different temperatures. *Trans. Nonferrous Met. Soc. China* **2018**, *28*, 177–185. [CrossRef]
100. Altenor, S.; Carene, B.; Emmanuel, E.; Lambert, J.; Ehrhardt, J.J.; Gaspard, S. Adsorption studies of methylene blue and phenol onto vetiver roots activated carbon prepared by chemical activation. *J. Hazard. Mater.* **2009**, *165*, 1029–1039. [CrossRef] [PubMed]

molecules

Article

Silica Monolith for the Removal of Pollutants from Gas and Aqueous Phases

Vanessa Miglio [1,2], Chiara Zaccone [1,2], Chiara Vittoni [1,2], Ilaria Braschi [2,3,*], Enrico Buscaroli [3], Giovanni Golemme [4], Leonardo Marchese [1,2] and Chiara Bisio [1,2,5,*]

1 Department of Sciences and Technological Innovation and Interdisciplinary Nano-SiSTeMI Centre, University of Eastern Piedmont A. Avogadro, Viale T. Michel 11, 15121 Alessandria, Italy; vanessa.miglio@uniupo.it (V.M.); chiara.zaccone@uniupo.it (C.Z.); chiara.vittoni@uniupo.it (C.V.); leonardo.marchese@uniupo.it (L.M.)
2 Interdisciplinary Nano-SiSTeMI Centre, University of Eastern Piedmont A. Avogadro, Viale T. Michel 11, 15121 Alessandria, Italy
3 Department of Agricultural and Food Sciences, University of Bologna, Viale G. Fanin 44, 40127 Bologna, Italy; enrico.buscaroli2@unibo.it
4 Department of Environmental Engineering, University of Calabria, Via P. Bucci 45A, 87036 Rende, Italy; giovanni.golemme@unical.it
5 CNR-SCITEC Istituto di Scienze e Tecnologie Chimiche "Giulio Natta", Via G. Venezian 21, 20133 Milano, Italy
* Correspondence: ilaria.braschi@unibo.it (I.B.); chiara.bisio@uniupo.it (C.B.); Tel.: +39-051-209-6533 (I.B.); +39-0131-36-0216 (C.B.)

Abstract: This study focused on the application of mesoporous silica monoliths for the removal of organic pollutants. The physico-chemical textural and surface properties of the monoliths were investigated. The homogeneity of the textural properties along the entire length of the monoliths was assessed, as well as the reproducibility of the synthesis method. The adsorption properties of the monoliths for gaseous toluene, as a model of Volatile Organic Compounds (VOCs), were evaluated and compared to those of a reference meso-structured silica powder (MCM-41) of commercial origin. Silica monoliths adsorbed comparable amounts of toluene with respect to MCM-41, with better performances at low pressure. Finally, considering their potential application in water phase, the adsorption properties of monoliths toward Rhodamine B, selected as a model molecule of water soluble pollutants, were studied together with their stability in water. After 24 h of contact, the silica monoliths were able to adsorb up to the 70% of 1.5×10^{-2} mM Rhodamine B in water solution.

Keywords: adsorption; toluene; rhodamine B; water stability of monolith

check for updates

Citation: Miglio, V.; Zaccone, C.; Vittoni, C.; Braschi, I.; Buscaroli, E.; Golemme, G.; Marchese, L.; Bisio, C. Silica Monolith for the Removal of Pollutants from Gas and Aqueous Phases. *Molecules* **2021**, *26*, 1316. https://doi.org/10.3390/molecules26051316

Academic Editor: Teofil Jesionowski

Received: 25 January 2021
Accepted: 24 February 2021
Published: 1 March 2021

1. Introduction

In recent years, chemical industries have been focusing on sustainable development approaches, promoting products and services that offer performance at lower costs, reducing significantly the environmental impact and improving the quality of life.

In connection to these approaches, mesoporous materials have been synthesized in the form of powder and extensively studied for various applications (i.e., catalysis, adsorption, etc.). Nevertheless, on an industrial level, the use of powders for environmental applications has been severely hindered due to their handling and recycling limitations. Therefore, in recent years, research has been focusing on developing materials on a macroscopic scale, aiming at facilitating their handling, recovery and reuse in order to expand the range of their applications to different fields [1].

To overcome problems related to the use of powders there are two possible options: (i) use pre-synthesized silica powders to form pellets, or (ii) directly synthesize silica monoliths. In the first case, it is often necessary to use one or different binders (such as methylcellulose) and to press the materials even under heating. This multistep time-consuming procedure could adversely affect the structure of mesoporous silica, thus causing performance alteration [2]. In the second case, the formation of silica monolith can be a convenient way to fully exploit the structural and functional properties of the material [3] by saving, at the same time, both reactants and time, in that a single-step synthesis is required.

Several studies have been carried out in the recent literature to obtain monolithic siliceous structures. Galarneau et al. synthetized macroporous silica monoliths with disordered mesoporosity, prepared by the spinodal decomposition method [4]. Nakanishi and coworkers synthesized silica gel monoliths with macro-mesoporous hierarchical structure via a spontaneous sol–gel process from silicon alkoxide using a structure-directing agent and a micellar swelling agent [5]. Kohns et al. obtained silica monoliths with submicrometric macropores, introducing urea as an agent to control the size of macropores, mesopores and skeleton thickness [6]. Fotoohi et al. prepared mesoporous silica monoliths using a simple one-pot sol–gel synthesis with subsequent atmospheric evaporation [1]. Roucher et al. synthesized self-supported macro-mesoporous SBA-15-Si (HIPE) monoliths by combining emulsion and cooperative templating mechanisms [7].

Silica monoliths have been used in applications mainly related to the field of chromatography for the preparation of columns [8]. However, recent examples report on their use for the removal of chemical compounds such as fatty acids, phenols, and sterols from wastewater effluents: in these cases, they are mainly exploited as composite [9] or biocomposite materials [10] after the inclusion of biomolecules or enzymes. Few examples of the use of silica monolith also concern the preparation of composite [5] or functionalized materials for CO_2 adsorption [4,11]. Organic-modified silica monoliths are also used for the sequestration of heavy metals and for the immobilization of enzymes [12,13].

VOCs (Volatile Organic Compounds) are a group of noxious organic compounds characterized by vapor pressure higher than 0.01 kPa at 293.15 K. The group includes some of the most common air pollutants released by chemical, petrochemical and allied industries [14] such as aliphatic, aromatic and chlorinated hydrocarbons, aldehydes, terpenes, alcohols, esters and ketones. Due to their toxicity, the removal of these compounds by means of adsorbent materials [15] is of environmental concern and a relevant issue for human health. Several recent studies focused on the use of mesoporous ordered silicas, functionalized with organic groups, for the adsorption of VOCs [16]. The toluene adsorption properties of different porous silicas (i.e., fumed silica, SBA-15 and commercial MCM-41) have been recently reported by our group [16] and we have shown that the porous architecture of different silicas have an important effect on the final adsorption properties. In the same conditions, we also tested the adsorption behaviour of siliceous zeolites and hypercross-linked polymers [17]. We derived that whereas siliceous zeolite adsorbs 21 Q% of toluene, HyperCross-linked Polymers (HCP_S) with flexible structures are able to adsorb more than 140 Q% of toluene.

Soluble dyes, such as Rhodamine B, Methylene blue, Congo red, Methyl Orange, etc., have been increasingly used for applications in different fields (i.e., food, paper, textiles, paints, pharmaceuticals, cosmetics, etc.) [18]. Their occurrence in industrial wastewater is worrying because they are harmful for ecosystems as well as animals and humans [19]. Although there are many treatment methods for removing dyes from aqueous solutions (e.g., chemical oxidation, photodegradation, membrane filtration), adsorption has been found to be a high-efficient and low-cost technology [20] and several sorbent materials such as activated carbons [21], clays [22], zeolites [23], activated alumina [24], etc., have been tested for this purpose. Among these, ordered mesoporous silicas are promising candidates due to their high specific surface area, uniform pore size and high pore volume.

For example, Rasalingam et al. showed that mesoporous MCM-48 and MCM-41 are able to remove, respectively, 87% and 81% of Rhodamine B from an aqueous solution at the concentration of 10^{-5} M [25]. They found that, in the used conditions, the MCM-41 sample (having a specific surface area of 1453 $m^2 \cdot g^{-1}$) is able to retain 8.2×10^{-4} $mol \cdot g^{-1}$ of Rhodamine

B. Others used MCM-41 (specific surface area of 1300.5 $m^2 \cdot g^{-1}$) to remove Rhodamine B at 25 °C and pH = 4 and in these conditions the solids absorb ca. 9.4×10^{-4} $mol \cdot g^{-1}$ [26].

In this work, for the first time, mesoporous silica monoliths without any surface functionalization are proposed for the removal of organic pollutants from water and for the reduction of volatile organic compounds (VOCs). Model molecules (i.e., Rhodamine B and toluene, respectively) are used to this purpose. Our aim is to study the adsorption capacities of a silica-based monolith intended as a possible candidate to overcome limitations related to the use of mesoporous silica in powder form (i.e., handling and recycling issues). We focused on the optimization of a synthesis procedure for the direct formation, in a single work phase, of cylindrical mesoporous silica monoliths and to the determination of their physico-chemical properties by using different experimental techniques. The spatial textural homogeneity of the synthesized monoliths and the reproducibility of the synthesis were assessed. Moreover, the monoliths were tested for the removal of toluene, chosen as a model molecule of aromatic hydrocarbons, from gas phase. A combination of FT-IR spectroscopy and volumetric analysis was adopted to study the adsorption process and gain knowledge on the interactions between adsorbent surface and toluene molecule. Finally, the ability of the same monoliths to remove Rhodamine B from aqueous solution was also studied by using UV-Visible spectroscopy. Fundamental parameters of the monoliths for real water applications (i.e., mechanical and textural stability under operating conditions) [27] were verified as well.

2. Results and Discussion

2.1. Physico-Chemical Characterization of Silica Monoliths

The morphology of the silica monoliths, named Mono-ICE (see Section 3.1), was studied by using scanning electron microscopy (SEM). The obtained micrographs of samples before (A) and after the calcination step (B) to remove the polyethylene oxide template (PEO, see Section 3.1), are reported in Figure 1.

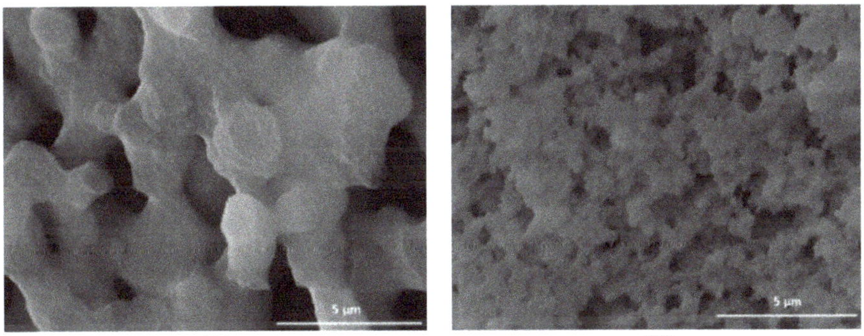

(A) (B)

Figure 1. SEM micrographs (20000×) of Mono-ICE before (**A**) and after calcination (**B**).

The Mono-ICE sample after the spinodal phase separation process (Figure 1A) is characterized by an irregular morphology, where the particle aggregation gives origin to interconnected macropores with a diameter of 4 µm and a skeleton thickness of 3 µm, on average, in accordance with what it has been reported in the reference article [4]. After the formation of the macroporous monoliths in acidic medium at low temperature, the ammonia treatment at 40 °C brings about the rearrangement of the weakly condensed silica network into denser nanodomains and the formation of larger mesopores, giving rise to the final hierarchical SBA-15 [4]. After the calcination of the organics (Figure 1B), the sample morphology appears deeply modified, with a strong reduction of macropore dimensions to 0.5–1 µm, on average.

The surface properties of the calcined silica monolith were then monitored by infrared spectroscopy (Figure S1 from Supplementary Materials). The FT-IR spectrum of Mono-ICE sample (Figure S1 from Supplementary Materials) is characterized by the presence of an intense band at 3743 cm^{-1}, due to the O-H stretching mode of isolated silanol groups present on the silica surface, and a broad band between 3700 and 3200 cm^{-1} with a maximum at 3521 cm^{-1} and a shoulder at ca. 3654 cm^{-1}, due to different surface silanol groups interacting with each other through hydrogen bonds. In the low frequency region, the spectrum shows three bands at 1983, 1863 and 1637 cm^{-1}, due to overtones and combination modes of the silica framework [28].

Textural properties of the monoliths were determined by N_2 adsorption–desorption isotherms at 77 K. In order to investigate whether the physico-chemical properties were homogeneous throughout the entire length of the cylindrical monoliths, several samples were singly divided into three parts of 1 cm in length each and of a weight of about 50 mg: the obtained samples were named Mono-ICE-Lateral A, Mono-ICE-Lateral B, Mono-ICE-Central C, respectively. Nitrogen adsorption and pore size distribution of the samples are shown in Figure 2A,B, respectively.

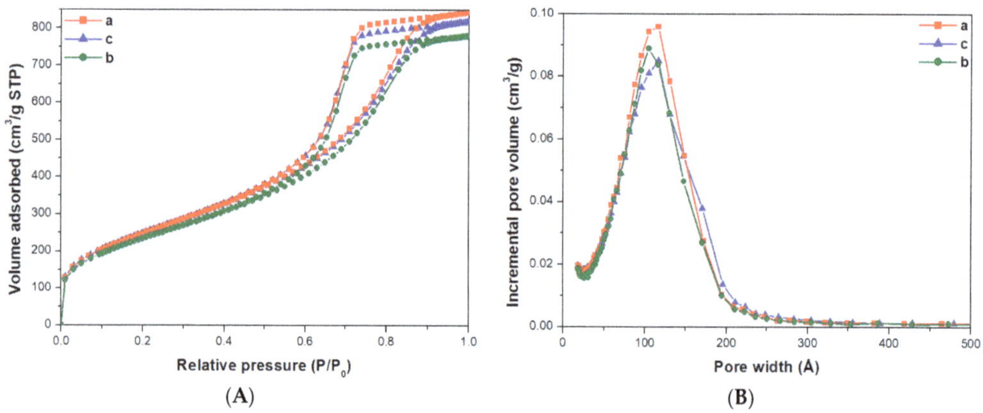

Figure 2. N_2 adsorption and desorption isotherms at 77K (**A**) and pores size distribution (**B**) of calcined Mono-ICE-Lateral A (a), Mono-ICE-Lateral B (b), and Mono-ICE-Central C (c).

Following the IUPAC classification, all the isotherms obtained (Figure 2A) are of type IVa, thus indicating a multilayer adsorption, which is typical of mesoporous solids. Hysteresis loops of type H2, due to disordered materials with a distribution of pore size and shape that is not well-defined, are found. All samples show a monomodal pore size distribution between 50 and 200 Å with maxima at around 110 Å (Figure 2B). The textural features of the three samples are similar, and this strongly suggests that the cylindrical monolith under investigation is structurally homogeneous along the entire length (Table 1).

To assess the reproducibility of the synthesis method, N_2 adsorption–desorption isotherms of monoliths from two different synthesis batches have been determined (Figure S2 from Supplementary Materials). The monoliths considered have similar textural properties, in terms of specific surface area and pore volume. In particular, Mono-ICE-1st repetition and Mono-ICE-2nd repetition have, on average, a surface area of 889 and 810 $m^2 \cdot g^{-1}$, respectively. By averaging the two values, the Mono-ICE monoliths have approximately a surface area of 850 $m^2 \cdot g^{-1}$, with a total pore volume of 1.2 $cm^3 \cdot g^{-1}$. Such values are quite close to those (700 $m^2 \cdot g^{-1}$, and pore size of 12 nm) reported by Galarneau et al. for the monoliths prepared with similar procedure [4].

2.2. Effect of the Water Treatment

The stability in water of the monolith samples (Mono-ICE-Lateral A, Mono-ICE-Lateral B and Mono-ICE-Central C) and the possible modifications on the textural properties after contacting with water were evaluated by N_2 physisorption analysis conducted at 77 K. To speed up the process, the three samples were soaked in warm water (T = 50 °C) for 36 h [29]. Samples will be hereafter named Mono-ICE-Lateral A-36h, Mono-ICE-Lateral B-36h and Mono-ICE-Central C-36h, respectively.

For the sake of brevity, only the results about Mono-ICE-Lateral A sample before and after the water treatment are shown in Figure 3. Similar results were obtained for Mono-ICE-Lateral B and Mono-ICE-Lateral C samples (Figure S3 from Supplementary Materials).

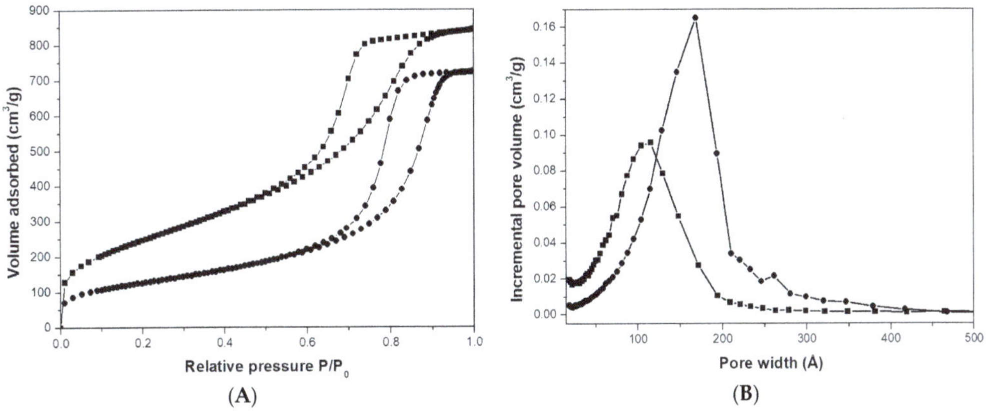

Figure 3. N_2 adsorption and desorption isotherms at 77 K (**A**) and pores size distribution (**B**) of calcined Mono-ICE-Lateral A (■) and Mono-ICE-Lateral A-36h (•).

The isotherms of the material before and after treatment in water at 50 °C show similar shapes, but at low pressure ($P/P_0 < 0.1$) the N_2 adsorption reduced to around 50% after the treatment in water (Figure 3A) because of the strong reduction in the amount of micropores (Figure 3B). The textural properties of the soaked material are indeed modified: (i) the surface area decreases from 909 to 455 $m^2 \cdot g^{-1}$ (Table 1); (ii) the pore size increases from 115 to 169 Å (Figure 3B); (iii) the total pore volume slightly decreases from 1.3 to 1.1 $cm^3 \cdot g^{-1}$ (Table 1). These changes are probably the effect of the hydrolysis and condensation of those parts of silica surrounding the micropores and characterized by a small radius of curvature, leaving a smoother surface inside the pores and enlarging the pore size [30]. Similar effects were also evidenced for other silica samples [29,31].

Table 1. Specific Surface Area and Total Pore Volume of Calcined Mono-ICE-Lateral A, Mono-ICE-Lateral B and Mono-ICE-Central C Samples, Before and After 36 h of Water Treatment.

Sample	SSA_{BET} [1] ($m^2 \cdot g^{-1}$)		V_P [2] ($cm^3 \cdot g^{-1}$)	
	(Before)	**(After)**	**(Before)**	**(After)**
Mono-ICE Lateral A	909	455	1.3	1.1
Mono-ICE Central C	906	465	1.3	1.1
Mono-ICE Lateral B	853	553	1.2	1.3

[1] Brunauer-Emmet-Teller (BET) specific surface area (SSA); [2] Total pore volume by Barrett, Joyner, and Halenda (BJH) method.

2.3. Toluene Adsorption from the Gas Phase

Toluene adsorption on silica monoliths was studied from qualitative and quantitative point of view by using FT-IR spectroscopy and volumetric analysis.

The adsorption has been carried out on the Mono-ICE central sample after calcination. The IR spectra obtained after the admission of 30 mbar of toluene on the sample and subsequent gradual decrease of the vapour pressure are reported in Figure 4A.

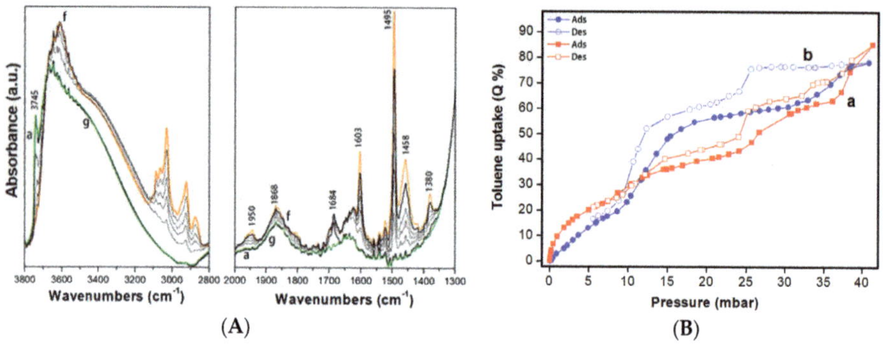

Figure 4. (**A**) FT-IR spectra of calcined Mono-ICE sample after outgassing at RT for 1 h (curve a, green), after dosages of 5, 10, 15, 20 mbar up to 30 mbar of toluene (curve f, orange), and after evacuation of toluene at RT for 30 min (curve g). (**B**) Toluene volumetric adsorption (full symbols) and desorption (empty symbols) isotherms of toluene at 35 °C on calcined Mono-ICE (a, ■), and MCM 41 (b, •) samples.

The admission of increasing toluene pressure on Mono-ICE silica samples (central part) (Figure 4A, curves b–f) results in a progressive disappearance of the band at 3745 cm^{-1}, due to isolated Si-OH species, and into the contemporary formation of a broad band centered at 3600 cm^{-1}. The progressive disappearance of the band related to isolated silanol species and to the formation of the band at lower frequencies is likely due to O–H$\cdots\pi$ interactions between silica silanols and toluene molecule [16].

Moreover, bands due to vibrations of the aromatic ring and the methyl group of toluene are also visible at lower wavenumbers. The vibrational modes of toluene adsorbed on the silica surface are described in Table 2.

Table 2. IR Bands Formed after the Adsorption of Toluene on Mono-ICE.

Frequency (cm^{-1})	Vibration Mode
3085, 3060, 3030	ν C-H aromatic ring
2922, 2871	ν_s and ν_{as} C-H toluene methyl group
1603	Quadrant stretching mode of mono-substituted ring C=C bond
1495	Semicircular stretching vibration mono-substituted aromatic ring
1460, 1377	Out-of-phase and in-phase deformations of methyl group

At higher pressure (above 5 mbar), the adsorption is likely driven by van der Waals interactions between the silica walls and the toluene molecules (host–guest interactions) and among toluene molecules (guest–guest interactions), as already observed elsewhere [16].

After decreasing the toluene pressure and subsequent evacuation at RT, the toluene has been fully desorbed: the band due to isolated SiOH species is completely restored, while the bands of toluene disappear (Figure 4A, curve g).

To gain information on the toluene uptake by the silica monolith, volumetric adsorption measurements at 35 °C were performed (Figure 4B). The isotherm of Mono-ICE sample (curve a) presents a complex shape where three regimes of adsorption are visible. In the adsorption branch, three different plateau are visible between: 5 and 10, 20 and 25, 30 and 35 mbar (with a toluene uptake of ca. 25, 40 and 50%, respectively). This interpretation is confirmed by the almost disappearing signal of the isolated silanol groups at 3745 cm^{-1} in the IR spectrum of Mono-ICE with toluene at 10 mbar (Figure 4A). On increasing the toluene pressure three adsorption steps are found, with a final toluene uptake of 85 Q%, or 1.00 $cm^3 \cdot g^{-1}$ of liquid toluene. The desorption isotherm forms three shallow hysteresis loops in the ranges 10–25, 25–35 and 35–40 mbar are due to a capillary condensation of the toluene molecules inside the pores. This complex behavior could be associated with the filling of the heterogeneous porosity of Mono-ICE sample. Although the calcined monolith showed a monomodal pore size distribution, pores with diameters ranging from 50 to 200 Å are visible, as described before (vide Figure 2).

The maximum toluene uptake is 85 Q% at 41 mbar, where

$$Q\% = \frac{m\ adsorbed\ toluene\ (\text{mg})}{100\ \text{mg}\ of\ sample} \tag{1}$$

The volumetric adsorption of toluene on silica monoliths has been compared to that of the reference mesoporous silica MCM-41 in the form of powder (Figure 4B, curve b) and already discussed in our recent publication [16]. For the sake of clarity, textural data of this sample are reported as supporting information (Table S1). Similarly to the Mono-ICE sample, the volumetric isotherm of toluene adsorbed on MCM-41 (Figure 4B, curve b) present three regimes of adsorption, however, only two hysteresis loops in the range 9–25 and 25–40 mbar, due to the toluene capillary condensation inside the pores, can be distinguished. The curve appears rapid until 9 mbar, and then the slope increases up to ca. 15 mbar, when the adsorption curve gradually tends to a first plateau at ca. 25 mbar with an uptake of ca. 59 Q%, which is associated with the filling of the fraction of mesopores with diameter lower than 40 Å (see pore size distribution, Figure S3B from Supplementary Materials). At pressures higher than 25 mbar, the slope increases again and progressively until 45 mbar, where the overall toluene uptake is ca. 78 Q%: the filling of the fraction of mesopores with dimensions from 40 to 80 Å likely occurs at this pressure.

It is here worth noticing that the overall toluene uptake for the two samples is similar, reaching 80–85 Q%, however the monolith adsorbs more at low pressure (<10 mbar). This effect should be correlated to the presence of micropores and narrow mesopores (ca. 20–30 Å) in Mono-ICE.

2.4. Rhodamine B Adsorption

The adsorption capacity of 1.5×10^{-2} mM Rhodamine B by the monoliths was studied by UV-Vis spectroscopy, at room temperature. In particular, the characteristic absorption band at 553 nm of Rhodamine B, due to the $\pi \rightarrow \pi^*$ transition of the chromophore unit [32], was observed after contact with the monolith (Figure 5) at increasing times (1 to 24 h).

Figure S6 from Supplementary Materials shows the adsorbed amount of Rhodamine B as a function of concentration of Rhodamine B solution over time.

Rhodamine B is a weak acid (pKa 4.2) with good solubility (34 $g \cdot L^{-1}$) in water. Its organic part is a cation in which the positive charge is shared by the two N atoms. At pH values larger than 4.2, the carboxylic group is predominantly deprotonated and the prevalent form of Rhodamine B is a Zwitterion (Figure S7 from Supplementary Materials). As reported in the literature, Rhodamine B molecules adsorb on the mesoporous silica materials through electrostatic, hydrogen, nonpolar and n–π bonding interaction. The effective adsorption sites on the silica surface are composed mainly of OH and/or oxygen bridges. At a pH between 5 and 6, Zeta potential of silica was negative, and the surfaces of the adsorbent was negatively charged. The positive moieties in the Rhodamine B zwitterion are attracted onto the surface silanolate groups present in silica materials through electrostatic forces. In addition, the residual surface hydroxyls of the mesoporous materials may also form H-bonds with the COO^- group present in the Rhodamine B zwitterion [25].

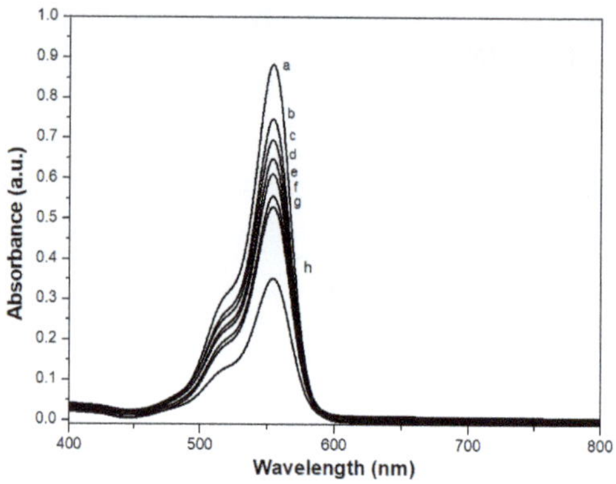

Figure 5. UV-Vis spectra, showing the intensity decrease of the maximum characteristic peak at 553 nm of 1.5×10^{-2} mM Rhodamine B in water solution (**a**), after 1 (**b**), 2 (**c**), 3 (**d**), 4 (**e**), 5 (**f**), 6 (**g**) and 24 (**h**) hours of contact with the calcined Mono-ICE, at room temperature.

It is reasonable to think that the Mono-ICE monolith, being entirely composed of silica, having a pH of 5.5 and a Zeta potential of −24 mV, behaves exactly in the same way.

Figure 6 shows the relative concentration of the Rhodamine B in aqueous solution after contact with Mono-ICE sample at room temperature, before and after it has undergone water treatment, over the time. The measurements have been replicated three times and average values and relative standard deviations are reported in the figure.

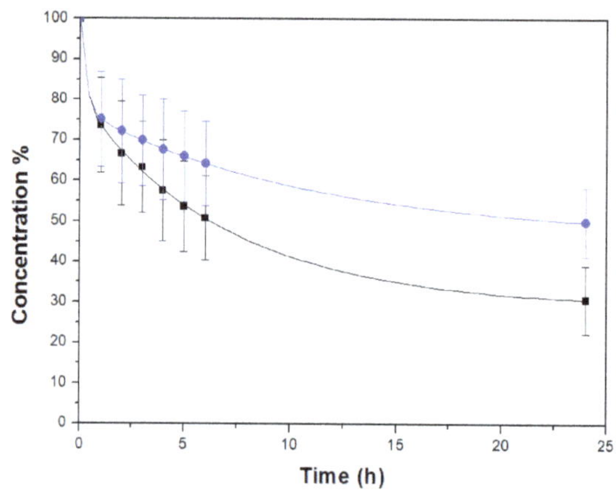

Figure 6. Concentration (%) decrease over time of 1.5×10^{-2} mM Rhodamine B water solution in the presence of calcined Mono-ICE before (■) and after water treatment (•). Error bars represent standard deviations calculated on averaging results collected from three replicated experiments.

The concentration of the Rhodamine B solution decreases over time: 70% of the dye is removed from the solution after 24 h of contact with Mono-ICE before the water treatment (Figure 6, black squares).

The adsorption of Rhodamine B on calcined silica monoliths was compared with that on MCM-41 silica powder, before and after water treatment (Figure S5 from Supplementary Materials). MCM-41 powder (before water treatment) absorbs 58% of Rhodamine B after the first hour of contact, and 92% after 24 h. Rhodamine B adsorbs more rapidly in MCM-41, compared to the Mono-ICE monolith, probably because it is a powder; in the Mono-ICE monolith Rhodamine B has to diffuse in the macropore system before it may reach the mesopores.

After 24 h of contact with Mono-ICE after the water treatment (Figure 6, black circles), the adsorbed amount of Rhodamine B solution decreased by 50%, due to the modification of textural properties (i.e., surface area decrease, the pore size increase, total pore volume decrease) already described (Figure S4B from Supplementary Materials and Table S1).

Similar changes of textural properties are reported for the MCM-41 powder after treatment in water at 50 °C for 36 h (Figure S4 from Supplementary Materials).

MCM-41 silica powder has better performance than the monolith due to the faster diffusion of the pollutant in the pores and having a larger specific surface area; however, the monolith shows excellent adsorption properties and, above all, since it is not synthesized in the form of powder, it is more usable from a technological point of view. As reported in the work of Rasalingam et al. the adsorption capacity of Rhodamine B on MCM-41 silica synthesized by them (having an SSA of 1453 $m^2 \cdot g^{-1}$) is 8.2×10^{-4} $mol \cdot g^{-1}$ [25]. The Mono-ICE monolith has an adsorption capacity of 1.07×10^{-6} $mol \cdot g^{-1}$. Moreover, unlike MCM-41 silica which is not stable in water, the monolith possesses perfect stability.

MCM-41 is adsorbing more Rhodamine B than Mono-ICE basically for two reasons: (i) the SSA of MCM-41 is larger; (ii) the reduced radius of curvature of the MCM-41 pores favours a stronger interaction with Rhodamine B. In addition, since MCM-41 has a larger tendency to dissolve in water than Mono-ICE because of its thinner walls [16,30], MCM-41 may develop a larger amount of silanol groups per unit surface.

Owing to its instability in water, together with the difficult handing of the powder, the advantages in the use of the MonoICE monolith become evident.

3. Materials and Methods

3.1. Materials

Silica monoliths were obtained by adapting the procedure described by Galarneau et al. [4]. In detail, a mixture of nitric acid (2.31 g, HNO_3 68 % w/w, Sigma-Aldrich 7697372, M.W. = 63.01 $g \cdot mol^{-1}$), polyethylene oxide-PEO (2.5 g, Sigma-Aldrich 25322683, M.W. > 20 kDa) and deionized water (24.5 g) was prepared and refrigerated at −19 °C for 1 h. Then, tetraethylorthosilicate-TEOS (20 g, $Si(OC_2H_5)_4$, Sigma-Aldrich 131903, M.W. = 208.33 $g \cdot mol^{-1}$), previously cooled at −19 °C for 1 h, was added to the mixture (molar ratios of the optimized final composition: 1 Si/0.60 PEO/0.26 HNO_3/14.21 H_2O).

Polyvinylchloride tubes (five of 8 mm in diameter and seven of 6 mm in diameter and 10 cm length) were then filled with the mixture and closed with a cap, paying particular attention to keep everything cold throughout the whole process by means of an ice bath. Following, the filled tubes were put in a vertical position, held in place by an appropriate support, in a 4 L water bath at 40 °C for 3 days.

The formed monoliths were then removed from the tubes and placed in a water bath at room temperature (RT). The water bath was regenerated every 30 min until neutral pH. The monoliths were then immersed in 1 L of 0.1 M ammonia aqueous solution (NH_4OH, Sigma-Aldrich 7664417, M.W. = 17.03 $g \cdot mol^{-1}$) in a Teflon bottle, and placed in an oven at 40 °C for 1 day, to catalyze the Ostwald ripening of the weakly condensed silica. Finally, the monoliths were recovered, further washed in a water bath until pH neutrality, dried at RT for 4 days and then calcined in O_2 flow at 550 °C for 6 h to remove the PEO still present. The obtained silica monoliths were named Mono-ICE (Figure 7).

A monolith was divided into 3 parts of 1 cm in length each, and weighed about 50 mg: 48.9 mg for Mono-ICE-Lateral A, 50.4 mg for Mono-ICE-Lateral B and 55.5 mg for Mono-ICE-Central C.

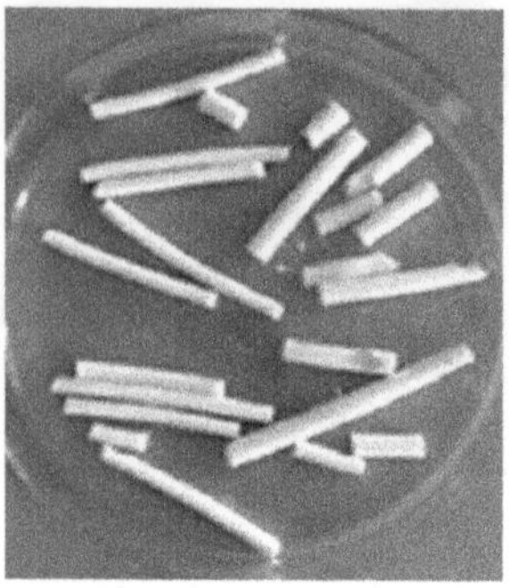

Figure 7. Mono-ICE samples.

3.2. *Stability in Water Treatments*

Considering that silica monoliths are soaked in water during their usage as pollutants removers from aqueous phase, the stability in water is as an important characteristic that the adsorbent must have.

To study the stability in water of silica monoliths (and the reference MCM41 silica powder), the solids (50 mg) were dispersed in 5 mL of water and the dispersion was heated at 50 °C for 36 h to speed possible degradation processes. Samples were then recovered and dried at 60 °C for 24 h.

3.3. *Characterization Techniques*

3.3.1. Scanning Electron Microscopy (SEM)

SEM images were acquired on a Quanta 200 scanning electron microscope (FEI, Eindhoven, The Netherlands) equipped with tungsten filament as electron source. Before the analysis, a conductive coating of gold was deposited on the monoliths by low-pressure plasma to avoid that insulating particles are electronically charged under the electron beam.

3.3.2. Infrared Spectra

Infrared spectra were collected by using a Thermo Electron Corporation (Whaltham, MA, USA) FT Nicolet 5700 spectrometer with 4 cm^{-1} resolution. Self-supporting pellets of Mono-ICE samples were obtained by grinding a piece of silica monolith and compressing the obtained powder with a mechanical press at ca. 7 tons·cm^{-2}. To perform the analysis, the obtained pellets were placed into an IR cell equipped with KBr windows permanently attached to a vacuum line (residual pressure $\leq 1 \times 10^{-3}$ mbar), allowing all treatments (and toluene adsorption/desorption experiments) to be carried out in situ. All the spectra were collected at beam temperature (ca. 35 °C) on samples previously dehydrated at RT for 30 min to completely remove the adsorbed water.

3.3.3. Nitrogen Adsorption Measurements

Nitrogen Adsorption Measurements were conducted at AlfatestLAB S.R.L (Cinisello Balsamo, Italy). Experiments were performed at 77 K in the pressure range between 0.01 and 1 P/P_0 of relative pressure using a 3Flex instrument (Micromeritics, Norcross, GA, USA). Prior to adsorption, the samples were outgassed and thermally treated as follows: 3 h at RT, 30 min at 50 °C, 30 min at 80 °C, 2 h at 110 °C and 3 h at 220 °C. The specific surface area of the samples was determined by the Brunauer–Emmett–Teller (BET) multipoint method in the range between 0.05 and 0.25 P/P_0. The pore size distribution was also calculated by applying the BJH (Barrett, Joyner, and Halenda) method on the adsorption branch (Thickness Curve: Halsey, Correction: Standard).

3.4. Toluene Adsorption

Volumetric Analysis. Toluene adsorption isotherms by gas phase were obtained at 35 °C by volumetric analysis of vapor sorption employing an Autosorb-iQ instrument (Quantachrome Instruments, Boynton Beach, FL, USA). Prior to adsorption, the samples were outgassed (final pressure 7×10^{-4} mbar) and thermally treated as follows: 30 min at 50 °C, 30 min at 80 °C, 2 h at 120 °C, 2 h at 150 °C and, finally, 12 h at 220 °C in order to remove completely adsorbed water.

3.5. Rhodamine B Adsorption

The Rhodamine B adsorption experiments were conducted by using a Lambda 900 UV-Visible spectrometer (Perkin Elmer, Waltham, MA, USA). Rhodamine B (CAS Number: 81-88-9, IUPAC name: [9-(2-carboxy phenyl)-6-diethylamino-3-xanthenylidene]-diethylammonium chloride, was purchased from Sigma Aldrich, St. Louis, MO, USA, (Analytical Standard). Before the experiment, a calibration line was obtained from five Rhodamine B standard solutions (2×10^{-2}, 5×10^{-3}, 2.5×10^{-3}, 1.25×10^{-3}, and 6×10^{-4} mM), resulting in the equation A = 85.702 × C, with R^2 = 0.99995.

For the contact tests, 317 mg of sample was placed in a sealable glass vial with 63.4 mL of Rhodamine B solution 1.46×10^{-2} mM (pH = 5.5). The bottles have been subsequently sealed with teflon-lined caps and placed on a mechanical stirrer (300 rpm) at RT. At different interval times (i.e., 1, 2, 3, 4, 5, 6, 24 h), an aliquot of each solution (ca. 1 mL) was then withdrawn and directly analyzed. Each experiment was repeated 3 times, every time using a fresh monolith.

Z potential of the monolith at the pH value of the Rhodamine B solution (pH = 5.5) was −24.6 mV.

4. Conclusions

In this work a synthesis procedure of mesoporous silica monoliths suitable for being used as adsorbents of organic pollutants in water (Rhodamine B), or in the gaseous phase (toluene, as a model of VOCs), was adopted.

Textural properties were found to be homogeneous over the entire length of the monolith, which has an average surface area of ca. 850 $m^2 \cdot g^{-1}$ and a total pore volume of 1.2 $cm^3 \cdot g^{-1}$. After a water treatment (36 h at 50 °C), the specific surface area remained approximately 500 $m^2 \cdot g^{-1}$, and the total pore volume 1.1 $cm^3 \cdot g^{-1}$.

From a comparison between Mono-ICE and MCM-41 silica powder, it was found that both samples adsorb comparable amounts of toluene, and that the silica monolith performs better at low pressure (<10 mbar).

Monoliths before and after the water treatment were tested as adsorbents for Rodamine B in aqueous solution. Although the water treatment reduced the specific surface area, the treated material was still able to adsorb 50% of the Rhodamine B with respect to 70% of the control sample.

The silica monolith prepared in this study shows slightly worse performances in the adsorption of Rhodamine B when compared to the silica MCM-41 powder, however it displays several advantages including greater ease in handling and recovering.

Supplementary Materials: The following are available online: Figure S1. FT-IR spectra of self-supported pellets of Mono-ICE calcined sample after treatment in vacuum at beam temperature (b.t.) for 30 min; Figure S2. Comparison between the first repetition (Frame A) and the second repetition (Frame B) of N2 adsorption and desorption isotherms of Mono-ICE-A (a), Mono-ICE-B (b) and Mono-ICE-C (c); Figure S3. N2 adsorption and desorption isotherms (Frame A) and pore size distribution (Frame B) of Mono-ICE-A-36h (a), Mono-ICE-B-36h (b) and Mono-ICE-C-36h (c). Figure S4. N_2 adsorption and desorption isotherms (Frame A) and pores size distribution (Frame B) of MCM-41 -before and after water treatment at 50 °C for 36 h; Figure S5. Concentration (%) decrease over time of 1.5×10^{-2} mM Rhodamine Rhodamine B in water solution in the presence of commercial MCM-41 powder. Table S1. Main textural features of MCM-41 silica.

Author Contributions: Conceptualization, V.M., C.Z. and C.B.; methodology, C.Z.; validation, C.V., and V.M.; investigation, V.M., C.V., C.Z.; resources, C.B.; data curation, V.M., C.V. and C.Z.; writing—original draft preparation, V.M., C.V., and C.Z.; writing—review and editing, C.B., E.B., I.B., G.G. and L.M.; supervision, C.B. and L.M.; project administration, C.B.; funding acquisition, C.B. and L.M. All authors have read and agreed to the published version of the manuscript.

Funding: This research was funded by the University of Bologna "2° *Bando "Proof of Concept d'Ateneo" (PoC UNIBO)—Finanziamento a supporto dello sviluppo di invenzioni brevettate" and by the Italian Ministry of the Economic Development "Bando per la realizzazione di programmi di valorizzazione dei brevetti tramite il finanziamento di progetti di Proof of Concept (PoC) delle Università italiane, degli Enti Pubblici di Ricerca (EPR) italiani e degli Istituti di ricovero e cura a carattere scientifico (IRCCS)."* Decreto direttoriale del 27 settembre 2019, registrato alla Corte dei Conti il 10 ottobre 2019, Reg. n. 1-968.

Data Availability Statement: Not available.

Conflicts of Interest: The authors declare no conflict of interest.

Sample Availability: Samples of the compounds are available from the authors.

References

1. Fotoohi, B.; Kazemzad, M.; Mercier, L. Additive-Free Synthesis of Robust Monolithic Mesoporous Silica Support Used in Catalysis. *Ceram. Int.* **2018**, *44*, 20199–20210. [CrossRef]
2. Michels, N.-L.; Mitchell, S.; Pérez-Ramírez, J. Effects of Binders on the Performance of Shaped Hierarchical MFI Zeolites in Methanol-to-Hydrocarbons. *ACS Catal.* **2014**, *4*, 2409–2417. [CrossRef]
3. Tomašić, V.; Jović, F. State-of-the-Art in the Monolithic Catalysts/Reactors. *Appl. Catal. A: Gen.* **2006**, *311*, 112–121. [CrossRef]
4. Galarneau, A.; Abid, Z.; Said, B.; Didi, Y.; Szymanska, K.; Jarzębski, A.; Tancret, F.; Hamaizi, H.; Bengueddach, A.; Di Renzo, F.; et al. Synthesis and Textural Characterization of Mesoporous and Meso-/Macroporous Silica Monoliths Obtained by Spinodal Decomposition. *Inorganics* **2016**, *4*, 9. [CrossRef]
5. Amatani, T.; Nakanishi, K.; Hirao, K.; Kodaira, T. Monolithic Periodic Mesoporous Silica with Well-Defined Macropores. *Chem. Mater.* **2005**, *17*, 2114–2119. [CrossRef]
6. Kohns, R.; Haas, C.P.; Höltzel, A.; Splith, C.; Enke, D.; Tallarek, U. Hierarchical Silica Monoliths with Submicron Macropores as Continuous-Flow Microreactors for Reaction Kinetic and Mechanistic Studies in Heterogeneous Catalysis. *React. Chem. Eng.* **2018**, *3*, 353–364. [CrossRef]
7. Roucher, A.; Bentaleb, A.; Laurichesse, E.; Dourges, M.-A.; Emo, M.; Schmitt, V.; Blin, J.-L.; Backov, R. First Macro-Mesocellular Silica SBA-15-Si(HIPE) Monoliths: Conditions for Obtaining Self-Standing Materials. *Chem. Mater.* **2018**, *30*, 864–873. [CrossRef]
8. Rodrigues, D.; Rocha-Santos, T.A.P.; Freitas, A.C.; Gomes, A.M.P.; Duarte, A.C. Strategies Based on Silica Monoliths for Removing Pollutants from Wastewater Effluents: A Review. *Sci. Total Environ.* **2013**, *461–462*, 126–138. [CrossRef] [PubMed]
9. El-Safty, S.A.; Shahat, A.; Rabiul Awual, M. Efficient Adsorbents of Nanoporous Aluminosilicate Monoliths for Organic Dyes from Aqueous Solution. *J. Colloid Interface Sci.* **2011**, *359*, 9–18. [CrossRef]
10. Duarte, K.; Justino, C.I.L.; Pereira, R.; Panteleitchouk, T.S.L.; Freitas, A.C.; Rocha-Santos, T.A.P.; Duarte, A.C. Removal of the Organic Content from a Bleached Kraft Pulp Mill Effluent by a Treatment with Silica-Alginate-Fungi Biocomposites. *J. Environ. Sci. Health Part A* **2013**, *48*, 166–172. [CrossRef]
11. Chen, C.; Yang, S.-T.; Ahn, W.-S.; Ryoo, R. Amine-Impregnated Silica Monolith with a Hierarchical Pore Structure: Enhancement of CO_2 Capture Capacity. *Chem. Commun.* **2009**, 3627. [CrossRef]
12. Hou, C.; Ghéczy, N.; Messmer, D.; Szymańska, K.; Adamcik, J.; Mezzenga, R.; Jarzębski, A.B.; Walde, P. Stable Immobilization of Enzymes in a Macro- and Mesoporous Silica Monolith. *ACS Omega* **2019**, *4*, 7795–7806. [CrossRef]
13. AlSuhaimi, A.O.; AlMohaimadi, K.M.; AlAlawi, B.N.; Ali, I. A Novel Porous Silica Monolith Functionalized with 5-Amino-1,10-Phenanthroline for SPE of Metal Ions in Groundwater Samples Prior to Their Analysis Using ICP-MS. *Anal. Methods* **2018**, *10*, 2337–2346. [CrossRef]
14. Hernández, M.A.; Velasco, J.A.; Asomoza, M.; Solís, S.; Rojas, F.; Lara, V.H. Adsorption of Benzene, Toluene, and *p* -Xylene on Microporous SiO_2. *Ind. Eng. Chem. Res.* **2004**, *43*, 1779–1787. [CrossRef]
15. Benkhedda, J.; Jaubert, J.-N.; Barth, D.; Perrin, L. Experimental and Modeled Results Describing the Adsorption of Toluene onto Activated Carbon. *J. Chem. Eng. Data* **2000**, *45*, 650–653. [CrossRef]
16. Vittoni, C.; Gatti, G.; Braschi, I.; Buscaroli, E.; Golemme, G.; Marchese, L.; Bisio, C. Toluene Adsorption by Mesoporous Silicas with Different Textural Properties: A Model Study for VOCs Retention and Water Remediation. *Materials* **2020**, *13*, 2690. [CrossRef] [PubMed]
17. Paul, G.; Begni, F.; Melicchio, A.; Golemme, G.; Bisio, C.; Marchi, D.; Cossi, M.; Marchese, L.; Gatti, G. Hyper-Cross-Linked Polymers for the Capture of Aromatic Volatile Compounds. *ACS Appl. Polym. Mater.* **2020**, *2*, 647–658. [CrossRef]
18. Chen, J.; Zhu, X. Magnetic Solid Phase Extraction Using Ionic Liquid-Coated Core–Shell Magnetic Nanoparticles Followed by High-Performance Liquid Chromatography for Determination of Rhodamine B in Food Samples. *Food Chem.* **2016**, *200*, 10–15. [CrossRef] [PubMed]

19. Dubey, S.K.; Yadav, R.; Chaturvedi, R.K.; Yadav, R.K.; Sharma, V.K.; Minhas, P.S. Contamination of Ground Water as a Consequence of Land Disposal of Dye Waste Mixed Sewage Effluents: A Case Study of Panipat District of Haryana, India. *Bull. Env. Contam. Toxicol.* **2010**, *85*, 295–300. [CrossRef]
20. Cheng, Z.-L.; Li, Y.; Liu, Z. Study on Adsorption of Rhodamine B onto Beta Zeolites by Tuning SiO_2/Al_2O_3 Ratio. *Ecotoxicol. Environ. Saf.* **2018**, *148*, 585–592. [CrossRef]
21. Nassar, M.M.; El-Geundi, M.S. Comparative Cost of Colour Removal from Textile Effluents Using Natural Adsorbents. *J. Chem. Technol. Biotechnol.* **2007**, *50*, 257–264. [CrossRef]
22. Chen, H.; Wang, A. Kinetic and Isothermal Studies of Lead Ion Adsorption onto Palygorskite Clay. *J. Colloid Interface Sci.* **2007**, *307*, 309–316. [CrossRef] [PubMed]
23. Wang, S.; Li, H.; Xie, S.; Liu, S.; Xu, L. Physical and Chemical Regeneration of Zeolitic Adsorbents for Dye Removal in Wastewater Treatment. *Chemosphere* **2006**, *65*, 82–87. [CrossRef]
24. Yeh, R.Y.; Liu, R.L.; Chiu, H.; Hung, Y. Comparative Study of Adsorption Capacity of Various Adsorbents for Treating Dye Wastewaters. *Int. J. Environ. Stud.* **1993**, *44*, 259–284. [CrossRef]
25. Rasalingam, S.; Peng, R.; Koodali, R.T. An Insight into the Adsorption and Photocatalytic Degradation of Rhodamine B in Periodic Mesoporous Materials. *Appl. Catal. B Environ.* **2015**, *174–175*, 49–59. [CrossRef]
26. Lee, C.-K.; Liu, S.-S.; Juang, L.-C.; Wang, C.-C.; Lin, K.-S.; Lyu, M.-D. Application of MCM-41 for Dyes Removal from Wastewater. *J. Hazard. Mater.* **2007**, *147*, 997–1005. [CrossRef]
27. Thiruvenkatachari, R.; Vigneswaran, S.; Naidu, R. Permeable Reactive Barrier for Groundwater Remediation. *J. Ind. Eng. Chem.* **2008**, *14*, 145–156. [CrossRef]
28. Vittoni, C.; Sacchetto, V.; Costenaro, D.; Mastroianni, S.; Hinsch, A.; Marchese, L.; Bisio, C. Gelation of Solvent-Free Electrolyte Using Siliceous Materials with Different Size and Porosity for Applications in Dye Sensitized Solar Cells. *Sol. Energy* **2016**, *124*, 101–113. [CrossRef]
29. Gatti, G.; Olivas Olivera, D.F.; Paul, G.; Bagatin, R.; Carati, A.; Tagliabue, M.; Bisio, C.; Marchese, L. On the Adsorption of Toluene on Amorphous Mesoporous Silicas with Tunable Sorption Characteristics. *Dalton Trans.* **2019**, *48*, 11781–11790. [CrossRef]
30. Galarneau, A.; Nader, M.; Guenneau, F.; Di Renzo, F.; Gedeon, A. Understanding the Stability in Water of Mesoporous SBA-15 and MCM-41. *J. Phys. Chem. C* **2007**, *111*, 8268–8277. [CrossRef]
31. Hahn, M.W.; Copeland, J.R.; van Pelt, A.H.; Sievers, C. Stability of Amorphous Silica-Alumina in Hot Liquid Water. *ChemSusChem* **2013**, *6*, 2304–2315. [CrossRef] [PubMed]
32. Olivero, F.; Carniato, F.; Bisio, C.; Marchese, L. Promotion of Förster Resonance Energy Transfer in a Saponite Clay Containing Luminescent Polyhedral Oligomeric Silsesquioxane and Rhodamine Dye. *Chem. Asian J.* **2014**, *9*, 158–165. [CrossRef] [PubMed]

molecules

Article

Volcanic Rock Materials for Defluoridation of Water in Fixed-Bed Column Systems

Wondwosen Sime Geleta [1,2], Esayas Alemayehu [3,4,*] and Bernd Lennartz [2,*]

1 School of Chemical Engineering, Jimma Institute of Technology, Jimma University, P.O. Box 378, Jimma, Oromia, Ethiopia; wondeto@gmail.com
2 Faculty of Agricultural and Environmental Sciences, University of Rostock, Justus-Von-Liebig-Weg 6, 18059 Rostock, Germany
3 Faculty of Civil and Environmental Engineering, Jimma Institute of Technology, Jimma University, P.O. Box 378, Jimma, Oromia, Ethiopia
4 Africa Center of Excellence for Water Management, Addis Ababa University, 1176 Addis Ababa, Ethiopia
* Correspondence: esayas16@yahoo.com (E.A.); bernd.lennartz@uni-rostock.de (B.L.); Tel.: +251-91-701-7002 (E.A.); +49-381-498-3180 (B.L.)

Abstract: Consumption of drinking water with a high concentration of fluoride (>1.5 mg/L) causes detrimental health problems and is a challenging issue in various regions around the globe. In this study, a continuous fixed-bed column adsorption system was employed for defluoridation of water using volcanic rocks, virgin pumice (VPum) and virgin scoria (VSco), as adsorbents. The XRD, SEM, FTIR, BET, XRF, ICP-OES, and pH Point of Zero Charges (pH_{PZC}) analysis were performed for both adsorbents to elucidate the adsorption mechanisms and the suitability for fluoride removal. The effects of particle size of adsorbents, solution pH, and flow rate on the adsorption performance of the column were assessed at room temperature, constant initial concentration, and bed depth. The maximum removal capacity of 110 mg/kg for VPum and 22 mg/kg for VSco were achieved at particle sizes of 0.075–0.425 mm and <0.075 mm, respectively, at a low solution pH (2.00) and flow rate (1.25 mL/min). The fluoride breakthrough occurred late and the treated water volume was higher at a low pH and flow rate for both adsorbents. The Thomas and Adams–Bohart models were utilized and fitted well with the experimental kinetic data and the entire breakthrough curves for both adsorbents. Overall, the results revealed that the developed column is effective in handling water containing excess fluoride. Additional testing of the adsorbents including regeneration options is, however, required to confirm that the defluoridation of groundwater employing volcanic rocks is a safe and sustainable method.

Keywords: adsorption; breakthrough curve; defluoridation; up-flow mode; volcanic rocks

Citation: Geleta, W.S.; Alemayehu, E.; Lennartz, B. Volcanic Rock Materials for Defluoridation of Water in Fixed-Bed Column Systems. *Molecules* **2021**, *26*, 977. https://doi.org/10.3390/molecules26040977

Academic Editor: Giorgio Vilardi
Received: 7 December 2020
Accepted: 8 February 2021
Published: 12 February 2021

1. Introduction

Credible evidence from scientific literature substantiates both beneficial and detrimental effects of fluoride on human health with only a narrow range between intake associated with these effects [1,2]. Consumptions of fluoride in low concentrations (<1.0 mg/L) is an essential micronutrient for the healthy development of bone and dental enamel [3]; however, it leads to the development of fluorosis if it is consumed beyond the permissible limit (>1.5 mg/L) [4].

In many parts of the world, groundwater sources are the single largest supply of drinking water. For many rift communities, it may be the only economically viable option for drinking water. In the Ethiopian rift valley, about 40% of deep and shallow wells are contaminated with up to 26 mg/L of fluoride [5,6]. The weathering of primary rocks and leaching of fluoride-containing minerals in soils yield fluoride-rich groundwater in the Ethiopian Rift, which is generally associated with a low calcium content and high bicarbonate concentrations [7,8].

Globally, more than 200 million people, including Ethiopia, rely on groundwater with a fluoride concentration above the permissible level [3,4,9]. According to the Central Statistical Agency of Ethiopia report [10], 3.8% of the population is affected by high-level fluoride concentrations (>1.5 mg/L) in groundwater, which is used for drinking purposes. In general, fluorosis turns out to be the most widespread geochemical-based disease in the East African rift, affecting more than 80 million people [11–14]. Thus, due to the health effect of high fluoride in groundwater, it is essential to reduce excess fluoride concentrations to the allowable limit (<1.5 mg/L).

So far, various technologies such as coagulation/precipitation, electro-coagulation, membrane separations, ion exchange, and adsorption had been attempted for efficient defluoridation of groundwater [15–18]. Some of the shortfalls of these techniques include expensiveness, fouling issues, regular maintenance, and complicated operational procedures. In comparison to the techniques mentioned above, the adsorption methodology is still one of the most widely applied methods, taking the lead of high removal efficiency, cost-effectiveness, ease of operation, simplicity of design, and availability of large varieties of adsorbents [19,20].

Various adsorbents have been investigated and reported for the removal of excess fluoride from water in an effective manner. Some of the widely employed adsorbents are La (III)-Al (III)-activated carbon modified by chemical route [21], biomaterial functionalized cerium nanocomposite [22], Quaternized Palm Kernel Shell (QPKS) [23], bone char and activated alumina [24], bone char [25], renewable biowaste [26], $MgFe_2O_4$–chitosan–CaAl nanohybrid [27], carbon nanotube composite [15], Neem Oil-Phenolic Resin Treated Biosorbent [17], etc. However, many of these suffer from either time-consuming synthesis procedure, high processing costs, availability of raw materials, or short lifespan, which makes them impractical to be applied in the rift valleys that are essentially impacted by high fluoride concentration in water [1]. Consequently, efforts have been made to obtain easily accessible and long-lasting, low-cost, and efficient adsorbents that may be applied for the purification of water in low-income countries such as Ethiopia.

In recent years, volcanic rocks (VPum and VSco) have received significant interest for pollutant removal due to their valuable properties such as high surface area, low-cost, easy accessibility, good mechanical resistance, and availability in large quantities [28]. The source of these rocks is volcanic magma that formed during volcanic eruptions. Pumice (VPum) is a finely porous rock frothy with air bubbles; Scoria (VSco) is a rough rock that seems like furnace slag [28]. VPum is often formed from rhyolite magma [28], it can also develop from trachytic or dacitic magma. Due to its high porosity and low specific gravity, it has been used for water and wastewater treatment processes [29]. VSco is a vesicular pyroclastic rock with basaltic compositions, reddish-brown to black, denser than VPum, somewhat porous with high surface area and strength. Both volcanic rocks are found in abundance in Europe (Italy, etc.), Central America, Southeast Asia (Indonesia, etc.), and East Africa (Ethiopia, Eritrea, etc.) [28,29]. Although several studies have been conducted on the application of volcanic rocks for pollutants-laden wastewaters [28–31], very little research has been directed to the defluoridation of groundwater using volcanic rocks.

Previously, defluoridation research has been conducted on batch experiments using natural adsorbents [6,32]. The sorption capacity of adsorbents gained from batch equilibrium is valuable in giving basic information about the effectiveness of the adsorbents. Nevertheless, the data obtained from batch studies may not be appropriate for continuous processes where the contact time for the achievement of an equilibrium might be insufficient [33]. Consequently, studies by different authors [34–36] reveal that continuous processes mode (fixed-bed column set-up) yields reliable information about the breakthrough time, appropriate adsorption conditions, and the stability of the adsorption performance which can then be used to evaluate the potential of prepared adsorbents for industrial applications [1]. Therefore, there is an interest to conduct adsorption studies in a flow-through system.

The primary objectives of the current work were to (i) investigate the fluoride sorption capacity of VPum and VSco in fixed-bed column set-up, (ii) compare the adsorption properties of both adsorbents with each other, (iii) assess the fluoride adsorption mechanisms with respect to varying solution pH, adsorbent particle size, and flow rate, (iv) deeper analyze the adsorption processes employing mathematical models such as the Adams–Bohart and Thomas model, and (v) finally verify the suitability of the models for the design of flow-through systems for the removal of fluoride from aqueous solutions.

2. Results and Discussions

2.1. Characterization of Adsorbents

2.1.1. Crystalline Structures and Material Properties and Experimental Conditions

The crystalline phases of VPum and VSco were characterized using the X-ray diffraction (XRD) instrumental technique. The mineralogical composition of the adsorbents was characterized by matching the X-ray diffractogram (Figure 1a (VPum), b (VSco)) with the database of the X'pert HighScore Plus software package (Version: 2.2b (2.2.2)). The results showed that the main crystalline phases in VSco were Silicon Oxide (SiO_2), Albite low ($Na(AlSi_3O_8)$), whereas Hematite (Fe_2O_3) and Silicon Oxide (SiO_2) and Albite high ($Na(AlSi_3O_8)$) are the dominant components of VPum. The presence of crystalline phases in VPum samples can be ascribed to the peaks at 2θ = 24.9°, 27.6°, 27.7°, 37.7°, 41.9°, 58.0°, 64.9°and 65.0°, while that of VSco sample appeared at 2θ = 22.2°, 23.9°, 23.9°, 23.9°, 28.2°, 30.0°, 33.9°, and 35.8°. The detected dome in both samples between 2θ = 10° and 40° is an indication of amorphous material. The amorphous phase(s) present in the adsorbents was estimated by the calibration method. This method makes use of the integrated counts associated with the amorphous and crystalline fraction (Equation (1)) [37].

$$C_m(\%) = \left[\frac{C_{pa}}{A_{pa} + C_{pa}}\right] \times 100 \quad (1)$$

where C_m is the measured crystallinity, C_{pa} and A_{pa} are the integrated peak areas for the crystalline and amorphous components, respectively. The results revealed that the presence of the amorphous phase (s) in VPum and VSco is 89% and 68%, respectively.

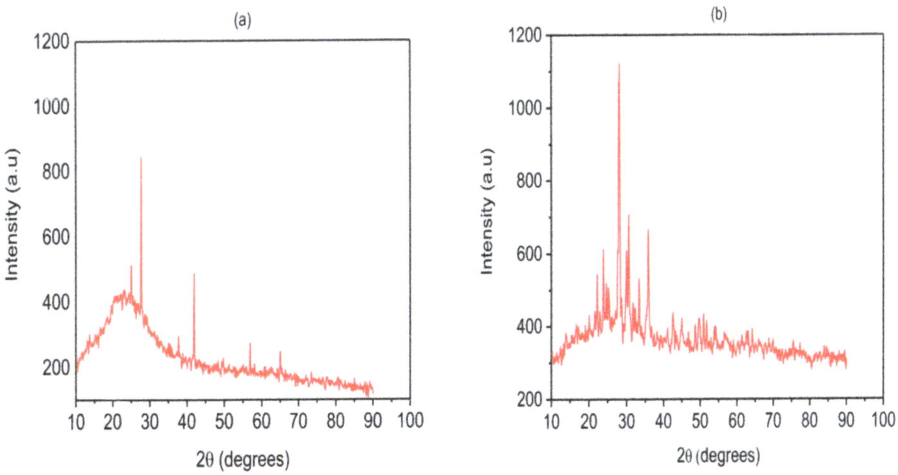

Figure 1. XRD patterns for (**a**) virgin pumice (VPum) and (**b**) virgin scoria (VSco).

The greater fraction of amorphous phase(s) in VPum compared with VSco possibly origins from simultaneous rapid cooling and depressurization of high-temperature volcano lava. The depressurization produces bubbles by lowering the boiling point of the lava. The simultaneous cooling then freezes the bubbles in the matrix of VPum. Due to rapid cooling, crystals do not have enough time to grow. A similar observation has been reported from the XRD analysis of pumice in previous studies [38,39].

Additionally, the results of material properties and experimental conditions were summarized in Table 1 as shown below.

Table 1. Material properties and experimental conditions.

Parameters	Virgin Scoria (VSco)			Virgin Pumice (VPum)		
Particle Size (mm)	**<0.075**	**0.075–0.425**	**0.425–2.00**	**<0.075**	**0.075–0.425**	**0.425–2.00**
Mass of adsorbents, m_{ads} (gm)	737.90	763.90	680.50	376.40	265.90	186.40
Bulk density, ρ_b (gm cm^{-3})	1.43	1.48	1.32	0.73	0.52	0.36
Particle density, ρ_s (gm cm^{-3})	2.37	2.33	1.94	1.61	1.31	0.64
Void volume, V_v (cm^3)	203.70	187.20	164.70	281.10	311.80	222.30
Total porosity, ε	0.40	0.36	0.32	0.55	0.61	0.43
Flow rate, Q (cm^3 min^{-1})	1.25	2.50	3.75	1.25	2.50	3.75
Empty Bed Contact Time, $_{EBCT}$ (min)	412.00	206.00	137.33	412.00	206.00	137.33
Filter (Superficial) velocity, V_f (cm min^{-1})	0.02	0.05	0.07	0.02	0.05	0.07
Effective (Interstitial) velocity, V_I (cm min^{-1})	0.06	0.13	0.23	0.05	0.08	0.17

2.1.2. Chemical Composition

The chemical analysis revealed that the major elements in VPum and VSco, as determined by ICP-OES (Table S1), are Si, Al, and Fe. Other elements were present in comparatively smaller quantities or below the detection limit of the instrument. In our previous study [28], the XRF measurement (Table S1) indicated that the oxides of Si, Fe, and Al were the major constituents of both VPum and VSco.

However, the chemical composition of the adsorbents might changes in time due to weathering processes. Consequently, representative samples have to be checked for possible changes induced due to weathering.

2.1.3. Fourier Transform Infrared (FTIR) Analysis

The FTIR spectrums of VPum (Figure 2a) and VSco (Figure 2b) at wavelengths ranging from 4000 to 400 cm^{-1} are shown in Figure 2. Due to the symmetric stretching vibration of Si-O-Si, the absorption band at ~1045.75 cm^{-1} can be assigned to the characteristic peak of $(SiO_4)^{2-}$ groups in the FTIR spectrum of VPum [39], whereas the band located at ~1011.5 in the FTIR spectrum of VSco can be belongs to the asymmetric stretching vibration of T-O-Si, T = Si or Al [40]. In the FTIR spectrum of VPum, the peaks at ~781 and ~695.25 belong to bending vibrations of Si-O-Si bond [38], whereas the band shown in the FTIR spectrum of VSco at ~759 is related to the stretching vibration of 6-fold coordinated Al(VI)-OH and 6-fold coordinated Al(VI)-O [41]. The small peaks shown in the FTIR spectrum of VSco at ~572 and ~539.25 can be attributed to the symmetric stretching of Si-O-Si and Al-O-Si [40,42], whereas the small band at ~465 belongs to bending vibrations of Si-O-Si and O-Si-O [42]. Certain peaks like the broadening peak at ~3602.5 cm^{-1} in the FTIR spectrum of VPum and sharper peak at ~2369.75 cm^{-1} in the FTIR spectrum of VSco belongs to the asymmetric stretching vibration of -OH bond can be allocated to adsorbed water molecules, whereas the peak at ~1645.75 cm^{-1} in VPum can be allocated to the bending vibration of H-O-H bond [38,39,42]. The most characteristic difference observed between the FTIR spectrum of VPum and FTIR spectra of VSco concerning the band attributed to the asymmetric stretching vibration of -OH bond. This band that is appeared as a broad band at about ~3602.5 cm^{-1} in the FTIR spectrum of VPum becomes sharper and shifts to lower frequencies (~2369.75 cm^{-1}) in the FTIR spectrum of VSco indicating that there is a high water content in VPum and could be correlated with less mechanical strength than VSco. Similar observations have been reported for a different system [42].

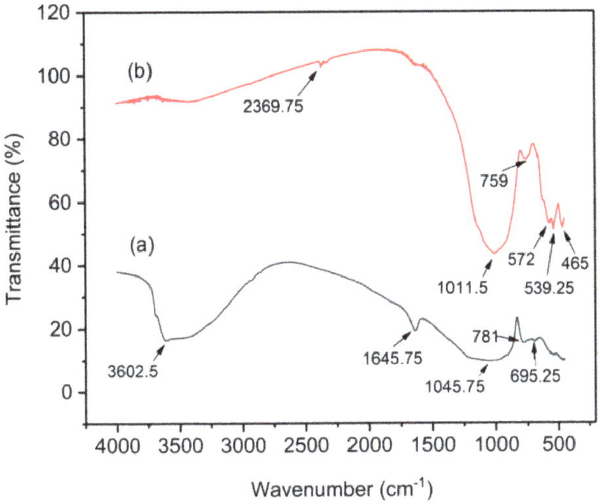

Figure 2. Fourier-transform infrared (FTIR) for (**a**) VPum and (**b**) VSco.

2.1.4. Scanning Electron Microscope (SEM) Analysis

The VPum (Figure 3a) and VSco (Figure 3b) SEM micrographs allowed direct observation of the surface morphology of the adsorbents with a magnification of $\times 100$. The structure of VPum showed that the surface of VPum had an interconnected porous surface [38,43], while VSco had an irregular shape and fibrous cavities (or pores). In addition, it may be said that these pores in VSco were either closed or in open forms (pores) [44]. As seen from the micrographs of the adsorbents, VPum had an interconnected inner porous surface (as indicated in Figure 3a, red-colored), while VSco (Figure 3b) is dominated by the dead-end pores. Consequently, the interconnected internal pore structure in VPum allows for better fluoride accessibility and, hence, better adsorption capacity than VSco.

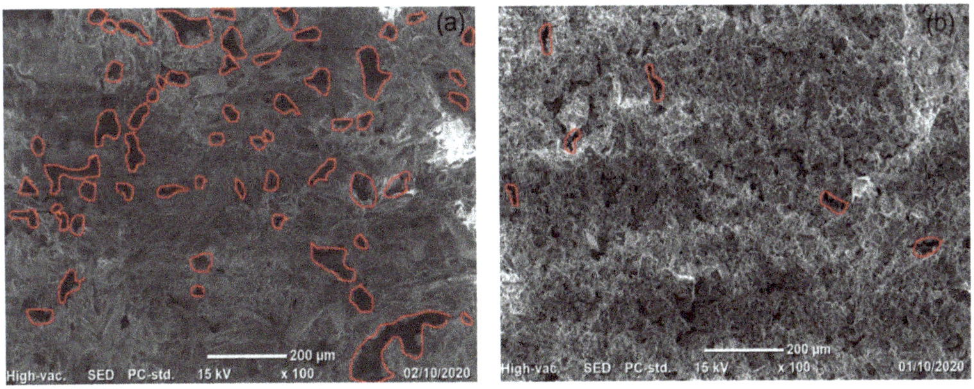

Figure 3. SEM micrographs for (**a**) VPum and (**b**) VSco.

2.1.5. pH and Point of Zero Charges (pH_{PZC})

The pH of the rock samples in water was found to be 6.65 and 7.20 for VPum and VSco, respectively. The point of zero charges (pH_{PZC}) of the adsorbents was identified as 6.85 for VPum and 6.98 for VSco at the intersection of the graph of the initial pH vs. the final pH (Figure 4). The slight difference observed in the adsorbents pH_{pzc} is related to their different characteristics. As can be seen from Table S1, the two volcanic rocks

(VPum and VSco) have different chemical compositions, which also influence the surface charge of the adsorbents. This is in agreement with previous studies [30,45], showing the effect of chemical composition on the zeta-potential of different materials. Below these values (pH < 6.85 for VPum and <6.98 for VSco), the surface of the adsorbents is positively charged. Thus, if the pH < pH_{PZC}, fluoride could possibly be adsorbed onto the surface of the adsorbents by coulombic attraction [6,46,47]. In addition, the curve for the blank experiment (for blank electrolyte solution 0.01 M NaCl) of both adsorbents is shown in Figure 4. As seen from the blank curve (Figure 4), a pH change without adding the adsorbents was obtained, which confirmed the sorbent dosing is not the only factor to fluctuate the pH of the solution.

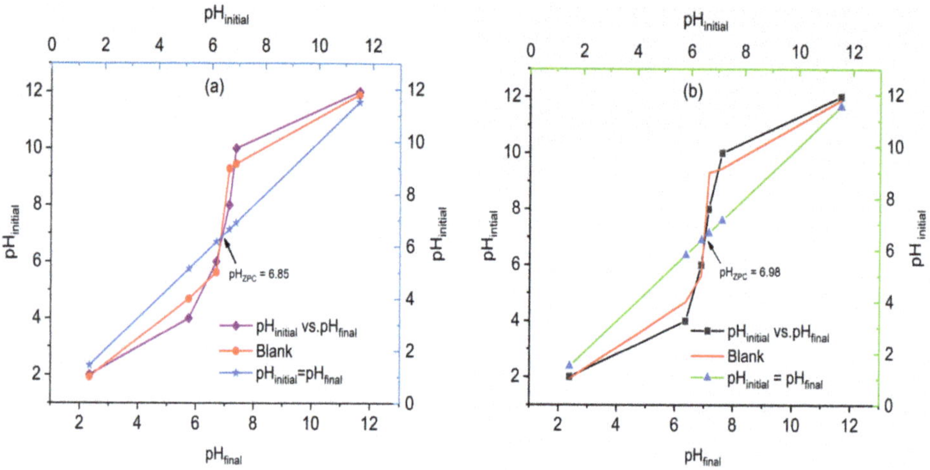

Figure 4. Determination of pH point of zero charges (pH_{PZC}) for (**a**) VPum and (**b**) VSco.

2.2. *Effect of Adsorbents Particle Size*

The effect of the particles size on the breakthrough behavior of fluoride was investigated for both VSco and VPum with grain size classes of silt to medium sand (<0.075, 0.075–0.425, 0.425–2.00 mm), while maintaining the same initial fluoride concentration (10 mg/L), bed depth (10cm), initial flow rate (1.25 mL/min), as well as solution pH (2.00) (Figure 5a (VPum), b (VSco)). As seen from Figure 5a (VPum) and b (VSco), on reducing the particle size from medium (0.425–2.00 mm) to silt (<0.075 mm) the breakthrough and exhaustion time noticeably increased for VSco, while the breakthrough and exhaustion time was high for VPum at a fine particle size (0.075–0.425 mm). The resulting breakthrough and removal of fluoride parameters are tabulated in Table 2. As can also be seen from Table 2, the amount of total adsorbed fluoride (q_{tot}) and the uptake of fluoride was high at silt (<0.075 mm) and fine (0.075–0425 mm) particle size for VSco and VPum, respectively. The smaller particle sizes provide large surface areas and/or sorption sites are more readily available. The results showed that the reduction of particle size of an adsorbent is a significant controlling factor in the fluoride–VSco system (at a particle size of <0.075 mm the fluoride uptake was high). A similar effect was observed for VPum (at a particle size of 0.075–0.425 mm the fluoride sorption capacity was high). However, the effect of particle size on the adsorption capacity is more pronounced for VSco than VPum. That means the pore spaces are more readily available in VPum as compared to VSco, showing that the pore space of VPum is a continuum (skeletal structure) while the pore space of VSco is dominated by dead-end pores. This infers VPum loses its internal porosity at the smallest particle size (<0.075 mm) since the continuum pore space (skeletal structure) is damaged when compared to the fine particle size (0.075–0425 mm) and resulting in smaller internal pore surface areas; consequently, the removal capacity of the adsorbent decreased. On the other hand, the pore space is not readily available in VSco (i.e., the internal pore space of VSco is dominated by dead-end pores). VSco at the smallest particle size (<0.075 mm) is, therefore, expected to have a large surface area, which leads to higher removal capacity compared to the fine particle size (0.075–0425 mm). A similar observation was reported for both adsorbents based on a batch adsorption experiment [28], and a similar remark was also drawn for pumice in the previous study [38]. Moreover, the BET specific surface area (S_{BET}) of the adsorbents was determined. As expected, VPum (3.50 m^2/g) has a larger surface area than VSco (2.49 m^2/g). Thus, all experiments other than the effect of particle sizes were conducted at a particle size of <0.075 mm for VSco and 0.075–0.425 mm for VPum.

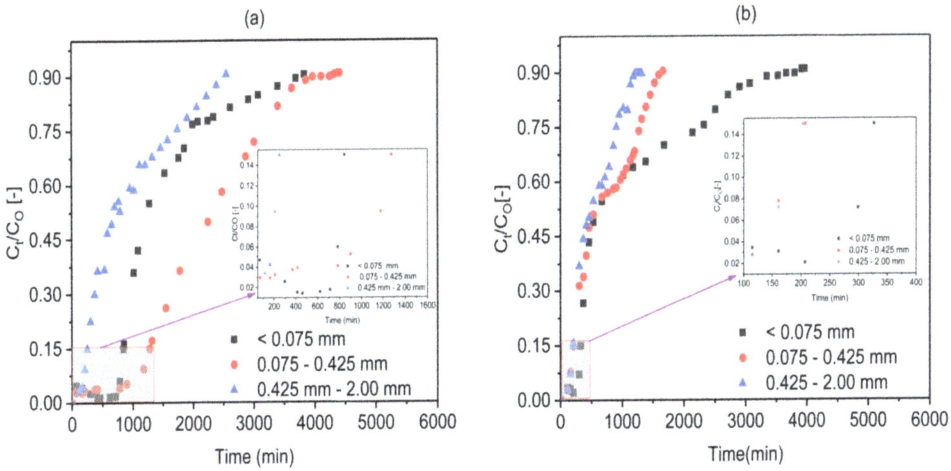

Figure 5. Effect of particle sizes on the breakthrough behavior of fluoride in (**a**) VPum and (**b**) VSco at (pH 2.00; influent fluoride concentration 10 mg/L (C_O: 10 mg/L); flow rate 1.25 mL/min (Q_O: 1.25 mL/min; bed depth 10 cm).

Table 2. Fixed-bed column parameters obtained for fluoride adsorption onto VPum and VSco.

VPum.	H (cm)	C_O (mg/L)	Q_O (mL/min)	pH	Particles size (P_{size}) (mm)	t_b (min)	t_e (min)	V_b (mL)	V_e (mL)	MTZ (cm)	EBCT (min)	q_{tot} (mg)	q_e (mg/kg)
	10	10	1.25	2.00	<0.075	816	1623	1019.64	2033.41	4.99	412	20.28	59.6
	10	10	1.25	2.00	0.075–0.425	1206	2339	1507.50	2923.70	4.84	412	29.24	109.9
	10	10	1.25	2.00	0.425–2.00	235	1013	293.23	1265.89	7.68	412	12.67	67.9
	10	10	1.25	4.00	0.075–0.425	278	500	347.50	625	4.44	412	6.25	23.51
	10	10	1.25	6.00	0.075–0.425	135	315	168.75	393.75	5.71	412	3.94	14.81
	10	10	2.50	2.00	0.075–0.425	215	634	538.47	1585.16	6.60	206	7.93	29.8
	10	10	3.75	2.00	0.075–0.425	75	359	282.69	1346.42	7.90	137	4.49	16.89
VSco	**H (cm)**	**C_O (mg/L)**	**Q_O (mL/min)**	**pH**	**Particles size (P_{size}) (mm)**	**t_b (min)**	**t_e (min)**	**V_b (mL)**	**V_e (mL)**	**MTZ (cm)**	**EBCT (min)**	**q_{tot} (mg)**	**q_e (mg/kg)**
	10	10	1.25	2.00	<0.075	415	1286	518.03	1607.60	6.77	412	16.08	22
	10	10	1.25	2.00	0.075–0.425	199	760	248.99	849.80	7.38	412	9.50	12.4
	10	10	1.25	2.00	0.425–2.00	231	591	288.17	739.12	6.10	412	7.39	10.9
	10	10	1.25	4.00	<0.075	296	487	370	608.75	3.92	412	6.09	8.2
	10	10	1.25	6.00	<0.075	227	393	283.75	491.25	4.22	412	4.91	6.7
	10	10	2.50	2.00	<0.075	185	445	462.95	1113.19	5.84	206	5.57	7.5
	10	10	3.75	2.00	<0.075	69	249	256.82	931.87	7.24	137	3.10	4.2

t_b = breakthrough time, t_e = exhaustion time, V_b = total effluent volume at breakthrough time, V_e = total effluent volume at exhaustion time MTZ = Mass Transfer Zone, EBCT = Empty Bed Contact Time, q_{total} = total amount of fluoride adsorbed from the column, q_e = equilibrium fluoride uptake per kg of adsorbent.

2.3. *Effect of Influent pH*

The influent solution's pH can noticeably affect the anions sorption on the adsorbents by changing the degree of ionization, the ion speciation, and the adsorbent's surface charge. Therefore, the effect of solution pH on adsorption of fluoride using VPum and VSco was investigated at different pH (2.00, 4.00, and 6.00) by a separate set of fixed-bed adsorption columns. The breakthrough curves obtained for both adsorbents are shown in Figure 6a,b for a fixed inlet flow rate of 1.25 mL/min, influent fluoride concentration of 10 mg/L, column bed depth of 10 cm, and a particle size of <0.075 mm for VSco and 0.075–0.425 mm for VPum.

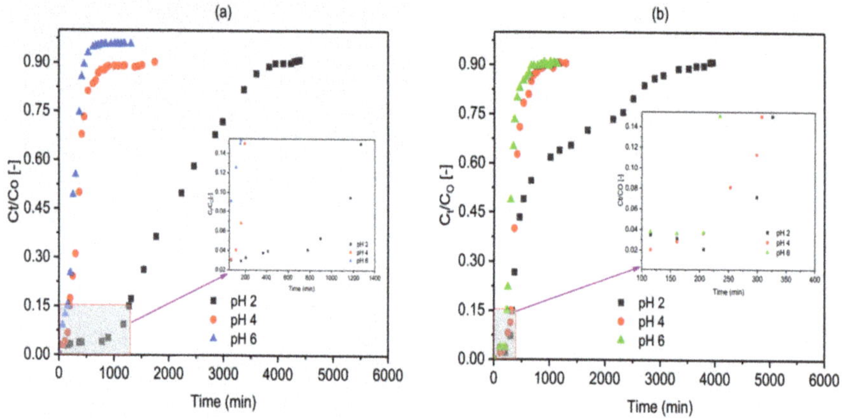

Figure 6. Effect of pH on the breakthrough behavior of fluoride in (**a**) VPum: 0.075–0.425 mm and (**b**) VSco: <0.075 mm (C_O: 10 mg/L; Q_O: 1.25 mL/min; bed depth 10 cm).

As can generally be observed from Figure 6a, b, the adsorption capacity of the adsorbents noticeably increased with decreasing pH. As can also be seen from Table 2 (VPum and VSco), the total amount of fluoride adsorbed (q_{tot}) was high for VPum (29.24 mg) and 16.08 mg for VSco at lower pH of 2, and the breakthrough time decreased from 1206 to 135 min for VPum and 415 to 227 min for VSco with an increase in pH from 2 to 6. The volume of water treated at the breakthrough time was higher at pH of 2.00 (1507.5 mL for VPum and 518.03 mL for VSco) than 4.00 (347.5 mL for VPum and 370 mL for VSco) and 6.00 (168.75 mL for VPum and 283.75 mL for VSco). This concludes the occurrence of the breakthrough time was longer, the amount of fluoride adsorbed, and treated water volume was high for a pH of 2.00. As pH varies, surface charge also varies; the sorption of charged species is affected. Therefore, the performance of adsorbents for better adsorption at low pH may be the result of the presence of a large number of H^+ ions at low pH values, and hence neutralize the negatively charged adsorbent surface [48], consequently dropping the interference of the adsorption of fluoride. In addition, this reality can be elucidated based on the pH value at the point of zero charges of the adsorbents (pH_{PZC} = 6.85 (VPum) and 6.98 (VSco)).

Moreover, the decrease in the adsorption of fluoride at pH 4.00 and 6.00 could also be due to the decrease in the number of H^+ or electrostatic repulsion of fluoride by negatively charged adsorbent surface [47,49].

Hence, the sorption of fluoride ions is due to an electrostatic phenomenon and surface complexation that perform independently or together for the adsorption of fluoride ions on the adsorbents. The removal mechanism at pH < pH_{PZC} is presumably due to columbic attraction of fluoride by positive surface charges (Equation (2)) and/or ligand exchange reactions of fluoride with surface hydroxyl groups (Equation (3)).

$$MOH_2^+ + F^- \leftrightarrow MOH_2^+ - - - F^- \quad (2)$$

$$MOH_2^+ + F^- \leftrightarrow MF + H_2O \quad (3)$$

where, M represents Fe, Al, Si, Ca, Mg, etc.

In this study, an increment in the final pH (pH ~7.20) was observed after the completion of the adsorption experiment, which is consistent with the columbic or ligand exchange type adsorption mechanism shown in Equations (2) and (3) [47]. This can be further explained by the capacity of the adsorbents to maintain a neutral pH after adsorption [6,50]. The capacity to maintain neutral effluent solution pH could be from the amphoteric nature of oxides in both adsorbents (Al_2O_3, Fe_2O_3, TiO_2, etc.) when compared with the effect of basic metallic oxides (Table S1). This type of observations were reported for the removal of pollutant in a previous study [6]. Furthermore, the elemental compositions of exchangeable cations also play a critical role in fluoride uptake during defluoridation [51]. There might be a slight increase in electro conductivities of the final solutions, which might not influence the adsorption process [52]. However, additional testing of the effluent solution for various compounds may be required to draw definite conclusions.

It is noted that the effect of pH on the adsorption capacity may be due to the shared impact of pH on the nature of the adsorbent surface, the existence of the adsorbed pollutant (fluoride ion), and the added acid and base to the working solution to adjust its pH. In this study, the optimum and effective removal of fluoride takes place at a pH of 2.00; hence, all experiments other than the effect of pH were conducted at a pH of 2.00 for both adsorbents.

2.4. Effect of Flow Rate

The effect of flow rate on the adsorption of fluoride using VPum and VSco was examined at the flow rates of 1.25, 2.50, and 3.75 mL/min whereas the bed depth (10 cm), influent solution pH (2.00), initial fluoride concentration (10 mg/L), and adsorbents particle size (<0.075 mm (Vsco) and 0.075–0.425 mm (VPum)) were held constant. As indicated in Figure 7a (VPum) and b (VSco), the breakthrough curves become steeper and shifted to the origin with an increasing flow rate while the breakthrough time decreased. The use of a high flow rate decreases the contact time of fluoride in the solution with the adsorbents, thereby allowing earlier breakthroughs to occur. Additionally, increasing the flow rate from 1.25 to 3.75 mL/min decreased the volume of water treated from 1507.5 to 282.69 mL and 518.03 to 256.03 mL for VPum and VSco, respectively (Table 2).

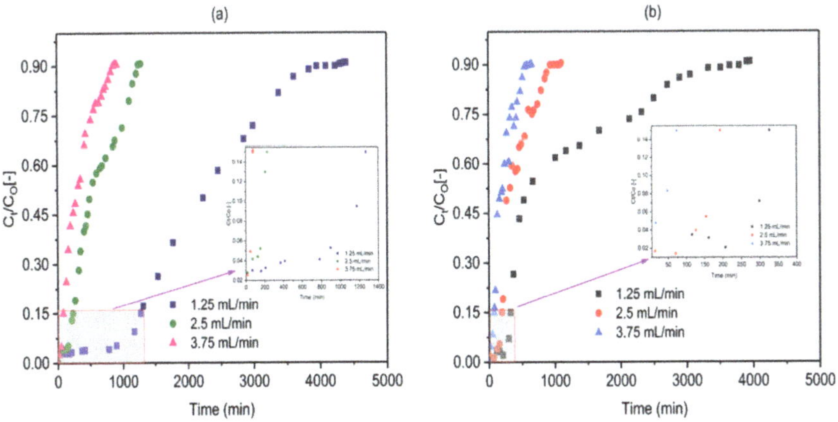

Figure 7. Effect of influent flow rate on the breakthrough behavior of fluoride in (**a**) VPum: 0.075–0.425 mm and (**b**) VSco: <0.075 mm (pH 2; C_O: 10 mg/L; bed depth 10 cm).

This was further supported by Mass Transfer Zone (MTZ) (Table 2) which increased with increasing flow rate. The total fluoride adsorbed (q_{tot}) increased from 4.49 mg to 29.24 mg for VPum and from 3.10 mg to 16.08 mg for VSco (Table 2) as the flow rate decrease from 3.75 to 1.25 mL/min. This results in the increase of the adsorption performance of the column from 17 to 110 mg/kg and 4.2 to 22 mg/kg for VPum and VSco, respectively (Table 2). The increase in sorption efficiency at a low flow rate shows that the adsorbates have sufficient time to penetrate and diffuse deeply into the pores of the adsorbents; hence, intraparticle mass transfer controls the sorption process. This was also verified by MTZ (Table 2) or unused bed, which decreased with decreasing flow rate. In general, at the lower flow rate, the contact time between the adsorbent and the fluoride was higher, resulting in an increased breakthrough time and treated water volume for the continuous column adsorption system. A similar type of observation was reported by various authors for fixed-bed column systems [53–55]. In this study, the optimum and effective removal of fluoride takes place at a flow rate of 1.25 mL/min; so, all experiments other than the effect of flow rate were performed at a flow rate of 1.25 mL/min.

2.5. Application of the Thomas Model

The values of the Thomas model parameters, K_T and q_o for both adsorbents shown in Table 3 for different experimental parameters were found from the non-linear optimization techniques according to Equation (18). The non-linear plots of the experimental (designated as exp.) and simulated (designated as cal.) breakthrough curves based on the Thomas model for VPum (a) and VSco (b) at different particle sizes (Figure S1), influent pH (Figure S2), and influent volumetric flow rate (Figure S3) were provided in Supplementary Materials. The results of K_T, q_o, and correlation coefficient (R^2) are shown in Table 3 for VPum and VSco. From the results, it can be seen that the values of R^2 range from 0.897 to 0.993 for VPum and 0.901 to 0.973 for VSco.

Table 3. Thomas model parameters for fluoride adsorption onto VPum and VSco.

VPum	H (cm)	C_o (mg/L)	Q (mL/min)	pH	P_{size} (mm)	K_T (L/min.mg) ($\times 10^4$)	$q_{0(cal.)}$ (mg/kg)	$q_{e(exp.)}$ (mg/kg)	R^2
	10	10	1.25	2.00	<0.075	2.199	48.6	59.6	0.950
	10	10	1.25	2.00	0.075–0.425	1.440	109.9	109.9	0.993
	10	10	1.25	2.00	0.425–2.00	1.840	57.9	67.9	0.897
	10	10	1.25	4.00	0.075–0.425	8.289	17.83	23.51	0.977
	10	10	1.25	6.00	0.075–0.425	12.099	13.08	14.81	0.995
	10	10	2.50	2.00	0.075–0.425	3.565	55.81	29.8	0.953
	10	10	3.75	2.00	0.075–0.425	5	45.82	16.9	0.962
VSco	**H (cm)**	**C_o (mg/L)**	**Q (mL/min)**	**pH**	**P_{size} (mm)**	**K_T (L/min.mg) ($\times 10^4$)**	**$q_{0(cal.)}$ (mg/kg)**	**$q_{e(exp.)}$ (mg/kg)**	**R^2**
	10	10	1.25	2.00	<0.075	1.23	19	22	0.901
	10	10	1.25	2.00	0.075–0.425	2.213	13.4	12.4	0.931
	10	10	1.25	2.00	0.425–2.00	3.310	10.1	10.9	0.956
	10	10	1.25	4.00	<0.075	10.140	7.06	8.2	0.973
	10	10	1.25	6.00	<0.075	15.520	5.42	6.7	0.965
	10	10	2.50	2.00	<0.075	5.263	13.3	7.5	0.929
	10	10	3.75	2.00	<0.075	7.220	11.2	4.2	0.944

The high values of R^2 indicate there were no significant disparities between the experimental data points and calculated data by the Thomas model for all particle sizes, influent solution pH, and influent volumetric flow rate. The observed differences between the experimental data and calculated data from the Thomas model may be due to the characteristic attribute weakness in the model. The Thomas model does not consider the external (film) and intra-particle diffusions in the adsorption system and, therefore, proposes adsorbate–adsorbent surface reactions to control the adsorption rate, hence the breakthrough. However, nearly all adsorption operations are typically not limited to surface reaction kinetics, but are also controlled by external and/or intra-particle diffusion [56,57]. Thus, the perceived disparities in this study indicate external and/or intra-particle mass transfer may be the rate-controlling steps in fluoride adsorption in a fixed-bed column

onto the adsorbents. Similar observations were drawn with the kinetic study of fluoride under fixed-bed conditions onto modified pumice [57]. As shown in Table 3, the value of the Thomas rate constant (K_T) increased as the influent flow rate increased but the value of the maximum solid-phase concentration (q_o) decreased. A related type of investigation on Thomas constants for different systems was reported by various authors [58,59].

2.6. Application of the Adams–Bohart Model

The parameter values of the Adams–Bohart model, K_{AB} and N_O, as depicted in Table 4 for Vpum and Vsco were similarly determined by non-linear regression analysis according to Equation (19).

Table 4. Adams–Bohart model parameters for fluoride adsorption onto VPum and VSco.

VPum	H (cm)	C_o (mg/L)	Q (mL/min)	pH	P_{size} (mm)	K_{AB} (L/min.mg) (×10⁴)	N_0 (mg/L)	R^2
	10	10	1.25	2.00	<0.075	2.187	35.55	0.950
	10	10	1.25	2.00	0.075–0.425	1.439	56.85	0.993
	10	10	1.25	2.00	0.425–2.00	1.741	20.77	0.911
	10	10	1.25	4.00	0.075–0.425	8. 289	9.22	0.995
	10	10	1.25	6.00	0.075–0.425	12. 099	6.76	0.953
	10	10	2.50	2.00	0.075–0.425	3.565	29.68	0. 962
	10	10	3.75	2.00	0.075–0.425	5.000	22.74	0.995
VSco	**H(cm)**	**C_o (mg/L)**	**Q (mL/min)**	**pH**	**P_{size} (mm)**	**K_{AB} (L/min.mg) (×10⁴)**	**N_o (mg/L)**	**R^2**
	10	10	1.25	2.00	<0.075	1.233	27.27	0.886
	10	10	1.25	2.00	0.075–0.425	2.213	17.68	0.886
	10	10	1.25	2.00	0.425–2.00	3.310	13.41	0.956
	10	10	1.25	4.00	< 0.075	10.145	10.13	0.969
	10	10	1.25	6.00	< 0.075	15.518	7.77	0.980
	10	10	2.50	2.00	< 0.075	5.263	19.72	0.929
	10	10	3.75	2.00	< 0.075	7.221	15.36	0.944

From the results presented in Table 4, it can be realized that the values of R^2 range from 0.911 to 0.993 for Vpum and 0.886 to 0.969 for VSco. The high values of R^2 designate the applicability of the Adams–Bohart model for describing the entire sorption mechanisms of fluoride onto VPum and VSco under a continuous fixed-bed flow process.

In a similar fashion with the Thomas model, the comparison of the non-linear plots of the experimental and calculated breakthrough curve, based on the Adams–Bohart model, are generally in good agreement for VPum (a) and VSco (b) at different particle sizes (Figure S4), influent solution pH (Figure S5), and influent flow rate (Figure S6) respectively. Only minor disparities were noticed at lower pH (2.00) and particle sizes of 0.075–0.425 mm and <0.075 mm for VPum and VSco, respectively. As seen in Table 4, the values of the kinetic constants were affected by the influent flow rate and increased with increasing flow rate. This presented that external mass transfer in the entire fluoride adsorption mechanisms in the fixed-bed column dominates the overall system kinetics [57,60]. In general, both the Adams–Bohart and the Thomas models could predict very well the entire region of the breakthrough curves for the fluoride-VSco and fluoride-VPum systems. In addition, both the Adams–Bohart model (Equation (19)) and the Thomas model (Equation (18)) are mathematically the same and, therefore, gave similar fit quality.

2.7. Comparison of Different Adsorbents on Fluoride Removal

A comparison has been made between volcanic rocks (VPum and VSco) used in this study and previously reported adsorbents for fluoride removal in a fixed-bed column system. The results for some adsorbents are presented in Table 5.

Table 5. Comparison of other adsorbents with VPum and VSco.

Adsorbents	Surface Area (m^2g^{-1})	Bed Depth (cm)	Fluoride in (mgL^{-1})	Adsorption Capacity (mg F^- g^{-1})	Adsorption Capacity per Surface Area ($mg.m^{-2}$)	References
Cement paste	NA*	20	15	0.149	-	[61]
aluminum modified iron oxide	NA	10.5	4	0.139	-	[62]
Acid-treated bentonite (GHB)	24.5	28	2.85	0.169	0.0069	[63]
MnO_2-coated Tamarind Fruit Shell	NA	6	2	0.883	-	[64]
kanuma mud	144.01	10	20	1.560	0.0108	[49]
Activated alumina (Grade OA-25)	250	10	5	0.74	0.0029	[65]
VPum	3.5	10	10	0.110	0.0314	This study
VSco	2.49	10	10	0.022	0.0088	This study

NA*: Not available.

As can be seen from these results in Table 5, the natural VPum used is comparable to cement paste and aluminum modified iron oxide in terms of defluoridation capacity. Values of adsorption capacity per unit surface area are, however, higher for VPum than those of acid-treated bentonite (GHB), kanuma mud, and activated alumina; and higher for VSco than acid-treated bentonite (GHB) and activated alumina (Grade OA-25) (Table 5). Both adsorbents are available in abundance in all parts of the world and are readily available in approximately 1/3 of Ethiopia's total area and are, hence, favored adsorption materials because of very low supply costs. The adsorbents used are primarily part of the natural environment. However, to improve the specific surface area and hence the defluoridation capacity, surface modification of the natural volcanic rocks may be appropriate.

3. Materials and Methods

3.1. Materials

In this study, rock samples were collected from volcanic cones (VPum: 8°10′ N 39°50′ E; VSco: 8°33′ N 39°16′ E) of the Main Rift Valley area of Oromia Regional State, East Showa Zone, Ethiopia, around 50–100 km East of Addis Ababa. The rocks are readily available in approximately 1/3 of the country's total area and are, thus, a preferred adsorption material because of very low supply costs [29,48,66].

3.2. Preparations of Adsorbents

The rock samples (VPum and VSco) were washed repeatedly with deionized water until all water-soluble compounds and dust were removed, and thereafter dried at 55 °C for 48 h [30,67]. After cooling samples down to room temperature, they were crushed with a mortar and sieved using different mesh sizes: silt (<0.075 mm), fine sand (0.075–0.425 mm), and medium sand (0.425–2.00 mm) [28,68]. All prepared samples were packed in air-tight plastic bags and stored at a cool and dry place for further use.

3.3. Preparations of Adsorbate

All glassware and bottles were thoroughly washed and rinsed with deionized water before usage. Chemicals used were analytical grade reagents and a fluoride stock solution (1000 mg/L) was prepared freshly by dissolving 2.21 g of anhydrous NaF (Merck KGaA, Darmstadt, Germany) in 1000 mL of deionized water. The synthetic fluoride solution of desired concentrations was made by diluting the stock solution. 0.1 M of NaOH and/or 0.1 M HCl solutions were used to adjust the pH values of the fluoride solution utilized in the column experimental experiments.

3.4. Adsorbent Characterizations

3.4.1. Crystalline Structures

The crystalline structures of the adsorbents were analyzed by an X-ray diffractometer (XRD-7000, Drawell, Shanghai, China) with Cu Kα as a radiation source (1.54056 Å) generated at 30 kV and 25 mA instrument. The diffractograms were gained with a step width of 2θ and a scan rate of 0.04°/min.

3.4.2. Chemical Composition

The elemental composition of the adsorbents was analyzed using inductively coupled plasma-optical emission spectroscopy (ICP-OES). X-ray fluorescence (XRF) spectroscopy was used to obtain information on the oxide contents of the adsorbents.

3.4.3. Fourier Transform Infrared (FTIR) Analysis

FTIR spectra of the samples were run on KBr pellets. The spectra were recorded over a range of 5000 to 400 cm^{-1} at a resolution of 0.1 cm^{-1} in a PerkinElmer spectrometer (UNSW Sydney, Australia) using a lithium tantalite ($LiTaO_3$) detector.

3.4.4. Scanning Electron Microscope (SEM) Analysis

A scanning electron microscope (SEM) (JCM-6000plus, Version 0.2, Peabody, MA, USA), operated at 15 kV, was used to determine the morphologies of VPum and VSco. The characteristics of the adsorbents were compared.

3.4.5. Determination of pH and Point of Zero Charges (pH_{PZC})

The pH of the adsorbents was determined using a pH meter in a 1:10 adsorbent/water ratio as per the standard method [6]. The pH at the point of zero charges (pH_{PZC}) of the adsorbents was examined based on the standard method. For this effect, 250 mL of 0.01 M NaCl solution as an electrolyte was positioned in a vessel, thermostated at 298 K, and N_2 was bubbled through the solution to stabilize the pH by preventing the dissolving of CO_2 from the air. In 6 Erlenmeyer flasks, 25 mL of the electrolyte was introduced and the pH was adjusted to the required value (2.00, 4.00, 6.00, 8.00, 10.00, and 12.00) by adding 0.1 M NaOH or 0.1 M HCl. The same procedure and method were performed for blank electrolyte solution (0.01 M NaCl). In each beaker, 0.25 g of the rock samples were added and shaken for 48 h. The suspension was subsequently filtrated and the final pH was determined. The point of zero charges (pH_{PZC}) was found at the intersection point by plotting the initial pH versus the final pH.

3.4.6. BET Analysis

The specific surface area (S_{BET}) of the adsorbents was measured using a nitrogen gas adsorption-desorption technique at 77k using surface analyzer equipment (Micrometrics/Gemini-2372). The samples were degassed at 300 °C under vacuum for at least 6 h before analysis. The Brunauer-Emmett-Teller (BET) equation was used to obtain a specific surface area (S_{BET}). The S_{BET} values of the two adsorbents (VPum and VSco) are compared.

3.5. Column Adsorption Experimental Set-Up and Procedures

Continuous fixed-bed column adsorption studies were carried out to assess the dynamic behavior of fluoride removal by using VPum and VSco. Continuous flow adsorption experiments were conducted in a small-scale cylindrical column of 8.1 cm internal diameter and 10 cm height with an empty bed volume of 515 cm^3. The column was filled with a weighted amount of adsorbent of different particle sizes (silt: <0.075 mm, fine sand: 0.075–0.425 mm, and medium sand: 0.425–2.00 mm). The same particle size was used if controlling parameters such as pH and flow rate were tested. The bed was conditioned with deionized water (pH: 7.00–7.30) for 12 h (overnight) to ensure a closely packed adsorbent and to avoid the potential occurrence of voids, channeling, or cracking, which can significantly affect the performance of the column.

A synthetic fluoride solution with a concentration of 10 mg/L was pumped to a packed bed column in up-flow mode to avoid channeling caused by gravity. The influent volumetric flow rate varied between experiments but was held constant in a given experiment using an adjustable peristaltic pump (MS-REGLO, Labortechnik-Analytic, Zürich, Switzerland). The experiments were conducted at room temperature (25 ± 2 °C). The effluent column sample was collected using an automatic fraction collector (RFI, MA-RON GmbH, Germany). The constant flow rate was verified by collecting and quantifying the effluent solution at regular time intervals. The column operation was stopped when the concentrations of the fluoride in the effluent exceeded 90% of its initial concentrations. Ion chromatography (930 Compact IC Flex, Metrohm, Switzerland) was used to quantify fluoride concentrations. The instrument uses 3.2 mmol/L Na_2CO_3/1.0 mmol/L $NaHCO_3$ as eluent, Metrosep A Supp 5–150/4.0 column, and a standard conductivity detector to measure the conductivity of the effluent solutions. The fluoride concentration was measured in the calibration range of 0.2–200 mg/L, contains inline dilution, inline dialysis, eluent degasser, CO_2 suppressor, and chemical suppressor. Suppression in IC maximizes the detection sensitivity of fluoride ions while reducing the background conductivity of the eluent.

The maximum tolerable breakthrough concentration (C_b) was 1.5 mg/L (15% of the influent initial concentration of 10 mg/L), which is recommended by WHO [4] as a maximum acceptable level for drinking water.

The effect of experimental parameters such as particles size (silt: <0.075 mm, fine sand: 0.075–0.425 mm, and medium sand: 0.075–0.425 mm), influent solution pH (2.00, 4.00, and 6.00), and influent volumetric flow rate (1.25, 2.50, and 3.75 mL/min) on breakthrough behavior and amount of fluoride removed were examined.

3.6. *Modeling and Analysis of Fixed-Bed Column Data*

A fixed-bed column adsorption performance is well described through the breakthrough curve concept [53]. The time of solute breakthrough and the shape of the breakthrough curve are important indicators for the operational adsorption processes; the breakthrough curve is directly linked to the viability and economics of the adsorption process [54,69]. The breakthrough patterns and according parameters are dependent on the operating conditions of the fixed-bed column such as adsorbent particle size, influent flow rate, and pH of the influent solution. Nevertheless, the pH value may not influence the breakthrough curve in a situation such as when using strongly basic anion exchangers. The primary and significant attribute is the sorbent selectivity to the pollutant, as well as the dynamic exchange capacity and full dynamic capacity of the column [70]. To investigate the performance of the column and to scale-up, the determination of breakthrough parameters is crucial. The breakthrough curves expressed in terms of the ratio of effluent to influent adsorbate concentration (C_t/C_o) as a function of time or effluent volume for a given height of the bed reflects the absorbed fluoride from the solution. Time equivalent to stoichiometric capacity (exhaustion time) and time equivalent to usable capacity (breakthrough time) is shown in Equations (4) and (5), respectively [54,59].

$$t_e = \int_{t=0}^{t=t_{total}} \left(1 - \frac{C_t}{C_o}\right) dt \tag{4}$$

$$t_b = \int_{t=0}^{t_b} \left(1 - \frac{C_b}{C_o}\right) dt \tag{5}$$

where t_e is exhaustion time (min), t_b is the breakthrough time (min) at which $C_t = C_b$ (mg/L) (for the present system, C_b = 1.5 mg/L).

The total value of fluoride adsorbed (q_{total}: mg) from the column for a given feed concentration and the flow rate was obtained from the area (A) under the breakthrough curve by integrating the adsorbed fluoride concentration C_{ad} ($C_{ad} = C_o - C_t$) (mgL^{-1}) versus t (min) and can be obtained from Equation (6) [55,71].

$$q_{total} = \frac{QA}{1000} = \frac{Q}{1000}\int_{t=0}^{t=t_{total}} C_{ad}dt \tag{6}$$

where t_{total}, and Q are the total flow time until saturation of the bed (min), and volumetric flow rate (mL/min), respectively.

Equilibrium fluoride uptake (q_{e}: mg kg^{-1}) (maximum capacity of the column) in the column is calculated by Equation (7) as the total amount of fluoride adsorbed (q_{total}) per kilogram of dry adsorbent mass (m) at the end of the total flow time [71].

$$q_{eq} = \frac{q_{total}}{m} \tag{7}$$

The effluent volume (V_e) and treated effluent volume or breakthrough volume (V_b) of solution can be found from Equations (8) and (9), respectively.

$$V_e = Qt_e \tag{8}$$

$$V_b = Qt_b \tag{9}$$

where, V_e is the total effluent volume at exhaustion time (mL), V_b, total effluent volume at the breakthrough time (mL), Q is the volumetric flow rate (mL/min), t_e and t_b are exhaustion and breakthrough time (min), respectively.

The Mass Transfer Zone (MTZ) or unused bed length (H_{UNB}) can be obtained from Equation (10) [54,59].

$$MTZ = H_T\left(\frac{t_e - t_b}{t_e}\right) \tag{10}$$

where H_T is total bed height (cm), t_e (min) is exhaustion time, and t_b is breakthrough time (min).

The Empty Bed Contact Time (EBCT), which measures the critical depth and the contact time between the solid phase adsorbent and the liquid phase, can be obtained from Equation (11).

$$EBCT = \frac{V_B}{Q} \tag{11}$$

where V_B is the volume of a fixed bed (mL) and Q is the flow rate (mL/min).

The bulk density (ρ_b: $gm.cm^{-3}$), which measures the adsorbent compaction status, and the particle density (ρ_p: $gm.cm^{-3}$) of the adsorbent can be obtained from Equations (12) and (13), respectively [72].

$$\rho_b = \frac{m_{ads}}{V_t} \tag{12}$$

$$\rho_p = \frac{m_{ads}}{V_t - V_v} \tag{13}$$

where m_{ads} is the dry mass of adsorbent (mg), and V_t is the bulk volume (cm^3) which includes the volume of adsorbent (V_B:cm^3) and the pore space between the adsorbent particles or void volume (V_v:cm^3).

The void volume (V_v:cm^3) of the adsorbent can be found from Equation (14) [72].

$$V_v = \frac{W_{Sat} - W_{dry}}{\rho_w} \tag{14}$$

where W_{dry} is the weight of dry adsorbent (g), W_{sat} is the weight of saturated adsorbent (g), and ρ_w is the density of water (g cm^{-3}).

The total porosity of the adsorbent (ε_b) can be obtained from Equation (15) [72].

$$\varepsilon_b = 1 - \frac{\rho_b}{\rho_p} \tag{15}$$

The filter (superficial) velocity (V_f:cm min^{-1}) and effective (interstitial) velocity (V_I:cm min^{-1}) can be found from Equations (16) and (17), respectively [72].

$$V_f = \frac{Q}{A} \tag{16}$$

$$V_I = \frac{Q}{A \times \varepsilon_b} \tag{17}$$

where A is the cross-sectional area of the fixed-bed (cm^2) and Q is the flow rate (cm^{3}min^{-1}).

3.7. *Fixed–Bed Column Breakthrough Curve Modeling*

The successive operation of a small scale column towards industrial applications can be well elucidated with the help of some models. Various models have been reported for predicting the breakthrough performance in fixed-bed adsorption [57,73]. In this study, the two most important and widely used mathematical models, the Thomas model and Adams–Bohart model, have been applied to the column experimental data for describing the dynamic behavior of fluoride adsorption using VPum and VSco in a fixed-bed column filter.

3.7.1. Thomas Model

The Thomas model [74] is one of the most extensively employed kinetic models to predict fixed-bed column performance. In addition to the prediction of the breakthrough curve for the effluent, the model can be used to determine the maximum uptake of adsorbate and adsorption rate constant [74]. The non-linear form of the Thomas model can be described as follows Equation (18), [75].

$$\frac{C_t}{C_o} = \frac{1}{1 + \exp\left[K_T q_o \frac{m}{Q} - K_T C_o t\right]} \tag{18}$$

where C_o (mg/L) is the initial solute concentration, C_t (mg/L) is the solute concentration at the time, t, Q (L/min) is the volumetric flow rate, q_o (mg/kg) is the maximum solid-phase concentration of solute (maximum column adsorption capacity), K_T is the Thomas rate constant (L/min mg), and m (kg) is the packed dry mass of the adsorbent in a fixed-bed.

3.7.2. Adams–Bohart Model

The Adams–Bohart model [76] was developed for the analysis of the dynamics of fixed-bed based on the assumption that the adsorption rate is proportional to both the residual adsorbent and adsorbate concentration. The nonlinear form of the Adams–Bohart model (Equation (19)) [77], was used for the prediction of breakthrough curves and model parameters.

$$\frac{C_t}{C_o} = \frac{1}{1 + \exp\left[K_{AB} N_o \frac{Z}{v} - K_{AB} C_o t\right]} \tag{19}$$

Where K_{AB} (L/mg min) is the kinetic constant, v (mL/min) is the linear flow rate, Z (cm) is a column bed depth, and N_O (mg/L) is the saturation concentration (adsorption capacity of the adsorbent per unit volume of the bed), and time t (min) ranges from the start to fluoride breakthrough point. The linear flow rate was determined by Equation (20).

$$v = \frac{Q}{A} \tag{20}$$

where Q (cm^3/min) is the volumetric flow rate, and A (cm^2) is the cross-sectional area of the bed [60,62,78].

4. Conclusions

In this study, the removal of fluoride from aqueous solutions was examined in a continuous fixed-bed adsorption column system using VPum and VSco. The characterizations investigations were performed using XRD, SEM, FTIR, BET, XRF, and ICP-OES equipment to reveal the mechanisms of adsorption and the suitability of the adsorbents for fluoride removal. The pH_{PZC} is 6.98 for VSco and 6.85 for VPum. The effects of process parameters such as adsorbent particle size, influent pH, and influent volumetric flow-rate on the performance of the adsorption process in a column were evaluated. The maximum removal capacity of 110 mg/kg for VPum and 22 mg/kg for VSco were achieved at a particle size of 0.075–0.425 mm and <0.075 mm, respectively, at lower solution pH (2.00) and flow rate (1.25 mL/min). The increase in adsorbent particle size, solution pH, and flow rate decreases the breakthrough and saturation time of the column bed and, consequently, lowers the amount of fluoride removal by VSco. The breakthrough and exhaustion time on VPum was high at a particle size of 0.075–0.425 mm, at lower solution pH and flow rate similar to that of VSco. Thus, in order to attain optimum performance, suitable experimental parameters are significant for the operation of the adsorption column. The Thomas and Adams–Bohart models were applied to the experimental data to estimate the breakthrough curves and to determine fixed-bed column kinetic parameters. Both the Adam–Bohart and the Thomas models could predict very well the entire region of the breakthrough curves for the fluoride-VSco and fluoride-VPum system. The results show that VPum and VSco could be used in a fixed-bed adsorption column for the removal of excess fluoride from water. The supply cost of the two adsorbents is very low; nevertheless, an overall cost analysis of the purification system is very important as it has implications for the feasibility (technical and economic) of the adsorption method. Additional testing of the adsorbents including representative samples test for possible compositional, mineralogical, and textural changes in time due to weathering, leaching test, competitive ions effects, and regeneration options is required to confirm that the defluoridation of groundwater employing volcanic rocks is a safe and sustainable method.

Supplementary Materials: Table S1: Elemental composition and oxides content of VPum and VSco, Figure S1: Experimental (exp.) and simulated (cal.; Thomas model) breakthrough curves of fluoride at different particle sizes for (a) VPum and (b) VSco (pH 2.00; C_O: 10 mg/L; Q_O: 1.25 mL/min; bed depth 10 cm), Figure S2: Experimental and simulated (Thomas model) breakthrough curves of fluoride at different pH for (a) VPum: 0.075–0.425 mm (b) VSco: <0.075 mm (CO: 10 mg/L; QO: 1.25 mL/min; bed depth 10 cm), Figure S3: Experimental and simulated (Thomas model) breakthrough curves of fluoride at different influent flow rate for (a) VPum: 0.075–0.425 mm and (b) VSco: <0.075 mm (pH 2.00; C_O: 10 mg/L; bed depth 10 cm), Figure S4: Experimental (exp.) and simulated (cal.; Adams–Bohart model) breakthrough curves of fluoride at different particle sizes for (a) VPum and (b) VSco (pH 2.00; C_O: 10 mg/L; Q_O: 1.25 mL/min; bed depth 10 cm), Figure S5: Experimental and simulated (Adams–Bohart model) breakthrough curves of fluoride at different pH for (a) VPum: 0.075–0.425 mm and (b) VSco: <0.075 mm (C_O: 10 mg/L; Q_O: 1.25 mL/min; bed depth 10 cm), Figure S6: Experimental and simulated (Adams–Bohart model) breakthrough curves of fluoride at different flow rate for (a) VPum: 0.075–0.425 mm and (b) VSco: <0.075 mm (pH 2.00; C_O: 10 mg/L; bed depth 10 cm).

Author Contributions: W.S.G. prepared the adsorbents, designed and conducted the adsorption experiments, analyzed the data, and prepared the manuscript; E.A. supervised the research work, updated and reviewed the manuscript; B.L. supervised the research work, reviewed and edited the article. All authors have read and agreed to the published version of the manuscript.

Funding: This research received no external funding.

Institutional Review Board Statement: Not applicable.

Informed Consent Statement: Not applicable.

Data Availability Statement: The data used in this study can be available from the authors on reasonable request.

Acknowledgments: The first author is greatly thankful to the German Academic Exchange Service (DAAD) for providing the scholarship during the study.

Conflicts of Interest: The authors declare no conflict of interest.

Sample Availability: Samples of the adsorbents (VPum and VSco) are available from the authors.

References

1. Kumari, U.; Mishra, A.; Siddiqi, H.; Meikap, B.C. Effective Defluoridation of Industrial Wastewater by Using Acid Modified Alumina in Fixed-Bed Adsorption Column: Experimental and Breakthrough Curves Analysis. *J. Clean. Prod.* **2021**, *279*, 123645:1–123645:17. [CrossRef]
2. Kumari, U.; Behera, S.K.; Siddiqi, H.; Meikap, B.C. Facile Method to Synthesize Efficient Adsorbent from Alumina by Nitric Acid Activation: Batch Scale Defluoridation, Kinetics, Isotherm Studies and Implementation on Industrial Wastewater Treatment. *J. Hazard. Mater.* **2020**, *381*, 120917. [CrossRef]
3. Ye, Y.; Yang, J.; Jiang, W.; Kang, J.; Hu, Y.; Ngo, H.H.; Guo, W.; Liu, Y. Fluoride Removal from Water Using a Magnesia-Pullulan Composite in a Continuous Fixed-Bed Column. *J. Environ. Manag.* **2018**, *206*, 929–937. [CrossRef] [PubMed]
4. WHO World Health Organization. *Guidelines for Drinking-Water Quality*, 4th ed.; WHO World Health Organization: Geneva, Swizerland, 2011; ISBN 978-92-4-154815-1.
5. Rango, T.; Vengosh, A.; Jeuland, M.; Whitford, G.M.; Tekle-haimanot, R. Biomarkers of Chronic Fluoride Exposure in Groundwater in a Highly Exposed Population. *Sci. Total Environ.* **2017**, *596–597*, 1–11. [CrossRef] [PubMed]
6. Fufa, F.; Alemayehu, E.; Lennartz, B. Defluoridation of Groundwater Using Termite Mound. *Water. Air. Soil Pollut.* **2013**, *224*, 1552:1–1552:15. [CrossRef]
7. Rango, T.; Bianchini, G.; Beccaluva, L.; Tassinari, R. Geochemistry and Water Quality Assessment of Central Main Ethiopian Rift Natural Waters with Emphasis on Source and Occurrence of Fluoride and Arsenic. *J. Afr. Earth Sci.* **2010**, *57*, 479–491. [CrossRef]
8. Demelash, H.; Beyene, A.; Abebe, Z.; Melese, A. Fluoride Concentration in Ground Water and Prevalence of Dental Fluorosis in Ethiopian Rift Valley: Systematic Review and Meta-Analysis. *BMC Public Health* **2019**, *19*, 1298. [CrossRef]
9. Cai, J.; Zhang, Y.; Pan, B.; Zhang, W.; Lv, L.; Zhang, Q. Efficient Defluoridation of Water Using Reusable Nanocrystalline Layered Double Hydroxides Impregnated Polystyrene Anion Exchanger. *Water Res.* **2016**, *102*, 109–116. [CrossRef]
10. CSAE Central Statistical Agency of Ethiopia. *Drinking Water Quality in Ethiopia—Results from the 2016 Ethiopia Socioeconomic Survey*; CSAE Central Statistical Agency of Ethiopia: Addis Ababa, Ethiopia, 2017.
11. Kut, K.M.K.; Sarswat, A.; Srivastava, A.; Pittman, C.U., Jr.; Mohan, D. A Review of Fluoride in African Groundwater and Local Remediation Methods. *Groundw. Sustain. Dev.* **2016**, *2–3*, 190–212. [CrossRef]
12. Shen, J.; Schäfer, A.I. Factors Affecting Fluoride and Natural Organic Matter (NOM) Removal from Natural Waters in Tanzania by Nano Filtration/Reverse Osmosis. *Sci. Total Environ.* **2015**, *527–528*, 520–529. [CrossRef]
13. Smedley, P.L.; Nkotagu, H.; Pelig-Ba, K.; MacDonald, A.M.; TylerWhittle, R.; Whitehead, E.J.; Kinniburgh, D.G. *Fluoride in Groundwater from High-Fluoride Areas of Ghana and Tanzania*; Commissioned Report CR/02/316; British Geological Survey Groundwater Systems and Water Quality Programme: Nottingham, UK, 2002.
14. Mohan, S.; Kumar, V.; Singh, D.K.; Hasan, S.H. Synthesis and Characterization of RGO/ZrO2 Nanocomposite for Enhanced Removal of Fluoride from Water: Kinetics, Isotherm, and Thermodynamic Modeling and Its Adsorption Mechanism. *RSC Adv.* **2016**, *6*, 87523–87538. [CrossRef]
15. Araga, R.; Kali, S.; Sharma, C.S. Coconut-Shell-Derived Carbon/Carbon Nanotube Composite for Fluoride Adsorption from Aqueous Solution. *Clean Soil Air Water* **2019**, *47*, 1–9. [CrossRef]
16. Waghmare, S.S.; Arfin, T. Fluoride Removal from Water by Various Techniques: Review. *Int. J. Innov. Sci. Eng. Technol.* **2015**, *2*, 560–571.
17. Manna, S.; Saha, P.; Roy, D.; Adhikari, B.; Das, P. Fixed Bed Column Study for Water Defluoridation Using Neem Oil-Phenolic Resin Treated Plant Bio-Sorbent. *J. Environ. Manag.* **2018**, *212*, 424–432. [CrossRef]
18. Gill, T.; Tiwari, S.; Kumar, P.A. A Review on Feasibility of Conventional Fluoride Removal Techniques in Urban Areas. *Int. J. Environ. Res. Dev.* **2014**, *4*, 179–182.
19. Liu, J.; Yue, X.; Lu, X.; Guo, Y. Uptake Fluoride from Water by Starch Stabilized Layered Double Hydroxides. *Water* **2018**, *10*, 745. [CrossRef]
20. Yadav, K.K.; Gupta, N.; Kumar, V.; Khan, S.A.; Kumar, A. A Review of Emerging Adsorbents and Current Demand for Defluoridation of Water: Bright Future in Water Sustainability. *Environ. Int.* **2018**, *111*, 80–108. [CrossRef]
21. Su, T.; Song, Y.; Lan, X.; Gao, W. Optimization for Removal e Ffi Ciency of Fl Uoride Using La (III)–Al (III)-Activated Carbon Modi Fi Ed by Chemical Route. *Green Process. Synth.* **2020**, *9*, 405–415. [CrossRef]
22. Nehra, S.; Raghav, S.; Kumar, D. Biomaterial Functionalized Cerium Nanocomposite for Removal of Fluoride Using Central Composite Design Optimization Study. *Environ. Pollut.* **2020**, 113773:1–113773:32. [CrossRef]

23. Abu Bakar, A.H.; Abdullah, L.C.; Mohd Zahri, N.A.; Alkhatib, M. Column Efficiency of Fluoride Removal Using Quaternized Palm Kernel Shell (QPKS). *Int. J. Chem. Eng.* **2019**, 5743590:1–5743590:13. [CrossRef]
24. Kennedy, A.M.; Arias-Paic, M. Fixed-Bed Adsorption Comparisons of Bone Char and Activated Alumina for the Removal of Fluoride from Drinking Water. *J. Environ. Eng.* **2020**, *146*, 04019099. [CrossRef]
25. Alkurdi, S.S.A.; Al-Juboori, R.A.; Bundschuh, J.; Hamawand, I. Bone Char as a Green Sorbent for Removing Health Threatening Fluoride from Drinking Water. *Environ. Int.* **2019**, *127*, 704–719. [CrossRef] [PubMed]
26. Rajkumar, S.; Murugesh, S.; Sivasankar, V.; Darchen, A.; Msagati, T.A.M.; Chaabane, T. Low-Cost Fluoride Adsorbents Prepared from a Renewable Biowaste: Syntheses, Characterization and Modeling Studies. *Arab. J. Chem.* **2019**, *12*, 3004–3017. [CrossRef]
27. Ghanbarian, M.; Ghanbarian, M.; Mahvi, A.H.; Tabatabaie, T. Enhanced Fluoride Removal over MgFe2O4–Chitosan–CaAl Nanohybrid: Response Surface Optimization, Kinetic and Isotherm Study. *Int. J. Biol. Macromol.* **2020**, *148*, 574–590. [CrossRef]
28. Alemayehu, E.; Lennartz, B. Virgin Volcanic Rocks: Kinetics and Equilibrium Studies for the Adsorption of Cadmium from Water. *J. Hazard. Mater.* **2009**, *169*, 395–401. [CrossRef] [PubMed]
29. Aregu, M.B.; Asfaw, S.L.; Khan, M.M. Identification of Two Low—Cost and Locally Available Filter Media (Pumice and Scoria) for Removal of Hazardous Pollutants from Tannery Wastewater. *Environ. Syst. Res.* **2018**, *7*, 10:1–10:14. [CrossRef]
30. Asere, T.G.; Mincke, S.; De Clercq, J.; Verbeken, K.; Tessema, D.A.; Fufa, F.; Stevens, C.V.; Du Laing, G. Removal of Arsenic (V) from Aqueous Solutions Using Chitosan–Red Scoria and Chitosan–Pumice Blends. *Int. J. Environ. Res. Public Health* **2017**, *14*, 895. [CrossRef]
31. Asere, T.G.; Verbeken, K.; Tessema, D.A.; Fufa, F.; Stevens, C.V.; Du Laing, G. Adsorption of As(III) versus As(V) from Aqueous Solutions by Cerium-Loaded Volcanic Rocks. *Environ. Sci. Pollut. Res.* **2017**, *24*, 20446–20458. [CrossRef]
32. Birhane, M.; Abebe, A.; Alemayehu, E.; Mengistie, E. Efficiency of Locally Available Filter Media on Fluoride and Phosphate Removal for Household Water Treatment System. *Chin. J. Popul. Resour. Environ.* **2014**, *12*, 110–115. [CrossRef]
33. Viswanathan, N.; Meenakshi, S. Enriched Fluoride Sorption Using Alumina/Chitosan Composite. *J. Hazard. Mater.* **2010**, *178*, 226–232. [CrossRef] [PubMed]
34. Monash, P.; Pugazhenthi, G. Removal of Crystal Violet Dye from Aqueous Solution Using Calcined and Uncalcined Mixed Clay Adsorbents. *Sep. Sci. Technol.* **2010**, *45*, 94–104. [CrossRef]
35. Patel, H. Fixed—Bed Column Adsorption Study: A Comprehensive Review. *Appl. Water Sci.* **2019**, *9*, 45:1–45:17. [CrossRef]
36. US EPA United States Environmental Agency. *Control of Organic Substances in Water and Wastewater*; Document No.: U.S. EPA-600/8-83-011; U.S. Environmental Protection Agency: Washington, DC, USA, 1983.
37. Rowe, M.C.; Brewer, B.J. AMORPH: A Statistical Program for Characterizing Amorphous Materials by X-Ray Diffraction 1. *Comput. Geosci.* **2018**, *120*, 24–31. [CrossRef]
38. Li, X.; Yang, W.; Zou, Q.; Zuo, Y. Investigation on Microstructure, Compostion, and Cytocompatibility of Natural Pumice for Potential Biomedical Application. *Tissue Eng. Part C* **2010**, *16*, 427–434. [CrossRef]
39. Sepehr, M.N.; Amrane, A.; Karimaian, K.A.; Zarrabi, M.; Ghaffari, H.R. Potential of Waste Pumice and Surface Modified Pumice for Hexavalent Chromium Removal: Characterization, Equilibrium, Thermodynamic and Kinetic Study. *J. Taiwan Inst. Chem. Eng.* **2014**, *45*, 635–647. [CrossRef]
40. Pirsaheb, M.; Mohammadi, H.; Sharafi, K.; Asadi, A. Fluoride and Nitrate Adsorption from Water by Fe(III)-Doped Scoria: Optimizing Using Response Surface Modeling, Kinetic and Equilibrium Study. *Water Sci. Technol. Water Supply* **2017**, *18*, 1117–1132. [CrossRef]
41. Djobo, J.N.Y.; Tchadjié, L.N.; Tchakoute, H.K.; Kenne, B.B.D.; Elimbi, A. Synthesis of Geopolymer Composites from a Mixture of Volcanic Scoria and Metakaolin. *J. Asian Ceram. Socities* **2014**, *2*, 387–398. [CrossRef]
42. Panias, D.; Giannopoulou, I.P.; Perraki, T. Effect of Synthesis Parameters on the Mechanical Properties of Fly Ash-Based Geopolymers. *Colloids Surfaces A Physicochem. Eng. Asp.* **2007**, *301*, 246–254. [CrossRef]
43. Asgari, G.; Roshani, B.; Ghanizadeh, G. The Investigation of Kinetic and Isotherm of Fluoride Adsorption onto Functionalize Pumice Stone. *J. Hazard. Mater.* **2012**, *217–218*, 123–132. [CrossRef]
44. Moradi, M.; Mansouri, A.M.; Azizi, N.; Amini, J. Adsorptive Removal of Phenol from Aqueous Solutions by Copper (Cu)—Modified Scoria Powder: Process Modeling and Kinetic Evaluation. *Desalin. Water Treat.* **2015**, *57*, 1–15. [CrossRef]
45. Souza, R.F.; Brandão, P.R.G.; Paulo, J.B.A. Effect of Chemical Composition on the Zeta -Potential of Chromite. *Miner. Eng.* **2012**, *36–38*, 65–74. [CrossRef]
46. Ayoob, S.; Gupta, A.K. Performance Evaluation of Alumina Cement Granules in Removing Fluoride from Natural and Synthetic Waters. *Chem. Eng. J.* **2009**, *150*, 485–491. [CrossRef]
47. Salifu, A.; Petrusevski, B.; Ghebremichael, K.; Modestus, L.; Buamah, R.; Aubry, C.; Amy, G.L. Aluminum (Hydr) Oxide Coated Pumice for Fluoride Removal from Drinking Water: Synthesis, Equilibrium, Kinetics and Mechanism. *Chem. Eng. J.* **2013**, *228*, 63–74. [CrossRef]
48. Alemayehu, E.; Thiele-Bruhn, S.; Lennartz, B. Adsorption Behaviour of Cr (VI) onto Macro and Micro-Vesicular Volcanic Rocks from Water. *Sep. Purif. Technol.* **2011**, *78*, 55–61. [CrossRef]
49. Chen, N.; Zhang, Z.; Feng, C.; Li, M.; Chen, R.; Sugiura, N. Investigations on the Batch and Fixed-Bed Column Performance of Fluoride Adsorption by Kanuma Mud. *Desalination* **2011**, *268*, 76–82. [CrossRef]
50. Alemayehu, E.; Lennartz, B. Adsorptive Removal of Nickel from Water Using Volcanic Rocks. *Appl. Geochem.* **2010**, *25*, 1596–1602. [CrossRef]

51. Obijole, O.A.; Gitari, M.W.; Ndungu, P.G. Mechanochemically Activated Aluminosilicate Clay Soils and Their Application for Defluoridation and Pathogen Removal from Groundwater. *Int. J. Environ. Res. Public Health* **2019**, *16*, 654. [CrossRef]
52. Sepehr, M.N.; Zarrabi, M.; Kazemian, H.; Amrane, A.; Yaghmaian, K.; Ghaffari, H.R. Removal of Hardness Agents, Calcium and Magnesium, by Natural and Alkaline Modified Pumice Stones in Single and Binary Systems. *Appl. Surf. Sci.* **2013**, *274*, 295–305. [CrossRef]
53. Aksu, Z.; Çağatay, S.S.; Gönen, F. Continuous Fixed Bed Biosorption of Reactive Dyes by Dried Rhizopus Arrhizus: Determination of Column Capacity. *J. Hazard. Mater.* **2007**, *143*, 362–371. [CrossRef]
54. Golie, W.M.; Upadhyayula, S. Continuous Fixed-Bed Column Study for the Removal of Nitrate from Water Using Chitosan/Alumina Composite. *J. Water Process Eng.* **2016**, *12*, 58–65. [CrossRef]
55. Paudyal, H.; Pangeni, B.; Inoue, K.; Kawakita, H.; Ohto, K.; Alam, S. Adsorptive Removal of Fluoride from Aqueous Medium Using a Fixed Bed Column Packed with Zr (IV) Loaded Dried Orange Juice Residue. *Bioresour. Technol.* **2013**, *146*, 713–720. [CrossRef] [PubMed]
56. Ghorai, S.; Pant, K.K. Equilibrium, Kinetics and Breakthrough Studies for Adsorption of Fluoride on Activated Alumina. *Sep. Purif. Technol.* **2005**, *42*, 265–271. [CrossRef]
57. Salifu, A. Laboratory-Scale Column Filter Studies for Fluoride Removal with Aluminium (Hydr) Oxide Coated Pumice, Regeneration and Disposal. In *Fluoride Removal from Groundwater by Adsorption Technology: The Occurrence, Adsorbent Synthesis, Regeneration and Disposal*; CRC Press/Balkema: Leiden, The Netherlands, 2017; pp. 127–129. ISBN 9780815392071.
58. Han, R.; Wang, Y.; Yu, W.; Zou, W.; Shi, J.; Liu, H. Biosorption of Methylene Blue from Aqueous Solution by Rice Husk in a Fixed-Bed Column. *J. Hazard. Mater.* **2007**, *141*, 713–718. [CrossRef]
59. Yagub, M.T.; Sen, T.K.; Afroze, S.; Ang, H.M. Fixed-Bed Dynamic Column Adsorption Study of Methylene Blue (MB) onto Pine Cone. *Desalin. Water Treat.* **2015**, *55*, 1026–1039. [CrossRef]
60. Quintelas, C.; Pereira, R.; Kaplan, E.; Tavares, T. Removal of Ni (II) from Aqueous Solutions by an Arthrobacter Viscosus Biofilm Supported on Zeolite: From Laboratory to Pilot Scale. *Bioresour. Technol.* **2013**, *142*, 368–374. [CrossRef]
61. Kang, W.H.; Kim, E.I.; Park, J.Y. Fluoride Removal Capacity of Cement Paste. *Desalination* **2007**, *202*, 38–44. [CrossRef]
62. García-sánchez, J.J.; Solache-ríos, M.; Martínez-miranda, V.; Morelos, C.S. Removal of Fluoride Ions from Drinking Water and Fluoride Solutions by Aluminum Modified Iron Oxides in a Column System. *J. Colloid Interface Sci.* **2013**, *407*, 410–415. [CrossRef]
63. Ma, Y.; Shi, F.; Zheng, X.; Ma, J.; Gao, C. Removal of Fluoride from Aqueous Solution Using Granular Acid-Treated Bentonite (GHB): Batch and Column Studies. *J. Hazard. Mater.* **2011**, *185*, 1073–1080. [CrossRef] [PubMed]
64. Sivasankar, V.; Ramachandramoorthy, T.; Chandramohan, A. Fluoride Removal from Water Using Activated and MnO2-Coated Tamarind Fruit (Tamarindus Indica) Shell: Batch and Column Studies. *J. Hazard. Mater.* **2010**, *177*, 719–729. [CrossRef] [PubMed]
65. Ghorai, S.; Pant, K.K. Investigations on the Column Performance of Fluoride Adsorption by Activated Alumina in a Fixed-Bed. *Chem. Eng. J.* **2004**, *98*, 165–173. [CrossRef]
66. Alemayehu, E.; Melak, F.; Sharma, S.K.; Van Der Bruggen, B.; Lennartz, B. Use of Porous Volcanic Rocks for the Adsorptive Removal of Copper. *Water Environ. J.* **2017**, *31*, 4–11. [CrossRef]
67. Kwon, J.-S.; Yun, S.-T.; Lee, J.-H.; Kim, S.-O.; Jo, H.Y. Removal of Divalent Heavy Metals (Cd, Cu, Pb, and Zn) and Arsenic (III) from Aqueous Solutions Using Scoria: Kinetics and Equilibria of Sorption. *J. Hazard. Mater.* **2010**, *174*, 307–313. [CrossRef]
68. Liu, C.; Evett, J.B. *Soil Properties-Testing, Measurament, and Evalution*; Banta Book Company: Upper Saddle River, NJ, USA, 2003; ISBN 0-13-093005-9.
69. Chen, I.-P.; Kan, C.-C.; Futalan, C.; Calagui, M.J.; Lin, S.-S.; Tsai, W.-C.; Wan, M.-W. Batch and Fixed Bed Studies: Removal of Copper (II) Using Chitosan-Coated Kaolinite Beads from Aqueous Solution. *Sustain. Environ. Res.* **2015**, *25*, 73–81.
70. De Gennaro, B.; Aprea, P.; Liguori, B.; Galzerano, B.; Peluso, A.; Caputo, D. Zeolite-Rich Composite Materials for Environmental Remediation: Arsenic Removal from Water. *Appl. Sci.* **2020**, *10*, 6939. [CrossRef]
71. Chen, S.; Yue, Q.; Gao, B.; Li, Q.; Xu, X.; Fu, K. Adsorption of Hexavalent Chromium from Aqueous Solution by Modified Corn Stalk: A Fixed-Bed Column Study. *Bioresour. Technol.* **2012**, *113*, 114–120. [CrossRef]
72. Worch, E. *Adsorption Technology in Water Treatment: Fundamentals, Processes, and Modeling*; Walter de Gruyter and Co.KG: Berlin, Germany, 2012; ISBN 9783110240221.
73. Mohan, S.; Singh, D.K.; Kumar, V.; Hasan, S.H. Modelling of Fixed Bed Column Containing Graphene Oxide Decorated by MgO Nanocubes as Adsorbent for Lead (II) Removal from Water. *J. Water Process Eng.* **2017**, *17*, 216–228. [CrossRef]
74. Thomas, H.C. Heterogeneous Ion Exchange in a Flowing System. *J. Am. Chem. Soc.* **1944**, *66*, 1664–1666. [CrossRef]
75. Singh, T.P.; Ghosh, S.; Cb, M. Adsorption of Fluoride from Industrial Wastewater in Fixed Bed Column Using Java Plum (Syzygium Cumini). *Asian J. Pharm. Clin. Res.* **2016**, *9*. [CrossRef]
76. Bohart, G.S.; Adams, E.Q. Some Aspects of the Behavior of Charcoal with Respect to Chlorine. *J. Am. Chem. Society* **1920**, *42*, 523–544. [CrossRef]
77. Chu, K.H. Breakthrough Curve Analysis by Simplistic Models of Fixed Bed Adsorption: In Defense of the Century-Old Bohart-Adams Model. *Chem. Eng. J.* **2020**, *380*, 122513:1–122513:8. [CrossRef]
78. Han, R.; Zou, L.; Zhao, X.; Xu, Y.; Xu, F.; Li, Y.; Wang, Y. Characterization and Properties of Iron Oxide-Coated Zeolite as Adsorbent for Removal of Copper (II) from Solution in Fixed Bed Column. *Chem. Eng. J.* **2009**, *149*, 123–131. [CrossRef]

molecules MDPI

Article

Application of Natural Clinoptilolite for Ammonium Removal from Sludge Water

Stephan Wasielewski [1,*], Eduard Rott [1], Ralf Minke [1] and Heidrun Steinmetz [2]

1 Institute for Sanitary Engineering, Water Quality and Solid Waste Management (ISWA), University of Stuttgart, Bandtaele 2, 70569 Stuttgart, Germany; eduard.rott@iswa.uni-stuttgart.de (E.R.); ralf.minke@iswa.uni-stuttgart.de (R.M.)
2 Faculty of Civil Engineering, University of Kaiserslau-tern, Paul-Ehrlich-Str. 14, 67663 Kaiserslautern, Germany; heidrun.steinmetz@bauing.uni-kl.de
* Correspondence: stephan.wasielewski@iswa.uni-stuttgart.de; Tel.: +49-711-685-65425

Abstract: Sludge water (SW) arising from the dewatering of anaerobic digested sludge causes high back loads of ammonium, leading to high stress (inhibition of the activity of microorganisms by an oversupply of nitrogen compounds (substrate inhibition)) for wastewater treatment plants (WWTP). On the other hand, ammonium is a valuable resource to substitute ammonia from the energy intensive Haber-Bosch process for fertilizer production. Within this work, it was investigated to what extent and under which conditions Carpathian clinoptilolite powder (CCP 20) can be used to remove ammonium from SW and to recover it. Two different SW, originating from municipal WWTPs were investigated (SW1: c_0 = 967 mg/L NH_4-N, municipal wastewater; SW2: c_0 = 718–927 mg/L NH_4-N, large industrial wastewater share). The highest loading was achieved at 307 K with 16.1 mg/g (SW1) and 15.3 mg/g (SW2) at 295 K. Kinetic studies with different specific dosages (0.05 $g_{CLI}/mg_{NH4\text{-}N}$), temperatures (283–307 K) and pre-loaded CCP 20 (0–11.4 mg/g) were conducted. At a higher temperature a higher load was achieved. Already after 30 min contact time, regardless of the sludge water, a high load up to 7.15 mg/g at 307 K was reached, achieving equilibrium after 120 min. Pre-loaded sorbent could be further loaded with ammonium when it was recontacted with the SW.

Keywords: ammonia; ammonium recovery; Freundlich; intraparticle diffusion; isoelectric state; Langmuir; pseudo-second-order; Temkin; zeolite; high-strength wastewater; sludge liquor

Citation: Wasielewski, S.; Rott, E.; Minke, R.; Steinmetz, H. Application of Natural Clinoptilolite for Ammonium Removal from Sludge Water. *Molecules* **2021**, *26*, 114. https://doi.org/10.3390/molecules26010114

Academic Editors: Chiara Bisio and Monica Pica

Received: 1 December 2020
Accepted: 23 December 2020
Published: 29 December 2020

Publisher's Note: MDPI stays neutral with regard to jurisdictional claims in published maps and institutional affiliations.

1. Introduction

In view of the world's population growing from 7.6 billion in 2017 to estimated 9.4–10.2 billion people by 2050, but also in consideration of rising living standards, correlating with increasing meat consumption, an increase of food requirements by 50% between 2012 and 2050 is to be expected [1]. This increased demand cannot be satisfied by the utilization of new farmland alone, since most of it is not developed, too remote from potential markets, susceptible to pest infestation or new cultivated would compete with the conservation of important ecosystems. Furthermore, potential arable land is limited to a small number of countries. Rather, increasing productivity and efficiency in agricultural production must contribute to meet the increased demand [2], resulting in a greater need for nutrients, especially nitrogen fertilizers.

Nowadays, the nutrition of half the world's population is ensured by the Haber-Bosch process, which enables the synthesis of ammonia (NH_3) for fertilizer production [3]. However, the production of NH_3 requires a high amount of energy (10 kWh/kg NH_3) [4]. Dawson and Hilton [5] calculated that 1.1% of the world's energy consumption can be attributed to the production of fertilizers; 90% of it due to the production of nitrogen fertilizers.

On the other hand, ammonium has severe negative environmental impacts. Because of its eutrophication potential, ammonium contributes to the growth of biomass in water bodies. Under alkaline conditions, which can occur during the day if intensive photosynthesis takes place, ammonium dissociates to ammonia, which has a toxic effect on aquatic fauna even in low concentrations.

In wastewater treatment rejected sludge water from the dewatering process of anaerobically treated sludge causes significant additional ammonium loads for the biological treatment step [6], resulting in additional need for energy and space. Additionally, external carbon sources might be necessary for a stable nitrogen elimination process. Instead of elimination and high-energy expenditure, the ammonium should be recovered to partially substitute the increasing worldwide demand for NH_3.

Recovery methods such as air stripping, bioelectrochemical systems, membrane separation, and ion exchange have been thoroughly investigated. However, these methods require additional chemicals as well as energy, and ammonia losses due to volatilization can occur [7].

The zeolite clinoptilolite (CLI) is known to be a very good ion exchanger, as it consists of a three-dimensional tetrahedral structure formed of AlO_4^- and SiO4, connected by a shared oxygen atom. The micropores formed by this structure are fine enough to allow entry and exchange of cations and water molecules [8]. This ability is based on the substitution of SiO_4 by AlO_4^-, leading to a negative charge in the structure, which has to be compensated by exchangeable cations such as Na^+, K^+, Ca^{2+}, and Mg^{2+} [9]. In a previous study with carpathian Clinoptilolite powder (CCP 20), 21.0 meq/100 g Na^+, 49.3 meq/100 g K^+, 65.6 meq/100 g Ca^{2+}, and 3.3 meq/100 g Mg^{2+} were exchanged with 136.9 meq/100 g NH_4^+ [10].

As soon as the exchange capability for ammonium is exhausted, CLI is proposed to be utilized as a slow-release fertilizer in agriculture [11] or regenerated by the use of sodium chloride, sodium carbonate, sodium bicarbonate, or sodium hydroxide solutions [12–15].

In a study investigating the adsorption of ammonium from different highly concentrated wastewaters, it was shown that elimination from leachate of a sewage sludge landfill (c_0 = 11.12–115.16 mg/L NH_4-N) was 10–20% lower than from a matrix-free solution [16]. It has been demonstrated, that ammonium from swine manure (c_0 = 0.43 M/L $NH_4^+ \approx$ 6150 mg/L NH_4-N) can be removed by means of CLI, but the load is reduced from 10 mg/g (matrix-free ammonium solution) to 2 mg/g (manure) due to the cations contained [17]. Furthermore, dilution of leachate from a municipal landfill (c_0 = 2292 mg/L NH_4-N) does not improve ammonium adsorption [18]. In addition, organic compounds that are not removed by activated carbon interfere with the sorption of ammonium from leachate (c_0 = 820 mg/L NH_4^+ = 637 mg/L NH_4-N) at CLI; the load increases when leachate is pretreated with activated carbon [19]. The results of these studies show that the composition of the medium to which CLI is applied might be decisive for the adsorption effect. Since the adsorption of ammonium from sludge water, has not been sufficiently investigated, the adsorption process in this complex medium is not well understood and the technical implementation is uncertain.

The objective of this study was to develop a deeper understanding of the factors influencing the sorption of ammonium from sludge water on powdered natural clinoptilolite.

2. Materials and Methods

2.1. Zeolite Samples and Chemicals

Since preliminary studied Slovakian CLI CCP 20 (CCP = Carpathian clinoptilolite powder) showed favorable sorption properties [10], it was employed for this study. It was obtained from the supplier Labradorit GmbH (Berlin, Germany), in a ground and sieved form (particle size smaller than 20 μm). The CLI was dried at 378 K for 24 h before use. No other pretreatment was conducted. The CCP 20 mainly consisted of Si (35.5% (*wt*/*wt*)), Al (5.4%), K (2.0%), Ca (1.6%), Fe (1.0%), Na (0.4%), Mg (0.3%), Ti (0.1%), Ba (0.08%), and Pb (0.001%), whereas Cr, Ni, As, Rb, Cd, Cs, Ba, Hg, and Tl were below the limit of detection.

NH_4Cl (p.a.), NaOH (p.a.) and HCl (32%, p.a.) were obtained from VWR International (Radnor, PA, USA).

2.2. Sludge Water and Matrix-Free Solution

Sludge water from two different WWTPs with high concentrations of ammonium nitrogen (see Table 1) were investigated. The sludge water originated from the dewatering process of anaerobically stabilized sludge by means of a chamber filter press. During normal operation, this sludge water is recycled to the main treatment process.

Table 1. Physical properties and constituents of the examined sludge waters.

Parameter	Unit	SW1	SW2	NH_4Cl
pH	–	7.9	8.0	5.3
Conductivity (298 K)	mS/cm	7.10	7.98	9.60
SS	mg/L	195	71–212	–
COD	mg/L	517	340–362	–
COD dissolved	mg/L	304	279–300	–
NH_4-N	mg/L	967	718–927	1000
NO_3-N	mg/L	<2	<1	–
NO_2-N	mg/L	<0.10	3.49	–
PO_4-P	mg/L	45.7	0.75–4.4	–
F^-	mg/L	n.d.	1.49	–
Cl^-	mg/L	174	266–292	–
Br^-	mg/L	<0.30	0.93	–
SO_4^{2-}	mg/L	~30	< 100	–

n.d.: not detected.

SW1 originated from a WWTP treating mainly municipal wastewater using the activated sludge process with upstream denitrification. Iron chloride sulphate is employed for phosphate precipitation. Primary sludge from mechanical treatment, secondary sludge from the biological stage and precipitated sludge from phosphorus elimination are admixed and anaerobically digested. Before dewatering, the digested sludge is pre-thickened and subsequently dewatered by means of a chamber filter press. For the investigations, the sludge water was taken from the outlet of the chamber filter press.

SW2 originated from a WWTP with a high industrial wastewater share. The wastewater is treated by trickling filters and a downstream denitrification. A mixture of aluminum and iron salts is employed as precipitant for phosphate elimination. Sludge from primary and secondary treatment as well as from phosphate elimination are pre-thickened and subsequently anaerobically digested. Afterwards, the digested sludge is dewatered with a chamber filter press, the outlet of which was sampled to gain SW2 for the investigations. From SW2, several samples were examined, which were taken at different times.

Matrix-free ammonium chloride solution (c_0 = 1000 mg/L NH_4-N) was prepared by dissolving NH_4Cl in distilled water.

Table 1 summarizes the physical properties and constituents of the examined sludge waters.

Both sludge waters showed only minor differences in composition except of COD (chemical oxygen demand) and PO_4-P. The COD concentration in SW1 was about twice as high as in SW2, but the dissolved COD concentration (filtered by 0.45 μm nylon membrane) was similar. Presumably, the retention of COD-causing particles in the sludge dewatering of SW1 was less efficient than in SW2. In addition, the concentration of PO_4-P in SW1 was considerably higher (factor 10).

In Table 2 the elementary composition of the examined sludge waters is listed.

Table 2. Elementary composition of the examined slugde waters.

Element	Unit	SW1	SW2
B	mg/L	0.5	0.294–0.390
Na	mg/L	126	129–132
Mg	mg/L	49.5	52.0–62.4
Al	mg/L	0.472/0.823	0.163–1.05
K	mg/L	234	87.5–91.0
Ca	mg/L	65	137–146
Cr	mg/L	0.002	<0.0001
Fe	mg/L	7.6	13.6
Ni	mg/L	0.15	0.030–0.042
Cu	mg/L	0.015	0.012–0.026
Zn	mg/L	n.d.	0.036–0.060
As	mg/L	<0.2	<0.0001
Se	mg/L	0.12	n.d.
Rb	mg/L	0.163	0.070
Sr	mg/L	0.075	0.721–0.770
Cd	mg/L	<0.02	<0.0001
Cs	mg/L	0.024	<0.003
Ba	mg/L	<0.02	0.012
Hg	mg/L	<0.02	<0.0001
Tl	mg/L	<0.02	<0.0001
Pb	mg/L	<0.005	0.002

n.d.: not detected.

The investigated sludge waters contained very low concentrations of heavy metals. They differed in their concentration of potassium and calcium, whereas the sodium and magnesium concentrations were almost identical.

Cations competing with ammonium ions for sorption sites such as K^+ and Na^+ were present in lower concentrations (K^+: SW1: 234 mg/L = 6.0 mmol/L; SW2: 87.5–91.0 mg/L = 2.2–2.3 mmol/L; Na^+: SW1: 126 mg/L = 10.3 mmol/L; SW2: 129–132 mg/L = 5.6–5.7 mmol/L) as ammonium (SW1: 967 mg/L NH_4-N = 67.7 mmol/L; SW2: 718–913 mg/L NH_4-N = 51.3–65.2 mmol/L). Accordingly, ammonium was present in multiple excess (SW1: $[NH_4^+]/[K^+] \approx 5$; $[NH_4^+]/[Na^+] \approx 10$; SW2: $[NH_4^+]/[K^+] \approx$ 234–30; $[NH_4^+]/[Na^+] \approx$ 9–12).

The ion ratios of ammonium to potassium and sodium were more favorable (high ammonium surplus) than in leachate from a sewage sludge landfill ($[NH_4^+]/[Na^+] \approx 2.7$; $[NH_4^+]/[K^+] \approx 14.6$ [16]) and leachate ($[NH_4^+]/[K^+] \approx 2.2$; $[NH_4^+]/[Na^+] \approx 0.9$ [19]), whereas a larger ammonium surplus was set in synthetic wastewater ($[NH_4^+]/[K^+] \approx 17.3$; $[NH_4^+]/[Na^+] \approx 6.9$ [20]).

2.3. *Experimental Design*

2.3.1. Isoelectric State of CLI and pH-Dependent Adsorption

Both sludge waters were adjusted to pH values ranging from 2 to 12 by HCl or NaOH prior to the experiment. A fixed specific sorbent mass at a ratio of 0.1 g CLI per mg NH_4-N was employed. Sorbent (20 g) and solution (200 mL) were stirred for 20 h on a magnetic stirrer (400 rpm) at room temperature (295 K) in closed bottles, subsequently membrane-filtered (0.45 μm pore size), and then the pH value as well as the ammonium concentration were determined. The initial pH values of the solutions were compared with those of the filtrates. The isoelectric state is the point at which both pH values are identical.

Sodium ions, which are added to the sample by NaOH to adjust the pH, compete with ammonium for the sorption sites in the CLI. In order to show the influence of the sodium ions without additionally pH value, equimolar NaCl was added. All experiments were conducted as triplicates. The titration curves of NH_4Cl solution, SW1 and SW2 are also depicted in Appendix A.

2.3.2. Isothermal Adsorption

Since sludge water originates from the digestion tower, operated moistly at mesophilic conditions, the influence of temperature on CLI loading was investigated. For this purpose, temperatures of 307 K (34 °C, mesophilic conditions in the digestion tower, considering small heat losses), 295 K (22 °C, room temperature) and 283 K (10 °C) were tested.

Since the ammonium concentration in the sludge water could not be changed, the amount of sorbent mass m (g) was varied instead. Thus, different quantities ranging from 2 g to 48 g sorbent were mixed with 200 mL sludge water V_p (mL) and stirred at a constant temperature (283 K, 295 K, and 307 K) on a magnetic stirrer at 400 rpm. After 20 h, the residual ammonium concentration c_{eq} (mg/L) as well as the pH in the filtrate were determined. Since the pH barely varied between the different dosages and a competing adsorption by Na^+ or H_3O^+ cations as well as a dilution due to the pH adjustment was to be avoided, a pH correction was not conducted. One experimental approach without sorbent for each examined pH was used to determine unwanted ammonium elimination, e.g., by stripping or adsorption onto parts of the glass apparatus. All experiments were conducted as triplicates. The ammonium concentration in the filtrate of that approach is expressed as c_B (mg/L). The loading q_{eq} (mg/g) of the sorbent mass was determined by Equation (1).

$$q_{eq} = \frac{(c_0 - (c_0 - c_B) - c_{eq}) \times (\frac{V_P}{1000})}{m} \quad (1)$$

2.3.3. Adsorption Kinetics

The influence of temperature (283 K–307 K; constant specific sorbent ratio of 0.1 $g_{CLI}/mg_{NH4\text{-}N}$; non pre-loaded CLI), the influence of the specific sorbent ratio (0.05–0.2 g CLI per mg NH_4-N; constant temperature 295 K; non pre-loaded CLI), as well as the influence of pre-load (0–11.4 $mg_{NH4\text{-}N}/g_{CLI}$; constant sorbent ratio 0.1 $g_{CLI}/mg_{NH4\text{-}N}$; constant temperature 307 K) were investigated in kinetic experiments. Furthermore, a matrix-free ammonium chloride solution (c_0 = 1000 mg/L NH_4-N) was examined as well. The sorption properties of CCP 20, including isoelectric state, pH-dependent elimination, isothermal adsorption, and thermodynamic properties were already investigated with similar matrix-free solution by Wasielewski et al. [10].

CCP 20 was mixed with the sample (1.5 L) on a magnetic stirrer. All experiments were conducted as triplicates. At periodic intervals, an aliquot (10 mL) was taken and immediately membrane-filtered (nylon membrane, 0.45 µm pore size) to prevent further contact between sorbent (CLI) and sample (NH_4Cl solution or sludge water). Subsequently, the ammonium concentration was measured in the filtrate and the time-dependent loading of the sorbent $q(t)$ (mg/g) was calculated. Since it is known from published studies that the adsorption kinetics strongly depend on the stirring speed [21–23], a high rotation frequency of 800 rpm was chosen to determine the maximum possible adsorption kinetic values. Due to sampling during the test, the total volume was continuously reduced. However, it can be assumed that during the sampling no change in the ratio of the sorbent mass to the volume of the solution occurred due to the homogeneously mixed conditions.

2.4. Adsorption Models

2.4.1. Freundlich Model

The nonideal, reversible adsorption of a heterogenous surface is described by the empirical Freundlich model [24]. It is not possible to calculate a complete loading, as the sorbent sites can be occupied in several layers. The loading of the sorbent $q_{eq,F}$ (mg/g) can be calculated by exponentiation of the corresponding equilibrium concentration c_{eq} (mg/L) with the factor $1/n$ (–), as described by Equation (2).

$$q_{eq,F} = K_F \, c_{eq}^{\frac{1}{n}} \quad (2)$$

Calculation methods for determining the constants K_F and *1/n* with the help of nonlinear regression or linearization are given, e.g., by Ho et al. [25]. In this study, the linearization was done by plotting log q_{eq} versus log c_{eq}. The gradient of the graph corresponds to n, while the tenth power of the intercept represents K_F.

2.4.2. Langmuir Model

A monomolecular layer of adsorbate on the available sorption sites is assumed according to the adsorption model of Langmuir [26]. Thus, the properties of the sorbent sites are identical and equivalent, so that a determination of the maximum adsorption capacity is possible. The loading of sorbent is calculated according to Equation (3), where K_L (L/mg) is the Langmuir constant and q_{max} (mg/g) the maximum capacity.

$$q_{eq,L} = \frac{q_{max} K_L c_{eq}}{1 + K_L c_{eq}} \quad (3)$$

The constants can either be deduced from linear or nonlinear regression based on measurement results. By plotting c_{eq}/q_{eq} vs. c_{eq}, $1/q_{eq}$ vs. $1/c_{eq}$, q_{eq} vs. q_{eq}/c_{eq}, or q_{eq}/c_{eq} vs. q_{eq}, a linear relationship for Equation (3) can be deduced [27]. Table 3 lists the four possible linear forms for determining Langmuir constants. In this study, only the type of isotherm with the highest coefficient of determination r^2 is listed. The coefficient of determination r^2 of the nonlinear form of the Langmuir isotherm and the experimentally determined loads q_{eq} and the arithmetical average loads $\overline{q_{eq}}$ were calculated according to Equation (4).

$$r^2 = \frac{\sum \left(q_{eq,L} - \overline{q_{eq}}\right)^2}{\sum \left(q_{eq,L} - \overline{q_{eq}}\right)^2 + \sum \left(q_{eq,L} - q_{eq}\right)^2} \quad (4)$$

Table 3. Linear forms of the Langmuir isotherm (according to [28]).

Type	Linear Form	Plot	K_L	q_{max}
I	$\frac{c_{eq}}{q_{eq}} = \frac{1}{q_{max}K_L} + \frac{c_{eq}}{q_{max}}$	$\frac{c_{eq}}{q_{eq}}$ vs. c_{eq}	slope/intercept	1/slope
II	$\frac{1}{q_{eq}} = \left[\frac{1}{q_{max}K_L}\right]\frac{1}{c_{eq}} + \frac{1}{q_{max}}$	$\frac{1}{q_{eq}}$ vs. $\frac{1}{c_{eq}}$	intercept/slope	1/intercept
III	$q_{eq} = q_{max} - \left[\frac{1}{K_L}\right]\frac{q_{eq}}{c_{eq}}$	q_{eq} vs. $\frac{q_{eq}}{c_{eq}}$	1/slope	intercept
IV	$\frac{q_{eq}}{c_{eq}} = q_{max}K_L - c_{eq}K_L$	$\frac{q_{eq}}{c_{eq}}$ vs. q_{eq}	(−1) slope	intercept/slope

2.4.3. Temkin Model

The isothermal loading of sorbents according to Temkin ([29] in [30]) is extended by the temperature parameter. Accordingly, the adsorption enthalpy is linearly proportional to the loading on the sorbent [31]. The form of the isotherm used in this work is taken from Ho et al. [25] (Equation (5)), where R is the universal gas constant (8.314459 J/(mol K)), T the temperature (K), b_T (1/mol), and A_T (L/mg) the Temkin isothermal constants.

$$q_{eq,T} = \frac{RT}{b_T} ln(A_T c_{eq}) \quad (5)$$

The linearized form of the Temkin isotherm is shown in Equation (6).

$$q_{eq,T} = \frac{RT}{b_T} ln(A_T) + \frac{RT}{b_T} ln(c_{eq}) \quad (6)$$

In a plot of ln c_{eq} vs. q_{eq}, the term RT/b_T is represented by the slope, whereas the intersection with the ordinate represents the term $RT\ ln(A_T)/b_T$. Subsequently, b_T and A_T can be deduced.

2.4.4. Thermodynamic Calculations

Energy adsorption or release, i.e., temperature increase or decrease, can be observed during the adsorption process. The standard free energy ΔG^0 (kJ/mol) can be calculated according to the following Equation (7)

$$\Delta G^0 = -RTln(K_d) \tag{7}$$

where K_d is the thermodynamic equilibrium constant, here the Freundlich constant (L/g). According to Milonjic [32], it should be noted that K_d must be dimensionless. Therefore, the use of the temperature-dependent equilibrium constant K_F must be corrected by a factor of 1000 g/L (density of water) into its dimensionless form. The relationship of the other thermodynamic parameters such as change in enthalpy ΔH^0 (kJ/mol) and change in standard entropy ΔS^0 (J/(mol K)) can be derived by means of the Gibbs–Helmholtz Equation (8).

$$\Delta G^0 = \Delta H^0 - T\Delta S^0 \tag{8}$$

From the plot of the logarithmic equilibrium constant K_d against the reciprocal value of the temperature *1/T* (Van't–Hoff diagram), a linear correlation can be derived. Here, the gradient corresponds to the quotient of the negative change in the free standard enthalpy ΔH^0 and the universal gas constant *R*. Furthermore, the quotient of the change of the free molar standard entropy ΔS^0 and the universal gas constant can be derived from the axis section.

Endothermic adsorption is described by a positive value of ΔH^0, meaning energy is absorbed by the adsorption process. A negative value indicates exothermic adsorption, meaning energy is being released. A spontaneous (exergonic) adsorption is expressed by negative ΔG^0, while negative ΔS^0 indicates a random adsorption behavior.

2.5. Kinetic Models

2.5.1. Intraparticle Diffusion

A mathematical description of the diffusion process is provided by the intraparticle diffusion model (ID). It presumes a correlation between the loading rate k_{ID} (mg/(min$^{0.5}$ g)) and the square root of the contact time *t* (min) ([33] in [34]). However, McKay et al. [35] extended this model by the constant C (mg/g), which is proportional to the thickness of the boundary layer as well as the initial adsorption by it. The time-dependent loading of the sorbent $q(t)_{ID}$ (mg/g) can be calculated by Equation (9).

$$q(t)_{ID} = k_{ID}t^{0.5} + C \tag{9}$$

To determine the loading rate k_{ID}, $q(t)$ versus $t^{0.5}$ is plotted. The slope of the resulting graph corresponds to k_{ID} while the intersection with the ordinate corresponds to C. Sole intraparticle diffusion occurs when the graph intersects the origin (C = 0). If a multistage diffusion process is present, two or more partial lines passing into each other can be approximated to the existing empirical measuring points of $q(t)$.

2.5.2. Pseudo-Second-Order

The time-dependent loading of the sorbent can be described by the pseudo-second-order (PSO) model according to Ho and McKay [36]. However, it is not possible to deduce the prevailing adsorption kinetic processes when using this model. It offers a macroscopic view of the adsorption process, based on the assumption that the adsorption rate is dependent on the loading of the ion exchange material at a certain point in time and its equilibrium state. The differential form of the PSO, i.e., as the differential of the load $q(t)$ (mg/g) at any time *t*, is given in Equation (10)

$$\frac{dq_{t,PSO}}{dt} = k_2(q_e - q_t)^2 \tag{10}$$

where k_2 is the pseudo-second-order rate (mg/(g min)) and q_e (mg/g) the load at equilibrium. From the integration of Equation (10) with the boundary conditions $q(t) = 0$ at $t = 0$ and $q(t) = q(t)$ at $t = t$, four different linear forms of the PSO model can be obtained (Table 4).

Table 4. Linear forms of PSO model (according to [37]).

Type	Linear Form	Plot	k_2	q_e
I	$\frac{t}{q(t)} = \frac{1}{k_2 q_e^2} + \frac{t}{q_e}$	$\frac{t}{q(t)}$ vs t	$(\text{slope})^2/\text{intercept}$	1/slope
II	$\frac{1}{q(t)} = \left[\frac{1}{k_2 q_e^2}\right]\frac{1}{t} + \frac{1}{q_e}$	$\frac{1}{q(t)}$ vs $\frac{1}{t}$	$(\text{intercept})^2/\text{slope}$	1/intercept
III	$q(t) = q_e - \left[\frac{1}{k_2 q_e}\right]\frac{q(t)}{t}$	$q(t)$ vs $\frac{q(t)}{t}$	$(-1)/(\text{slope} \times \text{intercept})$	intercept
IV	$\frac{q(t)}{t} = k_2 q_e^2 - k q_e q(t)$	$\frac{q(t)}{t}$ vs $q(t)$	$(\text{slope})^2/\text{intercept}$	−intercept/slope

In this study, only the type with the highest coefficient of determination r^2 (Equation (4)) is listed. All calculations in this study were conducted using Microsoft Excel 2019.

2.6. Analytical Methods

Ammonium was measured according to German standard DIN 38406-5 [38]. At a pH of about 12.6, ammonium cations and ammonia contained in the sample react with hypochlorite ions and salicylate ions in the presence of sodium pentacyanonitrosylferrate (2-)(nitroprusside sodium) as a catalyst to form a blue dye. The required hypochlorite ions are formed in the alkaline medium by hydrolysis of the dichloroisocyanuric acid ions. The spectral absorbance of the blue dye at 655 nm wavelength is linearly proportional to the ammonium concentration.

For determination of pH, probes (SenTix 950 + Multi 3430, WTW, Weilheim, Germany) were used.

F^-, Cl^-, NO_2-N, NO_3-N, Br^-, SO_4^{2-}, and PO_4-P were analyzed according to ISO 10304-1 [39] using the Dionec ICS-110 ion chromatograph (Thermo Fischer Scientific, Waltham, MA USA). Before the determination, the sample was filtered through a C18 cartridge (Strata C18-E (55 µm, 70 Å), Phenomenex, Torrance, USA) and diluted if necessary.

To determine the elementary composition of the sludge waters, 44 mL of sample were admixed with 2 mL HCl (32%), 3 mL HNO_3 (65%), 1 mL H_2O_2 (30%) and digested by a microwave (Start, MLS GmbH, Leutkirch, Germany) with a selected program run of 10 min at 443 K and a subsequent cooling phase of 20 min.

To determine the chemical elements of the zeolite, 0.3 to 0.5 g of the CLI were weighed and mixed with 6 mL HNO_3 (65%), 4 mL HF (48%), and 2 mL HCl (32%). The mixture was digested by microwave with a selected program run of 10 min at 383 K, then 5 min at 413 K, and finally 9 min at 463 K. Together with the cooling phase, the digestion lasted 64 min.

Heavy metals were analyzed by inductively coupled plasma mass spectrometry (Nexion 2000, Perkin Elmer, Waltham, MA, USA).

3. Results and Discussion

3.1. Isoelectric State of CLI and pH-Dependent Adsorption

In Figure 1 the final pH of SW1 and SW2 filtrates are plotted as a function of the initial pH after contact with CCP 20. The arbitrary pH value of SW1 was 7.9 and that of SW2 was 8.0. In the alkaline range, no considerable change in the pH was observed. This can be attributed to the decrease in ammonium uptake, as uncharged NH_3 is formed at pH >8, which is not adsorbed by the CCP 20. As a result, no cations are eluted that could lead to a change in the pH.

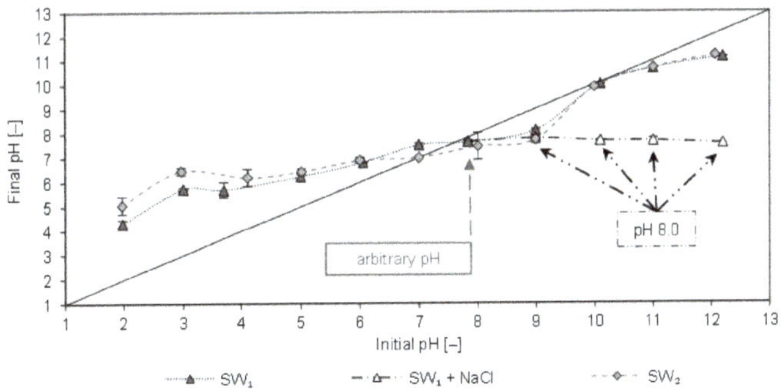

Figure 1. Final pH of the filtrates after 20 h contact with CCP 20 (SW1: c_0 = 967 mg/L NH_4-N; SW2: c_0 = 927 mg/L NH_4-N; T = 295 K; sorbent ratio 0.1 $g_{CLI}/mg_{NH4\text{-}N}$) as a function of the initial pH (adjusted pH of the solution before contact with CLI).

The pH increased in the acidic range (2–6), which can be attributed to the removal of NH_4^+ and the elution of cations (e.g., Na^+, K^+, Ca^{2+}, and Mg^{2+}). The isoelectric state (pH_{ISO}) of CCP 20 with both sludge waters occurred at pH values of 8 and 10. The same values of pH_{ISO} were also determined after contact of CCP 20 with matrix-free NH_4Cl solution [10]. Hence, an influence of the sludge water matrix on the pH_{ISO} is not detectable. Furthermore, Figure 1 shows the pH after spiking SW1 with equimolar NaCl instead of NaOH; the latter was added to the sample to adjust the pH. When spiked with NaCl (23–168 mmol/L), the pH remained at approx. 8.

Figure 2 depicts the elimination of ammonium from SW1 and SW2 by CCP 20 as a function of the initial pH. Furthermore, the degree of dissociation α of the NH_4^+/NH_3 system is plotted over the initial pH.

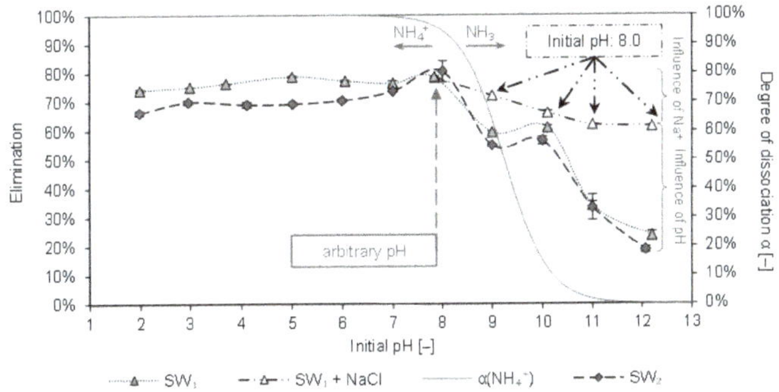

Figure 2. Elimination of ammonium from sludge water by CCP 20 (SW1: c_0 = 967 mg/L NH_4-N; SW2: c_0 = 927 mg/L NH_4-N; T = 295 K; sorbent ration 0.1 $g_{CLI}/mg_{NH4\text{-}N}$) after 20 h contact time as a function of different initial pH of the sludge water (adjusted pH before contact with CCP 20); in addition, the influence of an equimolar amount of Na^+ (from NaCl instead of NaOH to raise the pH) on the elimination is shown; degree of dissociation of ammonium in grey.

In the pH-range from 2 to 8 a consistent elimination between 66% and 81% was determined. A decrease in the elimination could be observed with pH values above 8. For SW1, the elimination decreased from 79% at pH 8 to 58% at pH 9 (59% at pH 10) and dropped to 24% at pH 12.2. At the same pH, the elimination from SW2 decreased from 81% to 55% (57% at pH 10) and finally to 19% (at pH 12). The batches with NaCl instead of NaOH indicate that the influence of sodium ions competing for sorption sites on the elimination is inferior to the influence of the pH. Although the elimination already declined from 79% to 62% due to the addition of Na^+ (from NaCl), the equimolar amount of NaOH, which increased the pH to 12.2, led to an even lower elimination of only 24%.

When comparing these results with results from similar experiments with matrix-free solution (pH 9: 73% elimination; pH 12.2: 20% elimination [10]), only a slightly negative influence of the SW matrix becomes apparent. Table 2 shows that both SW1 and SW2 had constituents (e. g., K^+, Na^+) that compete with ammonium for adsorption sites

Comparable studies with leachate have reported an elimination of 68% at pH 7 [22]. At pH 7 a higher loading (q = 17.7 mg/g) was achieved from swine liquid manure as compared to the arbitrary pH 8.2 (q = 12.5 mg/g) [17]. Furthermore, investigations with artificial swine wastewater stated that most ammonium was removed at pH 7, also showing a strong decrease in sorption with increasing pH [39]. On the contrary, the sorption of ammonium from drinking water was not affected by pH in the range of 5–9 [40].

The results here reveal that at a pH of 7, ammonium is eliminated to a high degree. In contrast to the literature results, a high degree of elimination could be realized with CCP 20 at arbitrary pH (7.9 or 8.0). A higher pH, however, should be avoided.

3.2. Isothermal Adsorption

Figure 3 displays the equilibrium loading q_{eq} of CCP 20 and the corresponding equilibrium concentration c_{eq} after 20 h contact time with SW1 and SW2 at different temperatures (283 K, 295 K, and 307 K). The lines represent the Freundlich equation, which gained the highest degree of determination of all tested isothermal equations (Freundlich, Langmuir, Temkin). The coefficients of the isothermal equations and their coefficient of determination are listed in Table 5. From the high concordance with the Freundlich isotherm, it can be deduced, that CLI has a heterogeneous surface, resulting in non-ideal sorption. Furthermore, the sorption process from sludge water is non-ideal, e.g., possible multiple occupancy of a sorption site as well as not all sorption sites are occupied.

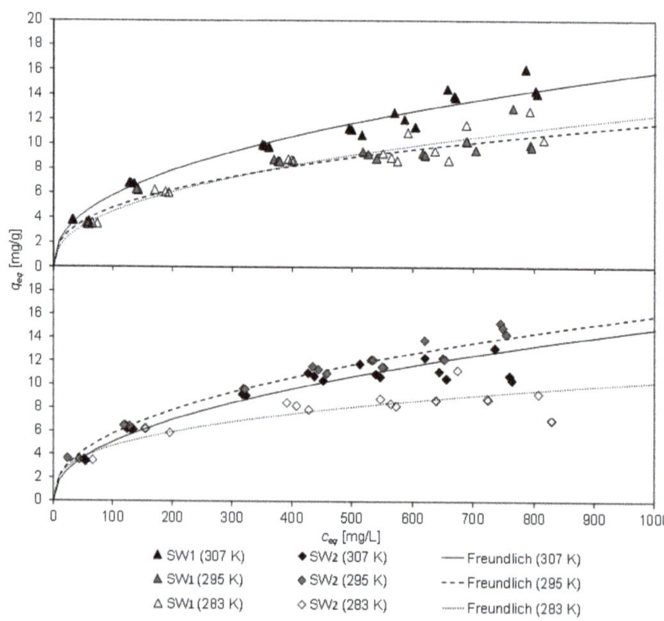

Figure 3. Equilibrium load q_{eq} and equilibrium concentration c_{eq} of CCP 20 and Freundlich isotherm of CCP 20 after 20 h contact time with SW1 or SW2 at different temperatures (SW1: c_0 = 967 mg/L NH_4-N; initial pH 8.1; final pH 7.6–8.6. SW2: c_0 = 866–913 mg/L NH_4-N; initial pH 7.9; final pH 7.2–8.0).

Table 5. Coefficients of isothermal adaptation according to Freundlich, Langmuir and Temkin for CCP 20 after 20 h contact time with SW1 or SW2 at different temperatures (SW1: c_0 = 967 mg/L NH_4-N; initial pH 8.1; final pH 7.6–8.6. SW2: c_0 = 866–913 mg/L NH_4-N; initial pH 7.9; final pH 7.2–8.0).

	Temperature	Freundlich			Langmuir			Temkin		
	T	K_F	$1/n$	r^2	K_L	q_{max}	r^2	A_T	b_T	r^2
	[K]	[L/g]	[−]	[−]	[L/mg]	[mg/g⁻]	[−]	[L/mg]	[J g /(mg mol)]	[−]
SW1	283	0.590	0.440	0.9087	0.005	12.38	0.5351	0.0495	846	0.8989
	295	0.811	0.385	0.9006	0.007	11.42	0.5202	0.0716	954	0.9180
	307	0.777	0.436	0.9603	0.005	17.31	0.5377	0.0631	735	0.9173
SW2	283	0.998	0.337	0.6701	0.006	11.68	0.5137	0.1188	1209	0.6828
	295	0.759	0.440	0.9711	0.005	16.64	0.5386	0.0670	724	0.9311
	307	0.596	0.464	0.8964	0.005	15.08	0.5257	0.0549	798	0.9322

The pH of the filtrate of SW1 (initial pH 8.1) changed independently to values between 7.6 and 8.6 during contact, and to 7.2 to 8.0 for SW2 (initial pH 7.9), whereby the former occurred with low sorbent masses and the latter was determined in a blind test without sorbent. As a result, the pH value dropped slightly due to the sorption process.

The highest equilibrium load of CCP 20 with ammonium from SW1 was 16.1 mg/g at 307 K. However, this was considerably lower by 13% than observed for matrix-free solution (NH_4Cl) with a similar concentration of 1000 mg/L NH_4-N (q_{eq} = 18.8 mg/g [10]). From SW2, the highest equilibrium load was 15.3 mg/g at 295 K (18% lower than from matrix-free solution). The load was lower at 307 K, but this was probably due to an alteration in the ammonium concentration in SW2 between the tests.

The minor loading of CCP 20 compared to matrix-free solution could be ascribed to the constituents in the sludge water, which interfere with the sorption process and thus lead to a reduction of the adsorption capacity. In particular, the deeper sorption sites in the framework of the CLI are probably more difficult to access. Access pores may be blocked or sorption sites may be occupied by other constituents in the sludge water. Furthermore, blocking of zeolite pores could be caused e.g., by solids. In addition, the viscosity of the sludge water changes depending on the temperature, which could also influence the loading of the CLI.

A decrease in the adsorption capacity of zeolite due to the wastewater matrix has been reported in several publications [16,17]. By using swine manure (c_0 = 7700 mg/L NH_4^+) instead of NH_4Cl solution (c_0 = 7700 mg/L NH_4^+), the uptake capacity of the tested CLI decreased from 10 mg/g to 2 mg/g [17]. To a similar extent as was found in this paper, a decrease of 10–20% of the adsorption capacity when using leachate from a sewage sludge landfill (c_0 = 115.16 mg/L NH_4-N) instead of NH_4Cl solution (c_0 = 119.48 mg/L NH_4-N) was reported [16]. The authors attributed this to ions such as Na^+, K^+, Mg^{2+}, and Ca^{2+} with ion ratios of ammonium to potassium of $[NH_4^+]/[K^+] \approx 14.6$ and sodium of $[NH_4^+]/[Na^+] \approx 2.7$ in the leachate. However, in both sludge waters investigated ammonium was present in multiple excess (SW1: $[NH_4^+]/[K^+] \approx 5$; $[NH_4^+]/[Na^+] \approx 10$; SW2: $[NH_4^+]/[K^+] \approx 24$–30; $[NH_4^+]/[Na^+] \approx 10$–12) and thus ammonium dominated the sorption process.

The determined ion ratio of ammonium to potassium and sodium in SW1 and SW2 was higher than reported by Wang et al. [16], but the difference in load between sludge water and matrix-free solution [10] was within the same magnitude. Obviously, the capacity of CCP 20 is influenced by both, the cations contained in the sludge water and by the matrix of the sludge water.

3.3. *Thermodynamic Properties*

Table 6 lists the determined thermodynamic state variables free reactivity enthalpy ΔG^0, free standard enthalpy ΔH^0, and molar standard entropy ΔS^0 after 20 h contact time of CCP 20 with SW1 and SW2.

Table 6. Thermodynamic properties of CCP 20 after 20 h contact time with SW1 or SW2 (SW1: c_0 = 967 mg L^{-1} NH_4-N; Initial pH 8.1; Final pH 7.6–8.6. SW2: c_0 = 866–913 mg L^{-1} NH_4-N; Initial pH 7.9; Final pH 7.2–8.0).

Sorptive	Temperature	Free Reaction Enthalpy	Free Standard Enthalpy	Molar Standard Entropy
	T	ΔG^0	ΔH^0	ΔS^0
[−]	[K]	[kJ/mol]	[kJ/mol]	[J/(K mol)]
SW1	283	−15.0	8.5	84.3
	295	−16.4		
	307	−17.0		
SW2	283	−16.2	−15.5	2.6
	295	−16.3		
	307	−16.3		

In the examined temperature range (283–307 K), an exergonic, i.e., voluntary sorption process of ammonium to CLI, can be deduced from to the negative free reaction enthalpy ΔG^0. The free standard enthalpy ΔH^0 was positive for SW1, i.e., an endothermic reaction was present. In contrast to this, an exothermic reaction was observed for SW2. The positive molar standard entropy ΔS^0 indicates a directed process. Independent of the matrix, the reaction of ammonium with CLI is voluntary and directed.

From experiments with CLI and matrix-free solution, an exergonic reaction was also reported with ΔG^0 ranging from −2.8662 to 0.22 kJ/mol [41], −0.79 to 1.63 kJ/mol [42], and −0.22 to 1.60 kJ/mol [43]. In this study, the values of ΔG^0 ranged from −15 to −17 kJ/mol. The much lower values regarding ΔG^0 of this study can be attributed to the smaller particle size and therefore short diffusion pathways of cations into the CLI. For their experiments, Alshameri et al. [41], Gunay [42], and Karadag et al. [43] used zeolites with larger particle sizes such as 0.063–0.074 mm, 0.3–0.6 mm, and 1.0–1.4 mm.

On the contrary to the results published by other researchers (ΔH^0: −49.384, −22.34, −5.43, −15.38 kJ/mol [41–44]), which indicate that adsorption of ammonium is exothermic, a slightly endothermic adsorption from SW1 (ΔH^0: 8.5 kJ/mol (SW1)) was found. However, adsorption from SW2 was exothermic (ΔH^0: −15.5 kJ/mol).

Furthermore, results reported with negative values of ΔS^0 (−0.1561, −43.03, −49.34, −74.42 kJ/(mol K) [41–44]) indicate decreasing ammonium uptake due to increasing randomness. In contrast to this, a strongly directed adsorption process, as indicated by positive ΔS^0 values ranging between 2.6 J/(K mol)$^{-1}$ (SW2) and 84.3 J/(K mol)$^{-1}$ (SW1), was achieved.

3.4. *Kinetic Studies*

3.4.1. Influence of Temperature on Kinetics

In Figure 4, the loading of CCP 20 with ammonium from sludge water at various temperatures (283 K to 307 K) after different contact times (up to 180 min) is depicted. The sorption kinetics at the investigated temperatures are fit to the ID model.

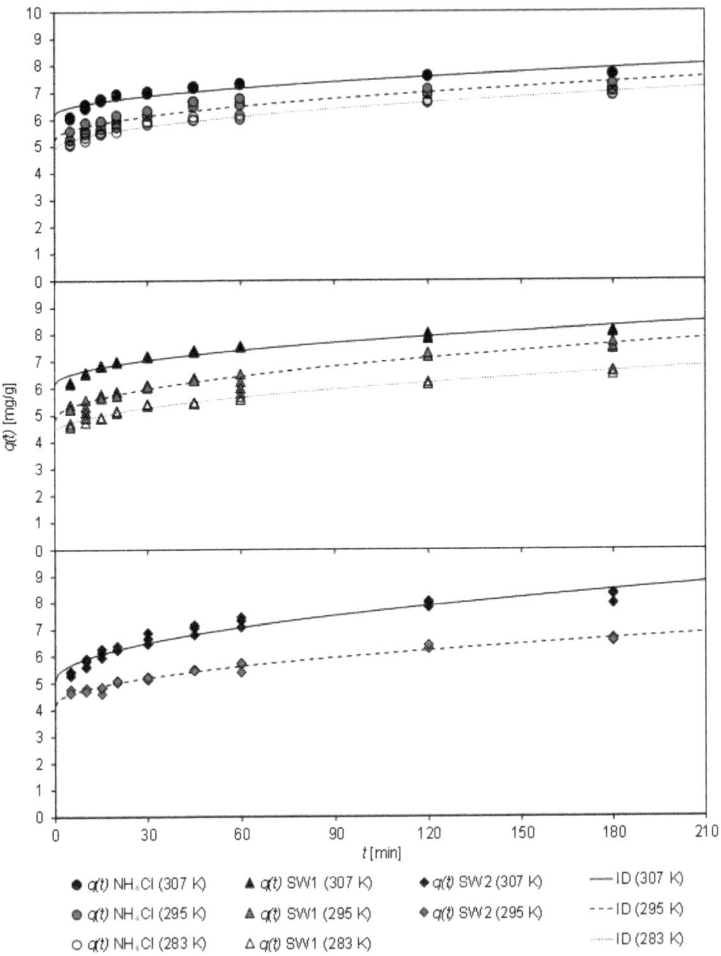

Figure 4. Loading $q(t)$ of CCP 20 as a function of contact time t at different temperatures (283–307 K) and fit to the ID model (specific sorbent ratio: 0.1 g_{CLI}/mg_{NH4-N}; NH4Cl solution: c_0 = 1000 mg/L NH4-N; initial pH 5.3; final pH 6.3–7.0; SW1: c_0 = 967 mg/L NH4-N; initial pH 7.6; final pH 7.4–8.3; SW2: c_0 = 913 mg/L NH4-N; initial pH 7.6; final pH 6.5–8.2).

Over the entire temperature range, a rapid adsorption of ammonium by CCP 20 was observed. Within the first five minutes, a high loading occurred, independently of the sludge water matrix.

Table 7 shows the coefficients of the kinetic fit according to both, PSO and ID model, the latter of which achieved a higher coefficient of determination.

Table 7. Coefficients of the sorption kinetics according to the PSO and ID models of CCP 20 at different temperatures (283–307 K, specific sorbent ratio: 0.1 g_{CLI}/mg_{NH4-N}; NH4Cl solution: c_0 = 1000 mg/L NH4-N; initial pH 5.3; final pH 6.3–7.0; SW1: c_0 = 967 mg/L NH4-N; initial pH 7.6; final pH 7.4–8.3; SW2: c_0 = 913 mg/L NH4-N; initial pH 7.6; final pH 6.5–8.2).

Temperature	Sorptive	Pseudo-Second-Order			Intraparticle Diffusion		
T [K]	[−]	k_2 [g/(mg min)]	q_e [mg/g]	r^2 [−]	k_{ID} [mg/(min$^{0.5}$ g)]	C [mg/g]	r^2 [−]
283		0.065	6.60	0.6853	0.159	4.88	0.9642
295	NH_4Cl	0.064	6.99	0.8319	0.159	5.25	0.8852
307		0.090	7.53	0.9153	0.129	6.16	0.8467
283		0.056	6.23	0.6116	0.172	4.32	0.5236
295	SW_1	0.043	7.19	0.7486	0.208	4.84	0.9566
307		0.070	7.87	0.8641	0.161	6.16	0.9228
295	SW_2	0.041	6.34	0.7231	0.185	4.19	0.9704
307		0.037	7.89	0.8370	0.252	5.12	0.9386

From the PSO model, an increase of k_2 from 0.065 g/(mg min) to 0.090 g/(mg min) could be derived with increasing temperature of the matrix-free NH_4Cl solution. For SW1, k_2 first decreased with increasing temperature from 0.056 g/(mg min) (at 283 K) to 0.043 g/(mg min) (295 K), but then increased to 0.070 g/(mg min) (307 K). With SW2, k_2 remained almost unchanged with 0.041 g/(mg min) (at 295 K) and 0.037 g/(mg min) (at 307 K). Thus, the sludge-water matrix caused a reduction of the sorption rate k_2. Temperature only had a minor effect on the sorption rate in the range of 283–295 K. The equilibrium load q_e was similar for all samples, with a higher value for q_e being obtained with increasing temperature.

For all three matrices, an increase in the initial sorption C was observed with increasing temperature, whereas the sorption rate k_{ID} was only slightly affected. In the case of the matrix-free solution, k_{ID} decreased from 0.159 mg/(min$^{0.5}$ g) at 283 K to 0.129 mg/(min$^{0.5}$ g) at 307 K. In contrast, k_{ID} increased slightly with SW1 between 283 K and 295 K. This low coefficient of determination of the sorption kinetics at 283 K indicates that no ID was present at low temperatures. Nevertheless, it was also shown that the kinetics at 295 K slowed down. Here, k_{ID} decreased from 0.208 mg/(min$^{0.5}$ g) to 0.161 mg/(min$^{0.5}$ g).

Regardless of the sorption kinetics, it should be recognized that ammonium uptake was mainly affected by initial sorption, which in turn was greater at higher temperatures. After the CCP 20 was in contact with the matrix-free solution or sludge waters for 30 min at 307 K, a load between 6.88 mg/g (SW2) and 7.15 mg/g (SW1) of was determined.

Similar conclusions regarding a decrease of the sorption rate as a result of an increase in temperature were described according to the PSO model (NH_4Cl solution) [43]. Furthermore, it was concluded that this was an exothermic process, which was slower due to increased temperatures. On the other hand, it was found that k_2 was reduced and k_{ID} increased due to increased temperature [23]. Consequently, the equilibrium was reached later. However, the values reported by Erdoğan and Ülkü [23] for k_{ID} with 4.8 10^{-3} mg/(min$^{0.5}$ g) at 298 K and 5.4 10^{-3} mg/(min$^{0.5}$'g) at 313 K were about a factor of 30 lower than the values obtained with CCP 20. This is probably ascribed to the larger particle sizes of the CLI (0.85–2.00 mm [23]) investigated.

3.4.2. Influence of the Specific Dosage on Kinetics

Figure 5 displays the loading $q(t)$ as a function of the contact time t of CCP 20 at different specific sorbent dosages (0.05–0.2 g_{CLI}/mg_{NH4N}) after contact with NH_4Cl solution as well as SW1 and SW2. In addition, the fit of the ID model, which had gained the highest coefficient of determination, is shown. A load between 3.07 mg/g and 7.25 mg/g was realized after only 5 min, regardless of the matrix. After 30 min contact time, depending on the specific sorbent addition (0.05–0.2 g_{CLI}/mg_{NH4N}), a loading of between 4.15 mg/g and 7.56 mg/g was obtained with the NH_4Cl solution. However, after 60 min, the load increased to values between 4.30 mg/g and 8.34 mg/g and after 120 min to between 4.41 mg/g and 8.66 mg/g.

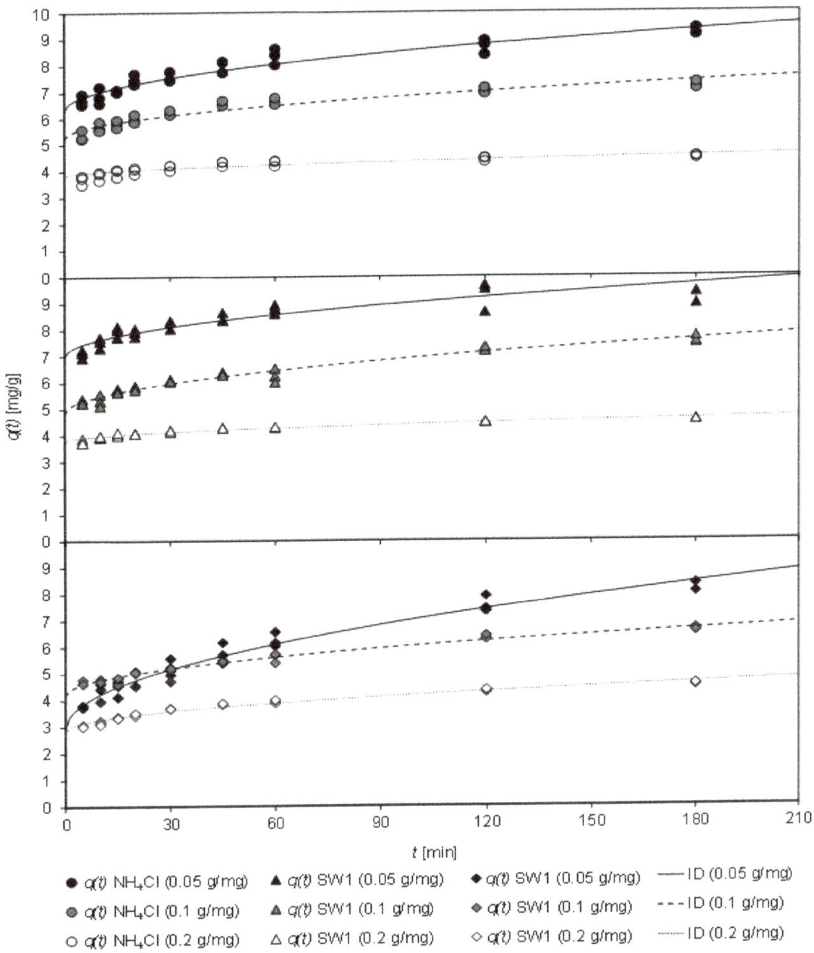

Figure 5. Loading $q(t)$ of CCP 20 as a function of the contact time t after different specific sorbent ratios (0.05–0.2 $g_{CLI}/mg_{NH4\text{-}N}$) aligned with the ID model (T = 295 K; NH4Cl solution: c_0 = 1000 mg/L NH4-N; initial pH 5.3; final pH 6.3–7.0; SW1: c_0 = 967 mg/L NH4-N; initial pH 7.6; final pH 7.5–8.5; SW2: c_0 = 718–913 mg/L NH4-N; initial pH 7.6; final pH 7.7–8.2).

Table 8 displays the coefficients of the sorption kinetics according to the PSO model and the ID model. The high coefficients of determination of the ID model indicate the limitation of the sorption rate by intraparticle diffusion.

Table 8. Coefficients of the sorption kinetics according to the PSO and ID models of CCP 20 after different sorbent loads (T = 295 K; NH4Cl solution: c_0 = 1000 mg/L NH4-N; initial pH 5.3; final pH 6.3–7.0; SW1: c_0 = 967 mg/L NH4-N; initial pH 7.6; final pH 7.5–8.5; SW2: c_0 = 718–913 mg/L NH4-N; initial pH 7.6; final pH 7.7–8.2.

Specific Sorbent Dosage	**Sample**	**Pseudo-Second-Order**			**Intraparticle Diffusion**		
m_S		k_2	q_e	r^2	k_{ID}	C	r^2
[g_{CLI}/mg_{NH4N}]		[g/(mg min)]	[mg/g]	[−]	[mg/(min$^{0.5}$ g)]	[mg/g]	[−]
0.05	NH$_4$Cl	0.041	8.84	0.7704	0.225	6.31	0.9119
0.1		0.064	6.99	0.8319	0.159	5.25	0.8852
0.2		0.154	4.45	0.7605	0.058	3.78	0.7595
0.05	SW1	0.048	9.26	0.7849	0.202	7.00	0.8543
0.1		0.043	7.19	0.7486	0.208	4.84	0.9566
0.2		0.188	4.45	0.8355	0.063	3.77	0.9134
0.05	SW2	0.008	8.56	0.8305	0.410	2.93	0.9500
0.1		0.020	6.78	0.7309	0.185	4.19	0.9704
0.2		0.062	4.35	0.8263	0.135	2.84	0.9647

The coefficients of the PSO model and the ID model indicate that the sorbent dosage has a decisive influence on the rate of ammonium uptake. In the investigated samples, k_2 increased with increasing sorbent dosage, i.e., the sorption equilibrium was achieved earlier. With the lowest specific sorbent dosage of 0.05 $g_{CLI}/mg_{NH4\text{-}N}$, k_2 was between 0.008 g/(mg min) (SW2) and 0.048 g/(mg min) (SW1) and rose disproportionately with a specific dosage of 0.2 $g_{CLI}/mg_{NH4\text{-}N}$ to 0.062 g/(mg min) (SW2), 0.154 g/(mg min) (NH_4Cl) or even up to 0.188 g/(mg min) (SW1). In contrast, the load q_e decreased from 9.26 mg/g (SW2) to 4.35 mg/g (SW1) with increasing sorbent dosage. The values for k_{ID} became smaller with increasing sorbent loading, i.e., the sorption equilibrium was achieved earlier. Thus, k_{ID} values of 0.410 mg/(min$^{0.5}$ g) (SW2), 0.202 mg/(min$^{0.5}$ g) (SW1), and 0.225 mg/(min$^{0.5}$ g) (NH_4Cl) where reached at a dosage of 0.05 $g_{CLI}/mg_{NH4\text{-}N}$, which then decreased to 0.135 mg/(min$^{0.5}$ g) (SW2) and 0.063 mg/(min$^{0.5}$ g) (SW1) and 0.058 mg/(min$^{0.5}$ g) (NH_4Cl), respectively, at a dosage of 0.2 $g_{CLI}/mg_{NH4\text{-}N}$. The constant C, which is proportional to the thickness of the boundary layer and represents the initial sorption, was also reduced from 6.31 mg/g to 3.78 mg/g (NH_4Cl), from 7.00 mg/g to 3.77 mg/g (SW1), and from 4.19 mg/g to 2.84 mg/g (SW2) when a larger sorbent dosage (0.05–0.2 $g_{CLI}/mg_{NH4\text{-}N}$) was applied. This can be ascribed to the fact that with higher specific dosage, more sorption sites are provided, resulting in a rapid sorbent equilibrium but lower load.

During the contact time investigated, only a slight influence of the sample matrix on the load has been observed. Neither was the uptake rate k_2 or k_{ID} influenced (except for SW2 at a specific dosage of 0.05 $g_{CLI}/mg_{NH4\text{-}N}$). However, regardless of the matrix, CCP 20 was in equilibrium after 120 min. In contrast, it was reported that for ammonium from matrix-free sorption solution one hour of contact time, but for that from the leachate 2.5 h were needed to achieve equilibrium [16]. The authors ascribed this to interfering cations such as Na^+, K^+, Mg^{2+}, and Ca^{2+} ($[NH_4^+]/[Na^+] \approx 2.7$; $[NH_4^+]/[K^+] \approx 14.6$) in the leachate investigated. However, the high stoichiometric excess of ammonium in the sludge waters investigated in this work was larger (SW1: $[NH_4^+]/[K^+] \approx 5$; $[NH_4^+]/[Na^+] \approx 10$; SW2: $[NH_4^+]/[K^+] \approx 24$–30; $[NH_4^+]/[Na^+] \approx 10$–12). Hence, the interfering of competing cations can be assumed as low. Therefore, no significant difference between matrix-free solution and sludge water could be determined in this experiment.

3.4.3. Influence of the Pre-load on Sorption Kinetics

During one sorption process, the CCP 20 may not be completely loaded, e.g., if the lowest possible residual concentration is to be achieved by a higher dosage of sorbent. In a process cascade, this partially pre-loaded sorbent could be returned and recontacted with sludge water. Similarly, a partial regeneration of the sorbent can result in partially loaded sorbent.

Figure 6 shows the loading $q(t)$ of partially loaded CCP 20 after contact with NH_4Cl solution, SW1, and SW2 as a function of the contact time t. For pre-loading, the sorbent was brought into contact with the sample for 30 min (q_1) or 60 min (q_2) at 307 K (based on the results from Section 3.4.1). In addition, the fit of the ID model is shown, which achieved the highest coefficient of determination (Table 9).

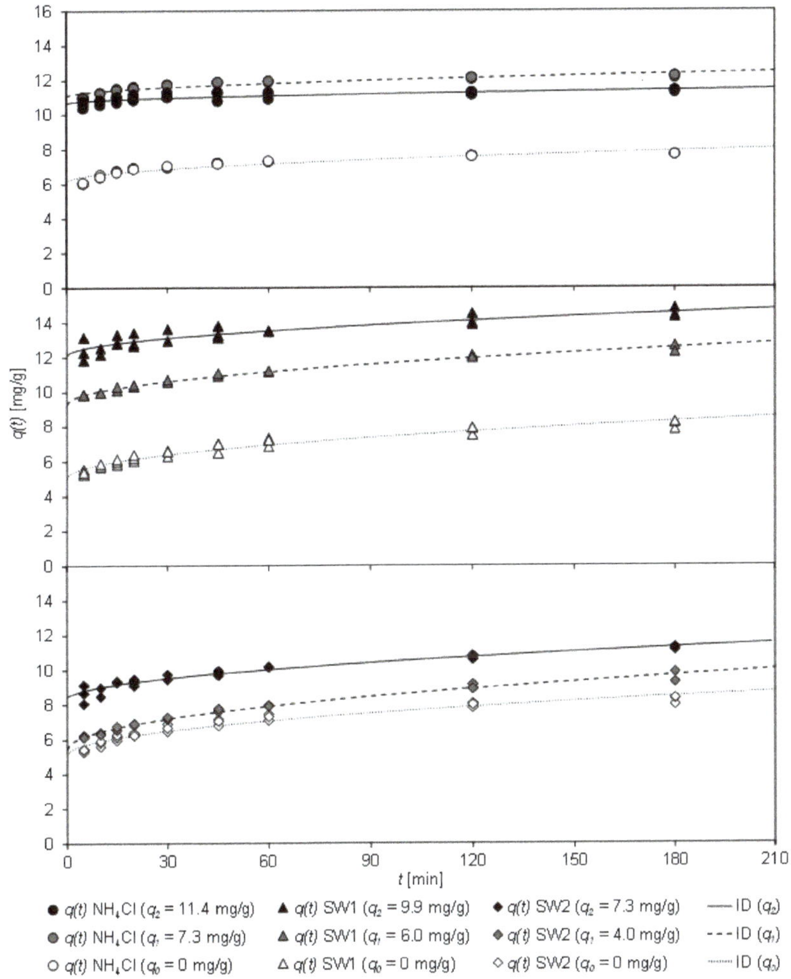

Figure 6. Loading $q(t)$ of CCP 20 as a function of the contact time t of differently pre-loaded CCP 20 (q_0–q_2 = 0–11.4 mg/g) and fit with the ID model (specific sorbent dosage: 0.1 $g_{CLI}/mg_{NH4\text{-}N}$; T = 307 K; NH4Cl solution: c_0 = 1000 mg/L NH4-N; initial pH 5.3; final pH 6.0; SW1: c_0 = 967 mg/L NH4-N; initial pH 7.9; final pH 7.6; SW2: c_0 = 775–913 mg/L NH4-N; initial pH 7.6; final pH 6.5–8.4).

Table 9. Coefficients of the sorption kinetics according to the PSO and ID models of CCP 20 with different pre-loads (specific sorbent dosage: 0.1 g_{CLI}/mg_{NH4-N}; T = 307 K; NH4Cl solution: c_0 = 1000 mg/L NH4-N; initial pH 5.3; final pH 6.0; SW1: c_0 = 967 mg/L NH4-N; initial pH 7.9; final pH 7.6; SW2: c_0 = 775–913 mg/L NH4-N; initial pH 7.6; final pH 6.5–8.4).

Pre-Load		Matrix	Pseudo-Second-Order			Intraparticle Diffusion		
			k_2	q_e	r^2	k_{ID}	C	r^2
[mg/g^{-1}]		[−]	[g/(mg min)]	[mg/g]	[−]	[mg/(min$^{0.5}$ g)]	[mg/g]	[−]
q_0	0		0.090	7.53	0.9153	0.129	6.16	0.8467
q_1	7.3	NH_4Cl	0.135	12.08	0.8943	0.096	11.07	0.8302
q_2	11.4		0.161	11.37	0.7261	0.054	10.69	0.5191
q_0	0		0.070	7.87	0.8641	0.161	6.16	0.9228
q_1	6.0	SW1	0.042	12.00	0.7325	0.243	9.30	0.9853
q_2	9.9		0.050	14.32	0.6985	0.182	12.11	0.7495
q_0	0		0.037	7.89	0.8370	0.252	5.12	0.9386
q_1	4.0	SW2	0.027	9.06	0.7603	0.316	5.46	0.9846
q_2	7.3		0.043	10.85	0.5479	0.229	8.25	0.5210

As Figure 6 shows, the loading of CCP 20 increased with increasing contact time. Unloaded sorbent is marked as q_0, pre-loaded sorbent as q_1 and q_2, respectively. Pre-loaded CCP 20 (q_1 and q_2) adsorbed additional ammonium on contact with the sample. A considerable change in the load $q(t)$ occurred for the matrix-free NH_4Cl solution. However, CCP 20 pre-loaded with 11.4 mg/g was already close to the sorption equilibrium, so that no significant increase in loading could be determined.

In contrast, with both wastewater matrices of the sludge waters SW1 and SW2 even at the highest pre-load ($q_2(SW1)$ = 9.9 mg/g and $q_2(SW2)$ = 7.3 mg/g), an increase of loading was achieved.

Table 9 shows the coefficients of the kinetic fit according to the PSO and ID models, the latter achieving higher coefficients of determination.

Within 5 min contact time, a high initial loading (C = 5.12–6.16 mg/g) of the unloaded sorbent (q_0) was achieved, independent of the sample matrix. When partially pre-loaded (q_1), the loading of the sorbent was considerably increased in all tests; depending on the pre-loading (approx. 80% for NH_4Cl, almost 100% for SW1 and approx. 60% for SW2). In the first pre-loading step (q_1), the initial loading C of NH_4Cl increased significantly from 6.16 mg/g (q_0) by about 80% to 11.07 mg/g (q_1). For SW1, it also increased considerably from 6.16 mg/g (q_0) to 9.30 mg/g (q_1), but for SW2 only slightly from 5.12 mg/g (q_0) to 5.46 mg/g (q_1). In case of the highest preload (q_2), the initial load C of matrix-free NH_4Cl was lower (10.69 mg/g), but in case of sludge water it increased (SW1: 12.11 mg/g; SW2: 8.25 mg/g). The decreasing values for k_{ID} of SW1 (q_1–q_2: 0.243–0.182 mg/(min$^{0.5}$ g)) and SW2 (q_1–q_2: 0.316–0.229 mg/(min$^{0.5}$ g)) indicate that the sorbent reached equilibrium faster due to the partial pre-loading. After 30 min, preloaded (q_2) CCP 20 was loaded to an extend between 9.45 mg/g (SW2) and 13.63 mg/g (SW1).

The fit of the sorption kinetics from matrix-free NH_4Cl solution by means of the PSO model reveals that k_2 increased with increasing pre-load of the sorbent, i.e., the sorbent achieved equilibrium faster. However, k_2 of unloaded CCP 20 (q_0) attained 0.090 g/(mg min), which increased to 0.135 g/(mg min) when partially loaded (q_1) and finally to 0.161 g/(mg min) with the highest pre-load (q_2), which was almost in equilibrium. On the contrary, the k_2 values for SW1 (q_0–q_2: 0.042–0.070 g/(mg min)) and SW2 (q_0–q_2: 0.027–0.043 g/(mg min)) do not allow clear conclusions due to their wide variation.

Nevertheless, it has been ascertained that the sorption equilibrium is achieved faster from matrix-free NH_4Cl solution than from sludge water. This can be ascribed to cations competing for sorption sites, but also to organic matter or particles contained in the sludge water and the slower diffusion of ammonium to deeper sorption sites. In a process cascade, in which a high CLI dosage has to achieve the lowest possible residual concentration, the partially loaded CLI could be brought into contact with sludge water again in order to use its capacity to the full extent. In order to achieve the highest possible loading of the sorbent, up to three sorption phases should be carried out, each lasting a maximum of 30 min.

4. Conclusions

From experiments with high strength sludge waters with ammonium concentrations from 718 mg/L NH_4-N to 967 mg/L NH_4-N by means of Carpathian clinoptilolite, the following boundary conditions can be derived with which the highest possible loading of the sorbent CCP 20 can be achieved:

- the pH value should be in a range of 2 to 8 (or the arbitrary pH if below 8),
- the temperature of 307 K should be preferred over lower temperatures (e.g., 283 K or 295 K),
- the choice of a low specific sorbent dose (e.g., <0.1 $g_{CLI}/mg_{NH4\text{-}N}$) is advantageous,
- pre-loaded sorbent should be recontacted with the sludge water several times (up to three times)
- the contact time of (pre-loaded) sorbent should be at least 30 min.

However, other boundary conditions may be relevant, depending on the objectives of the treatment, e.g., high loading of the sorbent, shortest possible contact time, low effluent concentrations. For the design as well as the implementation of the process, the required contact time is of major importance. In the experiments it could be shown that a high loading of the clinoptilolite can be achieved already after 30 min. Therefore, it can be deduced that the necessary equipment for the treatment of the relatively small partial flow of the sludge water compared to the main wastewater flow would only require minor construction and plant engineering upgrades.

Based on the found conditions, it is of interest for future investigations under which parameters the clinoptilolite can be regenerated and possibly reused. Furthermore, it should not be omitted that the liquid resulting from the regeneration is still usable or the recovered ammonium is available in a usable form.

Author Contributions: Conceptualization, S.W.; Investigation, S.W.; Methodology, S.W.; Visualization, S.W.; Writing—original draft, S.W.; Writing—review & editing, E.R., R.M. and H.S. All authors have read and agreed to the published version of the manuscript.

Funding: This research was funded by Federal Ministry of Economic Affairs and Energy following a resolution of the German Parliament.

Data Availability Statement: All data comes from the author.

Acknowledgments: We thank the Federal Ministry of Economic Affairs and Energy following a resolution of the German Parliament for sponsoring this research work (ZF4045511). The CLI was provided by Fluidtec, Kempenich, Germany.

Conflicts of Interest: The authors declare no conflict of interest.

Appendix A Titration Curve of SW1, SW2, and NH_4Cl Solution

During anaerobic digestion, large quantities of methane and CO_2 are formed, resulting in high concentrations of hydrogen carbonate in the sludge water. By lowering the pH, the hydrogen carbonate is converted into CO_2, which then outgasses.

The base capacity, i.e., the amount of OH^- required to adjust the pH to 8.2, corresponds to the amount of free CO_2 in the solution. The acid capacity, i.e., the amount of H_3O^+ necessary to adjust the pH to 4.3, corresponds to the amount of HCO_3^-. In addition, other compounds such as ammonium, borate, phosphate, sulfate, nitrate, etc. may react, therefore the titration method can only be applied to sludge water to a limited extent [45].

The amounts of OH^-or H_3O^+ required to adjust the pH of SW1, SW2 and NH_4Cl solution are shown in Figure A1.

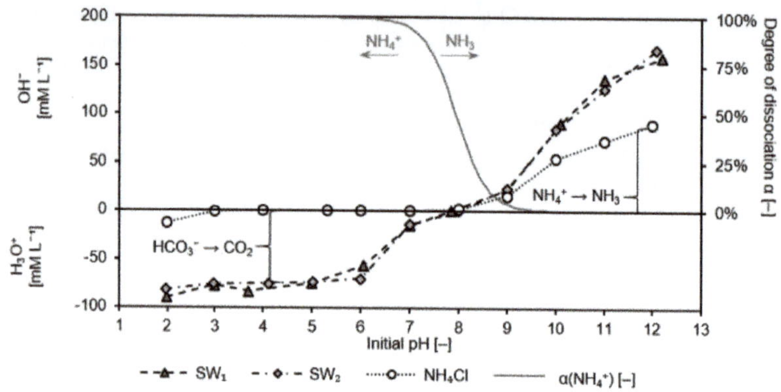

Figure A1. Quantity of OH^- (NaOH) and H_3O^+ (HCl) to adjust the pH of SW1, SW2, and NH_4Cl solution as well as the degree of dissociation of ammonium depending on the pH.

Only a small amount of H_3O^+ (pH 4: 0.09 mmol/L; pH 3: 1.2 mmol/L; pH 2: 12.7 mmol/L) was required to lower the pH of NH_4Cl (arbitrary pH 5.3), as no hydrogen carbonate was present. For an increase of the pH, a larger amount of OH^- was required due to the conversion of $NH_4{}^+$ into NH_3 (pH 7: 0.3 mmol/L; pH 8: 15.8 mmol/L; pH 12: 90 mmol/L). Since no other substances were present in the matrix-free solution, a stoichiometric transformation of OH^- for the conversion of $NH_4{}^+$ into NH_3 can be assumed when adjusting the pH value.

For both sludge waters SW1 and SW2, almost the same amount of H_3O^+ and OH^- was required. However, the amount of H_3O^+ and OH^- required for pH adjustment of NH_4Cl solution was much lower. Obviously, a large amount of hydrogen carbonate buffering the pH was present. For example, 74.5 mmol/L of H_3O^+ was needed to set up a pH of 5 for SW1 and 73.8 mmol/L of H_3O^+ for SW2. To adjust a pH of 10, 90.9 mmol/L and 85 mmol/L OH^-, respectively, were required. Due to the high concentration of interfering ions, it is not possible to determine the exact acid/base capacity and hydrogen carbonate concentration in the sludge waters. Based on the data, it can be estimated at approx. 75–80 mmol/L. The hydrogen carbonate is opposed by a sufficiently large amount of ammonium cations (55–69 mmol/L) as counter ion.

References

1. United Nations. *World Population Prospects: The 2017 Revision, Key Findings and Advance Tables*; Working Paper No. ESA/P/WP/248; United Nations: New York, NY, USA, 2017.
2. FAO. *The Future of food and Agriculture—Trends and Challenges*; FAO: Rome, Italy, 2017; ISBN 978-92-5-109551-5.
3. Erisman, J.W.; Sutton, M.A.; Galloway, J.; Klimont, Z.; Winiwarter, W. How a century of ammonia synthesis changed the world. *Nat. Geosci.* **2008**, *1*, 636–639. [CrossRef]
4. Appl, M. Ammonia. In *Ullmann's Encyclopedia of Industrial Chemistry: Ammonia*; Wiley-VCH Verlag GmbH & Co. KGaA: Weinheim, Germany, 2000; ISBN 9783527306732.
5. Dawson, C.J.; Hilton, J. Fertiliser availability in a resource-limited world: Production and recycling of nitrogen and phosphorus. *Food Policy* **2011**, *36*, S14–S22. [CrossRef]
6. Janus, H.M.; van der Roest, H.F. Don't reject the idea of treating reject water. *Water Sci. Technol.* **1997**, *35*, 27–34. [CrossRef]
7. Sengupta, S.; Nawaz, T.; Beaudry, J. Nitrogen and Phosphorus Recovery from Wastewater. *Curr. Pollut. Rep.* **2015**, *1*, 155–166. [CrossRef]
8. Englert, A.H.; Rubio, J. Characterization and environmental application of a Chilean natural zeolite. *Int. J. Min. Process* **2005**, *75*, 21–29. [CrossRef]
9. Weatherley, L.R.; Miladinovic, N.D. Comparison of the ion exchange uptake of ammonium ion onto New Zealand clinoptilolite and mordenite. *Water Res.* **2004**, *38*, 4305–4312. [CrossRef] [PubMed]
10. Wasielewski, S.; Rott, E.; Minke, R.; Steinmetz, H. Evaluation of Different Clinoptilolite Zeolites as Adsorbent for Ammonium Removal from Highly Concentrated Synthetic Wastewater. *Water* **2018**, *10*, 584. [CrossRef]

11. Cyrus, J.S.; Reddy, G.B. Sorption and desorption of ammonium by zeolite: Batch and column studies. *J. Env. Sci. Health* **2011**, *46*, 408–414. [CrossRef] [PubMed]
12. Guaya, D.; Valderrama, C.; Farran, A.; Armijos, C.; Cortina, J.L. Simultaneous phosphate and ammonium removal from aqueous solution by a hydrated aluminum oxide modified natural zeolite. *Chem. Eng. J.* **2015**, *271*, 204–213. [CrossRef]
13. Karadag, D.; Tok, S.; Akgul, E.; Turan, M.; Ozturk, M.; Demir, A. Ammonium removal from sanitary landfill leachate using natural Gördes clinoptilolite. *J. Hazard. Mater.* **2008**, *153*, 60–66. [CrossRef]
14. Zhang, H.; Li, A.; Zhang, W.; Shuang, C. Combination of Na-modified zeolite and anion exchange resin for advanced treatment of a high ammonia-nitrogen content municipal effluent. *J. Colloid Interface Sci.* **2016**, *468*, 128–135. [CrossRef] [PubMed]
15. Du, Q.; Liu, S.; Cao, Z.; Wang, Y. Ammonia removal from aqueous solution using natural Chinese clinoptilolite. *Sep. Purif. Technol.* **2005**, *44*, 229–234. [CrossRef]
16. Wang, Y.; Liu, S.; Xu, Z.; Han, T.; Chuan, S.; Zhu, T. Ammonia removal from leachate solution using natural Chinese clinoptilolite. *J. Hazard. Mater.* **2006**, *136*, 735–740. [CrossRef] [PubMed]
17. Montegut, G.; Michelin, L.; Brendle, J.; Lebeau, B.; Patarin, J. Ammonium and potassium removal from swine liquid manure using clinoptilolite, chabazite and faujasite zeolites. *J. Env. Manag.* **2016**, *167*, 147–155. [CrossRef]
18. Martins, T.H.; Souza, T.S.O.; Foresti, E. Ammonium removal from landfill leachate by Clinoptilolite adsorption followed by bioregeneration. *J. Env. Chem. Eng.* **2017**, *5*, 63–68. [CrossRef]
19. Farkas, A.; Rozic, M.; Barbaric-Mikocevic, Z. Ammonium exchange in leakage waters of waste dumps using natural zeolite from the Krapina region, Croatia. *J. Hazard. Mater.* **2005**, *117*, 25–33. [CrossRef]
20. Ji, Z.Y.; Yuan, J.S.; Li, X.G. Removal of ammonium from wastewater using calcium form clinoptilolite. *J. Hazard. Mater.* **2007**, *141*, 483–488. [CrossRef]
21. Lin, L.; Lei, Z.; Wang, L.; Liu, X.; Zhang, Y.; Wan, C.; Lee, D.; Tay, J.H. Adsorption mechanisms of high-levels of ammonium onto natural and NaCl-modified zeolites. *Sep. Purif. Technol.* **2013**, *103*, 15–20. [CrossRef]
22. Temel, F.A.; Kuleyin, A. Ammonium removal from landfill leachate using natural zeolite: Kinetic, equilibrium, and thermodynamic studies. *Desalination Water. Treat.* **2016**, *57*, 23873–23892. [CrossRef]
23. Erdoğan, B.C.; Ülkü, S. Ammonium sorption by Gördes clinoptilolite rich mineral specimen. *Appl. Clay Sci.* **2011**, *54*, 217–225. [CrossRef]
24. Freundlich, H. Over the adsorption in solution. *J. Phys. Chem.* **1906**, *57*, 385–470.
25. Ho, Y.-S.; Porter, J.F.; McKay, G. Equilibrium isotherm studies for the sorption of divalent metal ions onto peat: Copper, nickel and lead single component systems. *Water Air Soil Pollut.* **2002**, *141*, 1–33. [CrossRef]
26. Langmuir, I. The Adsorption of Gases on Plane Surfaces of Glass, Mica and Platinum. *J. Am. Chem. Soc.* **1918**, *40*, 1361–1403. [CrossRef]
27. Kinniburgh, D.G. General purpose adsorption isotherms. *Env. Sci. Technol.* **1986**, *20*, 895–904. [CrossRef]
28. Chen, X. Modeling of Experimental Adsorption Isotherm Data. *Information* **2015**, *6*, 14–22. [CrossRef]
29. Temkin, M.I.; Pyzhev, V. Kinetics of ammonia synthesis on promoted iron catalyst. *Acta Physicochim. USSR* **1940**, *12*, 217–222.
30. Dada, A.O.; Olalekan, A.P.; Olatunya, A.M.; Dada, O. Langmuir, Freundlich, Temkin and Dubinin–Radushkevich Isotherms Studies of Equilibrium Sorption of Zn^{2+} Unto Phosphoric Acid Modified Rice Husk. *IOSR-JAC* **2012**, *3*, 38–45. [CrossRef]
31. Aharoni, C.; Ungarish, M. Kinetics of activated chemisorption. Part 2—Theoretical models. *J. Chem. Soc. Faraday Trans. 1 Chem. Condens. Phases* **1977**, *73*, 456. [CrossRef]
32. Milonjic, S. A consideration of the correct calculation of thermodynamic parameters of adsorption. *J. Serb. Chem. Soc.* **2007**, *72*, 1363–1367. [CrossRef]
33. Weber, W.J.; Morris, J.C. Kinetics of Adsorption on Carbon from Solution. *J. Sanit. Eng. Div.* **1963**, *89*, 31–60.
34. Qiu, H.; Lv, L.; Pan, B.; Zhang, Q.; Zhang, W. Critical review in adsorption kinetic models. *J. Zhejiang Univ. Sci. A* **2009**, *10*, 716–724. [CrossRef]
35. McKay, G.; Otterburn, M.S.; Sweeney, A.G. The removal of colour from effluent using various adsorbents, III. Silica: Rate Processes. *Water Res.* **1980**, *14*, 15–20. [CrossRef]
36. Ho, Y.-S.; McKay, G. A Comparison of Chemisorption Kinetic Models Applied to Pollutant Removal on Various Sorbents. *Process Saf. Environ. Prot.* **1998**, *76*, 332–340. [CrossRef]
37. Ho, Y.-S. Second-order kinetic model for the sorption of cadmium onto tree fern: A comparison of linear and non-linear methods. *Water Res.* **2006**, *40*, 119–125. [CrossRef] [PubMed]
38. Deutsches Institut für Normung e.V., 38406–5. In *German Standard Methods for the Examination of Water, Waste Water and Sludge*; cations (group E); Determination of Ammonia-Nitrogen (E 5); Beuth Verlag GmbH: Berlin, Germany, 1983.
39. Deutsches Institut für Normung e.V., 10304–1. In *Water Quality—Determination of Dissolved Anions by Liquid Chromatography of Ions —Part 1: Determination of Bromide, Chloride, Fluoride, Nitrate, Nitrite, Phosphate and Sulfate*; Beuth Verlag GmbH: Berlin, Germany, 2009.
40. Li, M.; Zhu, X.; Zhu, F.; Ren, G.; Cao, G.; Song, L. Application of modified zeolite for ammonium removal from drinking water. *Desalination* **2011**, *271*, 295–300. [CrossRef]
41. Alshameri, A.; Yan, C.; Al-Ani, Y.; Dawood, A.S.; Ibrahim, A.; Zhou, C.; Wang, H. An investigation into the adsorption removal of ammonium by salt activated Chinese (Hulaodu) natural zeolite: Kinetics, isotherms, and thermodynamics. *J. Taiwan Inst. Chem. Eng.* **2014**, *45*, 554–564. [CrossRef]

42. Gunay, A. Application of nonlinear regression analysis for ammonium exchange by natural (Bigadic) clinoptilolite. *J. Hazard. Mater.* **2007**, *148*, 708–713. [CrossRef]
43. Karadag, D.; Koc, Y.; Turan, M.; Armagan, B. Removal of ammonium ion from aqueous solution using natural Turkish clinoptilolite. *J. Hazard. Mater.* **2006**, *136*, 604–609. [CrossRef]
44. Tosun, I. Ammonium removal from aqueous solutions by clinoptilolite: Determination of isotherm and thermodynamic parameters and comparison of kinetics by the double exponential model and conventional kinetic models. *Int. J. Env. Res. Public Health* **2012**, *9*, 970–984. [CrossRef]
45. Deutsches Institut für Normung e.V., 38409–7. In *German Standard Methods for the Examination of Water, Waste Water and Sludge—General Measures of Effects and Substances (Group H)—Part 7: Determination of Acid and Base Capacity (H 7)*; Beuth Verlag GmbH: Berlin, Germany, 2005.

molecules

MDPI

Article

Investigating the Aging Effects of Biochar on Soil C and Si Dissolution and the Interactive Impact on Copper Immobilization

Shaojun Jiang [1,2], Jiachen Wu [1], Lianxin Duan [1,2], Sheng Cheng [1,2], Jian Huang [1,2] and Tao Chen [1,2,*]

1 School of the Environment, South China Normal University, Guangzhou 510006, China; shaojunj93@163.com (S.J.); 15261826395@163.com (J.W.); duanlianxin1993@163.com (L.D.); ShengC@m.scnu.edu.cn (S.C.); 18371807641@163.com (J.H.)

2 Guangdong Provincial Key Laboratory of Chemical Pollution and Environmental Safety & MOE Key Laboratory of Theoretical Chemistry of Environment, South China Normal University, Guangzhou 510006, China

* Correspondence: tao.chen@m.scnu.edu.cn; Tel./Fax: +86-20-39310187

Academic Editors: Farid Chemat, Hussein Znad, Chiara Bisio and Monica Pica

Received: 1 June 2020; Accepted: 16 September 2020; Published: 21 September 2020

check for updates

Abstract: Aging tests were used to investigate the long-term effects of BC on the immobilization of Cu, and the soil silicon dissolution of three types soils (black soil, (BS), vegetable garden soil (VS) and red soil (RS)). Litchi branch biochars (BC) at 10% (*w*/*w*) were incubated with three Cu (400 mg/kg) contaminated soils. The effect on soil properties of pH, soil organic carbon (SOC), dissolved organic carbon (DOC) and available silicon content were investigated, along with the speciation distribution of Cu. The results indicated that SOC, DOC, and available silicon content (except, BC300) increased with the application of BCs. On the other hand, the DTPA (diethylenetriaminepentaacetic acid) extractable Cu content in BS, VS and RS soils were reduced by 4–12%, 18–25%, and 12–19%, respectively. The Cu availability in all soils first increased, and then decreased during the aging process. The sum of the other four fractions, including the carbonate fraction and the inert component increased by 4–4.5% (BS), 1.4–2.1% (VS), and 0.5–1% (RS) respectively, over the long-term process. Moreover, during the whole aging process, the soil properties (such as pH, SOC, DOC and available silicon content) were almost stable. This study demonstrates that BCs, especially those produced at a higher temperature, are superior to those been produced at 300 °C in immobilizing Cu and releasing available silicon in soils. However, the remediation efficiencies were restricted by the soil type contamination status and remediation time.

Keywords: BC; sequential extraction; copper; soil remediation; carbon-silicon interaction

1. Introduction

Copper pollution in agricultural soils can occur from copper mining and directly or indirectly from anthropogenic sources [1]. This can cause crop yields to fall, as well as allow copper to enter the food chain. Although copper is an essential trace element for the human body, long-term excessive intake can cause copper to accumulate in the body and endanger health [1–4]. Due to its potential toxicity, persistence and irreversibility by the United States Environmental Protection Agency (USEPA) has listed Cu as priory control pollutant [5–7]. With increased public awareness of health and safety, soil contamination by Cu has gained attention, and requires proper remediation.

In-situ remediation of Cu-contaminated soils is regarded as an effective, practicable, and environmentally friendly measure [8]. Biochar (BC) is the solid, carbon (C)-rich and silicon (Si)-rich product of heating biomass in a low oxygen environment (pyrolysis) [9–12]. Due to its highly porous micro-structure, active functional groups, high pH; rich Si content; surface area; and cation exchange capacity (CEC), BC can effectively immobilize contaminants. It does this by adsorption, ion exchange, surface complexation and precipitation. A number of studies have highlighted that BC further reduces the risk of heavy metals contamination to humans and the surrounding ecosystem [13–18]. Recent studies have reported the successful applications of BC in soil remediation, in both the short and long-term. For instance, Rees et al. [19] observed a reduction in extractable cadmium (Cd), copper (Cu (II)), and other metals in soils after the addition of 80% coniferous and 20% hardwood BC after seven days. Uchimiya et al. [20] also found a reduction in Cu (II) and Zn (II), using toxicity characteristic leaching procedure (TCLP) extraction in a contaminated soil 7 day incubation experiment after addition of cottonseed hull BC. Likewise, Bian et al. [21] used wheat straw BC to treat agriculture land and consistently and significantly increased soil pH, total organic carbon and observed reductions in Cd (II) and Pb (II) concentrations, following $CaCl_2$ and DTPA extractions over a three year period. Li et al. [22] demonstrated the link between BC type and long-term (three years) immobilization of Cd and Cu in acidic paddy soils, and recommended that the readily oxidized BC be applied to the soil to reduce the risk of Cd and Cu exposure. These short-term and long-term studies demonstrate that the application of BC in soil remediation is feasible. However, the data on long-term effects of BC on the immobilization of heavy metals appear to be inconsistent. In lightly contaminated agriculture soil, Lucchini et al. [23] demonstrated that BC caused small changes in metal fractionation in soil, but no significant changes in total metal concentrations in soil or plants were observed after three years. Over three consecutive harvests, hardwood BC treatments increased the uptake of metals in ryegrass in the last two harvests compared to the first harvest [24]. These studies highlight that the effects of BC on immobilizing heavy metals in soils need to be investigated in greater detail. Furthermore, the long-term immobilization of heavy metals in soils is affected by the aging process of BC [22,25]. BC undergoes a slowed growth rate with uncontrolled in-situ oxidation in soils, which leads to the formation of carboxylic, carbonylic, phenolic and other oxygen-containing functional groups on the surface of aged BC [22,25–29]. As, such, the aging process is not synchronized in soils of different types [27,30,31].

Lots of studies have focused on the organic component of BCs after adding them to soil, because of their function in carbon sequestration and environmental remediation, such as the adsorption of contaminants by oxygen-containing functional groups [12,32,33]. However, the function and effects of BC on the inorganic components are unclear, especially for BC that have higher mineral content (silicon-containing component). On the one hand, BC acts as a potential bio-available Si source and, more importantly, along with Si and Al complexes from soil, results in an increase in Fe and soil organic matter [34]. For instance, the adsorption of silicic acid on Al or Fe oxides may promote a lower phosphate-fixation in acidic soils, thus improving the rate of available P. Silicic acid dissolved in the soil solution can be adsorbed into soil minerals, particularly Fe and Al oxides/hydroxides [35]. There is a general lack of knowledge about the effects of BC aging on silicon flow and pollutant control in soil systems. Silicon-rich biochar is considered a rich source of silicon materials [36]. Therefore, BC can be redefined as a silicon-improving material in soil. Liu et al. [37] reported that the effect of wheat straw BC on soil Si in different regions. Unfortunately, the interaction between soil-available Si content and carbon components with biochar during the aging process in soils was not fully understood. In addition, the stable organo-mineral (C and Si) fraction produced via the aging process between BC and minerals plays a vital role in the stabilization of heavy metals.

Based on the expounded data, we propose the hypothesis that the interaction between soil-available Si content and carbon components in biochar during the aging process assists Cu immobilization immediately following biochar amendment. To test this hypothesis, this study was set up to explore the relationship between C and Si in soil, as well as to determine the availability of Cu in three different soils, during the aging process. The oxidizability under accelerated aging of BC under wet-dry conditions was simulated in the laboratory to gain insights into the aging process of BC. Specifically, this study aims to: (1) compare the effect of BC under different aging processes on three soil properties, and the mobility Cu in soil over the course of a year, and (2) determine the relationship between C and Si in the soil, with added BC, and as a result of stimulated aging.

2. Materials and Methods

2.1. Soils and Biochar

Three topsoil samples (0–20 cm) were collected from different site fields. one sample of black soil (BS) was collected from non-polluted agricultural fields near Harbin, China (45°40′ N and 126°35′ E); Two samples of red soil (RS) and vegetable garden soil (VS) were collected from agricultural fields (23°07′ N and 113°43′ E) and vegetable fields (23°02′ N and 113°43′ E) in Guangzhou, China. The sample of RS and VS were classified as clay soil (Ultisols), BS was classified as sandy loam soil (Histosols). After collection, the samples were air-dried, and passed through 2 mm and 0.147 mm nylon sieves before using for further characterization analysis.

Litchi branches were collected from an orchard in ZengCheng District, Guangzhou City, Guangdong Province, south China (23°41′ N and 113°81′ E). The BC preparation process followed the method outlined in a previous study [38]. The BCs obtained under oxygen-limited conditions of different temperatures were named BC300 and BC600, where the suffix number refers to the carbonization temperature. The properties of the soils and BC are listed in Tables S1 and S2.

2.2. Experimental Design

The evaluation of retention properties Cu^{2+} on BC300 and BC600, followed the method described by Lin et al. [39]. Briefly, an aliquot of 0.05 g BC sample was added to 50 mL centrifuges tubes (polypropylene) that contained 40 mL of 50 mg/L Cu^{2+} solution and 0.01M NaOH or HCl was used to adjust the solution pH. Then, the tubes were shaken at 250 rpm for 24 h at room temperature (25 °C ± 2) using a horizontal shaker. The amount of metal adsorbed was also calculated following the method described by Lin et al. [39].

The soil was air-dried and sieved to remove impurities, such as stones and organic residues. The total Cu content was adjusted to 400 mg/kg by adding $Cu(NO_3)_2$ solutions, mixed thoroughly, and aged at room temperature for 1.5 months. This study included nine treatments: BS0, VS0 and RS0 (control, containing BS, VS and RS without BC amendment), BS1, VS1 and RS1 (containing, BS, VS and RS with 10% (*w*/*w*) BC300 amendment), BS2, VS2 and RS2 (containing, BS, VS and RS with 10% (*w*/*w*) BC600 amendment). All treatments were performed in triplicate. The incubation experiment was conducted in a constant temperature room at 25 °C with 65% relative humidity, under a wet and dry (wet-dry) cycle for one year. The samples were checked regularly, and water added to compensate for water loss every 10 d to maintain 80% water holding capacity (WHC), then air-dried in the nature environment. During the incubation, soil samples (~50 g) were collected on days 0, 1, 3, 5, 7, 14, 21, 28, 60, 180 and 365 from each pot after thoroughly mixing the soils. Soil samples were air-dried and sieved for further analysis.

2.3. Chemical Analysis

Soil properties, including pH, organic matter, cation exchange capacity, available nitrogen and phosphorus contents and other physicochemical properties, were determined according to previously described methods [40]. Briefly, the pH (1:2.5 of soil:water), soil organic carbon (SOC) (K_2CrO_4-H_2SO_4 oil-bath heating), cation exchange capacity (CEC) (1 M ammonium acetate leaching at pH 7.0), total N (TN) (N/C soil analyzer, Flash, EA, Milan, Italy), available P (AP) (Olsen method) and available K (AK) (1 M ammonium acetate extraction) were determined. The dissolved organic carbon (DOC) was determined extraction of the soil (5 g) with distilled water (25 mL) and shaking for 2 h. The extract solution was filtered (0.45 μm) prior to DOC determination using a TOC analyzer (TOC V-CPH, Jena, Germany). The available Si (acetate buffer method) from soil using the method adopted from Yu et al. [41]. The total Cu, and plant-available Cu of soil were estimated according to the method by Lu [42]. Briefly, the total heavy metals in the soil were digested by HNO_3-HF-H_2O_2 and the plant-available Cu of soils were extracted from the treated soil using diethylenetriaminepentaacetic acid (DTPA, pH 7.3). The different forms of Cu metal in soil were extracted following the method of Tessier [43], and the methods are summarized in Table S3. All samples were centrifuged at 3000 rpm for 5 min, then, the suspensions were filtered through a 0.45 μm filter, acidified (with concentrated HNO_3) and Cu concentrations were determined using an atomic absorption spectrophotometer (AAS, iCE 3500, Thermo Scientific, MA, USA).

2.4. Data and Statistical Analysis

In the study, the following parameters were evaluated:

(1) Effects of BC application on an increase or decrease in Si or DOC content in soil calculated by Equation (1):

$$E_{bc} = \frac{[\text{Si content or DOC content in soil from objects with application of BC}]}{[\text{Si content or DOC content in soil from objects without the application of BC}]} \times 100\% \quad (1)$$

where, E_{bc}—percent increase or decrease in Si content or DOC content in soil under amendment with BC.

(2) The immobilization effectiveness (IE) of Cu in soil was calculated by Equation (2):

$$IE(Cu) = \left(1 - \frac{C_t}{C_o}\right) \times 100\% \quad (2)$$

where, C_t and C_o are the copper concentration in the leachate of the treated soil and control, respectively.

All the experiments were conducted in triplicate. Data are presented as means with standard deviations. Statistical analyses were performed using SPSS Statistics 19.0 software (IBM, Armonk, NY, USA). A one-way ANOVA with multiple comparisons by LSD test at $p < 0.05$ significance level was used to determine differences. Graphs were prepared using Origin 9.0 (OriginLab Corp, Northampton, MA, USA).

3. Results and Discussion

3.1. Soil Properties of Soil and BC

The soil properties are summarized in Table S1. Analysis of the soil properties and the pH shows that there were differences among the collected soil samples. The pH values followed the order BS (pH = 6.15) > VS (pH = 5.88) > RS (pH = 4.86). Likewise, the highest and lowest SOC were observed in BS and RS, respectively, BS and RS were collected from north and south China, respectively, indicating the positive correlation between latitudes for different sites. Moreover, they had the highest extractable cation contents of available silicon and BS compared with the other soil samples. However, the lowest content of available silicon and low CEC values were observed in RS and VS. Prior to spiking with Cu, these three soils were considered unpolluted, because the concentration of heavy metals, such as Cu, Pb, Cd and Zn concentration in soils all lower than the risk screening values for soil contamination of agricultural land (Table S1) [39]. In comparison, the DOC of BCs gradually decreased due increased degradation of carbon components at high temperatures. Noticeable, as shown in Table S2, the C content in BC gradually increases with the pyrolysis temperature increases, while the H, O, and N content continues to decrease, and the atomic ratios H/C, O/C, and (O + N)/C also decrease. These results indicate that biochar with a higher pyrolysis temperature has a higher degree of carbonization and a more complete π-conjugated aromatic structure [39]. Litchi BC morphology and the absorbent capacities have been discussed in the literature [38,39,44]: the observed increase in the p-conjugated system, and metal-retention capability. Therefore, this was not considered in the current study.

3.2. Sorption Behavior of Cu by BCs

The sorption capacities of Cu, by different BCs are presented in Figure 1. The maximum adsorption capacities for Cu increased with the rising pyrolysis temperature from 24.33 (for BC300) to 67.23 mg/g (for BC600), respectively. Notably, the maximum adsorption capacity between BCs of different temperatures different higher than previous studies (Table S4). This can be attributed to the pH, ash contents, functional groups and high CEC values of the BCs (Table S1). This result is consistent with other studies [45] and indicates the positive influence of BC inorganic compounds on metal immobilization in aqueous solutions. In addition, an amount of studies on the large surface area, pore structure and functional group of biochars indicated that metal immobilization was possible [12,13,22]. Interestingly, following BC treatment with sodium silicate, the Cu sorption capacities in BC300 + Si and BC600 + Si increased by 24% and 9%, respectively, compared to BC300 and BC600. Zhao et al. [46] from an adsorption study suggested that the addition of silicon not only enhanced the adsorption capacity of biochar, but also effectively resisted the decrease in the adsorption capacities on BCs after aging (H_2O_2 and H_2SO_4 oxidation). Previous studies have also shown that silica particles might affect C arrangement and structural composition by producing a Si-C coupling system in the biochar matrix [47]. As discussed above, BC modified with silicon shows potential for enhanced metal adsorption.

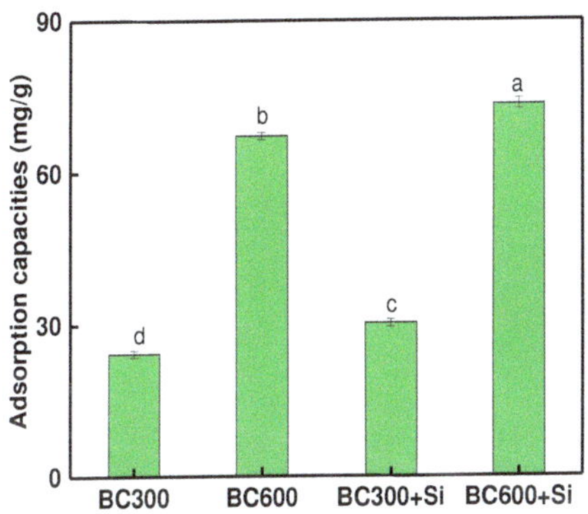

Figure 1. The adsorption capacities of Cu on different BC. BC300 + Si and BC600 + Si are BC modified with 5 M sodium silicate. Values are the mean ± standard deviations, different letters in the same column represent significant difference at $p < 0.05$ ($n = 3$, LSD test).

3.3. Effects of BC Amendment on Soil pH, DOC, and SOC

The impact of BC on soil pH, DOC and SOC, over one year is shown in Figures 2 and 3, Table 1 and Table S5. Compared with the control, the application of BC slightly elevated soil pH, and the influence was more obvious with the application of biochar that was higher in pH (Figure 2). The pH values of BS0, VS0 and RS0 showed a slow, slightly increasing trend at certain time points, during the whole incubation process, which may be related to the neutrality of the deionized water used for wet-dry cycle (Figure 2). After BC treatment and aging, the pH of BS1 and BS2 on day 5 decreased to 6.39 and 6.60 compared with the pH on day 1, and rapidly increase to 6.51 and 6.87, from day 3 to day 28, followed by a slow decrease during the whole aging process (Figure 2a).

Table 1. Soil organic carbon (SOC), as affected by biochar (BC) application after one-year of aging. BS, black soil. VS, Vegetable garden soil. RS, red soil.

Treatment	BS0	BS1	BS2	VS0	VS1	VS2	RS0	RS1	RS2
SOC (%)	3.32	5.07	4.38	1.53	3.27	2.37	0.90	2.41	1.85

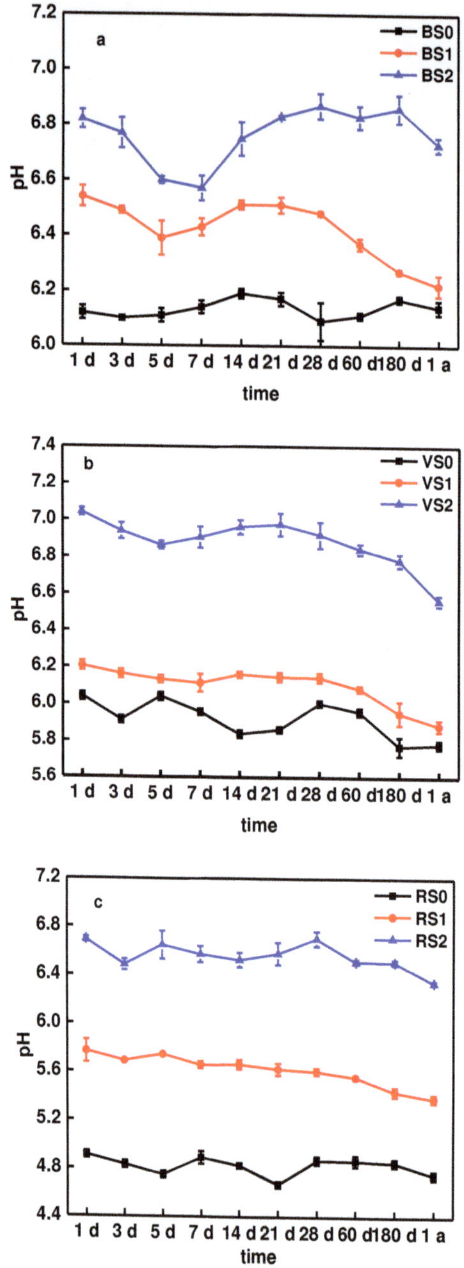

Figure 2. The change of pH in BS (**a**), VS (**b**) and RS (**c**) soils amended with biochars. Note: Error bars are standard deviations of the means (n = 3). (BS0: black soil, BS1: black soil + BC300, BS2: black soil + BC600, VS0: Vegetable garden soil, VS1: Vegetable garden soil + BC300, VS2: Vegetable garden soil + BC600, RS0, Red soil, RS1: Red soil + BC300, RS2: Red soil + BC600).

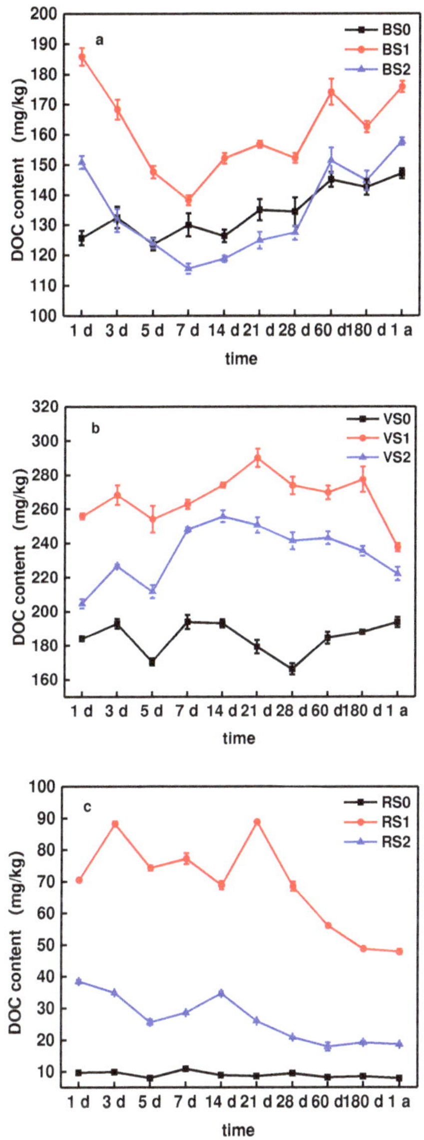

Figure 3. The change in DOC content in BS (**a**), VS (**b**) and RS (**c**) soils amended with biochars. Note: Error bars are standard deviations of the means (n = 3).

However, compared to BS the pH changes of VS and RS were different during aging process, mainly related to differences in soil types and properties. After biochar treatment and aging, the pH of VS1, VS2, RS1 and RS2 rapidly increased on day 1, then slowly decrease during the by one-year aging process (Figure 2b,c). Previous studies have shown that BC during the 72-day aging process had the effect of improving the soil pH, particularly in the early stage [48]. Wang et al. [33] showed that pH rapidly decreased from 0 to 30 days and slowly decreased from 31 to 95 days. Xu et al. [49] showed that pH (BC) value decreased from 8.20–10.7 to 7.5–9.7 after a 25-cycle aging test with BC, which can explain the soil slow decreased for pH during aging process (Figure 2).Furthermore, in a field experiment under a cucumber–sweet potato–rape rotation, Jiang et al. [44] showed that BC application can alter the soil acidity, while the natural aging of biochar in soil, cause the pH of soil to slowly decrease.

As SOC did not significantly change during the whole incubation process in any of the same treatments soil, we did not measure SOC at every time points, indicating that both types of biochar remained stable in the soil in the long term [22]. Changes in BC significantly increased the SOC content, compared with the control (Table 1). The increase in SOC was greater among VS and RS, compared to BS. Likewise, in soil treated with BC, DOC was significantly higher than in the untreated soil. The higher the pyrolysis temperature, the weaker the effect of improving soil DOC (Figure 3), which may due to the high DOC in BC300 (Table S1).

The DOC concentrations in the soil, in common with pH, showed variation between collection periods. However, there was consistent decreasing trend in VS and RS in the latter part of the aging process, compared to the early soil (Figure 3b,c). With treatment by BS, however, the DOC content in BS1 and BS2 decreased from 1day to 7 days, and then increased from day 7 until the end of the process. This could be because BC improves the soil environment and enhances the adsorption of soil to organic components in the initial stage, but the labile organic compound in soil and biochar samples may be released during incubation after 7 days (Figure 3a). A number of studies have also reported that BC application could significantly increase DOC content in soils and decreased during long term aging process [24,50,51]. Li et al. [22] showed that the contents of soil organic matter (SOM) and DOC were significantly higher in the soil treated with biochar. There was a slow change in SOM during the 3-year period; DOC contents, on the other hand, changed over time, increasing steadily up to harvest one, and then decreasing over harvest 2 and 3. BC can modify soil DOC, in both quality and quantity by means of releasing DOC into the soil solution and/or stimulating the generation of more DOC from the soil organic carbon pool [51]. Moreover, Jiang et al. [44] reported that there was a downwards trend in DOC content after the aging process of crop cultivation. This may be attributed to the mineralization of C that occurs during the growing season, in the presence of N from inorganic fertilizer.

3.4. *Effect of Biochar on Soil Available Silicon*

Silicon, as a chemical modification in its soluble form, can effectively decrease heavy metal bioavailability [52]. The role of Si in BC, on Cu mobility during the aging process of typical soils, such as BS, VS and RS, is not yet fully understood. The results of the concentrations of available Si in soil amended with BCs are shown in Figure 4. Figure 4 a show that BS1 decreased the available silicon by 11% and 12% compared with BS0 and BS2 in last stage, respectively. Due to the increase in available silicon content in the later stage of the aging process, BS2 is close to BS0 in the final stage. During the entire aging process, the available silicon content in BS1 was always lower than BS0.This is due to the BC300 having lower ash and SiO_2 content (Table S1) and potentially an equilibrium has been reached between the Si adsorbed by soil and the Si dissolved from BC surface [32]. Compared with VS0 and RS0, the treatments VS1 and RS1 also show the same trend during o1a aging process (Figure 4b,c).

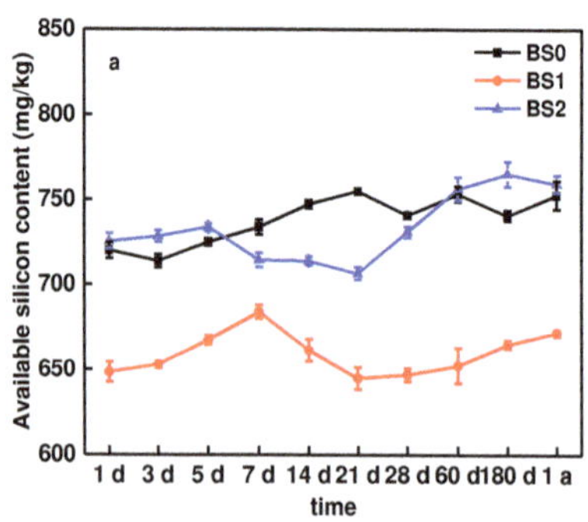

Figure 4. *Cont.*

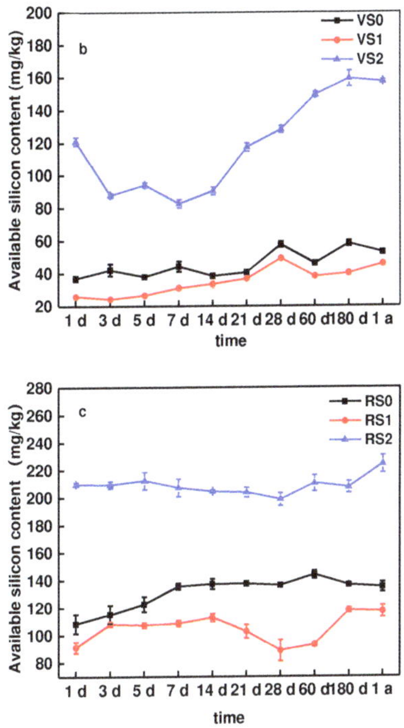

Figure 4. The change of available silicon content in BS (**a**), VS (**b**) and RS (**c**) soils amended with biochars. Note: Error bars are standard deviations of the means ($n = 3$).

Similarly, compared to VS0 and RS0, the available silicon of VS2 and RS2 significantly increased by 87–228% and 46–94%, respectively, with the application of BC600 (Table S5). This result is due to the fact that Si the major component of the ash and BC derived from BC600 had a higher Si content, thus resulting in enhanced Si available for the lower Si soils, such as VS and RS. Previous studies have shown that BC can be used as a Si pool. It supports the dynamic process of Si release from BC modified soil [11,12,32,52].

Figure 4 also shows the silicon dissolution of the experimental values for different soil-BC mixtures. For BC600, the higher improvement in silicon dissolution in a low-silicon soil (RS) compared to that in a high-silicon soil (VS) treated with Si-rich BC (BC600) was observed. However, the opposite result was reported by Wang et al. [32]. For Si-rich BCs, the higher improvement in silicon dissolution in a high-silicon soil (HSS) compared to that in a low-silicon soil treated with Si-rich BCs was observed [32]. According this phenomenon, we suggest that the interaction between the soil and the BC may have caused the change, the BC and soil properties major the influence factor.

3.5. DTPA-Extractable Cu in the Soil

As shown in Figure 5, BC amendment reduced the mobility of Cu contents in soils, as seen by the reduction of DTPA-extractable metal content. By comparison, BC600 was more effective at metal immobilization than BC300, and achieved the highest reduction in the DTPA-extractable metal content in the three soils. This is because the adsorption capacities were higher than other BCs (Table 2). In addition, a clear difference was found between treatments with different soils. Xiao et al. [45] showed that BCs were more effective at reducing DTPA-extractable metal contents and metal immobilization than straw materials, this led them to believe that adsorption might not be the only mechanism governing metal immobilization in soils. Similarly, He et al. [53] demonstrated that the BC soil oxidation reaction would significantly affect the performance of BC on metal mobility and speciation in soils. Compared to BS, this result indicates that the BC (BC300 and BC600) significantly affects the immobilization of Cu in RS and VS. The Cu immobilization rates were higher in BS during the one-year aging process, which might be due to a lower concentration of available Cu in VS and RS, and because the bond between Cu and soil in BS was not as strong (Figure 5 and Table 2).

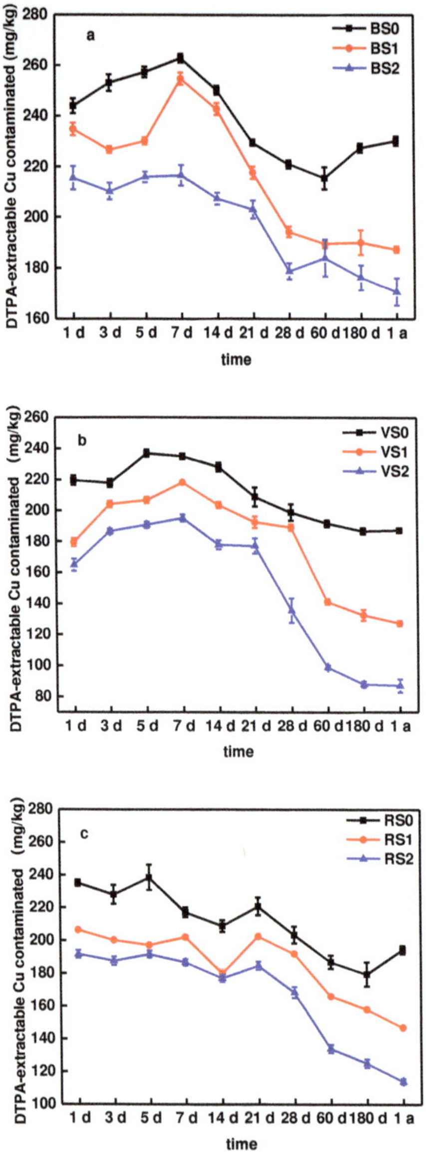

Figure 5. The change of DTPA-extractable Cu concentration in BS (**a**), VS (**b**) and RS (**c**) soils amended with biochars. Note: Error bars are standard deviations of the means ($n = 3$).

Table 2. The immobilization effectiveness (IE) of Cu in three amended soils with biochars (BCs). Values are the mean ± standard deviations, and the different letters in the same column represent that significant difference at $p < 0.05$ ($n = 3$, LSD test).

	Time	BS		VS		RS	
		BC300	BC600	BC300	BC600	BC300	BC600
IE(Cu,%)	1 d	4 ± 0.25 f	12 ± 0.85 h	18 ± 0.17 d	25 ± 0.53 d	12 ± 0.29 d	19 ± 0.24 e
	3 d	10 ± 0.53 d	17 ± 0.66 f,g	6 ± 0.54 i	14 ± 0.57 g	12 ± 0.37 d	18 ± 0.37 e,f
	5 d	11 ± 0.33 d	16 ± 0.33 f,g	13 ± 0.16 e	19 ± 0.33 f	17 ± 0.41 b	20 ± 0.76 d
	7 d	3 ± 0.21 g	18 ± 0.57 d,e	7 ± 0.21 h	17 ± 0.16 g	7 ± 0.34 e	14 ± 0.54 i
	14 d	3 ± 0.08 g	17 ± 0.49 e,f	11 ± 0.22 f	22 ± 0.24 e	14 ± 0.49 c	15 ± 0.37 h,i
	21 d	5 ± 0.29 e	12 ± 0.24 h	8 ± 0.47 g	15 ± 0.49 i	8 ± 0.29 e	16 ± 0.24 g,h
	28 d	12 ± 0.22 c	19 ± 0.57 c,d	5 ± 0.21 j	32 ± 0.21 c	6 ± 0.17 f	17 ± 0.24 f,g
	60 d	12 ± 0.25 c	15 ± 0.93 g	26 ± 0.53 c	48 ± 0.33 b	11 ± 0.29 d	28 ± 0.33 c
	180 d	16 ± 0.22 b	22 ± 0.24 b	29 ± 0.45 b	53 ± 0.08 a	12 ± 0.17 d	30 ± 0.54 b
	1 a	19 ± 0.12 a	26 ± 0.41 a	32 ± 0.39 a	53 ± 0.49 a	24 ± 0.37 a	41 ± 0.76 a

Beesley and Dickinson [54] showed that the application of hardwood-derived BC increased the concentrations of DOC, which has an adverse effect on the immobilization of Cu. Therefore, the DTPA-extractable Cu contents higher than other soils during one-year aging process is probably due to the increase in DOC (Figure 3) and SOC (Table S1) in the BS treatments (BS0, BS1 and BS2). In terms of the change trends, the Cu availability in nine treatments fluctuated sight, but generally showed an increase-decrease trend. For BS, the DTPA-extractable Cu concentration decreased from 215 mg/kg on day 1 to 170 mg/kg on one year with BC600 application (Figure 5a). The DTPA-extractable Cu concentration in BS1 fluctuated between 187 and 234 mg/kg, accounting for 47–58% of the BS total Cu on the incubation (Figure 5a). Compared to BS, the DTPA-extractable Cu concentration in VS and RS also have be similar on trend during one-year aging process (Figure 5). The changes in Cu availability in soils reflected that the transport and transformation processes between soil and BC were dynamic during the aging process.

Taking into consideration the pH value (Figure 2), DOC (Figure 3) and available silicon content (Figure 4), we suggest that the interaction between these factors and the soil DTPA-extractable Cu may have caused the change in Cu availability. Previous works [22,24,45,52] have also shown that an increase in soil pH, DOC and Si content after BC application could reduce the mobility of metal and promote its transformation to a more stable state. The aging experiment reported by Shen et al. [17] showed that the pH of the biochar decreased after accelerated aging due to the loss of dissolved alkaline minerals, indicating the pH could be a mobilization mechanism of the Cu during accelerated aging. Li et al. [51] study showed that a BC induced DOC modification in soil alters the speciation of metal contaminants by changing soil conditions, which suggests that the biochar-DOM interaction is an important mechanism in determining the mobility of heavy metal contaminants in biochar amended soils. Moreover, silicon from BC, as a chemical amendment in its soluble form, can effectively decrease heavy metal bioavailability [52]. Therefore, the pH values of soil, DOC, available silicon content and DTPA-extractable Cu concentration are dynamic during the aging process with BC application, indicating there was a close relationship between them.

3.6. Tessier Cu Fractions in the Soil

The Tessier method was used to determine the five Cu fractions in the soil. The five Cu fractions in the soil were significantly affected in all nine-soil treatments during one-year aging process, compared with the early aging stage (i.e., on day 1). However, a slight change occurred at each time points: in general, F1 and the sum of F1 and F2 showed decreasing trends. Hence in this study, we only evaluated and compared the five Cu fractions at the last stage (i.e., at one year). Figure 6 shows the fractionation of Cu in different treatments. We can see that the Cu bound to Fe and Mn oxide (F3), organic matter (F4) and the residual fraction (F5) covers over 68–89% of its total content in almost all the treatments.

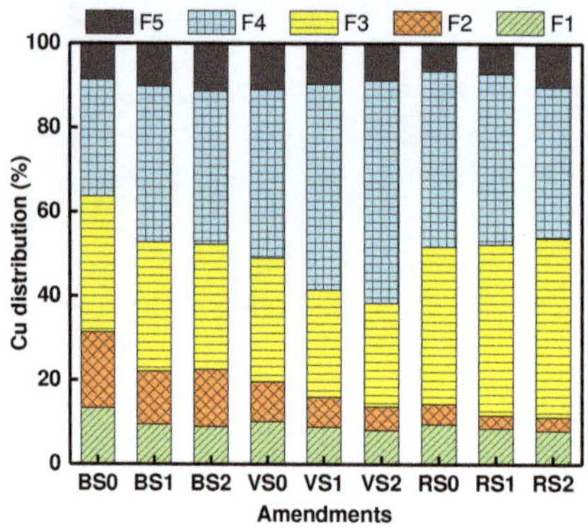

Figure 6. Cu fractionations of soils after 1a incubation with different BC amendments. F1, exchangeable; F2, carbonate bound; F3, Fe-Mn oxide bound; F4, organic matter bound; F5, residual fraction.

Comparing the three soils, F1 and the sum of F1 and F2 is higher in BS than in VS and RS. This also supports the high DTPA-extractable Cu concentration for BS (Figure 6). The sum of F1 and F2 of BS0, BS1 and BS2 were 31%, 22% and 23%, respectively, and higher than the VS and RS treatments; The phenomenon might occur because SOC (Table S1) and DOC (Figure 4) is high in the BS, both of which decreased the ability of BC to fixate Cu [54]. Although, Park et al. [10] reported that Cu is present as a more stable fraction complex in this soil, but we believe that Cu in artificially contaminated soil more easily complexes with DOC in the soil during the one-year aging, enabling high SOC and DOC content. It is well known that Pb in F3, F4, and F5 is an inert component. While, easily exchangeable Cu (F1) is directly related to its bioavailability, and Cu bound to carbonates (F2) can be converted to bioavailable component as soil pH decreases. Meng et al. [48] indicated that BC continuously immobilized the bioavailable metal fraction in the soil during simulation of the actual aging process over 8 years. Therefore, the sum of the other four fractions including the carbonate fraction, F3, F4 and F5 all significantly increased in the long-term process.

3.7. Limitation and Environmental Implication

As demonstrated above, biochar is capable of immobilizing Cu in soils. However, soil properties, such as pH, SOC and DOC, and the properties of BC and the aging time, all affect the of efficiency of metal immobilization. As shown in Figure 2, the application of BC facilitated the stabilization of pH in the soil, but in the long-term aging process, the soil pH decreased. Therefore, in order to maintain the long-term immobilization of Cu in soils by BC, it is appropriate to add new BC in the long-term amended stages of soil. BC also plays a significant role in various terrestrial biogeochemical processes of Si and C (Figures 3 and 4): however, the Si content in BC for amended pollution has not caused issues. In this study, we first considered the available silicon content, in relation to Cu immobilization during aging process. The results show that the release of available Si is important to the fixation of Cu, which is consistent with the experimental results (Figure 1).

Moreover, the effect of BC on soil Si in different soils was investigated, our results were contradictory to other studies, e.g., Wang et al. [32] with regards to whether the available Si content of soil increased or decreased after BC amendment. This may be due to feedstocks of BC and soil properties. More studies are needed to focus on the carbon-silicon biogeochemical cycles and the heavy metal transport distribution between soil and water ecosystems. On the other hand, our result indicate that BC continuously immobilized the bioavailable Cu fraction in the soil (Figures 5 and 6),and BC has been applied to heavy metal contaminated soils as a metal stabilizing method, but it is unclear when and in what form the fixed heavy metals are released again. Further studies should focus on combining the Si source with the heavy metal (BC or soil) in contaminated soil during the aging process.

4. Conclusions

In this study, the addition of BC facilitated the immobilization of Cu in three soils types. However, the efficiency was affected by both the properties of the BC and the incubation time, as well as soil type. BC may not perform well in Cu contaminated soils with high SOC and DOC content. In comparison, the addition of BC enhanced soil DOC and pH value in VS and RS and significantly decreased the available Si content in both BS1 and BS2 during the aging process. Moreover, all the treatments significantly increased the SOC content. Generally, BCs, especially those high in pH and high ash content, had a better performance on soil Cu immobilization, and BC application facilitated the transformation of Cu speciation from F1 or F2 into a stable fraction. Although BC application reduced the extractable Cu contents in DTPA-extracts, the remediation performance was not satisfactory, especially for BS soil treatment due to the high DTPA-extractable Cu concentration during the whole aging process. This study showed that field applications of BC in heavy-metal associated soil remediation must consider the soil type and region attributes into account, and a study with respect to the soil type and the Si source (BC or soil) in contaminated soil during the aging process are important when BC applications are evaluated.

Supplementary Materials: The following are available online, Figure S1: The locations of the soil sampling sites, Table S1: The properties of the soil and BC, Table S2: Elemental compositions and atomic ratios of biochars, Table S3: Extractants used in sequential extraction of metal in soil, Table S4 Comparison of sorption capacity of Cu with biochars derived from different materials, Table S5: Effects of the application of BC on an increase/decrease in silicon content/DOC content in soil.

Author Contributions: Data curation, S.J., J.W. and S.C.; Funding acquisition, T.C.; Investigation, S.J., J.W. and J.H.; Methodology, L.D.; Software, S.C.; Writing—original draft, S.J.; Writing—review & editing, L.D. and T.C. All authors have read and agreed to the published version of the manuscript.

Funding: This work was supported by grants from the Guangdong Provincial Science and Technology Program (2015B020237003) and Provincial Science and Technology Plan Project of Guangdong Province (2018SG00118).

Conflicts of Interest: The authors declare no conflict of interest.

References

1. Xu, X.; Huang, R.; Liu, J.; Shu, Y. Fractionation and release of Cd, Cu, Pb, Mn, and Zn from historically contaminated river sediment in Southern China: Effect of time and Ph. *Environ. Toxicol. Chem.* **2019**, *38*, 464–473. [CrossRef] [PubMed]
2. Qu, C.; Shi, W.; Guo, J.; Fang, B.; Wang, S.; Giesy, J.P.; Holm, P.E. China's soil pollution control: Choices and challenges. *Environ. Sci. Technol.* **2016**, *50*, 13181–13183. [CrossRef]
3. Niu, L.; Yang, F.; Xu, C.; Yang, H.; Liu, W. Status of metal accumulation in farmland soils across China: From distribution to risk assessment. *Environ. Pollut.* **2013**, *176*, 55–62. [CrossRef]
4. Lu, A.; Wang, J.; Qin, X.; Wang, K.; Han, P.; Zhang, S. Multivariate and geostatistical analyses of the spatial distribution and origin of heavy metals in the agricultural soils in Shunyi, Beijing, China. *Sci. Total Environ.* **2012**, *425*, 66–74. [CrossRef]
5. Chen, H.; Teng, Y.; Lu, S.; Wang, Y.; Wang, J. Contamination features and health risk of soil heavy metals in China. *Sci. Total Environ.* **2015**, *512–513*, 143–153. [CrossRef]
6. Chen, H.; Teng, Y.; Lu, S.; Wang, Y.; Wu, J.; Wang, J. Source apportionment and health risk assessment of trace metals in surface soils of Beijing metropolitan, China. *Chemosphere* **2016**, *144*, 1002–1011. [CrossRef]
7. Zhao, F.J.; Ma, Y.; Zhu, Y.G.; Tang, Z.; McGrath, S.P. Soil contamination in China: Current status and mitigation strategies. *Environ. Sci. Technol.* **2015**, *49*, 750–759. [CrossRef]
8. He, B.Y.; Yu, D.P.; Chen, Y.; Shi, J.L.; Xia, Y.; Li, Q.S.; Wang, L.L.; Ling, L.; Zeng, E.Y. Use of low-calcium cultivars to reduce cadmium uptake and accumulation in edible amaranth (*Amaranthus mangostanus* L.). *Chemosphere* **2017**, *171*, 588–594. [CrossRef] [PubMed]

9. Cao, X.D.; Ma, L.N.; Gao, B.; Bandara, W.H. Dairy-manure derived biochar effectively sorbs lead and atrazine. *Environ. Sci. Technol.* **2009**, *43*, 3285–3291. [CrossRef]
10. Park, J.H.; Choppala, G.K.; Bolan, N.S.; Chung, J.W.; Chuasavathi, T. Biochar reduces the bioavailability and phytotoxicity of heavy metals. *Plant Soil* **2011**, *348*, 439–451. [CrossRef]
11. Xiao, X.; Chen, B.; Zhu, L. Transformation, morphology, and dissolution of silicon and carbon in rice straw-derived biochars under different pyrolytic temperatures. *Environ. Sci. Technol.* **2014**, *48*, 3411–3419. [CrossRef]
12. Wang, Y.; Xiao, X.; Xu, Y.; Chen, B. Environmental effects of silicon within biochar (Sichar) and carbon-silicon coupling mechanisms: A critical review. *Environ. Sci. Technol.* **2019**, *53*, 13570–13582. [CrossRef] [PubMed]
13. Shen, Z.; Jin, F.; Wang, F.; McMillan, O.; Al-Tabbaa, A. Sorption of lead by Salisbury biochar produced from British broadleaf hardwood. *Bioresour. Technol.* **2015**, *193*, 553–556. [CrossRef]
14. Shen, Z.; McMillan, O.; Jin, F.; Al-Tabbaa, A. Salisbury biochar did not affect the mobility or speciation of lead in kaolin in a short-term laboratory study. *J. Hazard. Mater.* **2016**, *316*, 214–220. [CrossRef]
15. Shen, Z.; Som, A.M.; Wang, F.; Jin, F.; McMillan, O.; Al-Tabbaa, A. Long-term impact of biochar on the immobilisation of nickel (II) and zinc (II) and the revegetation of a contaminated site. *Sci. Total Environ.* **2016**, *542*, 771–776. [CrossRef]
16. Shen, Z.; Zhang, Y.; McMillan, O.; Jin, F.; Al-Tabbaa, A. Characteristics and mechanisms of nickel adsorption on biochars produced from wheat straw pellets and rice husk. *Environ. Sci. Pollut. Res. Int.* **2017**, *24*, 12809–12819. [CrossRef]
17. Shen, Z.; Hou, D.; Zhao, B.; Xu, W.; Ok, Y.S.; Bolan, N.S.; Alessi, D.S. Stability of heavy metals in soil washing residue with and without biochar addition under accelerated ageing. *Sci. Total Environ.* **2018**, *619–620*, 185–193. [CrossRef]
18. Shen, Z.; Tian, D.; Zhang, X.; Tang, L.; Su, M.; Zhang, L.; Li, Z.; Hu, S.; Hou, D. Mechanisms of biochar assisted immobilization of Pb(2+) by bioapatite in aqueous solution. *Chemosphere* **2018**, *190*, 260–266. [CrossRef]
19. Rees, F.; Simonnot, M.O.; Morel, J.L. Short-term effects of biochar on soil heavy metal mobility are controlled by intra-particle diffusion and soil pH increase. *Eur. J. Soil Sci.* **2014**, *65*, 149–161. [CrossRef]
20. Uchimiya, M.; Bannon, D.I.; Wartelle, L.H. Retention of heavy metals by carboxyl functional groups of biochars in small arms range soil. *J. Agric. Food Chem.* **2012**, *60*, 1798–1809. [CrossRef]
21. Bian, R.; Joseph, S.; Cui, L.; Pan, G.; Li, L.; Liu, X.; Zhang, A.; Rutlidge, H.; Wong, S.; Chia, C.; et al. A three-year experiment confirms continuous immobilization of cadmium and lead in contaminated paddy field with biochar amendment. *J. Hazard. Mater.* **2014**, *272*, 121–128. [CrossRef] [PubMed]
22. Li, H.; Ye, X.; Geng, Z.; Zhou, H.; Guo, X.; Zhang, Y.; Zhao, H.; Wang, G. The influence of biochar type on long-term stabilization for Cd and Cu in contaminated paddy soils. *J. Hazard. Mater.* **2016**, *304*, 40–48. [CrossRef]
23. Lucchini, P.; Quilliam, R.S.; Deluca, T.H.; Vamerali, T.; Jones, D.L. Does biochar application alter heavy metal dynamics in agricultural soil? *Agric. Ecosyst. Environ.* **2014**, *184*, 149–157. [CrossRef]
24. Karami, N.; Clemente, R.; Moreno-Jimenez, E.; Lepp, N.W.; Beesley, L. Efficiency of green waste compost and biochar soil amendments for reducing lead and copper mobility and uptake to ryegrass. *J. Hazard. Mater.* **2011**, *191*, 41–48. [CrossRef]
25. Ren, X.; Wang, F.; Zhang, P.; Guo, J.; Sun, H. Aging effect of minerals on biochar properties and sorption capacities for atrazine and phenanthrene. *Chemosphere* **2018**, *206*, 51–58. [CrossRef]
26. Dong, X.; Li, G.; Lin, Q.; Zhao, X. Quantity and quality changes of biochar aged for 5 years in soil under field conditions. *Catena* **2017**, *159*, 136–143. [CrossRef]
27. Lawrinenko, M.; Laird, D.A.; Johnson, R.L.; Jing, D. Accelerated aging of biochars: Impact on anion exchange capacity. *Carbon* **2016**, *103*, 217–227. [CrossRef]
28. Sorrenti, G.; Masiello, C.A.; Dugan, B.; Toselli, M. Biochar physico-chemical properties as affected by environmental exposure. *Sci. Total Environ.* **2016**, *563–564*, 237–246. [CrossRef]
29. Hale, S.; Hanley, K.; Lehmann, J.; Zimmerman, A.; Cornelissen, G. Effects of chemical, biological, and physical aging as well as soil addition on the sorption of pyrene to activated carbon and biochar. *Environ. Sci. Technol.* **2011**, *45*, 10445–10453. [CrossRef]
30. Liu, G.; Chen, L.; Jiang, Z.; Zheng, H.; Dai, Y.; Luo, X.; Wang, Z. Aging impacts of low molecular weight organic acids, (LMWOAs) on furfural production residue-derived biochars: Porosity, functional properties, and inorganic minerals. *Sci. Total Environ.* **2017**, *607–608*, 1428–1436. [CrossRef]

31. Zhang, X.; Sarmah, A.K.; Bolan, N.S.; He, L.; Lin, X.; Che, L.; Tang, C.; Wang, H. Effect of aging process on adsorption of diethyl phthalate in soils amended with bamboo biochar. *Chemosphere* **2016**, *142*, 28–34. [CrossRef] [PubMed]
32. Wang, Y.; Xiao, X.; Chen, B. Biochar impacts on soil silicon dissolution kinetics and their interaction mechanisms. *Sci. Rep.* **2018**, *8*, 8040. [CrossRef] [PubMed]
33. Wang, Y.; Xiao, X.; Zhang, K.; Chen, B. Effects of biochar amendment on the soil silicon cycle in a soil-rice ecosystem. *Environ. Pollut.* **2019**, *248*, 823–833. [CrossRef] [PubMed]
34. Farmer, V.C.; Delbos, E.; Miller, J.D. The role of phytolith formation and dissolution in controlling concentrations of silica in soil solutions and streams. *Geoderma* **2005**, *127*, 71–79. [CrossRef]
35. Bruun Hansen, H.C.; Raben-Lange, B.; Raulund-Rasmussen, K.; Borggaard, O.K. Monosilicate adsorption by ferrihydrite and goethite at pH 3–6. *Soil Sci.* **1994**, *158*, 40–46. [CrossRef]
36. Ma, J.F.; Tamai, K.; Yamaji, N.; Mitani, N.; Konishi, S.; Katsuhara, M.; Ishiguro, M.; Murata, Y.; Yano, M. A silicon transporter in rice. *Nature* **2006**, *440*, 688–691. [CrossRef]
37. Liu, X.; Li, L.; Bian, R.; Chen, D.; Qu, J.; Wanjiru Kibue, G.; Pan, G.; Zhang, X.; Zheng, J.; Zheng, J. Effect of biochar amendment on soil-silicon availability and rice uptake. *J. Plant Nutr. Soil Sci.* **2014**, *177*, 91–96. [CrossRef]
38. Liu, J.; Jiang, S.; Chen, D.; Dai, G.; Wei, D.; Shu, Y. Activation of persulfate with biochar for degradation of bisphenol A in soil. *Chem. Eng. J.* **2020**, *381*, 122637. [CrossRef]
39. Lin, N.; Zhang, H.; Jia, Z.Z.; Huang, R.L.; Shu, Y.H. Adsorption of Pb(II) by biochars derived from three types of biomass. *J. Agro-Environ. Sci.* **2016**, *35*, 992–998.
40. Lu, R. *Analytical Methods of Soil Agrochemistry*; China Agricultural Science and Technology Publishing House: Beijing, China, 1999; pp. 18–99.
41. Yu, H.Y.; Ding, X.; Li, F.; Wang, X.; Zhang, S.; Yi, J.; Liu, C.; Xu, X.; Wang, Q. The availabilities of arsenic and cadmium in rice paddy fields from a mining area: The role of soil extractable and plant silicon. *Environ. Pollut.* **2016**, *215*, 258–265. [CrossRef]
42. Lu, K.; Yang, X.; Gielen, G.; Bolan, N.; Ok, Y.S.; Niazi, N.K.; Xu, S.; Yuan, G.; Chen, X.; Zhang, X.; et al. Effect of bamboo and rice straw biochars on the mobility and redistribution of heavy metals (Cd, Cu, Pb and Zn) in contaminated soil. *J. Environ. Manag.* **2017**, *186*, 285–292. [CrossRef]
43. Tessier, A.; Campbell, P.G.C.; Bisson, M. Sequential extraction procedure for the speciation of particular trace elements. *Anal. Chem.* **1979**, *51*, 844–851. [CrossRef]
44. Jiang, S.; Liu, J.; Wu, J.; Dai, G.; Wei, D.; Shu, Y. Assessing biochar application to immobilize Cd and Pb in a contaminated soil: A field experiment under a cucumber-sweet potato-rape rotation. *Environ. Geochem. Health* **2020**, 1–12. [CrossRef]
45. Xiao, R.; Wang, P.; Mi, S.; Ali, A.; Liu, X.; Li, Y.; Guan, W.; Li, R.; Zhang, Z. Effects of crop straw and its derived biochar on the mobility and bioavailability in Cd and Zn in two smelter-contaminated alkaline soils. *Ecotoxicol. Environ. Saf.* **2019**, *181*, 155–163. [CrossRef]
46. Zhao, Z.; Nie, T.; Zhou, W. Enhanced biochar stabilities and adsorption properties for tetracycline by synthesizing silica-composited biochar. *Environ. Pollut.* **2019**, *254*, 113015. [CrossRef]
47. Ahmad, M.; Ahmad, M.; Usman, A.R.A.; Al-Faraj, A.S.; Abduljabbar, A.; Yong, S.O.; Al-Wabel, M.I. Date palm waste-derived biochar composites with silica and zeolite: Synthesis, characterization and implication for carbon stability and recalcitrant potential. *Environ. Geochem. Health* **2019**, *41*, 1687–1704. [CrossRef]
48. Zhuo, S.H.; Meng, W.; Xu, T.; Deng, Y.Y.; Lin, Z.B.; Wang, X.G. Transport and transformation of Cd between biochar and soil under combined dry-wet and freeze-thaw aging. *Environ. Pollut.* **2020**, *263*, 114449.
49. Xu, Z.; Xu, X.; Tsang, D.C.W.; Cao, X. Contrasting impacts of pre- and post-application aging of biochar on the immobilization of Cd in contaminated soils. *Environ. Pollut.* **2018**, *242*, 1362–1370. [CrossRef]
50. Wagner, A.; Kaupenjohann, M. Suitability of biochars (pyro- and hydrochars) for metal immobilization on former sewage-field soils. *Eur. J. Soil Sci.* **2014**, *65*, 139–148. [CrossRef]
51. Li, G.; Khan, S.; Ibrahim, M.; Sun, T.R.; Tang, J.F.; Cotner, J.B.; Xu, Y.Y. Biochars induced modification of dissolved organic matter (DOM) in soil and its impact on mobility and bioaccumulation of arsenic and cadmium. *J. Hazard. Mater.* **2018**, *348*, 100–108. [CrossRef]
52. Sui, F.; Wang, J.; Zuo, J.; Joseph, S.; Munroe, P.; Drosos, M.; Li, L.; Pan, G. Effect of amendment of biochar supplemented with Si on Cd mobility and rice uptake over three rice growing seasons in an acidic Cd-tainted paddy from central South China. *Sci. Total Environ.* **2019**, *709*, 136101. [CrossRef]

53. He, E.; Yang, Y.; Xu, Z.; Qiu, H.; Yang, F.; Peijnenburg, W.; Zhang, W.; Qiu, R.; Wang, S.Z. Two years of aging influences the distribution and lability of metal(loid)s in a contaminated soil amended with different biochars. *Sci. Total Environ.* **2019**, *673*, 245–253. [CrossRef]
54. Beesley, L.; Dickinson, N. Carbon and trace element fluxes in the pore water of an urban soil following green waste compost, woody and biochar amendments, inoculated with the earthworm *Lumbricus terrestris*. *Soil. Biol. Biochem.* **2011**, *43*, 188–196. [CrossRef]

Sample Availability: Samples of the compounds, such as Sodium silicate, biochars and soils, are available from the authors.

molecules

MDPI

Article

Competitive Interaction of Phosphate with Selected Toxic Metals Ions in the Adsorption from Effluent of Sewage Sludge by Iron/Alginate Beads

Hanna Siwek * and **Krzysztof Pawelec**

Department of Bioengineering, West Pomeranian University of Technology Szczecin, ul. J.Słowackiego 17, 71-434 Szczecin, Poland; krzysztof.pawelec@zut.edu.pl

* Correspondence: hanna.siwek@zut.edu.pl

Academic Editors: Chiara Bisio and Monica Pica

Received: 3 June 2020; Accepted: 26 August 2020; Published: 31 August 2020

check for updates

Abstract: Wastewater is characterized by a high content of phosphate and toxic metals. Many studies have confirmed the sorption affinity of alginate adsorbents for these ions. In this study, the adsorption of phosphate from effluent of sewage sludge on biodegradable alginate matrices cross-linked with Fe^{3+} ions (Fe_Alg) was investigated. Kinetics and adsorption isotherms were tested in laboratory conditions in deionized water (DW_P) and in the effluent (SW_P), and in the same solutions enriched in toxic metals ions—Cu^{2+}, Cd^{2+}, Pb^{2+}, and Zn^{2+} (DW_PM and SW_PM). Batch experiments were performed by changing the concentration of phosphate at constant metal concentration. Kinetics experiments indicated that the pseudo-second-order model displayed the best correlation with adsorption kinetics data for both metals and phosphate. The Freundlich equation provided the best fit with the experimental results of phosphate adsorption from DW_P and DW_PM, while the adsorption from SD_P and SD_PM was better described by the Langmuir equation. For tested systems, the affinity of the Fe_Alg for metal ions was in the following decreasing order: $Pb^{2+} > Cu^{2+} > Cd^{2+} > Zn^{2+}$ in DW_PM, and $Pb^{2+} > Cu^{2+} > Cd^{2+} > Zn^{2+}$ in SW_PM. The metals' enrichment of the DW_P solution increased the affinity of Fe_Alg beads relating to phosphate, while the addition of the metals of the SW_P solution decreased this affinity.

Keywords: phosphorus removal; toxic metals; adsorption; alginate beads; sewage sludge

1. Introduction

The growing world population causes intensification of food production, which generates an increasing use of fertilizers, including phosphate ones. The main industrial source of phosphorus (P) is phosphate, the resources of which, like oil and natural gas, are non-renewable. The results show that 70% of the global production of phosphate rock is currently produced from reserves which will be depleted within 100 years and combining this with increasing demand will result in a significant global production deficit, which by 2070 will be larger than the current production [1,2]. Phosphorus does not disappear like hydrocarbons after "consumption" but can be recycled. In accordance with the principles of sustainable development, attempts are made to search for alternative sources of phosphorus in the human food chain, e.g., in municipal waste. Municipal wastewaters may contain from 6 to 8 mg/L of total phosphorous [3,4]. The national P budgets in Central Europe show that municipal wastewater contains a P load that could theoretically replace 40% to 50% of the annually applied mineral P fertilizer in agriculture [5].

A particularly numerous group of methods for the recovery of phosphorus from wastewater and manure is the crystallization of struvite [6] or combustion [6–8]. An alternative method of phosphorus removal at various stages of wastewater and sewage sludge treatment can be adsorption of mineral phosphorus forms from the liquid phase. To this end, adsorbents containing iron compounds, mainly oxides and carbonates, e.g., allophane [9], neutralized red sludge containing 40–45% Fe_2O_3 [10], fly ash granulated and sintered at 1000–1350 °C [11] or goethite [12,13] were tested. Due to the high impact of pH and redox potential on the phosphorus binding process with the participation of minerals and iron-containing waste, these adsorbents have little application significance [14,15]. The second group of sorbents are adsorbents activated with multivalent metals, e.g., lanthanum-modified bentonite [16] or iron-modified: refined aspen wood fiber [17,18], coir pith [19], eggshell [20] and *Staphylococus xylosus* biomass [21]. All the adsorbents described above were tested only on a small scale, and their main shortage is difficult recovery after the phosphorus binding process. The third group of sorbents are hydrogel composites obtained on the basis of natural polysaccharides, e.g., alginates, the effective action of which has been repeatedly confirmed during purification of the aqueous environment from pigments, heavy metals and antibiotics. Alginate is a natural anionic polysaccharide extracted from algae, which is non-toxic and biodegradable. The major component of alginate is Na-salt of alginic acid with abundant free hydroxyl and carboxyl groups distributed along the backbone chain of the polymer. It can pass through an irreversible chemical process with polyvalent cations (except magnesium) to form a crosslinking bond, and finally forming the thermo-irreversible gel [22]. Alginate cross-linked with multivalent metals, e.g., calcium, forms structures with a large specific surface [23,24]. After chemical modification, it can be converted into adsorbents with anionic sorption capacity. By such modification using Fe(III) or Zr(IV) compounds, adsorbents for purifying water from oxygen anions containing Se(IV), Cr(VI) and As(V) [25,26] or P [27,28] compounds were obtained.

Wastewater is characterized by a high content of toxic metals, examples of metal contents in municipal wastewater are: 0.75 mgCu^{2+}/L, 1.13 mgZn^{2+}/L and 0.4 mgPb^{2+}/L, whereas in industrial wastewater: 2.13 mgCu^{2+}/L, 17.3 mgZn^{2+}/L, 6.1 mgPb/L and 0.1 mgCd^{2+}/L [29]. Many studies have confirmed the sorption affinity of alginate adsorbents for these metals [22] so the metals can compete in the phosphate adsorption process. The aim of this paper was to study the adsorption of phosphate in a multicomponent system on biodegradable alginate matrices cross-linked with Fe^{3+} ions (Fe_Alg). The tests were carried out in effluent water generated during the processing of sewage sludge from sewage treatment plants, and the tested effluent has been enriched with selected multivalent metals. To assess the competitiveness of effluent components for phosphate adsorption, analogous comparative studies were carried out in ion solutions prepared in deionized water.

2. Results

2.1. *Phosphate Adsorption Kinetics*

The kinetics of phosphate binding from solutions without and with addition of metal ions by Fe_Alg matrices showed that the amount of adsorbed phosphate increased over time (Figure 1). The main goal of the kinetic studies was to determine the time after which the tested systems reach the state of adsorption equilibrium, and thus how long the adsorption time will be in the study of adsorption isotherms.

In the examined systems, the equilibrium was established after 56 h, after which 70.1% and 59.2% phosphate were removed from the DW_P and DW_PM solutions and 76.5% and 63.3% from SW_P and SW_PM, respectively. It has to be noted that that from 53.54% (for SD_PM solution) to 92.83% (for SW_P solution) of the maximum sorption capacity is achieved within 8 h.

Kinetics experiments indicated that the pseudo-second-order model displayed the best correlation with phosphate adsorption kinetics data for all the tested systems. The adjusted determination coefficients for the linear t/a_t dependence as a function of time ranged from 0.996 to 0.999 (Table 1). The presence of metals in the DW_PM solution increased values of a_{eq} and V_o, while in the SW_PM solution, these values decreased. Values of adsorption phosphate at equilibrium state are consistent with the adsorption capacities obtained from the phosphate adsorption isotherms presented in this study. The values of k_2 calculated for the models describing adsorption phosphate from SW_P and SW_PM solutions were higher compared to k_2 values calculated for DW_P and DW_PM solutions.

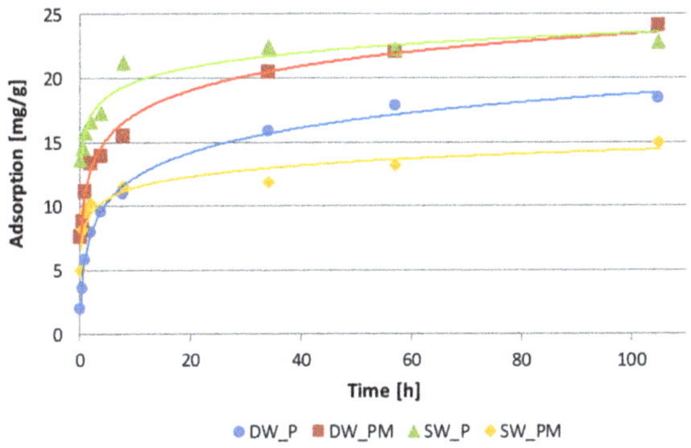

Figure 1. Adsorption kinetics of phosphate ions by alginate matrices cross-linked with Fe^{3+} ions. DW_P—deionized water enriched with phosphate, SW_P—effluent from a sludge enriched with phosphate, DW_PM—deionized water enriched with phosphate and metal ions, SW_PM—effluent from a sludge enriched with phosphate and metal ions.

Table 1. Estimated pseudo-second-order kinetic model parameters for phosphate adsorption by alginate matrices cross-linked with Fe^{3+} ions. DW_P—deionized water enriched with phosphate, SW_P—effluent from a sludge enriched with phosphate. DW_PM—deionized water enriched with phosphate and metal ions, SW_PM—effluent from a sludge enriched with phosphate and metal ions.

Solution	Adsorption at Equilibrium State	Rate Constant of Adsorption	Initial Adsorption Rate	Adjusted Coefficient of Determination
	a_{eq} (mgPO$_4$/g)	k_2 (g/mg h)	V_o (mg/g h)	$R_{adj}{}^2$
DW_P	18.83 ± 0.32	0.011 ± 0.002	4.17 ± 0.85	0.998
DW_PM	22.27 ± 0.51	0.028 ± 0.011	13.79 ± 5.28	0.996
SW_P	22.83 ± 0.12	0.047 ± 0.015	24.39 ± 7.79	0.999
SW_PM	15.11 ± 0.12	0.067 ± 0.022	15.41 ± 5.12	0.999

As pH is an important factor affecting the removal of ions from aqueous solution, the pH changes of the solutions during adsorption were studied. In the conducted experiment, the pH of effluent water was higher in comparison with deionized water, i.e., 7.0 and 5.8, respectively. During the process, the pH changed to a small extent, while the most significant changes were found for the DW_PM solution, in which the pH decreased by 0.6 (Figure 2). A greater decrease of the pH in DW_PM and SW_PM solutions compared to DW_P and SW_P indicates a competition for binding sites' metal ions and remaining protons on the alginate matrix.

Changes in the content of investigated metals in alginate capsules during the phosphate adsorption process showed that the most lead was removed from DW_PM, whereas the most copper from SW_PM solutions (Figure 3). The lowest adsorption was found for Zn^{2+} for the two tested systems. The equilibrium in the examined systems was established after 56 h. After that time, 70.6% Pb^{2+}, 56% Cu^{2+}, 14.1% Zn^{2+} and 28.84% Cd^{2+} were removed from DW_PM, and 70.4% Pb^{2+}, 84.5% Cu^{2+}, 26.6% Zn^{2+} and 51.7% Cd^{2+} from SW_PM. The sorption affinity of the studied metals for the adsorbent changed in the following order: $Pb^{2+} > Cu^{2+} > Cd^{2+} > Zn^{2+}$ in DW_PM, and $Cu^{2+} > Pb^{2+} > Cd^{2+} > Zn^{2+}$ in SW_PM.

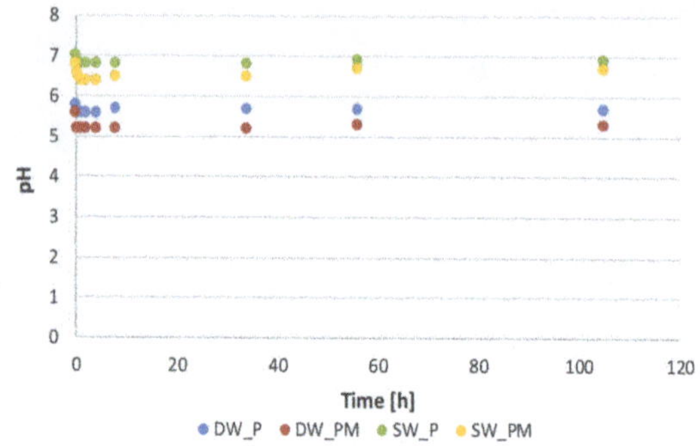

Figure 2. Changes in the pH of the tested systems during the adsorption process. DW_P—deionized water enriched with phosphate, SW_P—effluent from a sludge enriched with phosphate. DW_PM—deionized water enriched with phosphate and metal ions, SW_PM—effluent from a sludge enriched with phosphate and metal ions.

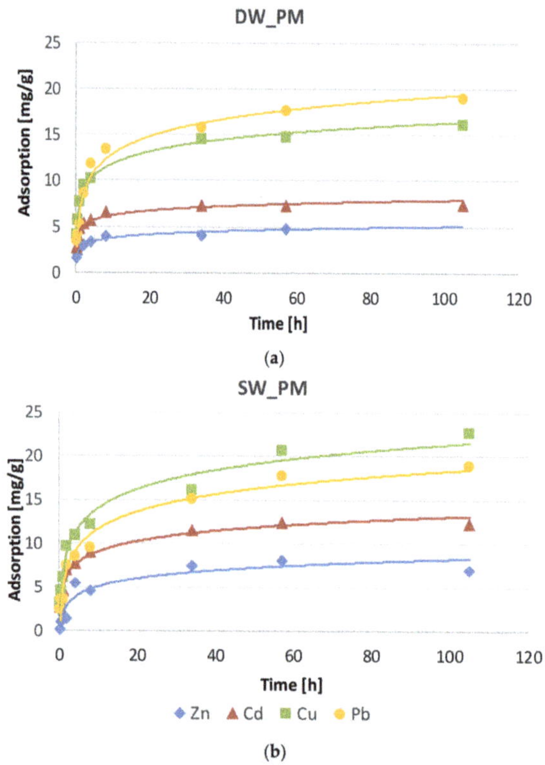

Figure 3. The kinetics of metal ion binding in the presence of phosphate by alginate matrices cross-linked with Fe^{3+} ions in: (**a**) deionized water enriched with phosphate and metal ions (DW_PM) and (**b**) effluent from a sludge enriched with phosphate and metal ions (SW_PM).

Kinetics of the metals binding by Fe_Alg beads in the presence of phosphate indicated that the pseudo-second-order model displayed the best correlation with adsorption kinetics data for all metals. The adjusted determination coefficients for the linear t/a_t dependence as a function of time ranged from 0.982 to 0.999 (Table 2). Many studies published in the literature also report pseudo-second-order kinetics for adsorption of: Cd^{2+} and Cu^{2+} by alginate–based attapulgit foams [30], Cu^{2+} ions by glutaraldehyde crosslinked humic acid-immobilized sodium alginate [31], Cd^{2+}, Cu^{2+}, and Pb^{2+} ions by alginate modified with the grafting of urea and biuret [32] and Pb^{2+} ions by biocharealginate beads [33]. The values of k_2 calculated for the models describing adsorption of all tested metals from the DW_PM solution were higher compared to k_2 values calculated for the SW_PM solution. It indicates that adsorption of metals was faster in the DW_PM solution. Among the examined metals, the binding process was the fastest in the case of Cd^{2+}, for which k_2 was 0.422 ± 0.075 g/mg h in DW_PM and 0.059 ± 0.022 g/mg h in SW_PM. This process was the slowest in the case of Pb^{2+}, for which k_2 was 0.018 ± 0.005 g/mg h in DW_PM and 0.010 ± 0.003 g/mg h in SW_PM.

Table 2. Estimated pseudo-second-order kinetic model parameters for metals adsorption by alginate matrices cross-linked with Fe^{3+} ions. DW_PM—deionized water enriched with phosphate and metal ions, SW_PM—effluent from a sludge enriched with phosphate and metal ions.

System	Metal	Adsorption at Equilibrium State	Rate Constant of Adsorption	Initial Adsorption Rate	Adjusted Coefficient of Determination
		a_{eq} (mgPO$_4$/g)	k_2 (g/mg h)	V_o (mg/g h)	R_{adj}^2
DW_PM	Pb	19.12 ± 0.39	0.018 ± 0.005	6.49 ± 1.89	0.997
	Zn	4.13 ± 0.03	0.422 ± 0.075	7.18 ± 1.28	0.999
	Cd	7.50 ± 0.03	0.166 ± 0.035	9.35 ±1.98	0.997
	Cu	15.08 ± 0.12	0.059 ± 0.011	13.48 ± 2.50	0.998
SW_PM	Pb	19.42 ± 0.56	0.011 ± 0.003	4.06 ± 1.02	0.993
	Zn	7.23 ± 0.16	0.059 ± 0.022	3.10 ± 1.15	0.999
	Cd	12.52 ± 0.19	0.042 ± 0.010	6.58 ± 1.59	0.997
	Cu	22.94 ± 1.03	0.010 ± 0.003	5.07 ± 1.63	0.982

During the adsorption process, the release of iron ions from Fe_Alg beads took place. The most iron was released by systems enriched with metals (Figure 4). After completion of the tests, the DW_PM system released on average about 35 times more iron compared to the DW_P system, while the SW_PM system released on average about 5 times more iron than the SW_P system. Iron ions were released into the solution mainly during the first four hours of adsorption, as was the ion binding on the adsorbent. A positive correlation with a high correlation coefficient (R = 0.894) was calculated between the amount of released phosphate ions and the amount of adsorbed metal ions in the DW_PM solution, no such correlation was found in the SW_PM solution (Figure 5).

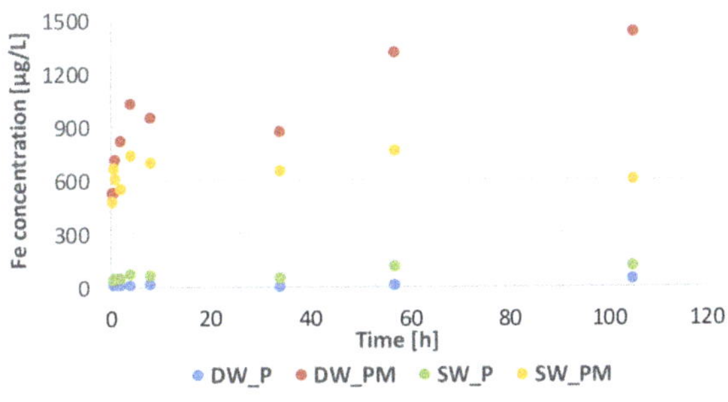

Figure 4. Changes in iron concentration in the tested solutions during the adsorption process. DW_P—deionized water enriched with phosphate, SW_P—effluent from a sludge enriched with phosphate, DW_PM—deionized water enriched with phosphate and metal ions, SW_PM—effluent from a sludge enriched with phosphate and metal ions.

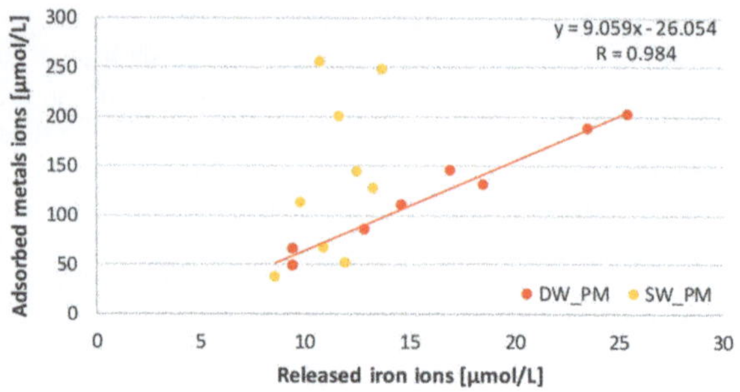

Figure 5. Relationship between the amount of iron released from alginate matrices cross-linked with Fe^{3+} ions and the total amount of metal ions adsorbed from a 1 L solution.

2.2. Phosphate Adsorption Isotherm

According to the Giles classification [34], the shape of the obtained phosphate adsorption isotherms onto Fe_Alg beads indicates that they are L-type isotherms of subgroup 1 (Figure 6). The isotherms are most commonly found in solute adsorption in aqueous solution and it was indicated that the adsorption occurs due to relatively weak forces, such as van der Waals forces and a low competition from solvent molecules [35]. The same form of isotherm was received by authors examining the process of phosphate adsorption on various adsorbents, e.g., goethite [12], wood fiber treated with carboxymethyl [18] and surfactant-modified natural zeolite [36].

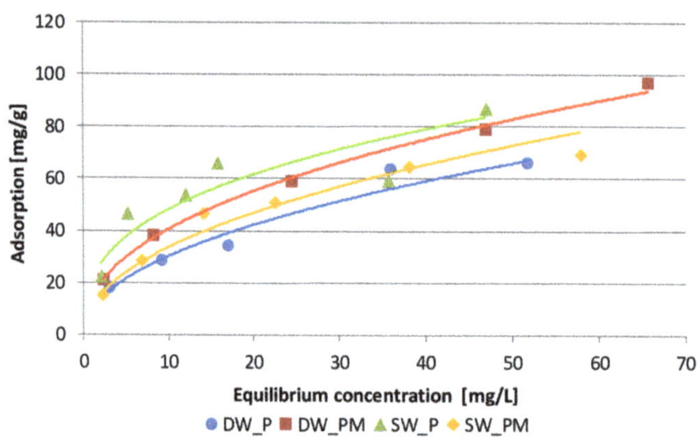

Figure 6. Freundlich isotherms of phosphate adsorption onto alginate matrices cross-linked with Fe^{3+} ions. DW_P—deionized water enriched with phosphate, SW_P—effluent from a sludge enriched with phosphate, DW_PM—deionized water enriched with phosphate and metal ions, SW_PM—effluent from a sludge enriched with phosphate and metal ions.

This isotherm type is usually described by the Freundlich or Langmuir models. The phosphate adsorption system from DW_P and DW_PM was better described by the Freundlich model, because the value of R_{adj}^2 was higher than the value of this parameter calculated for the Langmuir model (Table 3). This type of isotherm presumes that the multilayer of the adsorption process occurs on a heterogenous surface. The values k_F were higher in the isotherm describing adsorption in DW_PM, which indicates that in the equilibrium state, a higher sorption capacity of Fe_Alg beads relating to phosphate was in this system.

Table 3. Freundlich and Langmuir isotherm constants for phosphate adsorption onto alginate matrices cross-linked with Fe^{3+} ions. DW_P—deionized water enriched with phosphate, SW_P—effluent from a sludge enriched with phosphate, DW_PM—deionized water enriched with phosphate and metal ions, SW_PM—effluent from a sludge enriched with phosphate and metal ions.

Solution	Freundlich Model *			Langmuir Model **		
	$1/n_F$	k_F	$R_{adj}{}^2$	q_m ($mgPO_4$ g^{-1})	k_L (L $mgPO_4{}^{-1}$)	$R_{adj}{}^2$
DW_P	0.49 ± 0.06	9.98 ± 0.89	0.968	89.3 ± 21.9	0.055 ± 0.008	0.913
DW_PM	0.44 ± 0.01	14.5 ± 0.34	0.998	112 ± 32.0	0.063 ± 0.007	0.947
SW_P	0.42 ± 0.08	19.8 ± 3.26	0.898	100 ± 11.8	0.135 ± 0.005	0.994
SW_PM	0.48 ± 0.05	11.0 ± 0.95	0.964	83.5 ± 8.01	0.083 ± 0.003	0.993

* Freundlich $q_e = k_F C_e^{1/n_F}$; ** Langmuir $q_e = \frac{q_m k_L C_e}{1+k_L C_e}$

The phosphate adsorption from SW_P and SW_PM systems was better described by the Langmuir model: values of $R_{adj}{}^2$ were very high: 0.994 and 0.993, respectively. This model indicates that adsorption is proportional to the fraction of surface of the open adsorbent. The value k_L was higher in the isotherm describing adsorption in the SW_P solution, which indicates that in the equilibrium state, a higher affinity of Fe_Alg beads relating to phosphate was in the SW_P system. In the presence of metals, this affinity decreased. Comparable values of q_m in SW_P and SW_PM systems indicate a comparable number of active sites.

A comparison between the calculated parameters of adsorption isotherms and parameters calculated by other authors, who studied adsorption of phosphate ions on adsorbents containing iron, indicates comparable or higher sorption capacity of Fe_Alg beads relating to phosphate in the multicomponent solutions. For values of the k_F calculated for iron oxide tailing [15], goethite and alginate/goethite beads [37], Fe-Mn binary oxide [13] ranged from 11.7 to 27.2, while the calculated values of q_m ranged from 8.2 to 108 $mgPO_4$ g^{-1}.

3. Discussion

Adsorption studies in multicomponent solutions, such as wastewater and their processing products, are important for the design, optimization and operation of purification technologies in the aquatic environment. In particular, heavy metal ion adsorption in multicomponent systems has been recognized as a strong antagonistic removal process, in which the properties and concentrations of compounds affect the adsorption efficiency [38–40]. The research conducted on the competitiveness of phosphate and multivalent metal adsorption by Fe_Alg showed that phosphate adsorption in the DW_PM solution was greater compared to the SW_PM solution. Intensification of phosphate adsorption in multicomponent solutions may be caused by multivalent metals forming additional active centers on the adsorbent. The mechanism of metal and phosphate adsorption on alginate gel matrices is different. Carboxylate function groups of alginate are negatively charged in neutral and alkaline media and hence, have greater affinity to cations. The sorption of polyvalent metal ions onto alginate takes place via a specific ion exchange mechanism involving the replacement of other cations by metal ions [41,42]. Studies on adsorption of multivalent metals to calcium-crosslinked alginate beads have shown that the metal uptake capacity at low pH is attributed to an ionic exchange protons [43,44] and showed high affinity for polyvalent metal ions, especially within a low-concentration region [45]. Alginate acid, in combination with multivalent metals, creates structures with large specific surface area and anionic sorption capacity, which enables phosphate binding on alginate adsorbents. Metals in the form of hydrogel capsules are characterized by higher sorption capacity per unit of metal, which is a consequence of the development of the adsorbent's specific surface and large dispersion of metal cations on it [46], which can lead to the formation of more active sites on the adsorbent. Alginate is rich in carboxyl, hydroxyl and other active functional groups which can react with heavy metals through ion exchange or complex reaction [22,42]. The research carried out in this work showed that all the tested metals were adsorbed in solutions prepared in deionized water and in effluent water along with phosphate on Fe_Alg. During adsorption, partial exchange of the adsorbed metals with iron ions took place, which is indicated by a multiple increase in iron concentration in solutions containing DW_PM and SW_PM metals compared to DW_P and SW_P solutions. It suggests an ionic exchange process. This applies more to the DW_PM system, where a positive correlation was found between the amount of adsorbed metal ions and the amount of released iron ions. Studies on the relationship between the physical parameters of various metal ions, including toxic metal ions, and the binding affinity of these metal ions for alginate, have shown that Pb^{2+}, Cd^{2+}, Cu^{2+} and Zn^{2+} have higher affinity to alginates in comparison to Fe^{3+} [47,48]. In the DW_PM solution, the ionic exchange between metal ions and Fe^{3+} was four times greater than in the SW_PM solution, which is indicated by a much greater amount of released Fe^{3+} ions in the DW_PM solution. For the tested systems, the affinity of the Fe_Alg for metal ions was in the following decreasing order: $Pb^{2+} > Cu^{2+} > Cd^{2+} > Zn^{2+}$ in DW_PM, and $Cu^{2+} > Pb^{2+} > Cd^{2+} > Zn^{2+}$ in SW_PM. Metals' adsorption in the DW_PM solution is consistent with other reports, in which the adsorption of these metals on calcium-crosslinked alginate adsorbents [49,50], on alginate modified by the grafting of urea maintained [51] and on activated carbon-containing alginate adsorbent [52] was studied. An affinity of Fe_Alg for number of metal ions found in the study are consistent with the ability of sodium alginate to bind to multivalent cations following the sequence of $Pb^{2+} > Cu^{2+} > Cd^{2+} > Ba^{2+} > Sr^{2+} > Ca^{2+} > Co^{2+} > Ni^{2+} > Zn^{2+} > Mn^{2+}$ [53].

The second mechanism of metal binding in alginate is the formation of coordination complexes [42,53]. According to the metals' classification of Nieboer and Richardson [54], lead belongs to type B metals and is characterized by a high value of the covalent index, i.e., strong ability to accept electrons from the ligand donor atom, and high value of the ion index, which is a measure of the possibility of ionic bond formation. Lead ions have the highest values of these two parameters among environmentally important metals. As a result, lead ions in a neutral environment and in the presence of various ligands with donor atoms form one of the most stable complexes. This explains why lead adsorption was high and comparable in the both of the tested systems: DW_PM and SW_PM. The other examined metals: copper, cadmium and zinc, belong to the intermediate-type metals and their ions are characterized by lower value of the covalent index (copper has the largest one) and the ion index compared to lead ions. Analysis via Fourier transform infrared spectroscopy (FTIR) of the sodium alginate crosslinking with $CaCl_2$ after adsorption from tetra metallic mixture solution, Cu^{2+}, Zn^{2+}, Ni^{2+} and Cd^{2+}, showed that metal ions bind to carboxyl and hydroxyl groups. The adsorption experiments demonstrated that the beads exhibited a high removal efficiency for the selective adsorption of Cu^{2+}. It was due to better bond stability with Cu^{2+} compared to other metals [50]. Research conducted by Chen et al. [42] showed that the nature of lead uptake was the typical ion exchange between Ca^{2+} and Pb^{2+} at the carboxylate anionic site, whereas the copper uptake took place through the ion exchange between Ca^{2+} and Cu^{2+} as well as the formation of the coordination complex. These different binding mechanisms may cause a greater adsorption of copper than that of lead and a relatively low adsorption of phosphate in the SW_PM solution. The higher adsorption of metals, found in the SW_PM (pH 6.8) solution compared to DW_PM (pH 5.6), could have been influenced by the small pH difference between these solutions. Numerous experimental results indicate that pH is one of the most crucial factors influencing the efficiency of the metal ion uptake process. It is justifiable, as it is able to change both the surface properties of the bio-sorbent and the metal ion form in the bulk solution. The former factor is usually modeled by assuming the effect of competition between the metal ions and protons for the available binding sites or by applying the ion-exchange models, in which binding of the metal ion occurs in the reaction with protonated surface sites, accompanied by release of proton(s) [49]. Studies of ion exchange of various metals in alginate under different pH conditions showed that this exchange significantly decreased with decreasing pH [48]. The exception in the experiment was lead, the adsorption of which in the DW_PM and SW_PM solution was at a comparable level. Greater adsorption of the investigated metals of the intermediate type in the SW_PM solution may result from the presence of various types of compounds that, after adsorption, can form complex compounds with these metals. Intermediate-type metals, compared to B-type metals, have stronger tendency to complex with ligands other than water, and form complexes with donor atoms, such as oxygen, nitrogen and sulfur [54].

The research carried out showed that in the solution prepared based on distilled water, the addition of metals increased the adsorption of phosphate, which is indicated by the higher value of adsorption at equilibrium state in the DW_PM solution compared to DW_P, and the higher value of k_F in the Freundlich isotherm calculated for the phosphate adsorption process in these systems. The increase in phosphate adsorption in a solution containing only phosphate and metals compared to a solution containing only phosphate indicates a slight synergistic effect of metals on phosphate adsorption. This process was best described by the Freundlich model, which is characteristic of multilayer adsorption. The slower adsorption rate of phosphate in the DW_PM solution compared to the SW_PM solution could be related to the formation of subsequent layers. In the SW_PM metal-enriched effluent water, the adsorption of phosphate on Fe_Alg was lower compared to the adsorption in the SW_P system. The addition of metals reduced phosphate adsorption, which is indicated by the lower value of adsorption at equilibrium state in the SW_PM solution compared to SW_P, and the k_L parameter in the Langmuir isotherm calculated for the phosphate adsorption process in this system. This process was best described by the Langmuir model, which is characteristic of monolayer coverage and no later interaction between adsorbed molecules. Calculated reaction rate constants in the pseudo-second-order equation show that the process was much faster than for solutions containing only phosphate and metals (DW_P and DW_PM). This could be due to the formation of a single layer of adsorbed components, largely formed by metals, as indicated by their greater adsorption in the SW_PM solution compared to DW_PM. Results indicated that phosphate and multivalent metal ions in effluent water might show competitive adsorption. Less phosphate adsorption in this system could also be caused by the presence of compounds in effluent water that, in the presence of metals, compete with phosphate for active sites on the adsorbent surface. Metal ions studied at work can form complexes with donor atoms, such as oxygen, nitrogen and sulfur. These complexes could hinder the adsorption of phosphates in the SW_PM solution. As a result of the adsorption of Cu^{2+}, Zn^{2+} and Cd^{2+}, new active sites could be created, which had a greater affinity for other SW compounds of anionic nature then phosphate. Such anion behavior was observed during the study of the effect of the coexisting ions on the adsorption of fluorides on the alginate gelled with zirconium in wastewater. The presence of HCO_3^-, SO_4^{2-} and PO_4^{3-} had a large negative impact on fluoride removal. The decreased defluorination was attributed to the lower affinity of zirconium alginate for fluoride and a competition between the fluoride ions and the other anions [55].

The conducted research indicates that the possibility of using phosphate ion adsorption on alginate matrices to recover phosphorus from liquid waste in sewage treatment plants is limited. To limit the competitive effect of metals on the phosphate adsorption process, the method should be used in municipal wastewater treatment plants in small, non-industrial agglomerations. Treated effluent water must not contain toxic metals, such as lead and cadmium, which do not perform any biological functions. In the case of other metals, tests should be carried out to determine their limit concentrations, which do not pose a threat to the environment.

4. Materials and Methods

4.1. Materials

Alginate matrices, cross-linked with Fe^{3+} ions (Fe_Alg) with bead size 2.5–3.0 mm, were used as adsorbents. The hydrogel beads were produced with the injection method. Nine grams of sodium alginate powder (Keltone HV, ISP-Germany, Marl, Germany) was dispersed in 600 mL of deionized water to give a 1.5% *w*/*v* alginate solution. This solution was mixed with a mechanic stirrer until a transparent, viscous solution was obtained. In order to receive a solution containing fully hydrated polymer chains, the mixing process was carried out with use of a magnetic agitator for 24 h. Hydrogel capsules of alginate/Fe(III) were received by dosing the sodium alginate in a quantity of 5 mL each time with the classic 10 mL syringe (Polfa S.A., Lublin, Poland), ended up with a needle diameter 0.8 × 40 mm (TERUMO, Belgium, Levuen, Belgium), to 100 mL $FeCl_3$ 0.155 M solution, at 2–3-min intervals. Time of the reaction after injection of the polysaccharide solution was 70 min. During forming the beads, the $FeCl_3$ solution was stirred continuously at 600 rpm. The beads were rinsed with distilled water, the liquid was drained, and the adsorbent was sealed in a deionized water (DW) container.

The following salts were used to prepare the adsorbed ion solutions:

$CuSO_4$, $(CH_3COOH)_2Cd$, $Pb(NO_3)_2$, $ZnSO_4$ and KH_2PO_4 (EUROCHEM BGD, Tarnów, Poland). The solutions were prepared in DW and by enriching water effluent from a sludge thickener press (SW) with them, some selected qualitative indicators of SW are shown in Table 4. The SW solution comes from a mechanical-biological sewage treatment plant (Nowogard in Poland) with a capacity of 3400 m^3/d.

Table 4. General characteristics of the tested effluent from a sludge.

Water Quality Indicators	Value	Water Quality Indicators	Value
pH	7	N_Ammonia, mg/L	12.19
RedOx Potential Eh, mV	−36.6	N_Nitrite, mg/L	0.03
Color, mgPt/L	19 s	N_Nitrate, mg/L	7.43
turbidity, (Nephelometric Turbidity Units) NTU	65.8	Cu^{2+}, mg/L	0.0065
Hardness, $CaCO_3$ mg/L	220.4	Pb^{2+}, mg/L	0.0341
Alkalinity, mval/L	4.1	Zn^{2+}, mg/L	0.0592
Phosphate, mg/L	8.06	Cd^{2+}, mg/L	0.0025

4.2. Analytical Methods

Determinations of tested water quality indicators were carried out using standard methods [56]. Contents of total metals (Cu, Cd, Pb, Zn and Fe) in the solutions were determined by means of Atomic Absorption Spectroscopy using a spectrometer ThermoElemental, Solaar S, Walthman, USA with atomization occurring in acetylene/air flame. Phosphate concentration was measured by the molybdenum blue colorimetric method [57] using a two-beam spectrophotometer Techcomp UV/VIS 8500 at 890 nm wavelength.

4.3. Adsorption Kinetic Measurements

The phosphate adsorption kinetics on Fe_Alg beads were studied in 10 mg PO_4 L^{-1} solutions prepared on the basis of DW and SW without addition of DW_P and SW_P, as well as with the addition of metal ions: Cu^{2+}, Cd^{2+}, Pb^{2+} and Zn^{2+} (DW_PM and SW_PM) at the concentration of each metal ion of 10 mg/dm^3 and a constant ionic strength of 30 mmol against KCl, as proposed by Naira et al. [58]. A volume of 50 mL of prepared solutions and 0.01 g dry matter Fe_Alg matrices was added into the conical flasks with a capacity of 100 mL. Nine measuring series were prepared for each type of solution in duplicate. Changes in phosphate and metal contents (including iron) in the aqueous phase were tested in successive series after 0.25, 0.5, 1, 12, 4, 8, 34, 56 and 105 h. In parallel, the concentration of phosphate and metals in control samples without adsorbent was tested.

4.4. Phosphate Adsorption Experiments

Phosphate adsorption characteristics were studied in a static batch system proposed by Naira et al. [58]. The tests were carried out at 20 °C in DW and SW, which were enriched with phosphate(V) ions at a concentration of 10, 20, 30, 40, 60 and 80 mg L^{-1} (DW_P and SW_P), and metal ions: Cu^{2+}, Cd^{2+}, Pb^{2+} and Zn^{2+} at a concentration of 5 mg L^{-1} each (DW_PM and SW_PM), with a constant ionic strength of 30 mmol against KCl. A volume of 50 mL of prepared solutions and 0.01 g DS Fe_Alg matrices was added into the conical flasks with a capacity of 100 cm^3. Changes in the content of phosphate ions and metals in the solution were tested at equilibrium after 56 h. The mixtures were shaken on a laboratory shaker for two hours in the beginning and the end of the adsorption process. The experiment was carried out in triplicate.

4.5. Data Analysis

The linearized form of equations was used to determine the parameters of mathematical models describing the kinetics of the adsorption process studied. Kinetic data of phosphate sorption on Fe_Alg was described using the pseudo-second-order rate equation developed by Ho and McKay [59,60] (Equation (1)):

$$\frac{da_t}{dt} = k_2 \cdot \left(a_{eq} - a_t\right)^2 \tag{1}$$

The solution to this equation under boundary conditions $t = 0$ to $t = t$ and from $a_t = 0$ to $a_t = a_{eq}$ is Equation (2):

$$\frac{1}{a_t} = \frac{1}{k_2 a_{eq}^2} + \frac{t}{a_{eq}} \tag{2}$$

where k_2 is the second-order rate constant of adsorption, and a_{eq} is the amount of phosphate adsorbed at equilibrium [61].

The equilibrium data for the removal of phosphate in the present investigation were analyzed using the two-parameter model, Langmuir (Equation (3)) and Freundlich (Equation (4)) isotherms:

$$q_e = \frac{q_m k_L C_e}{1 + k_L C_e} \tag{3}$$

$$q_e = k_F C_e^{1/n_F} \tag{4}$$

where C_e is the equilibrium concentration of phosphate in the solution, q_e is the correspondent uptake capacity of the adsorbent, q_m is the maximum adsorption capacity, k_L is the affinity constant (Langmuir constant), k_F and n_F are Freundlich constants and $1/n_F$ is the heterogeneity factor.

The linear equations of Langmuir and Freundlich are represented as follows (Equations (5) and (6), respectively).

$$\frac{C_e}{q_e} = \frac{1}{k_L(q_m)} + \frac{C_e}{q_m} \tag{5}$$

$$log q_e = \log k_F + \frac{(log C_e)}{n_F} \tag{6}$$

Linear regression analysis has been used to determine the best-fit isotherm and the method of least squares has been used for finding parameters of the equations. For the fitting-degree of the isotherms and kinetic equation with the experimental data, the adjusted coefficient of determination, ${R_{adj}}^2$, was used. The ${R_{adj}}^2$ is a good tool in selecting models, which takes into account the experimental degrees of freedom ($n1$), where n is the number of data points [35]. All statistical analyses were performed with the software STATISTICA version 13.3.

Author Contributions: H.S. conceived and directed the study and partly wrote the paper; K.P. performed the experiments and partly wrote the paper. All authors have read and agreed to the published version of the manuscript.

Funding: This research received no external funding.

Acknowledgments: The authors would like to acknowledge the management of a sewage treatment plant in Nowogard, Poland for for their cooperation.

Conflicts of Interest: The authors declare no conflict of interest.

References

1. Cooper, J.; Lombardi, R.; Boardman, D.; Carliell-Marquet, C. The future distribution and production of global phosphate rock reserves. *Resour. Conserv. Recycl.* **2011**, *57*, 78–86. [CrossRef]
2. Cordell, D.; Drangert, J.-O.; White, S. The story of phosphorus: Global food security and food for thought. *Glob. Environ. Chang.* **2009**, *19*, 292–305. [CrossRef]
3. Cornel, P.; Schaum, C. Phosphorus recovery from wastewater: Needs, technologies and costs. *Water Sci. Technol.* **2009**, *59*, 1069–1076. [CrossRef] [PubMed]
4. Parsons, S.; Smith, J.A. Phosphorus Removal and Recovery from Municipal Wastewaters. *Elements* **2008**, *4*, 109–112. [CrossRef]
5. Egle, L.; Rechberger, H.; Krampe, J.; Zessner, M. Phosphorus recovery from municipal wastewater: An integrated comparative technological, environmental and economic assessment of P recovery technologies. *Sci. Total Environ.* **2016**, *571*, 522–542. [CrossRef]
6. Wzorek, Z.; Lenik, E.; Gorazda, K.; Wilkosz, A. Popioły ze spalania dpadów z przemysłu mięsnego i osadów ściekowych jako źródło fosforu. *Arch. Gospod. Odpadami Ochr. Środowiska* **2006**, *3*, 83–90.
7. Kasprzyk, M.; Gajewska, M.; Molendowska, S. Możliwość odzysku fosforu z odcieków, osadów ściekowych i popiołów po termicznym przekształcaniu osadów ściekowych. Possibilities of phosphorus recovery from effluents, sewage sludge and ashes from sewage sludge thermal processing. *Ecol. Eng.* **2017**, *18*, 65–78. (In Polish) [CrossRef]
8. Krüger, O.; Adam, C. Phosphorus in recycling fertilizers—Analytical challenges. *Environ. Res.* **2017**, *155*, 353–358. [CrossRef]
9. Gibbs, M.M.; Hickey, C.W.; Ozkundakci, D. Ocena trwałości i porównanie skuteczności czterech czynników inaktywujących P w zarządzaniu wewnętrznymi ładunkami fosforu w jeziorach: Inkubacje osadów. *Hydrobiologia* **2011**, *658*, 253–275. [CrossRef]
10. Yue, Q.; Zhao, Y.; Li, Q.; Li, W.; Gao, B.; Han, S.; Qi, Y.; Yu, H. Research on the characteristics of red mud granular adsorbents (RMGA) for phosphate removal. *J. Hazard. Mater.* **2010**, *176*, 741–748. [CrossRef]
11. Bus, A.; Baryła, A.; Karczmarczyk, A. Wybór materiału reaktywnego do usuwania fosforu z wód i ścieków na przykładzie kruszywa popiołoporytowego Pollytag. *Inżynieria Ekologiczna* **2014**, *39*, 33–41. [CrossRef]
12. Chitrakar, R.; Tezuka, S.; Sonoda, A.; Sakane, K.; Ooi, K.; Hirotsu, T. Phosphate adsorption on synthetic goethite and akaganeite. *J. Colloid Interface Sci.* **2006**, *298*, 602–608. [CrossRef] [PubMed]
13. Zhang, G.; Liu, H.; Liu, R.; Qu, J. Removal of phosphate from water by a Fe–Mn binary oxide adsorbent. *J. Colloid Interface Sci.* **2009**, *335*, 168–174. [CrossRef] [PubMed]
14. Tanada, S.; Kabayama, M.; Kawasaki, N.; Sakiyama, T.; Nakamura, T.; Araki, M.; Tamura, T. Removal of phosphate by aluminum oxide hydroxide. *J. Colloid Interface Sci.* **2003**, *257*, 135–140. [CrossRef]
15. Zeng, L.; Li, X.; Liu, J. Adsorptive removal of phosphate from aqueous solutions using iron oxide tailings. *Water Res.* **2004**, *38*, 1318–1326. [CrossRef]
16. Haghseresht, F.; Wang, S.; Do, D. A novel lanthanum-modified bentonite, Phoslock, for phosphate removal from wastewaters. *Appl. Clay Sci.* **2009**, *46*, 369–375. [CrossRef]
17. Eberhardt, T.L.; Min, S.-H.; Han, J.S. Phosphate removal by refined aspen wood fiber treated with carboxymethyl cellulose and ferrous chloride. *Bioresour. Technol.* **2006**, *97*, 2371–2376. [CrossRef] [PubMed]
18. Eberhardt, L.T. Biosorbents prepared from wood particles treated with anionic polymerand iron salt: Effect of particle size on phosphate adsorption. *Bioresour. Technol.* **2008**, *99*, 626–630. [CrossRef]
19. Krishnan, K.A.; Haridas, A. Removal of phosphate from aqueous solutions and sewage using natural and surface modified coir pith. *J. Hazard. Mater.* **2008**, *152*, 527–535. [CrossRef]
20. Mezenner, N.Y.; Bensmaili, A. Kinetics and thermodynamic study of phosphate adsorption on iron hydroxide-eggshell waste. *Chem. Eng. J.* **2009**, *147*, 87–96. [CrossRef]
21. Aryal, M.; Liakopoulou-Kyriakides, M. Equilibrium, kinetics and thermodynamic studies on phosphate biosorption from aqueous solutions by Fe(III)-treated Staphylococusxylosus biomass: Common ion effect. *Colloids Surf. A Physicochem. Eng. Asp.* **2011**, *387*, 43–49. [CrossRef]
22. Wang, B.; Wan, Y.; Zheng, Y.; Lee, X.; Liu, T.; Yu, Z.; Huang, J.; Ok, Y.S.; Chen, J.; Gao, B. Alginate-based composites for environmental applications: A critical review. *Crit. Rev. Environ. Sci. Technol.* **2018**, *49*, 318–356. [CrossRef]

23. Bartkowiak, A.; Hunkeler, D. Alginate–Oligochitosan Microcapsules: A Mechanistic Study Relating Membrane and Capsule Properties to Reaction Conditions. *Chem. Mater.* **1999**, *11*, 2486–2492. [CrossRef]
24. Hill, C.B.; Khan, E. A Comparative Study of Immobilized Nitrifying and Co-Immobilized Nitrifying and Denitrifying Bacteria for Ammonia Removal from Sludge Digester Supernatant. *Water Air Soil Pollut.* **2008**, *195*, 23–33. [CrossRef]
25. Min, J.H.; Hering, J.G. Arsenate sorption by Fe(III)-doped alginate gels. *Water Res.* **1998**, *32*, 1544–1552. [CrossRef]
26. Hering, J.G.; Min, H.J. Removal of selenite and chromate using Fe(III)doped alginate gels. *Water Environ. Res.* **1999**, *71*, 169–175.
27. Siwek, H.; Bartkowiak, A.; Włodarczyk, M.; Sobecka, K. Removal of Phosphate from Aqueous Solution Using Alginate/Iron (III) Chloride Capsules: A Laboratory Study. *Water Air Soil Pollut.* **2016**, *227*, 427. [CrossRef]
28. Yeon, K.H.; Park, H.; Lee, S.H.; Park, Y.M.; Lee, S.H.; Iwamoto, M. Zirconium mesostructure immobilized in calcium alginate for phosphate removel. *Korean J. Chem. Eng.* **2008**, *25*, 1040–1046. [CrossRef]
29. Chiban, M.; Soudani, A.; Sinan, F.; Persin, M. Wastewater treatment by batch adsorption method onto micro-particles of dried Withania frutescens plant as a new adsorbent. *J. Environ. Manag.* **2012**, *95*, S61–S65. [CrossRef]
30. Wang, Y.; Feng, Y.; Zhang, X.-F.; Zhang, X.; Jiang, J.; Yao, J. Alginate-based attapulgite foams as efficient and recyclable adsorbents for the removal of heavy metals. *J. Colloid Interface Sci.* **2018**, *514*, 190–198. [CrossRef]
31. Chen, J.H.; Liu, Q.L.; Hu, S.R.; Ni, J.C.; He, Y.S. Adsorption mechanism of Cu(II) ions from aqueous solution by glutaraldehyde crosslinked humic acid-immobilized sodium alginate porous membrane adsorbent. *Chem. Eng. J.* **2011**, *173*, 511–519. [CrossRef]
32. Benettayeb, A.; Guibal, E.; Morsli, A.; Kessas, R. Chemical modification of alginate for enhanced sorption of Cd(II), Cu(II) and Pb(II). *Chem. Eng. J.* **2017**, *316*, 704–714. [CrossRef]
33. Do, X.-H.; Lee, B.-K. Removal of Pb^{2+} using a biochar–alginate capsule in aqueous solution and capsule regeneration. *J. Environ. Manag.* **2013**, *131*, 375–382. [CrossRef]
34. Giles, C.H.; Smith, D.; Huitson, A. A general treatment and classification of the solute adsorption isotherm. I. Theoretical. *J. Colloid Interface Sci.* **1974**, *47*, 755–765. [CrossRef]
35. Piccin, J.S.; Cadaval, T.R.S.; de Pinto, L.A.A.; Dotto, G.L. Adsorption Isotherms in Liquid Phase: Experimental, Modeling, and Interpretations. In *Adsorption Processes for Water Treatment and Purification*; Bonilla-Petriciolet, A., Mendoza-Castillo, D., Reynel-Ávila, H., Eds.; Springer: Cham, Switzerland, 2017; pp. 19–51. [CrossRef]
36. Dionisiou, N.S.; Matsi, T.; Misopolinos, N.D. Phosphorus Adsorption–Desorption on a Surfactant-Modified Natural Zeolite: A Laboratory Study. *Water Air Soil Pollut.* **2013**, *224*, 1362–1363. [CrossRef]
37. Siwek, H.; Bartkowiak, A.; Włodarczyk, M. Adsorption of Phosphates from Aqueous Solutions on Alginate/Goethite Hydrogel Composite. *Water* **2019**, *11*, 633. [CrossRef]
38. Tovar-Gómez, R.; Moreno-Virgen, M.D.R.; Moreno-Pérez, J.; Bonilla-Petriciolet, A.; Hernández-Montoya, V.; Duran-Valle, C. Analysis of synergistic and antagonistic adsorption of heavy metals and acid blue 25 on activated carbon from ternary systems. *Chem. Eng. Res. Des.* **2015**, *93*, 755–772. [CrossRef]
39. Rojas-Mayorga, C.; Mendoza-Castillo, D.; Silvestre-Albero, J.; Bonilla-Petriciolet, A. Tailoring the adsorption behavior of bone char for heavy metal removal from aqueous solution. *Adsorpt. Sci. Technol.* **2016**, *34*, 368–387. [CrossRef]
40. Sousa, F.W.; Oliveira, A.G.; Ribeiro, J.P.; Rosa, M.D.F.; Keukeleire, D.; Nascimento, R.F.D. Green coconut shells applied as adsorbent for removal of toxic metal ions using fixed-bed column technology. *J. Environ. Manag.* **2010**, *91*, 1634–1640. [CrossRef]
41. Papageorgiou, S.K.; Katsaros, F.; Kouvelos, E.P.; Nolan, J.W.; Le Deit, H.; Kanellopoulos, N.K. Heavy metal sorption by calcium alginate beads from *Laminaria digitata*. *J. Hazard. Mater.* **2006**, *137*, 1765–1772. [CrossRef]
42. Chen, J.P.; Hong, L.; Wu, A.S.; Wang, L. Elucidation of Interactions between Metal Ions and Ca Alginate-Based Ion-Exchange Resin by Spectroscopic Analysis and Modeling Simulation. *Langmuir* **2002**, *18*, 9413–9421. [CrossRef]
43. Ngomsik, A.-F.; Bée, A.; Siaugue, J.-M.; Cabuil, V.; Cote, G. Nickel adsorption by magnetic alginate microcapsules containing an extractant. *Water Res.* **2006**, *40*, 1848–1856. [CrossRef] [PubMed]
44. Jung, W.; Jeon, B.-H.; Cho, D.-W.; Roh, H.-S.; Cho, Y.; Kim, S.-J.; Lee, D.S. Sorptive removal of heavy metals with nano-sized carbon immobilized alginate beads. *J. Ind. Eng. Chem.* **2015**, *26*, 364–369. [CrossRef]

45. Gotoh, T.; Matsushima, K.; Kikuchi, K.-I. Adsorption of Cu and Mn on covalently cross-linked alginate gel beads. *Chemosphere* **2004**, *55*, 57–64. [CrossRef] [PubMed]
46. Shin, E.W.; Karthikeyan, K.G.; Tshabalala, M.A. Orthophosphate Sorption onto Lanthanum-Treated Lignocellulosic Sorbents. *Environ. Sci. Technol.* **2005**, *39*, 6273–6279. [CrossRef]
47. Idota, Y.; Kogure, Y.; Kato, T.; Yano, K.; Arakawa, H.; Miyajima, C.; Kasahara, F.; Ogihara, T. Relationship between Physical Parameters of Various Metal Ions and Binding Affinity for Alginate. *Boil. Pharm. Bull.* **2016**, *39*, 1893–1896. [CrossRef]
48. Karagündüz, A.; Unal, D. New method for evaluation of heavy metal binding to alginate beads using pH and conductivity data. *Adsorption* **2006**, *12*, 175–184. [CrossRef]
49. Plazinski, W. Sorption of lead, copper, and cadmium by calcium alginate. Metal binding stoichiometry and the pH effect. *Environ. Sci. Pollut. Res.* **2012**, *19*, 3516–3524. [CrossRef]
50. Yang, N.; Wang, R.; Rao, P.; Yan, L.; Zhang, W.; Wang, J.; Chai, F. The Fabrication of Calcium Alginate Beads as a Green Sorbent for Selective Recovery of Cu(II) from Metal Mixtures. *Crystals* **2019**, *9*, 255. [CrossRef]
51. Park, H.G.; Kim, T.W.; Yoo, I.K. Activated carbon-containing alginate adsorbent for the simultaneous removal of heavy metals and toxicorganics. *Process Biochem.* **2007**, *42*, 1371–1377. [CrossRef]
52. Russo, R.; Malinconico, M.; Santagata, G. Effect of Cross-Linking with Calcium Ions on the Physical Properties of Alginate Films. *Biomacromolecules* **2007**, *8*, 3193–3197. [CrossRef] [PubMed]
53. Jiang, X.; An, Q.; Xiao, Z.-Y.; Zhai, S.-R.; Shi, Z. Versatile core/shell-like alginate@polyethylenimine composites for efficient removal of multiple heavy metal ions (Pb^{2+}, Cu^{2+}, $CrO4^{2-}$): Batch and fixed-bed studies. *Mater. Res. Bull.* **2019**, *118*, 110526. [CrossRef]
54. Nieboer, E.; Richardson, D.H. The replacement of the nondescript term 'heavy metals' by a biologically and chemically significant classification of metal ions. *Environ. Pollut. Ser. B Chem. Phys.* **1980**, *1*, 3–26. [CrossRef]
55. Zhou, Q.; Lin, X.; Qian, J.; Wang, J.; Luo, X. Porous zirconium alginate beads adsorbent for fluoride adsorption from aqueous solutions. *RSC Adv.* **2015**, *5*, 2100–2112. [CrossRef]
56. Helrich, K. Official methods of analysis. *Assoc. Off. Anal. Chem.* **1990**, *1*, 1–1230.
57. European Standard EN1189. *Water Quality-Determination of Phosphorus-Ammonium Molybdate Specrtometric Method*; European Committee for Standardization: Brussels, Belgium, 1996.
58. Nair, P.S.; Logan, T.J.; Sharpley, A.N.; Sommers, L.E.; Tabatabai, M.A.; Yuan, T.L. Interlaboratory Comparison of a Standardized Phosphorus Adsorption Procedure. *J. Environ. Qual.* **1984**, *13*, 591–595. [CrossRef]
59. Ho, Y.; McKay, G. Pseudo-second order model for sorption processes. *Process Biochem.* **1999**, *34*, 451–465. [CrossRef]
60. Ho, Y. Review of second-order models for adsorption systems. *J. Hazard. Mater.* **2006**, *136*, 681–689. [CrossRef]
61. Kumar, M.M.; Ramalingam, S.; Senthamarai, C.; Niranjanaa, M.; Vijayalakshmi, P.; Sivanesan, S. Adsorption of dye from aqueous solution by cashew nut shell: Studies on equilibrium isotherm, kinetics and thermodynamics of interactions. *Desalination* **2010**, *261*, 52–60. [CrossRef]

Sample Availability: Samples of the compounds are not available from the authors.

molecules

MDPI

Article

Sustainable Development of Enhanced Luminescence Polymer-Carbon Dots Composite Film for Rapid Cd^{2+} Removal from Wastewater

Mohammed Abdullah Issa * and Zurina Z. Abidin *

Department of Chemical and Environmental Engineering, Faculty of Engineering, University Putra Malaysia, Serdang 43400 UPM, Selangor, Malaysia

* Correspondence: mohbaghdadi1@yahoo.com (M.A.I.); zurina@upm.edu.my (Z.Z.A.)

Academic Editors: Chiara Bisio and Monica Pica

Received: 14 June 2020; Accepted: 10 July 2020; Published: 3 August 2020

Abstract: As a remedy for environmental pollution, a versatile synthetic approach has been developed to prepare polyvinyl alcohol (PVA)/nitrogen-doped carbon dots (CDs) composite film (PVA-CDs) for removal of toxic cadmium ions. The CDs were first synthesized using carboxymethylcellulose (CMC) of oil palms empty fruit bunch wastes with the addition of polyethyleneimine (PEI) and then the CDs were embedded with PVA. The PVA-CDs film possess synergistic functionalities through increasing the content of hydrogen bonds for chemisorption compared to the pure CDs. Optical analysis of PVA-CDs film was performed by ultraviolet-visible and fluorescence spectroscopy. Compared to the pure CDs, the solid-state PVA-CDs displayed a bright blue color with a quantum yield (QY) of 47%; they possess excitation-independent emission and a higher Cd^{2+} removal efficiency of 91.1%. The equilibrium state was achieved within 10 min. It was found that adsorption data fit well with the pseudo-second-order kinetic and Langmuir isotherm models. The maximum adsorption uptake was 113.6 mg g^{-1} at an optimal pH of 7. Desorption experiments showhe that adsorbent can be reused fruitfully for five adsorption-desorption cycles using 0.1 HCl elution. The film was successfully applied to real water samples with a removal efficiency of 95.34% and 90.9% for tap and drinking water, respectively. The fabricated membrane is biodegradable and its preparation follows an ecofriendly green route.

Keywords: biomass waste; *N* doped carbon dots; adsorption; Cd (II); mechanism; water remedy

1. Introduction

Water is the prime component of human survival. Nevertheless, environmental pollution caused by heavy metals owing to the inappropriate disposal of wastewater is becoming a worldwide concern [1,2]. Cd(II) metal is one of the most toxic substances that tend to accumulate in living organisms. According to the Disease Registry and Environmental Protection Agency (EPA), Cd(II) ranks seventh on the top 20 toxic chemicals priority list [3]. Long-term exposure to Cd(II) results in a series of harmful diseases in humans ecosystem damage [4]. Due to its toxic effects on human health, the separation and elimination of Cd^{2+} from aquatic media is still a hot research topic. Among the solutions, zeolites [5], polymeric compounds [6] and carbon-based materials [7,8] are the most widely used sorbents for the removal of chemical pollutants, due to their large surface area and tunable pore size.

Nowadays, carbon dots (CDs) have gained considerable attention for heavy metal removal from aqueous systems [9,10]. These CDs, with sizes less than 10 nm, exhibit outstanding properties like ease of preparation, non-toxic nature and excellent luminescence behavior. These features make CDs ideal to be applied for wastewater remediation, photocatalysis, cell imaging, cancer therapy and metal sensing [11–14]. Many protocols including an arc-discharge approach [15], electrochemical oxidation [16], microwave treatment [17] and a hydrothermal route [18] have been demonstrated for the production of CDs. Among all the reported protocols, the hydrothermal route is the most preferred technique for its simplicity, mild conditions and high fluorescence quantum yield (QY).

More recently, hydrothermal approaches using a variety of natural carbonaceous precursors have been successfully employed to obtain CDs [19–24]. In particular, the usage of lignocellulosic wastes was fruitfully utilized for their cost-effective and eco-friendly resources [25–31]. However, existing works have indicated several drawbacks like weak QY and harsh synthesis conditions [27,28]. One should note that the replacement and clustering degree of cellulose rings and the chain length, as well as the choice of additives, are the key factors influencing the final QY of CDs. Carboxymethylcellulose (CMC) is the main modification of cellulose in which carboxymethyl species replace several hydrogen atoms of the hydroxyl moieties on the cellulose framework [10,32]. These biocompatible resources enriched with ether and hydroxyl groups can accelerate the synthesis process through successive substitution reactions and enhance the removal uptake by providing adsorption sites [2,29]. Thus, full exploitation of these low-value wastes to recycle them and develop new value-added carbonaceous materials is of great interest. The encapsulation of CDs into solid-state film is still challenging as they show photo-quenching properties after the immobilization processes, make them unsuitable for application in a variety of fields [4].

Previous works have reported the workability of incorporating CDs with metallic compounds or polymers without affecting the optical performance [33,34]. There have been a few works explaining the integration of CDs into hydrogels for several functions. Among them, polyvinyl alcohol-based films were frequently used to prepare freestanding luminescence films for green CD encapsulation [35–38], due to their biocompatibility, non-toxicity and hydrophilicity [39–43]. Despite the strong optical efficiency that was achieved using CDs-based hybrid films, several existing drawbacks, including long equilibration times of more than 1 h [7,44,45] and low removal uptake [45–47], have restricted their Cd^{2+} removal applications. Therefore, it would be sensible to develop highly photoluminescence CDs to enhance their performance towards heavy metal removal from contaminated water. It is well known that surface doping of CDs with N can effectively tune its electronic structure, provides available sites and enhanced PL, which are great significance for efficient and selective recognition and capture of metal ions [10,48].

In this regard, herein, we report a versatile one-pot route for production CDs from CMC with the assistance of PEI dopant. The as-obtained CDs were then integrated with a PVA matrix to develop a (PVA-CDs) free-standing membrane, which was further applied for Cd^{2+} removal from wastewater. The physicochemical properties of bare CDs and PVA-CDs film were thoroughly studied using different analytical techniques including low and high-resolution transmission electron microscopy (TEM, HRTEM), X-ray diffraction (XRD), zeta potential (η), X-ray photoelectron spectra (XPS), Fourier transform infrared (FTIR) and UV/VIS spectroscopy (UV–Vis) and fluorescence spectra (PL). Impressively, the as-developed PVA-CDs film exhibited blue emission with enhanced QY up to 47% in comparison to the isolated CDs (greenish, 44%), which could be ascribed to a H bonding effect. The photo- and thermostability of the film resulted in no apparent reduction in the PL emission. The developed film was utilized for Cd^{2+} removal from water and emphasis was placed on the adsorption mechanism by studying the sorption kinetics and isotherms. Furthermore, desorption and regeneration studies were also conducted to evaluate the reusability of the film.

2. Results and Discussion

2.1. Synthesis of CDs and PVA-CDs

Figure 1 shows a schematic representation of CDs and PVA-CDs composite development. CMC, as a green and economic resource, is used as a carbon precursor for the production of fluorescent CDs. In contrast to other lignocellulosic materials reported in the literature [27,29], CMC of empty fruit bunches is enriched with ether and hydroxyl moieties which could play a major role in the enhancement of substitution degree and carbonization acceleration. Branched PEI was selected as a dopant for its feature to inject electron-rich N into the final framework of CDs and increasing QY with the ability to capture metal ions [9]. The as-formed CDs exhibit a yellow color, showing greenish emission under UV-lamp (365 nm), good water solubility and long shelf-life (up to 1 year) with no apparent agglomeration. These CDs were encapsulated with PVA polymer host to develop a solid-state hybrid film for heavy metal complexation in an attempt to overcome the separation issues of the liquid phase [4]. In comparison to the liquid CDs, the as-obtained film is transparent under daylight while they exhibited a bright blue emission under UV exposure. This transformation from greenish to blue could be due to the formation of hydrogen bonds from the numerous interactions of CD surface sites and the PVA substrate [41,49].

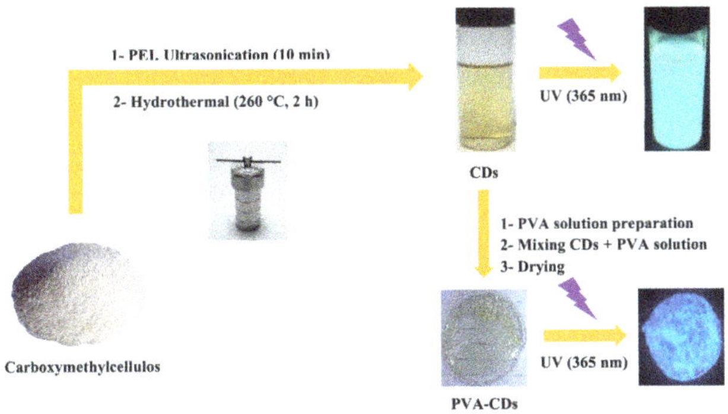

Figure 1. Schematic representation of the preparation of CDs and PVA-CDs film from CMC waste.

2.2. *Physical Structure*

Figure 2 show the TEM, HRTEM and XRD results of the CDs. TEM images (Figure 2a) reveal that the obtained CDs have a monodispersed spherical structure, with an average size of 4.1 nm. Based on the HRTEM images (inset of Figure 2a), these nanodots have high crystallinity domains, demonstrating the presence of a graphitic inner core. The well-resolved lattice fringes show an interplanar distance of 0.24 nm, which is close to the (100) facet of graphite, similar to previous reported studies [50–52]. Based on the particle size distribution histogram, which was calculated from one hundred nanoparticles (Figure 2b), the CDs have a size distribution ranging from 3 to 9 nm with an average diameter of about 4.2 ± 0.5 nm. The XRD data displayed a diffraction peak at a 2θ value of 23.8° attributed to the (002) lattice spacing of graphitic-like carbon, in line with other works [50–53]. The measured zeta potential of the CDs (Figure S1) found to be −8.71 mV at pH 7, implying a negatively-charged CDs edges.

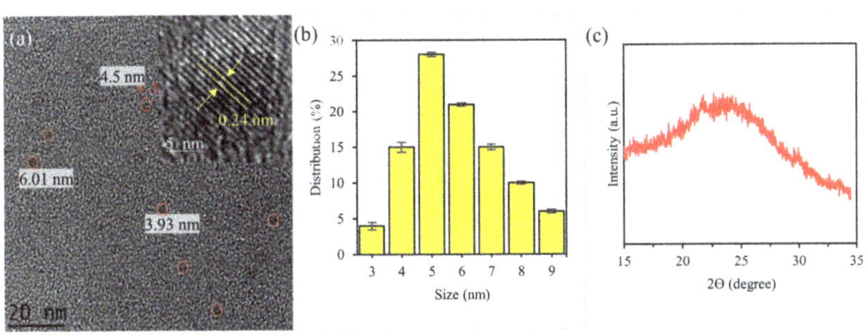

Figure 2. (**a**) TEM of CDs, (**b**) particle size distribution, (**c**) XRD patterns of CDs. Inset: HRTEM of CDs.

2.3. Chemical Structure

XPS was used to gain insight into the chemical composition of the used and formed materials. As presented in Figure S2, the XPS spectrum of CMC is mainly composed of C1s and O1s. Meanwhile, CDs indicated the presence of C1s, O1s and N1s, suggesting the formation of graphene CDs with the abundance of N. It can also be seen that the peaks of both C1s and O1s were remarkably higher than that of CMC, implying the occurrence of aromatization reaction with a high degree of surface oxidation. The high-resolution XPS survey of C1s (Figure 3a) displayed C-C, C-N, C-O and C=O at 284.6, 285.5, 287.4, and 287.9 eV, respectively. The dominant peak of sp^2 hybridized C-C (52.4%) suggests a high degree of graphitization. The N_1s band (Figure 3b) showed three peaks at 398.8, 399.7 and 400.8 eV associated with graphitic N, pyridinic N and N–H, respectively. The formation of graphitic and pyridinic N indicate the effective N-doping of CD. Meanwhile, non-graphitic amines like N-H distributed around CDs through Maillard chemistry reaction [54–60]. The O_1s spectrum (Figure 3c) showed two peaks at 530.7 and 531.9 eV corresponding to O-H and C=O, respectively. It is worth noting that the incorporation of active moietises including N-H and C=O can effectively enhance the QY of CDs, which provide more active sites for capturing a variety of analytes [4,13,48].

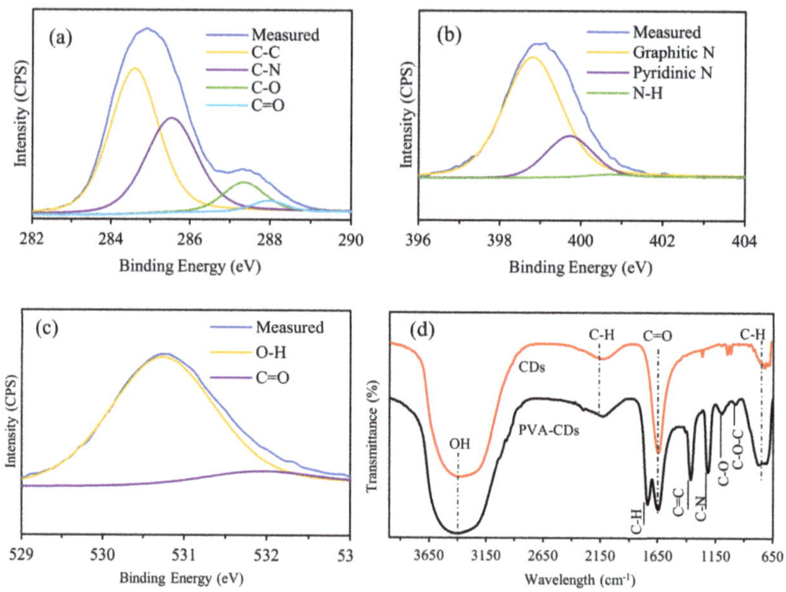

Figure 3. The high-resolution XPS survey of (**a**) C1s, (**b**) N1s and (**c**) O1s, (**d**) FTIR spectra of CDs and PVA-CDs.

The corresponding elemental compositions (in atomic ratios) are summarized Table S1. The atomic ratio C/N/O is found to be 67/12/21, demonstrating the successful doping of N-atom into the conjugated carbon skeleton domains of the CDs [61]. The high content of carbon compared to oxygen emphasizes a fruitful aromatization reaction with the elimination of O moieties. These active sites can tune the electronic structure, surface chemical characteristics and PL emission, which are vitally necessary for the selective sensing and removal of different metal ions [10,25].

Further confirmation about the chemical binding and surface functionalities was obtained using FTIR as presented in Figure 3d. In comparison to the spectra of CDs, new peaks at 1438 and 1738 cm^{-1} were formed, which are ascribed to hybridized C=C and C-H moieties, respectively [40]. Besides, there is a shift from 1266 to 1368 cm^{-1} and an increase in the intensity of the C-N band, implying a plausible involvement of graphitic N in the encapsulation process. Furthermore, the change in position and shape of the C–H bending from 725 to 772 cm^{-1} indicates the formation of hydrogen bonds between CDs and PVA polymer [40]. Moreover, an obvious increment and shift of the C-O, C-O-C and C=O bands from 1046 to 1096 cm^{-1}, 1017 to 979 cm^{-1} and 1641 cm^{-1} to 1647 cm^{-1}, respectively, were also observed. This provides evidence that both O species and topological traps were altered [49,62]. The above results are consistent with the XPS data, confirming the successful interaction between PVA host and CDs.

According to the XPS and FTIR data, it can be said that the abundance of H-containing functional groups on the surface of CDs plays a significant role to interact covalently with the PVA matrix [63]. In details, the presence of –OH, –COOH, and –NH implies that the H-containing groups are partly tethered on the edge of the CDs, rather than being completely encapsulated in the CDs [63]. From the FTIR spectra in Figure 3d, the appearance and enhancement of the C=C, C-N and C-H bands demonstrate the existence of chemical linkages between the CDs with the PVA host, forming a cross-linking network domain.

2.4. Optical Properties

UV-vis absorption and PL spectra were recorded to study the optical behavior of the liquid and solid-state CDs. The UV-Vis absorption spectra (Figure 3a) of both CDs and PVA-CDs film exhibit two absorption peaks centered at 294 and 340 nm, corresponding to the π-π^* transition and n-π^* transition of C=C and C=O, in line with previously reported data for carbon dot-based materials [10,40].

The PL spectra (Figure 3b) reveal that the excitation/emission maxima of PVA-CDs film were red-shifted by about 10 and 3 nm, respectively, compared to that of CDs. This may be due to the PVA environment, which can act as a surface passivation agent for CDs [64]. The QY of the developed PVA-CDs composite film was determined and found to be 47%, which is higher than that of CDs (inset of Figure 2b). The QY enhancement might be due to hydrogen bonding interactions in the PVA/CDs composite as suggested by the FTIR analysis (Figure 3b). This binding can produce plenty of physical crosslinking points for the CDs in the more confined environment of the PVA matrix, which provide a stabilization effect on the electron/hole pairs for more efficient radiative recombination, as reported in the literature [41,49,64]. The emission spectra (310–380 nm in range) of the CDs remained unshifted as the excitation wavelength was varied from 310–350 nm. Beyond 350 nm, a red-shift of the PL intensity was observed. The particle size selection and surface traps of CDs are the main factors responsible for the emission variation [27,51,65,66]. In contrast to the liquid CDs, PVA-CDs film exhibits only excitation-independent emission phenomena when exposed to varied excitation wavelengths (Figure 7c), which agrees with previous work [40]. This phenomenon is more favorable for a variety of applications as it prevents auto-luminescence [67].

To evaluate the photostability of the PVA-CDs composite film, the developed films were stored under daylight for one month and the PL spectra were recorded. The PL spectra (Figure S3a) display no apparent loss in PL intensity even after this duration of daylight illumination, retaining about 97% of their original PL emission. Moreover, the effect of varied temperate on the PL emission was also measured. As presented in Figure S3b, PL intensity retention of 97.8% was shown as the temperature increases from 25 to 65 °C. The enhanced QY and photostability of the PVA-CDs render them ideal for using these films in a wide range of applications including the detection and removal of metal ions.

2.5. Heavy Metal Complexation Test Using PVA-CDs Films

XPS and FTIR data confirm that the as-obtained PVA-CDs composite film is enriched with active sites. These sites can act as a bridge for the PVA-CDs to interact with heavy metals leading to metal complexation and rapid removal from wastewater. In this work, the selectivity of the PVA-CDs to adsorb heavy metals was tested in the presence of 50 ppm of Ni^{2+}, Pb^{2+}, Cd^{2+}, Zn^{2+} and Hg^{2+} at pH 7 for 10 min. The results reveal that the removal efficiencies of PVA-CDs film towards Ni^{2+}, Pb^{2+}, Cd^{2+}, Zn^{2+} and Hg^{2+} were about 46.37, 91.26, 61%, respectively (Figure S4). This suggests the more effective chelating kinetics of Cd^{2+} to bind with the functional moieties on the PVA-CDs surface compared to other metal ions [10,68].

2.6. Effect of pH and Contact Time on Cd^{2+} Removal

To optimize the analytical performance, the influence of pH and response time on Cd^{2+} adsorption using PVA-CDs film were evaluated. Figure 4a shows the % adsorption of Cd^{2+} using PVA-CDs and uncoated CDs at different pH values (3–11). The adsorption efficiency of Cd^{2+} for both PVA-CDs and uncoated CDs were found to rise in alkaline solutions, with a maximum % adsorption of 91 and 83, respectively. The low uptake in acidic medium could be ascribed to the release of H_3O^+ ions. The latter compete with Cd^{2+} for the surface sites of the adsorbent, leaving free metal in the suspension [69,70]. However, this effect is reduced with rising pH and the electrostatic attractions become dominant, thus enhancing the adsorption capacity [1]. Next the effect of contact time (from 1 to 25 min) on the % adsorption of Cd^{2+} was studied (Figure 4b). As shown, after only 10 min the PVA-CDs film reached the highest adsorption removal (91%). Meanwhile, an incubation period of 30 min was found sufficient to achieve an adsorption of 86% using uncoated CDs. In comparison to CDs, the PVA-CDs film is more efficient in terms of both adsorption time and removal efficiency. This enhancement in a remarkably short duration could be related to the formation of hydrogen bonds that facilitate provide more electrostatic exchange with Cd^{2+} and thus improve the % adsorption [40].

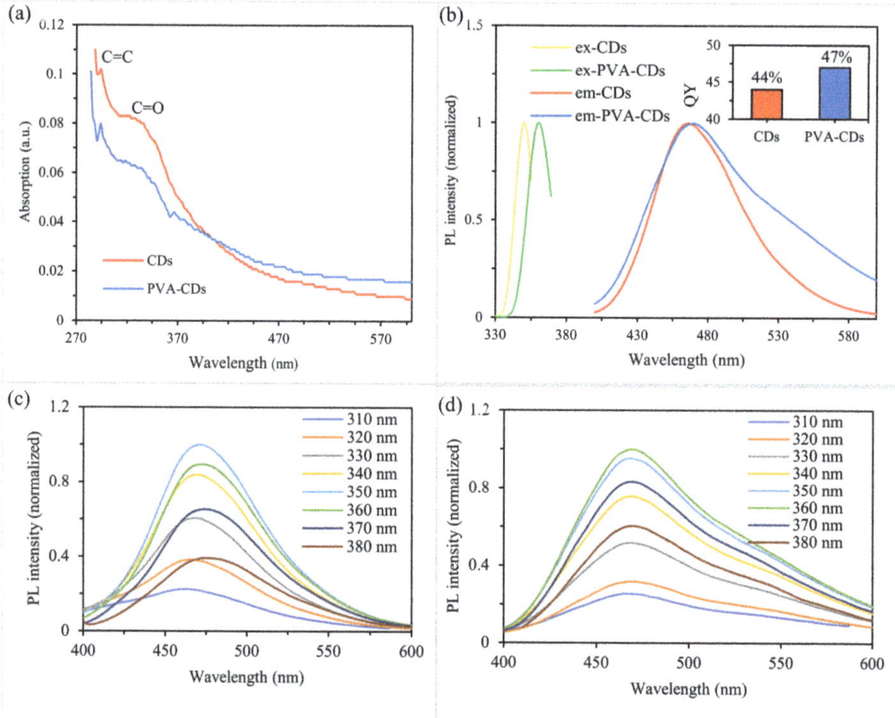

Figure 4. (**a**) UV-Vis absorption spectra of bare CDs and PVA-CDs film with inset photographs under daylight (left) and UV-light (right), (**b**) PL emission with inset showing QY, (**c**) excitation-dependent emission spectra of pared CDs and (**d**) excitation-dependent emission spectra of PVA-CDs film.

2.7. Adsorption Kinetics and Isotherm

To explain the adsorption of Cd^{2+} onto PVA-CDs, the first- and second-order kinetic models were evaluated, which can be described by the following equations:

First-order kinetic model:

$$\log(q_e - q_t) = \log q_e - (k_1/2.303)t \quad (1)$$

Second-order kinetic model:

$$t/q_t = 1/\left(k_2\, q_e^2\right) + t/q_e \quad (2)$$

where q_t and q_e (mg/g) are, respectively, the adsorption capacity at time t and equilibrium. k_1 (min^{-1}) and k_2 (g (mg min)$^{-1}$) are the rate constants of first- and second-order models, respectively.

The adsorption performance can be determined using Langmuir or Freundlich models using the following equations, respectively:

$$C_e/q_e = C/q_m + 1/(q_m\, k_L\,) \quad (3)$$

$$\log q_e = \log k_f + 1/n \times \log C_e \quad (4)$$

where q_e and C_e are the equilibrium adsorption uptake (mg/g) and equilibrium concentration of the metal ions (mg/L), respectively. q_m is the maximum adsorption capacity (mg/g) and K_L is Langmuir constant. Freundlich variable K_f describes the capacity of the absorption experiment and n indicates the isotherm nonlinearity.

Studying the adsorption kinetics is necessary for wastewater treatment as it affords substantial insights into the adsorbent nature and the dynamic process. Thus, the rate of Cd^{2+} removal on PVA-CDs was evaluated from the straight-line plots of ln($q_e - q_t$) versus (t) for first-order (Figure 4a) and t/q_t against (t) for pseudo-second-order models (Figure 4b). As shown, the pseudo-second-order kinetic model has better correlation parameter values compared to the first-order kinetics (Figure 5c, Figure S5a, Table 1), from which it can be concluded that the mechanism is probably chemisorption [4].

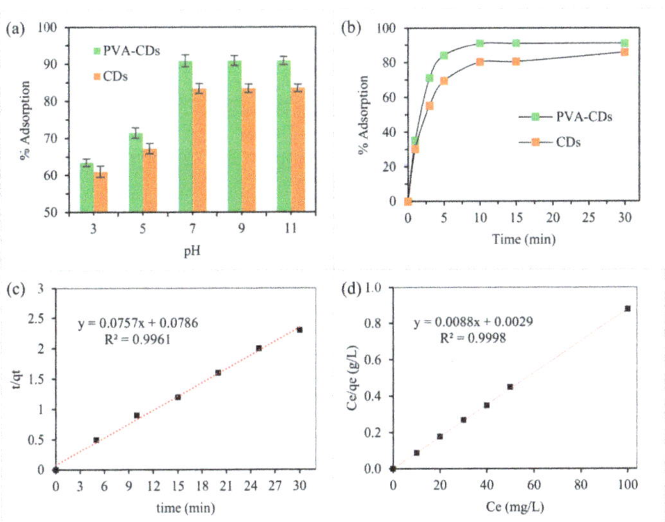

Figure 5. The effect of (**a**) pH and (**b**) contact time (pH = 7) on the adsorption efficiency for Cd^{2+} using CDs and PVA-CDs, (**c**) Pseudo-second-order model for adsorption Cd^{2+} onto PVA-CDs, (**d**) linearity graph of Langmuir isotherm.

Table 1. Kinetic models and isotherms parameters for Cd^{2+} adsorption onto PVA-CDs.

Pseudo-First Order			Pseudo-Second Order			Langmuir			Freundlich		
K_1	q_e	R^2	q_e	K_L	R^2	q_e	K_L	R^2	K_f	n	R^2
0.1	5.4	0.9853	13.2	0.075	0.9961	113.6	0.016	0.9998	4.64	3.1	0.9853

The adsorption isotherm is studied to calculate the adsorption uptake of an adsorbent and also to predict the interaction process. For instance, the Langmuir model proposes that the adsorption occurs by monolayer formation with no remarkable interaction between the adsorbent and adsorbate moieties, whereas Freundlich assumed that the adsorption capacity occurs on a heterogonous adsorbent surface [69]. In this regard, the adsorption isotherms (Figure 5d and Figure S5b) under the optimal conditions at room temperature were plotted. As presented in Table 1, it can be seen that Langmuir isotherm was a better fit than the Freundlich model according to the correlation coefficients (R^2). Based on Freundlich model investigation, n values are higher than one, confirming that the sorption reaction could be related to the electrostatic binding between Cd^{2+} and electron-rich species on the PVA-CDs surface [4,71]. The maximum adsorption uptake for cadmium onto N-CDs is determined to be 113.6 mg/g.

To compare the adsorption uptake of N-CDs with other adsorbents, a summary of adsorbent for Cd^{2+} removal is listed in Table 2.

Table 2. Comparison of several nanoadsorbents used for Cd^{2+} removal.

Adsorbent	Time (min)	q_m (mg/g)	Ref.
$NiFe_2O_4$/hydroxyapatite/graphene quantum dots	10	344.83	[1]
Chitosan-carbon dots	3	112.4	[4]
Poly (acrylic acid)-activated carbon nanocomposite	120	473.2	[7]
SnO_2 quantum dots decorated reduced graphene oxide	20	40.81	[46]
Carbon dots modified mesoporous organosilica	500	105.5	[44]
Carbon quantum dots/layered double hydroxide hybrid	20	12.6	[47]
Hollow calcite single crystals with CQDs	90	66.68	[45]
Multiwall carbon nanotubes	120	88.62	[72]
Nano-pumice	120	200	[73]
Nanoscale carbon black	120	31.7	[74]
N doped carbon dots integrated with PVA films	10	113.6	This work

It can be observed that the present nanoprobe has better performance, not only in terms of high adsorption capacity but also of needing less removal time to reach equilibrium. The Langmuir separation factor (RL) was also determined and found to be 0.61, emphasizing the favorable Cd^{2+} adsorption onto PVA-CDs.

2.8. Adsorption Mechanism

Efficiency improvements in heavy metal removal are mainly due to the PVA matrix, as previously reported [10,75–77]. Meanwhile, the reduction of Cd^{2+} on uncoated CDs could be explained on the basis of electrostatic attractions between Cd^{2+} centers and electron-rich functionalities around CDs. That means a metal-ligand complex is formed due to the donation effect of electrons from -COO^-, -OH and -N to the d-orbitals of metal, in line with previously reported works [48,78,79]. However, for the PVA-CDs membrane, a synergetic adsorption activity was noticed. The adsorption/desorption mechanisms of Cd^{2+} ions are illustrated in Figure 6. Recent investigations show that PVA-CDs film favored the rapid adsorption of Cd^{2+} by providing more -OH sites for target metal ions [70,75–77].

According to the Pearson assumption, the affinity of heavy metals towards the available sites is ascribed to the soft acid–soft base binding, where the PVA–CDs act as a soft base and Cd^{2+} as a soft acid [4,80].

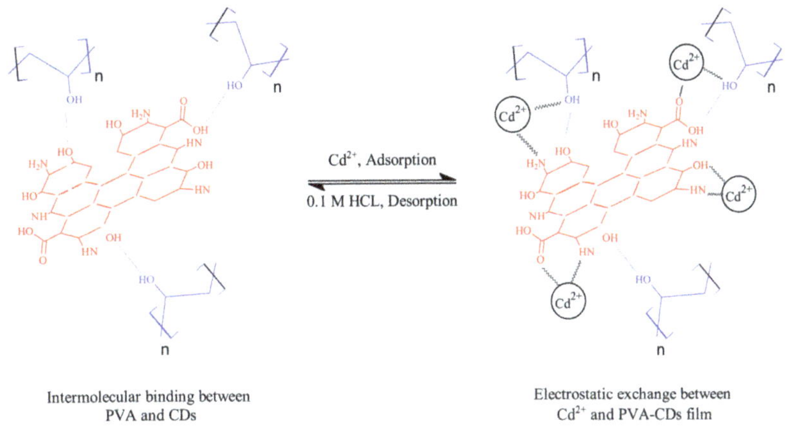

Figure 6. Schematic representation for the adsorption/desorption of Cd^{2+} by PVA-CDs nanocomposite.

To support the aforementioned assumption, several analyses were carried out. TEM images and particle size distribution (Figure S6a) show that there is a considerable increase in the nanocomplex size, with an average diameter of 8.6 nm, suggesting the agglomeration of the complex [81–83]. In comparison to the uncoated CDs (Figure S1), the surface charges of the hybrid film (Figure S6b) were neutralized with 5.08 eV after adding Cd^{2+} due to the electrostatic effect [48]. FTIR measurements (Figure S6c) prove that there is significant shift of the wavenumbers and increments of the characteristic peaks, particularly C-N, C=O, C-O and C-O-C, demonstrating the involvement of the -COO^- and -N-containing groups in Cd^{2+} complexation. All the above observations confirm the complex formation between Cd^{2+} and ligand spices of PVA-CDs.

2.9. Desorption and Recyclability

The regeneration and reusability performance of the adsorbent was studied to evaluate its activity for practical use. First, recycling experiments to recover the adsorbed Cd^{2+} were conducted using urea (0.1 M), HCl (0.1 M), EDTA (0.1 M) and NaCl (0.1 M). As shown in Figure 7a, the results revealed the superior effectiveness of HCl to recover Cd^{2+} from PVA-CDs, consistent with a previous study [84].

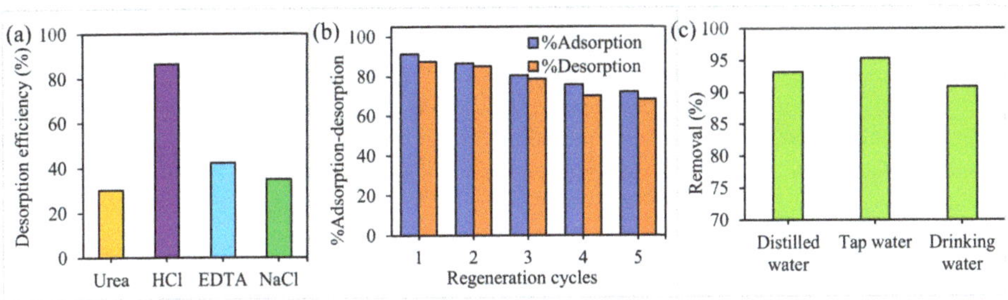

Figure 7. (**a**) Different de-sorbents to recover Cd^{2+} from PVA-CDs, (**b**) reusability study of PVA-CDs for five cycles using HCl and (**c**) real water sample test of PVA-CDs.

Next, regeneration tests were carried out using HCl as an eluent. As presented in Figure 7b, there is no substantial loss of % adsorption even after five adsorption-desorption cycles, in which only a 19.2% decline in the removal efficiency of the recycled material was shown. The loss in % adsorption might be assigned to the reduction of the adsorbent or the irreversible occupation of partial adsorption sites [85]. The good reusability emphasized that the adsorbent could be potentially used for practical contaminated water remediation.

2.10. Analysis of Real Water Samples

To further validate the applicability of the developed film, the removal of cadmium was tested in real water samples, including tap water and drinking water. The water samples were acquired from a biochemistry lab (UPM, Malaysia) and spiked with 50 μM Cd^{2+} ion solution. The values obtained were 95.34% and 90.9% in the case of tap and drinking water, respectively (Figure 7c). The results were compared with the same concentrations in distilled water with a % removal of 93.2%. These above data emphasized the reliability of the developed luminescence film for the recognition of Cd^{2+} in the environment.

3. Materials and Methods

3.1. Materials

CMC of oil palms empty fruit bunch wastes was collected from the Waris Nove Company (Pahang, Malaysia). Linear PEI (Mn ~5000) and quinine sulfate (quinine hemisulfate salt monohydrate) were purchased from Sigma-Aldrich (St. Louis, MO, USA). PVA (98% hydrolyzed, average molecular weight of 72,000 $gmol^{-1}$) was obtained from R&M Chemicals (Selangor, Malaysia). Deionized water (DI) has been used throughout all the experiments.

3.2. Preparation of CDs and PVA-CDs

CMC (1 g) was mixed with PEI (1 g) and then dispersed in DI (250 mL). Then, the mixture was heated in an autoclave at 260 °C for 2 h. The resultant supernatant was centrifuged at 12,000 rpm for 10 min to obtain free-dark carbonaceous material. The final suspension was purified through vacuum filtration (0.22 μm) to remove the precipitate. It was then placed into a dialysis membrane (1 kDa) so that the contents of salt ions were eliminated. Polyvinyl alcohol-CDs film (PVA-CDs film) was prepared by mixing N-CDs (1 mL, 1 mg/mL) with PVA solution (1 mL, 1 wt%) under continuous stirring. Then, the obtained hydrogel was poured on a glass substrate and dried at 80 °C for 1 h. Finally, the PVA-CDs composite film was peeled off from the glass surface to obtain a freestanding film and stored for further utilization.

3.3. Characterizations

Transmission electron microscopy (TEM) and high-resolution TEM (HRTEM) readings were carried out using a G2 F20 electron microscope (Tecnai, Hillsboro, OR, USA) with an acceleration voltage of 200 kV. X-ray powder diffraction (XRD) patterns were measured using a PANalytical (Malvern, UK) diffractometer with Cu-Kα radiation. Fourier transform infrared (FTIR) spectra (Thermo Nicolet, Midland, ON, Canada) were measured of 4 cm^{-1} resolution with KBr as a standard within the range from 650 to 4000 cm^{-1}. X-ray photoelectron spectra (XPS) (Physical Electronics PHI 5400 spectrometer, Mimos Semiconductors, Kuala Lumpur, Malaysia) were measured using Al-Ka radiation (hv = 1486.6 eV). Before deconvolution, charge correction was established using Origin 9.0 (OriginLab Corporation, Northampton, MA, USA) at C_{1s} by setting binding energies of C-C and C-H at 284.8 eV. The pH values were obtained by a PB-10 pH-meter (Beijing Sartorius Instruments Co. Ltd., Beijing, China) and zeta potential study of N-CDs was carried out using Zetasizer Nano ZS (Malvern, UK). UV–Vis spectra of all samples were recorded using a UV-1800 spectrophotometer (Shimadzu, BA29AP, Corston, UK). Fluorescence spectra were measured in quartz cuvettes with 1 cm path length using a LS

55 Fluorescence Spectrometer (PerkinElmer, South San Francisco, CA, USA) with a slit width of 15 and 5 nm for excitation and emission, respectively, and a scan rate of 240 nm/min.

3.4. Quantum Yield Determinations

Quinine sulfate in 0.1 M H_2SO_4 (QY = 54%) was used as a standard fluorophore. The QY was determined by comparing both absorption and PL emission of the prepared and used materials with that of quinine sulfate using the following Equation (5) [48]:

$$QY = Q_R \frac{I}{I_S} \frac{OD_S}{OD} \frac{\eta^2}{\eta_{S2}} \quad (5)$$

where (I) is integrated intensity, (OD) optical density and (η) is refractive index. The subscript S refers to the standard fluorophore of known QY.

3.5. Adsorption Studies

The feasibility of using the developed PVA–CDs films for heavy metal adsorption was evaluated. To test the selectivity, the hybrid films (0.1 g) were immersed in solutions of five selected metals: $NiCl_2$, $Pb(NO_3)^2$, $Cd(NO_3)^2$, $Zn(NO_3)^2$, $HgCl_2$ (30 mL, 50 ppm) at ambient temperature, pH = 7, then washed several times with DI and the supernatant then analyzed using atomic absorption spectrometry (AAS) to calculate the remaining metal ion in the solutions.

The kinetic tests were conducted under optimum conditions, agitating at 200 rpm. After a certain time (t) of adsorption, the residual Cd^{2+} concentration was measured using the aforementioned technique. For isotherm modeling, experiments were carried out at room temperature using various Cd^{2+} concentrations under optimal pH and contact time conditions. All the experiments were repeated three times. The equilibrium adsorption capacity, q_e (mg/g) and removal efficiency, %adsorption was calculated using the following equations:

$$qe = \frac{(Co - Ce)v}{m} \quad (6)$$

$$\%\text{Adsorption} = \frac{Co - Ce}{Co} \times 100 \quad (7)$$

where m (g) is the mass of adsorbent, v (L) is the volume of solution, Co and Ce (mg/ L) are the initial and equilibrium concentration of Cd^{2+} in solution, correspondingly.

4. Conclusions

Eco-friendly photoluminescent CDs were produced from CMC and PEI. The as-made CDs solution with an abundance with oxygen and nitrogen species displayed a high quantum yield for monodispersed particles with a great graphitization index. The CDs were integrated with PVA to develop PVA-CDs fluorescent film which was utilized for the removal of Cd^{2+} from aqueous media. The results show that the film exhibited a bright blue color, a higher QY of 47% and an improved Cd^{2+} removal percentage of 91.1 % compared to that of bare CDs (greenish, 44%, 80.4%, respectively). The equilibrium adsorption time was 10 and 30 min for PVA-CDs and CDs, respectively, at an optimal solution pH of 7. The adsorption process could be explained by pseudo-second-order kinetics and the Langmuir isotherm model. Under maximum conditions, an optimal adsorption uptake of 113.6 mg g^{-1} was achieved. Nanocomplexation is proposed to be the main reason for the removal of Cd^{2+}. The luminescent film was reusable for multiple desorption-adsorption cycles when regenerated with HCl solution. The feasibility of using the film was also tested for tap and drinking water with sufficient removal efficiencies. This work provides insights for developing an eco-friendly and rapid Cd^{2+} removal method with practical adsorbent membranes using sustainable and cost-effective carbon dots.

Supplementary Materials: The following are available online, Figure S1: Zeta potential of CDs, Figure S2: XPS spectrum of CMC, CDs and PVA-CDs, Figure S3: Photostability of PVA-CDs composite film at different aging times (a) and at different heating temperatures (b), Figure S4: Selectivity removal of Ni(II), Pb(II), Cd(II), Zn(II) and Hg(II) by using PVA-CDs composite film, Figure S5: Pseudo-first-order model (a) and Freundlich isotherm (b) for removal of Cd2+ onto PVA-CDs, Figure S6: (a) TEM image, (b) Zeta potential and (c) FTIR of PVA-CDs film in the presence of Cd2+. Inset: size distribution, Table S1: Elemental compositions of the EFB, CDs and PVA-CDs samples by XPS analysis.

Author Contributions: Conceptualization, M.A.I.; methodology M.A.I.; software, M.A.I. and Z.Z.A.; validation, M.A.I. and Z.Z.A.; formal analysis, M.A.I.; investigation, M.A.I.; resources, M.A.I. and Z.Z.A.; data curation, M.A.I. and Z.Z.A.; writing—original draft preparation M.A.I.; writing—review and editing, M.A.I. and Z.Z.A.; visualization, M.A.I.; supervision, Z.Z.A.; project administration, Z.Z.A.; funding acquisition, Z.Z.A. All authors have read and agreed to the published version of the manuscript.

Funding: This research was funded by University Putra Malaysia (UPM) Malaysia grant number GP-IPS/2017/9556800 And the APC was funded by University Putra Malaysia (UPM) Malaysia.

Conflicts of Interest: The authors declare that there is no conflict of interest regarding the publication of this work.

References

1. Kahrizi, P.; Mohseni-Shahri, F.S.; Moeinpour, F. Adsorptive removal of cadmium from aqueous solutions using NiFe2O4/hydroxyapatite/graphene quantum dots as a novel nano-adsorbent. *J. Nanostruct. Chem.* **2018**, *8*, 441–452. [CrossRef]
2. Wang, Y.; Wu, D.; Wei, Q.; Wei, D.; Yan, T.; Yan, T.; Yan, L.; Hu, L.; Du, B. Rapid removal of Pb (II) from aqueous solution using branched polyethylenimine enhanced magnetic carboxymethyl chitosan optimized with response surface methodology. *Sci. Rep.* **2017**, *7*, 1–11. [CrossRef] [PubMed]
3. Tchounwou, P.B.; Yedjou, C.G.; Patlolla, A.K.; Sutton, D.J. Heavy Metals Toxicity and the Environment. *Mol. Clin. Environ. Toxicol.* **2012**, *101*, 133–164.
4. Jlassi, K.; Eid, K.; Sliem, M.H.; Abdullah, A.M.; Chehimi, M.M.; Krupa, I. Rational synthesis, characterization, and application of environmentally friendly (polymer–carbon dot) hybrid composite film for fast and efficient UV-assisted Cd2+ removal from water. *Environ. Sci. Eur.* **2020**, *32*, 1. [CrossRef]
5. Figueiredo, H.; Quintelas, C. Tailored zeolites for the removal of metal oxyanions: Overcoming intrinsic limitations of zeolites. *J. Hazard. Mater.* **2014**, *274*, 287–299. [CrossRef]
6. Jlassi, K.; Abidi, R.; Benna, M.; Chehimi, M.M.; Kasak, P.; Krupa, I. Bentonite-decorated calix (4) arene: A new, promising hybrid material for heavy-metal removal. *Appl. Clay Sci.* **2018**, *161*, 15–22. [CrossRef]
7. Ge, H.; Wang, J. Ear-like poly (acrylic acid)-activated carbon nanocomposite: A highly efficient adsorbent for removal of Cd(II) from aqueous solutions. *Chemosphere* **2017**, *169*, 443–449. [CrossRef]
8. Tadesse, A.; Ramadevi, D.; Hagos, M.; Battu, G.; Basavaiah, K. Synthesis of nitrogen doped carbon quantum dots/magnetite nanocomposites for efficient removal of methyl blue dye pollutant from contaminated water. *RSC Adv.* **2018**, *8*, 8528–8536. [CrossRef]
9. Wang, J.; Chu, Y.; Qiu, F.; Li, X.; Wu, H.; Pan, J.; Niu, X.; Zhang, T.; Yang, D. One-pot simple green synthesis of water-soluble cleaner fluorescent carbon dots from cellulose and its sensitive detection of iron ion. *J. Clean. Prod.* **2017**. [CrossRef]
10. Issa, M.A.; Abidin, Z.Z.; Pudza, M.Y.; Zentou, H. Efficient removal of Cu(II) from aqueous systems using enhanced quantum yield nitrogen-doped carbon nanodots. *RSC Adv.* **2020**, *10*, 14979–14990. [CrossRef]
11. Ren, G.; Zhang, Q.; Li, S.; Fu, S.; Chai, F.; Wang, C.; Qu, F. One pot synthesis of highly fluorescent N doped C-dots and used as fluorescent probe detection for Hg^{2+} and Ag^{+} in aqueous solution. *Sens. Actuators B Chem.* **2017**, *243*, 244–253. [CrossRef]
12. Sun, X.; Lei, Y. Fluorescent carbon dots and their sensing applications. *Trends Anal. Chem.* **2017**, *89*, 163–180. [CrossRef]
13. Li, L.; Dong, T. Photoluminescence Tuning in Carbon Dots: Surface Passivation or/and Functionalization, Heteroatom Doping. *J. Mater. Chem. C* **2018**, *6*, 7944–7970. [CrossRef]
14. Ortega-Liebana, M.C.; Chung, N.X.; Limpens, R.; Gomez, L.; Hueso, J.L.; Santamaria, J.; Gregorkiewicz, T. Uniform luminescent carbon nanodots prepared by rapid pyrolysis of organic precursors confined within nanoporous templating structures. *Carbon* **2017**, *117*, 437–446. [CrossRef]

15. Xu, X.; Ray, R.; Gu, Y.; Ploehn, H.J.; Gearheart, L.; Raker, K.; Scrivens, W.A. Electrophoretic Analysis and Purification of Fluorescent Single-Walled Carbon Nanotube Fragments. *J. Am. Chem. Soc.* **2004**, *126*, 12736–12737. [CrossRef]
16. Li, H.; He, X.; Kang, Z.; Huang, H.; Liu, Y.; Liu, Y.; Liu, J.; Lian, S.; Tsang, C.H.A.; Yang, X.; et al. Water-Soluble Fluorescent Carbon Quantum Dots and Photocatalyst Design. *Angew. Chem. Int. Ed.* **2010**, *49*, 4430–4434. [CrossRef]
17. Purbia, R.; Paria, S. A simple turn on fluorescent sensor for the selective detection of thiamine using coconut water derived luminescent carbon dots. *Biosens. Bioelectron.* **2016**, *79*, 467–475. [CrossRef]
18. Sahu, S.; Behera, B.; Maiti, T.K.; Mohapatra, S. Simple one-step synthesis of highly luminescent carbon dots from orange juice: Application as excellent bio-imaging agents. *Chem. Commun.* **2012**, *48*, 8835–8837. [CrossRef]
19. Mehta, V.N.; Jha, S.; Basu, H.; Singhal, R.K.; Kailasa, S.K. One-step hydrothermal approach to fabricate carbon dots from apple juice for imaging of mycobacterium and fungal cells. *Sens. Actuators B Chem.* **2015**, *213*, 434–443. [CrossRef]
20. Qin, X.; Lu, W.; Asiri, A.M.; Al-Youbi, A.O.; Sun, X. Microwave-assisted rapid green synthesis of photoluminescent carbon nanodots from flour and their applications for sensitive and selective detection of mercury(II) ions. *Sens. Actuators B Chem.* **2013**, *184*, 156–162. [CrossRef]
21. Xu, J.; Lai, T.; Feng, Z.; Weng, X. Formation of fluorescent carbon nanodots from kitchen wastes and their application for detection of Fe3+. *Luminescence* **2014**, *30*, 420–424. [CrossRef]
22. Du, F.; Zhang, M.; Li, X.; Li, J.; Jiang, X. Economical and green synthesis of bagasse-derived fluorescent carbon dots for biomedical applications. *Nanotechnology* **2014**, *25*, 315702. [CrossRef]
23. Lu, W.; Quin, X.; Liu, S.; Chang, G.; Zhang, Y.; Luo, Y.; Asiri, A.M.; Al-Youbi, A.O.; Sun, X. Economical, green synthesis of fluorescent carbon nanoparticles and their use as probes for sensitive and selective detection of mercury(II) ions. *Anal. Chem.* **2012**, *84*, 5351–5357. [CrossRef]
24. Tyagi, A.; Tripathi, K.M.; Singh, N.; Choudhary, S.; Gupta, R.K. Green synthesis of carbon quantum dots from lemon peel waste: Applications in sensing and photocatalysis. *RSC Adv.* **2016**, *76*, 72423–72432. [CrossRef]
25. Issa, M.A.; Abidin, Z.Z.; Sobri, S.; Abdul-Rashid, S.; Mahdi, M.A.; Ibrahim, N.A.; Pudza, M.Y. Fabrication, characterization and response surface method optimization for quantum efficiency of fluorescent nitrogen-doped carbon dots obtained from carboxymethylcellulose of oil palms empty fruit bunch. *Chin. J. Chem. Eng.* **2019**, *28*, 584–592. [CrossRef]
26. Jayaweera, S.; Yin, K.; Hu, X.; Ng, W.J. Facile preparation of fluorescent carbon dots for label-free detection of Fe3+. *J. Photochem. Photobiol. A Chem.* **2019**, *370*, 156–163. [CrossRef]
27. Shen, P.; Gao, J.; Cong, J.; Liu, Z.; Li, C.; Yao, J. Synthesis of Cellulose-Based Carbon Dots for Bioimaging. *ChemistrySelect* **2016**, *1*, 1314–1317. [CrossRef]
28. Wu, Q.; Li, W.; Tan, J.; Wu, Y.; Liu, S. Hydrothermal carbonization of carboxymethylcellulose: One-pot preparation of conductive carbon microspheres and water-soluble fluorescent carbon nanodots. *Chem. Eng. J.* **2015**, *266*, 112–120. [CrossRef]
29. Wu, P.; Li, W.; Wu, Q.; Liu, Y.; Liu, S. Hydrothermal synthesis of nitrogen-doped carbon quantum dots from microcrystalline cellulose for the detection of Fe 3+ ions in an acidic environment. *RSC Adv.* **2017**, *7*, 44144–44153. [CrossRef]
30. Jayaweera, S.; Yin, K.; Hu, X.; Ng, W.J. Fluorescent N/Al Co-Doped Carbon Dots from Cellulose Biomass for Sensitive Detection of Manganese (VII). *J. Fluoresc.* **2019**, *29*, 1291–1300. [CrossRef]
31. Yang, G.; Wan, X.; Su, Y.; Zeng, X.; Tang, J. Acidophilic S-doped carbon quantum dots derived from cellulose fibers and their fluorescence sensing performance for metal ions in an extremely strong acid environment. *J. Mater. Chem. A* **2016**, *4*, 12841–12849. [CrossRef]
32. Wang, L.Y.; Wang, M.J. Removal of Heavy Metal Ions by Poly(vinyl alcohol) and Carboxymethyl Cellulose Composite Hydrogels Prepared by a Freeze-Thaw Method. *ACS Sustain. Chem. Eng.* **2016**, *4*, 2830–2837. [CrossRef]
33. Feng, T.; Ai, X.; An, G.; Yang, P.; Zhao, Y. Charge-Convertible Carbon Dots for Imaging-Guided Drug Delivery with Enhanced in Vivo Cancer Therapeutic Efficiency. *ACS Nano* **2016**, *10*, 4410–4420. [CrossRef] [PubMed]
34. Khan, I.; Kumar, H.; Mishra, G.; Gothwal, A.; Kesharwani, P.; Gupta, U. Polymeric Nanocarriers: A New Horizon for the Effective Management of Breast Cancer. *Curr. Pharm. Des.* **2017**, *23*, 5315–5326. [CrossRef]

35. Wang, W.; Lai, H.; Cheng, Z.; Kang, H.; Wang, Y.; Zhang, H.; Wang, J.; Liu, Y. Water-induced poly(vinyl alcohol)/carbon quantum dot nanocomposites with tunable shape recovery performance and fluorescence. *J. Mater. Chem. B* **2018**, *6*, 7444–7450. [CrossRef]
36. Hoang, Q.B.; Mai, V.T.; Nguyen, D.K.; Truong, D.Q.; Mai, X.D. Crosslinking induced photoluminescence quenching in polyvinyl alcohol-carbon quantum dot composite. *Mater. Today Chem.* **2019**, *12*, 166–172. [CrossRef]
37. Gamal El-Shamy, A. Novel hybrid nanocomposite based on Poly(vinyl alcohol)/carbon quantum dots/fullerene (PVA/CQDs/C60) for thermoelectric power applications. *Compos. Part. B Eng.* **2019**, *174*, 106993. [CrossRef]
38. Taspika, M.; Permatasari, F.A.; Nuryadin, B.W.; Mayangsari, T.R.; Aimon, A.H.; Iskandar, F. Simultaneous ultraviolet and first near-infrared window absorption of luminescent carbon dots/PVA composite film. *RSC Adv.* **2019**, *9*, 7375–7381. [CrossRef]
39. Xu, L.; Zhang, Y.; Pan, H.; Xu, N.; Mei, C.; Mao, H.; Zhang, W.; Cai, J.; Xu, C. Preparation and performance of radiata-pine-derived polyvinyl alcohol/carbon quantum dots fluorescent films. *Materials* **2020**, *13*, 67. [CrossRef]
40. Aziz, S.B.; Hassan, A.Q.; Mohammed, S.J.; Karim, W.O.; Kadir, M.F.Z.; Tajuddin, H.A.; Chan, N.N.M.Y. Structural and Optical Characteristics of PVA : C-Dot Composites : Tuning the Absorption of Ultra Violet (UV) Region. *Nanomaterials* **2019**, *9*, 216. [CrossRef]
41. Luminescent poly (vinyl alcohol)/carbon quantum dots composites with tunable water-induced shape memory behavior in different pH and temperature environments. *ACS Appl. Mater. Interfaces* **2016**, *8*, 34744–34754. [CrossRef]
42. Wang, Y.; Zhao, Y.; Zhang, F.; Chen, L.; Yang, O.; Liu, X. Fluorescent polyvinyl alcohol films based on nitrogen and sulfur co-doped carbon dots towards white light-emitting devices. *New. J. Chem.* **2016**, *40*, 8710–8716. [CrossRef]
43. Alhosseini, S.N.; Moztarzadeh, F.; Mozafari, M.; Asgari, S.; Dodel, M.; Samadikuchaksaraei, A.; Kargozar, S.; Jalali, N. Synthesis and characterization of electrospun polyvinyl alcohol nanofibrous scaffolds modified by blending with chitosan for neural tissue engineering. *Int. J. Nanomed.* **2012**, *7*, 25–34.
44. Wang, L.; Cheng, C.; Tapas, S.; Lei, J.; Matsuoka, M.; Zhang, J.; Zhang, F. Carbon dots modified mesoporous organosilica as an adsorbent for the removal of 2,4-dichlorophenol and heavy metal ions. *J. Mater. Chem. A* **2015**, *3*, 13357–13364. [CrossRef]
45. Yang, T.; He, R.; Nie, G.; Wang, W.; Zhang, G.; Hu, Y.; Wu, L. Creation of Hollow Calcite Single Crystals with CQDs: Synthesis, Characterization, and Fast and Efficient Decontamination of Cd(II). *Sci. Rep.* **2018**, *8*, 1–12. [CrossRef]
46. Dutta, D.; Thiyagarajan, S.; Bahadur, D. SnO2 quantum dots decorated reduced graphene oxide nanocomposites for efficient water remediation. *Chem. Eng. J.* **2016**, *297*, 55–65. [CrossRef]
47. Rahmanian, O.; Dinari, M.; Abdolmaleki, M.K. Carbon quantum dots/layered double hydroxide hybrid for fast and efficient decontamination of Cd(II): The adsorption kinetics and isotherms. *Appl. Surf. Sci.* **2018**, *428*, 272–279. [CrossRef]
48. Issa, M.A.; Abidin, Z.Z.; Sobri, S.; Rashid, S.; Mahdi, M.A.; Ibrahim, N.A.; Pudza, M.Y. Facile Synthesis of Nitrogen-Doped Carbon Dots from Lignocellulosic Waste. *Nanomaterials* **2019**, *9*, 1500. [CrossRef]
49. Hu, M.; Gu, X.; Hu, Y.; Deng, Y.; Wang, C. PVA/Carbon Dot Nanocomposite Hydrogels for Simple Introduction of Ag Nanoparticles with Enhanced Antibacterial Activity. *Macromol. Mater. Eng.* **2016**, *301*, 1352–1362. [CrossRef]
50. Wang, R.; Wang, X.; Sun, Y. One-step synthesis of self-doped carbon dots with highly photoluminescence as multifunctional biosensors for detection of iron ions and pH. *Sens. Actuators B Chem.* **2017**, *241*, 73–79. [CrossRef]
51. Da Silva Souza, D.R.; Caminhas, D.C.; Mesquita, J.P.; Pereira, F.V. Luminescent carbon dots obtained from cellulose. *Mater. Chem. Phys.* **2018**, *203*, 148–155. [CrossRef]
52. Zhou, J.; Zhou, H.; Tang, J.; Deng, S.; Yan, F.; Li, W.; Qu, M. Carbon dots doped with heteroatoms for fluorescent bioimaging: A review. *Microchim. Acta* **2017**, *184*, 343–368. [CrossRef]
53. Wu, F.; Su, H.; Wang, K.; Wong, W.K.; Zhu, X. Facile synthesis of N-rich carbon quantum dots from porphyrins as efficient probes for bioimaging and biosensing in living cells. *Int. J. Nanomed.* **2017**, *12*, 7375–7391. [CrossRef]

54. Sevilla, M.; Fuertes, A.B. The production of carbon materials by hydrothermal carbonization of cellulose. *Carbon* **2009**, *47*, 2281–2289. [CrossRef]
55. Yu, K.; Wang, J.; Song, K.; Wang, X.; Liang, C.; Dou, Y. Hydrothermal Synthesis of Cellulose-Derived Carbon Nanospheres from Corn Straw as Anode Materials for Lithium ion Batteries. *Nanomaterials* **2019**, *9*, 93. [CrossRef]
56. García-Bordejé, E.; Pires, E.; Fraile, J.M. Parametric study of the hydrothermal carbonization of cellulose and effect of acidic conditions. *Carbon* **2017**, *123*, 421–432. [CrossRef]
57. Li, R.; Wang, L.; Shahbazi, A. A Review of Hydrothermal Carbonization of Carbohydrates for Carbon Spheres Preparation. *Trends Renew. Energy* **2015**, *1*, 43–56. [CrossRef]
58. Falco, C.; Baccile, N.; Titirici, M.M. Morphological and structural differences between glucose, cellulose and lignocellulosic biomass derived hydrothermal carbons. *Green Chem.* **2011**, *13*, 3273–3281. [CrossRef]
59. Pires, N.R.; Santos, C.M.W.; Sousa, R.R.; de Paula, R.C.M.; Cunha, P.L.R.; Feitosa, J.P.A. Novel and fast microwave-assisted synthesis of carbon quantum dots from raw cashew gum. *J. Braz. Chem. Soc.* **2015**, *26*, 1274–1282. [CrossRef]
60. Bhattacharyya, S.; Ehrat, F.; Urban, P.; Teves, R.; Wyrwich, R.; Döblinger, M.; Feldmann, J.; Urban, A.S.; Stolarczyk, J.K. Effect of nitrogen atom positioning on the trade-off between emissive and photocatalytic properties of carbon dots. *Nat. Commun.* **2017**, *8*, 1–9. [CrossRef]
61. Arul, V.; Sethuraman, M.G. Facile green synthesis of fluorescent N-doped carbon dots from *Actinidia deliciosa* and their catalytic activity and cytotoxicity applications. *Opt. Mater.* **2018**, *78*, 181–190. [CrossRef]
62. Hao, Y.; Gan, Z.; Xu, J.; Wu, X.; Chu, P.K. Applied Surface Science Poly (ethylene glycol)/carbon quantum dot composite solid films exhibiting intense and tunable blue-red emission. *Appl. Surf. Sci.* **2014**, *311*, 490–497. [CrossRef]
63. Bai, J.; Ren, W.; Wang, Y.; Li, X. High-performance thermoplastic polyurethane elastomer/carbon dots bulk nanocomposites with strong luminescence. *High Perform. Polym.* **2020**, 1–11. [CrossRef]
64. Bandi, R.; Devulapalli, N.P.; Dadigala, R.; Gangapuram, B.R.; Guttena, V. Facile Conversion of Toxic Cigarette Butts to N,S-Codoped Carbon Dots and Their Application in Fluorescent Film, Security Ink, Bioimaging, Sensing and Logic Gate Operation. *ACS Omega* **2018**, *3*, 13454–13466. [CrossRef]
65. Zhang, Y.; He, Y.H.; Cui, P.P.; Feng, X.T.; Chen, L.; Yang, Y.Z.; Liu, X.G. Water-soluble, nitrogen-doped fluorescent carbon dots for highly sensitive and selective detection of Hg^{2+} in aqueous solution. *RSC Adv.* **2015**, *5*, 40393–40401. [CrossRef]
66. Ding, H.; Wei, J.S.; Zhang, P.; Zhou, Z.Y.; Gao, Q.Y.; Xiong, H.M. Solvent-Controlled Synthesis of Highly Luminescent Carbon Dots with a Wide Color Gamut and Narrowed Emission Peak Widths. *Small* **2018**, *14*, 1–10. [CrossRef]
67. Qin, J.; Zhang, L.; Yang, R. Solid pyrolysis synthesis of excitation-independent emission carbon dots and its application to isoniazid detection. *J. Nanoparticle Res.* **2019**, *21*. [CrossRef]
68. Patır, K.; Gogoi, S.K. Nitrogen-doped carbon dots as fluorescence ON-OFF-ON sensor for parallel detection of copper(II) and mercury(II) ions in solutions as well as in filter paper-based microfluidic device †. *Nanoscale Adv.* **2018**, *1*, 592–601. [CrossRef]
69. Biswas, S.; Meikap, B.C.; Sen, T.K. Adsorptive Removal of Aqueous Phase Copper (Cu2+) and Nickel (Ni2+) Metal Ions by Synthesized Biochar–Biopolymeric Hybrid Adsorbents and Process Optimization by Response Surface Methodology (RSM). *Water. Air. Soil Pollut.* **2019**, *230*. [CrossRef]
70. Alsuhybani, M.; Alshahrani, A.; Algamdi, M.; Al-Kahtani, A.A.; Alqadami, A.A. Highly efficient removal of Pb(II) from aqueous systems using a new nanocomposite: Adsorption, isotherm, kinetic and mechanism studies. *J. Mol. Liq.* **2020**, *301*, 112393. [CrossRef]
71. Wang, X.; Liang, X.; Wang, Y.; Wang, X.; Liu, M.; Yin, D.; Xia, S.; Zhao, J.; Zhang, Y. Adsorption of Copper (II) onto activated carbons from sewage sludge by microwave-induced phosphoric acid and zinc chloride activation. *Desalination* **2011**, *278*, 231–237. [CrossRef]
72. Mubarak, N.M.; Sahu, J.N.; Abdullah, E.C.; Jayakumar, N.S.; Ganesan, P. Microwave assisted multiwall carbon nanotubes enhancing Cd(II) adsorption capacity in aqueous media. *J. Ind. Eng. Chem.* **2015**, *24*, 24–33. [CrossRef]
73. Khorzughy, S.H.; Eslamkish, T.; Ardejani, F.D.; Heydartaemeh, M.R. Cadmium removal from aqueous solutions by pumice and nano-pumice. *Korean J. Chem. Eng.* **2014**, *32*, 88–96. [CrossRef]

74. Zhou, D.-M.; Wang, Y.-J.; Wang, H.-W.; Wang, S.-Q.; Cheng, J.-M. Surface-modified nanoscale carbon black used as sorbents for Cu(II) and Cd(II). *J. Hazard. Mater.* **2010**, *174*, 34–39. [CrossRef] [PubMed]
75. Dolgormaa, A.; Lv, C.-J.; Li, Y.; Yang, J.; Yang, J.-X.; Chen, P.; Wang, H.-P.; Huang, J. Adsorption of Cu(II) and Zn(II) ions from aqueous solution by gel/PVA-modified super-paramagnetic iron oxide nanoparticles. *Molecules* **2018**, *23*, 2982. [CrossRef]
76. Pour, Z.S.; Ghaemy, M. Removal of dyes and heavy metal ions from water by magnetic hydrogel beads based on poly(vinyl alcohol)/carboxymethyl starch-g-poly(vinyl imidazole). *RSC Adv.* **2015**, *5*, 64106–64118. [CrossRef]
77. Al-qudah, Y.H.F.; Mahmoud, G.A.; Khalek, M.A.A. Radiation crosslinked poly (vinyl alcohol)/acrylic acid copolymer for removal of heavy metal ions from aqueous solutions. *J. Radiat. Res. Appl. Sci.* **2014**, *7*, 135–145. [CrossRef]
78. Li, M.; Li, B.; Zhou, L.; Zhang, Y.; Cao, Q.; Wang, R.; Xiao, H. Fluorescence-sensitive adsorbent based on cellulose using for mercury detection and removal from aqueous solution with selective 'on-off' response. *Int. J. Biol. Macromol.* **2019**, *132*, 1185–1192. [CrossRef]
79. Abdullah, M.; Abidin, Z.Z. Coagulation Treatment of Palm Oil Mill Effluent using Plant-Based Tannin. *Pollut. Res.* **2018**, *37*, 788–793.
80. Wiśniewski, M.; Gauden, P.A. Pearson's hard-soft acid-base principle as a means of interpreting the reactivity of carbon materials. *Adsorpt. Sci. Technol.* **2006**, *24*, 389–402. [CrossRef]
81. Nair, R.V.; Thomas, R.T.; Sankar, V.; Muhammad, H.; Dong, M.; Pillai, S. Rapid, Acid-Free Synthesis of High-Quality Graphene Quantum Dots for Aggregation Induced Sensing of Metal Ions and Bioimaging. *ACS Omega* **2017**, *2*, 8051–8061. [CrossRef] [PubMed]
82. Minh, L.; Phan, T.; Hoon, S.; Phan, T.; Yeol, K.; Ha, S. Synthesis of fluorescent silicon quantum dots for ultra-rapid and selective sensing of Cr(VI) ion and biomonitoring of cancer cells. *Mater. Sci. Eng. C J.* **2018**, *93*, 429–436.
83. Stafiej, A.; Pyrzynska, K. Adsorption of heavy metal ions with carbon nanotubes. *Seperation Purif. Technol.* **2007**, *58*, 49–52. [CrossRef]
84. Cui, Y.W.; Li, J.; Du, Z.F.; Peng, Y.Z. Cr(VI) adsorption on red mud modified by lanthanum: Performance, kinetics and mechanisms. *PLoS ONE* **2016**, *11*, 1–16. [CrossRef]
85. Anand, S.R.; Bhati, A.; Saini, D.; Gunture Chauhan, N.; Khare, P.; Sonkar, S.K. Antibacterial Nitrogen-doped Carbon Dots as a Reversible 'fluorescent Nanoswitch' and Fluorescent Ink. *ACS Omega* **2019**, *4*, 1581–1591. [CrossRef]

Sample Availability: Samples of the compounds are available from the authors.

molecules MDPI

Article

$CuFe_2O_4$/Polyaniline (PANI) Nanocomposite for the Hazard Mercuric Ion Removal: Synthesis, Characterization, and Adsorption Properties Study

Saad S. M. Hassan [1,*], Ayman H. Kamel [1,*], Amr A. Hassan [1,2], Abd El-Galil E. Amr [3,4,*], Heba Abd El-Naby [1], Mohamed A. Al-Omar [3] and Ahmed Y. A. Sayed [3]

1 Chemistry Department, Faculty of Science, Ain Shams University, Abbasia, Cairo 11566, Egypt; amr_hassan@sci.asu.edu.eg (A.A.H.); hoba_science@hotmail.com (H.A.E.-N.)
2 Department of Chemistry, Virginia Commonwealth University, Richmond, VA 23284, USA
3 Pharmaceutical Chemistry Department, College of Pharmacy, King Saud University, Riyadh 11451, Saudi Arabia; malomar1@ksu.edu.sa (M.A.A.-O.); ahmedyahia009@gmail.com (A.Y.A.S.)
4 Applied Organic Chemistry Department, National Research Center, Dokki, Giza 12622, Egypt
* Correspondence: saadsmhassan@sci.asu.edu.eg (S.S.M.H.); ahkamel76@sci.asu.edu.eg (A.H.K.); aamr@ksu.edu.sa (A.E.-G.E.A.); Tel.: +20-12-2216-2766 (S.S.M.H.); +20-10-0074-3328 (A.H.K.); +966-565-148-750 (A.E.-G.E.A.)

Academic Editors: Chiara Bisio and Monica Pica
Received: 11 May 2020; Accepted: 9 June 2020; Published: 12 June 2020

Abstract: Copper ferrite nano-particles ($CuFe_2O_4$) were synthesized, characterized, modified with polyaniline to form $CuFe_2O_4$/PANI nano-composite. They were used as new adsorbents for the removal of the hazardous mercuric ions from aqueous solutions. High resolution transmission electron microscope (HR-TEM), X-ray diffraction (XRD), Fourier-transform infrared (FT-IR) and Brunauer–Emmett–Teller (BET) were used for the characterization of the synthesized $CuFe_2O_4$ nano-particles (NPs) in presence and absence of PANI nano-composite. The synthesized $CuFe_2O_4$NPs were of spherical shape with an average size of 10.8 nm. XRD analysis displayed crystal peaks for $CuFe_2O_4$NPs and amorphous peaks $CuFe_2O_4$/PANI nano-composite due to the existence of polyaniline layer. Contact time, adsorbent dose, solution pH, adsorption kinetics, adsorption isotherm and recyclability were studied. The method at the optimum conditions exhibited high performance with high mercury removal percentage of up to 99% with a maximum adsorption capacity 12.5 and 157.1 mg/g for $CuFe_2O_4$ and $CuFe_2O_4$/PANI, respectively. The adsorption processes were fitted to Langmuir isotherms. The adsorption behavior of $CuFe_2O_4$@PANI composite towards Hg^{2+} ions is attributed to the soft acid–soft base strong interaction between PANI and Hg(II) ions. High stability and enhanced re-usability are offered using $CuFe_2O_4$@PANI composite due to its enhanced removal efficiency. No significant removal decrease was noticed after five adsorption–desorption cycles. In addition, it possesses an easy removal from aqueous solutions by external magnetic field after adsorption experiments. These indicated the enhancement of polyaniline to the surface of $CuFe_2O_4$ toward the adsorption of mercury from aqueous solutions.

Keywords: $CuFe_2O_4$ nano-particles; $CuFe_2O_4$/PANI composite; mercury (II) removal; adsorption

1. Introduction

A clean water resource is a vital and necessary goal for the whole world. Toxic heavy metals like Hg, Pb, Cd, and Ni are considered the most dangerous environmental pollutants in the water, thus becoming of prior anxiety because of their toxicity and non-biodegradability to plants, animals and human [1,2]. Mercury is one of these heavy metals that can cause serious environmental and health problems as chronic and acute poisoning. It exists in different forms such as metallic Hg, Hg^+, Hg^{2+}, and organic mercury containing phenyl, methyl, and ethyl groups, etc. It causes different diseases such as Alzheimer's disease, amyotrophic lateral sclerosis, Parkinson's disease, and damaging of the immune system and kidneys. Mercury is considered as prior hazardous pollutant by the Agency for Toxic Substances and Disease Registry [3]. One of its natural sources is the volcanoes that produce almost half of the mercury emissions released in atmosphere. It is also produced from different industrial sources such as pharmaceuticals, chloralkali, plastic, textile, paint, rubber, paper, cement, electronic industry, coal combustion, fertilizers, oil refining, and rubber processing [4,5]. The other half is generated by humans by various means including 65% in combustion, 11% in the production of gold, 6.8% in the production of non-ferrous metal, 6.4% in the production of cement, 3.0% in the waste disposal including municipal waste, and 3.0% in the production of caustic soda [6]. According to the World Health Organization (WHO), 1 μg/L is the maximum permissible concentration of Hg(II)in drinking water [7]. According to the European Union (EU), the maximum acceptable level of Hg(II) is 5 μg/L for wastewater discharge [8–10]. River and lake water in the nearby industries may contain mercuric discharges which are fatal for aquatic as well as for human life. These discharges could accumulate in the stomach and remain non digestible resulting in the formation of cancerous diseases. Long-term exposure to mercury could cause serious damage to nerves, brain, kidney, lung irritation, eye irritation, skin rashes, vomiting, and diarrhea [11].Researchers have been used a lot of techniques to get rid of heavy metals in particularly mercury ion from waste water such as sorption and filtration [12], ion exchange [13,14], chemical precipitation [6], adsorption [15–17], solid phase extraction [18], and adsorption process using nano-materials [19–27]. The adsorption technique is the most effective and commonly used due to its high removal efficiency and cheapness.

Recently, there is a focus on the application of nano-materials in the removal of different environmental pollutants. This is based on their distinctive properties such as high surface area, high adsorption, and special photoelectric property. However, they are suffering from difficulty of their separation from aqueous solutions due to their small particle size which restricts the application in water treatment. So, it is preferable using magnetic nano-materials that can be easily separated from solution with external magnetic field [28–30].

Magnetic nano-materials possess adsorbent properties that qualify them for use as promising adsorbent materials, which open up a wide field for engineering separation applications. These magnetic nano-particles can be separated based on their nanostructures due to the easy direction of magnetization, which will vary depending on the arrangement of the atoms in the magnetic structure [31–33]. Applying a low density magnetic field stimulates the magnetization of the material and therefore makes the use of magnetic force possible, but when the magnetic field is cut off, the magnetization immediately decreases to zero. This last point is important for the release of particles after adsorption of the waste [34,35]. The main drawback of using magnetic nano-particles is the low potential pollutant removal ability. To invade this defect, the surface of magnetic nano-particles has been modified. The surface properties of nano-particles can be greatly enhanced after this modification. This is preferred through the Van der Waals interaction between the modified material and the reduced solvent shielding of the ions in the interlamellar environment.

Polyaniline (PANI) has attracted much attention because of its several unique properties [36–38]. It is highly stable in air and soluble in various solvents and exhibits dramatic changes in its electronic structure and physical properties in the protonated state. It also shows magnetic behavior because of its high spin density [39,40].

In the present work, modification of $CuFe_2O_4$ nano-particles (NPs) with polyaniline was used as a novel adsorbent for mercury removal in aqueous solutions. The nano-particles were synthesized, characterized, and used as an adsorbent for mercury removal under optimum conditions. The removal efficiency of the prepared adsorbents was investigated, and their adsorption and desorption behaviors towards mercury species were studied.

2. Results and Discussion

2.1. Adsorbent Characterization

2.1.1. X-ray Diffraction Pattern

The phase identification of $CuFe_2O_4$ and $CuFe_2O_4$/PANI nano-composites was illustrated by X-ray diffraction (XRD) as shown in Figure 1. All of the high intensity peaks are indexed and refined as tetragonal structure with I41/amd space group, which is consistent with standard Joint Committee on Powder Diffraction Standards (JCPDS) card no. 34-0425.The obtained XRD pattern exhibits good crystallinity for $CuFe_2O_4$. The reflection plans (101), (112), (211), (220), (303), and (224) coincide with the tetragonal spinel phase for $CuFe_2O_4$ with a characteristic peak appears at 2θ 35.5°. The reflection plans (010), (100), and (110) coincide with the amorphous phase of standard data for polyaniline. It is apparent that the broad diffraction peak centered at 2θ value 25.3° (110) in Figure 1 is the characteristic peak of the PANI layer. This can be ascribed to the periodicity parallel and perpendicular to the polymer chains, respectively [41]. The characteristic peak of $CuFe_2O_4$ still appears at 35.5° and little shift for the other peaks when doped with PANI. The average crystalline size of the prepared nano-composite was calculated using Scherrer's equation [42]:

$$\text{D} = 0.9\ \lambda/\beta\ cos\ \theta \tag{1}$$

where D is the average crystalline size, λ is the wavelength of CuKα,β is the full width at half maximum (FWHM) of most intense diffraction peak (211), and θ is the Bragg's angle. The average particle size is estimated to be 10.8 and 23.4 nm for $CuFe_2O_4$ and $CuFe_2O_4$/PANI nano-composites, respectively.

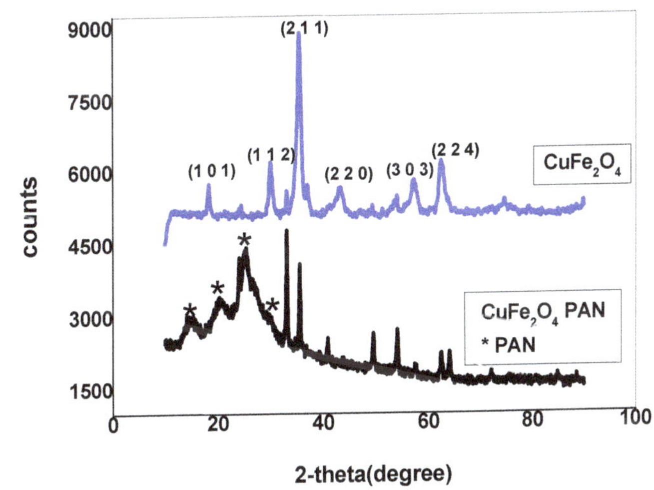

Figure 1. Pattern of both $CuFe_2O_4$ and $CuFe_2O_4$/PANI nano-composites.

2.1.2. High Resolution Transmission Electron Microscopy (HRTEM)

TEM images for $CuFe_2O_4$ nano-particles showed spherical shaped nano-particles with small agglomeration and nano sizes of 10.8 nm that coincides with the XRD result. The particles are dense and regularly distributed with clear boundary between neighboring particles as observed in Figure 2a. TEM images of $CuFe_2O_4$/PANI nano-composite revealed the light shell nature of PANI in which dark core copper ferrite particles are embedded as shown in Figure 2b.

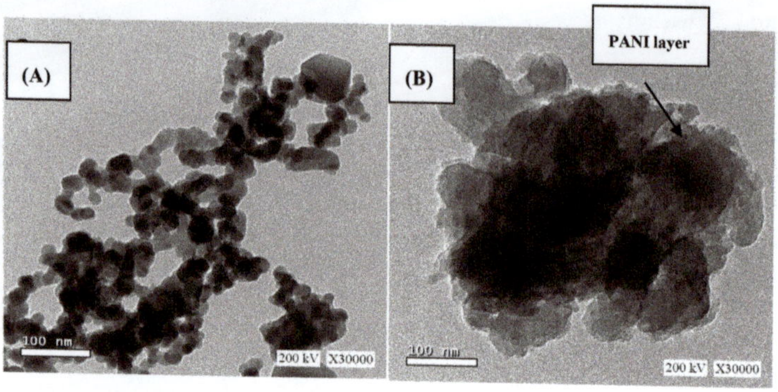

Figure 2. Images of (**A**) $CuFe_2O_4$ and (**B**) $CuFe_2O_4$/PANI nano-composites.

2.1.3. Brunauer–Emmett–Teller (BET)

The N_2 adsorption–desorption experiment at 77 K for $CuFe_2O_4$ and $CuFe_2O_4$/PANI nano-composites are shown in Figure 3. The figure shows an adsorption isotherm of the type IV with a hysteresis loop that is associated with capillary condensation within the mesoporous regions [43], with a hysteresis loop type H3, which is usually indicative of aggregates of platelet particles or adsorbents containing slit pores. The initial part of the isotherm (until $p/p^o \approx 0.4$) can be attributed to monolayer/multilayer adsorption because it follows the same path of desorption, which demonstrates weak adsorbate–adsorbent interactions. The hysteresis loop begins at $p/p^o = 0.4$ and it ends at $p/p^o = 0.95$; the hysteresis loop exhibits limited adsorption. This phenomenon is related to the presence of particles that are not rigidly joined together. The BET surface area and pore volume of the nano-composite are recorded in Table 1. The pore size of $CuFe_2O_4$ nano-particles was about 9.9 nm. It was regarded as a mesoporous material of surface area 44.7 m^2/g and pore volume of 0.11 cm^3/g. For $CuFe_2O_4$/PANI nano-composites, the BET surface area is lower, around 30.8 m^2/g, due to the lower cumulative volume of pores (0.06 cm^3/g).

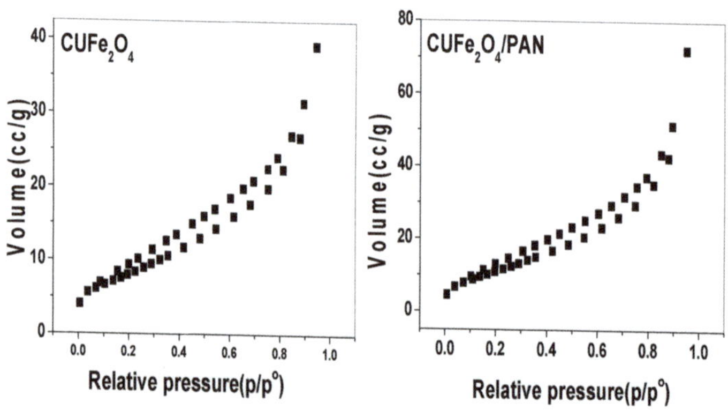

Figure 3. N_2 adsorption–desorption isotherms of coupled $CuFe_2O_4$ and $CuFe_2O_4$/PANI (polyaniline) nano-composites.

Table 1. General surface characteristics of $CuFe_2O_4$ and $CuFe_2O_4$/PANI nano-composites obtained by N_2 adsorption at 77 K.

Sample	Surface Area(m^2/g)	Average Pore Volume(cm^3/g)	Average Pore Diameter (nm)
$CuFe_2O_4$/PANI	30.8	0.06	17.8
$CuFe_2O_4$ NP	44.7	0.11	9.9

2.1.4. Fourier Transforms–Infrared Spectroscopy (FTIR)

The FTIR spectra of $CuFe_2O_4$ and $CuFe_2O_4$/PANI nano-composites are shown as supplementary material in Supplementary Materials Figure S1. $CuFe_2O_4$ spectrum has only a characteristic peak at 574.9 cm^{-1} of M–O bond while $CuFe_2O_4$/PANI spectrum has several characteristics peaks corresponding to polyaniline. These include peaks at 3419.1 cm^{-1} assigned for N–H stretching, 1561.1 cm^{-1} assigned to stretching vibration of C=C, 1469.7 cm^{-1} assigned to stretching vibration of C–C, 1298.7 cm^{-1} C–N stretching vibrations, 1135.8 cm^{-1} for C–H bending mode and 777 cm^{-1} assigned to the wagging of =C–H. Hence the obtained results confirm the presence of copper ferrite nano-particles doped PANI.

2.1.5. Thermal Analysis

Thermal-gravimetric analysis (TGA) of $CuFe_2O_4$/PANI is presented in Figure S2. It showed an overall weight loss of 35% in the range of 25–800 °C. A weight loss before 100 °C is noticed in the TGA curve due to residual water evaporation. Another weight loss is noticed within the ranging from 310 to 480 °C and 480 to 630 °C due to the thermal degradation of the lower and the higher weight PANI chains, respectively.

2.2. Adsorption Study

Nano-composite particles consisting of $CuFe_2O_4$ and that doped with PANI were prepared and tested as adsorbing substances to remove mercuric ions from aqueous solutions and some industrial waste water.

2.2.1. Effect of Mercury Concentration

The removal efficiency of mercury ions using $CuFe_2O_4$NPs was 82% beginning from 10 up to 120 μg/mL. The adsorption performed at pH value 7 for 30 mL of the adsorbent solution stirred for 120 min. While $CuFe_2O_4$/PANI nano-composites exhibit higher removal efficiency of 99.5% when varying the concentration of Hg^{2+} from 10 to 32 μg/mL. It begins to decrease to 92.3% upon increasing the concentration of mercury up to 120 μg/mL (Figure 4).

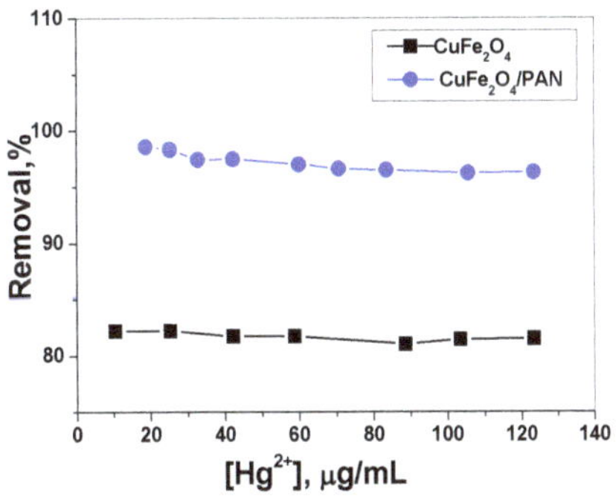

Figure 4. Effect of Hg^{2+} concentration on adsorption optimization using $CuFe_2O_4$ and $CuFe_2O_4$/PANI nano-composites sorbents; Conditions: (V = 30 mL, contact time = 2 h for $CuFe_2O_4$ and 1 h for $CuFe_2O_4$/PANI, adsorbent amount = 0.1 g $CuFe_2O_4$ and 0.2 g $CuFe_2O_4$/PANI and pH 7).

2.2.2. Effect of Contact Time

As shown in Figure 5, the concentration of Hg(II) ions was studied relative to the contact time of each adsorbent. It was found that the time required to obtain more than 80% of Hg(II) removal was 2 h for $CuFe_2O_4$. However, in case of $CuFe_2O_4$/PANI composite, the time required to achieve the equilibrium was one hour with a removal percentage of 99.5%. To examine the adsorption mechanism, kinetics is the vital feature. Pseudo first order and second order models were fitted as the practical kinetics data. The obtained results were presented in Table 2. The adsorption process for both $CuFe_2O_4$ and $CuFe_2O_4$/PANI composite obeyed the second order model.

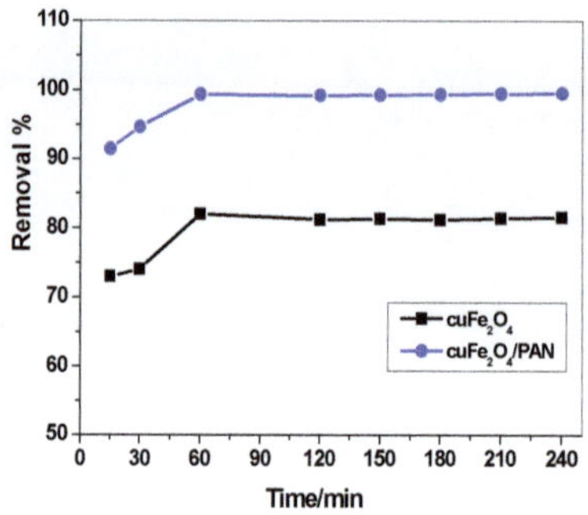

Figure 5. Effect of contact time on adsorption optimization using $CuFe_2O_4$ and $CuFe_2O_4$/PANI nano-composites sorbents (V = 30 mL, Hg^{2+} concentration = 25 µg/mL, adsorbent amount = 0.1 g $CuFe_2O_4$ and 0.2 g $CuFe_2O_4$/PANI and pH 7).

Table 2. Adsorption kinetics parameters.

Adsorbent	Pseudo-First Order			Second Order			
	$k_1 (min^{-1})$	q_{e1} (mg/g)	R^2	k^2 (g/(mg. min))	q_{e2} (mg/g)	q_e^{exp}	R^2
$CuFe_2O_4$	0.0056	1.571	0.942	5.3×10^{-3}	5.8922	7.1086	0.991
$CuFe_2O_4$/PANI	0.0732	2.2134	0.953	0.1121	8.3356	8.4123	0.998

2.2.3. Effect of Adsorbent Amount

To optimize the amount of adsorbent NPs, different amounts from each adsorbent in the range of 0.05 to 0.3 g were put in contact with 30 mL of 25 µg/mL Hg^{2+} solutions of pH 7 and 60 min contact time. As shown in Figure 6, it was observed that the maximum adsorption (i.e., 99.5% removal efficiency) was attained after using 0.2 and 0.1 g for $CuFe_2O_4$NPs and $CuFe_2O_4$/PANI composite, respectively.

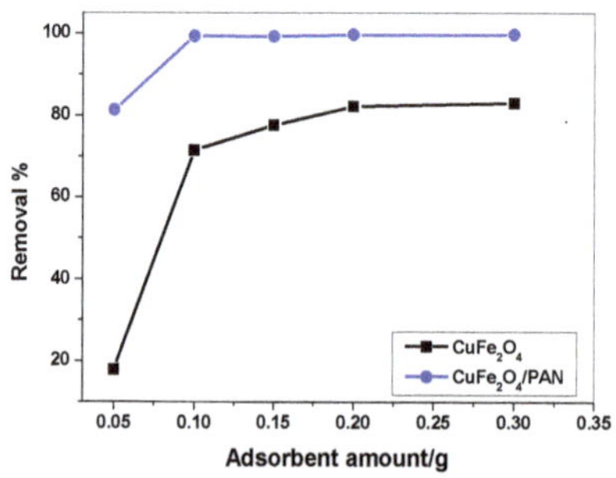

Figure 6. Effect of $CuFe_2O_4$ and $CuFe_2O_4$/PANI nano-composite sorbents amount on adsorption optimization (V = 30 mL, Hg^{2+} concentration = 25 µg/mL, contact time =2 h for $CuFe_2O_4$ and 1 h for $CuFe_2O_4$/PANI and pH 7).

2.2.4. Effect of pH

The pH is an essential parameter for Hg^{2+} adsorption due to its relevance to Hg speciation, as well as the interactions between Hg species and adsorbent surfaces. When the feed water pH was varied from 6.0 to 9.0, Hg^{2+} removal efficiency of $CuFe_2O_4$/polyaniline remained at ~99.5% (Figure 7A). For $CuFe_2O_4$NPs the removal percentage of Hg(II) ions became constant until pH reaches 7. This can be explained that at higher pH values, oxygen-containing groups (e.g., –OH) are ionized to –O–, forming negative charges on the $CuFe_2O_4$ surface.

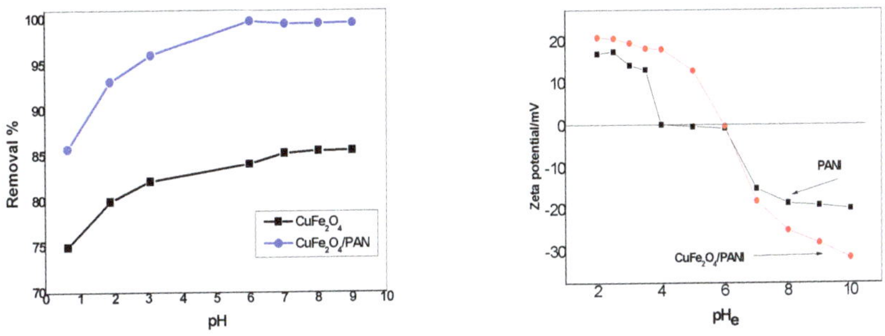

(A) (B)

Figure 7. (**A**) Effect of pH on adsorption optimization using $CuFe_2O_4$ and $CuFe_2O_4$/PANI nano-composite sorbents (V= 30 mL, Hg^{2+} concentration= 25 µg/mL, adsorbent amount = 0.1 g $CuFe_2O_4$ and 0.2 g $CuFe_2O_4$/PANI and contact time = 2 h for $CuFe_2O_4$ and 1 h for $CuFe_2O_4$/PANI). (**B**) Plots of the zeta potential as a function of pH for $CuFe_2O_4$/PANI and PANI.

Based on zeta potential results (Figure 7B),the Point of Zero Charge for both PANI and $CuFe_2O_4$/PANI composite, is around 4–6 and 6, respectively. $CuFe_2O_4$/PANI composite had net negative charges at pH > 6.0 and positive charges at pH < 6.0. At low pH values (e.g., pH < 5.0 for PANI-HCl), nitrogen atoms of imine groups were preferentially bound by protons, causing the PANI surfaces carrying positive charges.

2.3. Adsorption Isotherms

Langmuir (Equation (2), Freundlich (Equation (3), and Temkin (Equation (4) models were applied to calculate the sorption of Hg^{2+} ions for both $CuFe_2O_4$ and $CuFe_2O_4$/PANI nano-composite (Figures 8 and 9).

$$1/Q_t = 1/X_m b\, C_t + 1/X_m \tag{2}$$

$$Log\ Q_t = (1/n)\ log\ C_t + log\ k_F \tag{3}$$

$$Q_t = (RT/B_T)\ ln\ C_t + (RT/B_T)\ ln\ K_T \tag{4}$$

where: Q_t is adsorption capacity at equilibrium (mg/g), C_t is equilibrium concentration of the Hg^{2+} solution (µg/mL), t (min) is contact time, X_m (mg/g) is maximum monolayer adsorption capacity and b (L/mg) is the adsorption equilibrium constant. Relative adsorption capacities and sorption intensities n and K_f (mg/g), and the constants of Freundlich model, were calculated. Temkin constants, B_T (kJ/mol) and K_T (L/mg) whose are constants of heat of sorption and maximum binding energy were estimated. A 30 mL of different mercury concentrations ranging from 10 to 200 µg/mLwere tested under the optimum conditions and the adsorption was expressed by three equilibrium models: Langmuir, Freundlich, and Temkin to illustrate the adsorption capacity and adsorption behavior. The theory of Langmuir assumes that the adsorption occurs by monolayer on the surface of the adsorbent with the same adsorption sites (homogeneous surface), while Freundlich is an empirical theory at which the adsorption occurs by multilayer on the surface of the adsorbent with different adsorption sites (heterogeneous surface). Temkin assumed that there are indirect interactions between adsorbate molecules and the heat of adsorption of all molecules decrease linearly with increasing surface coverage [44]. The results are summarized in Table 3 and confirm the reasonable adsorption capacity of the used nano-composite material and follows Langmuir isotherm model.

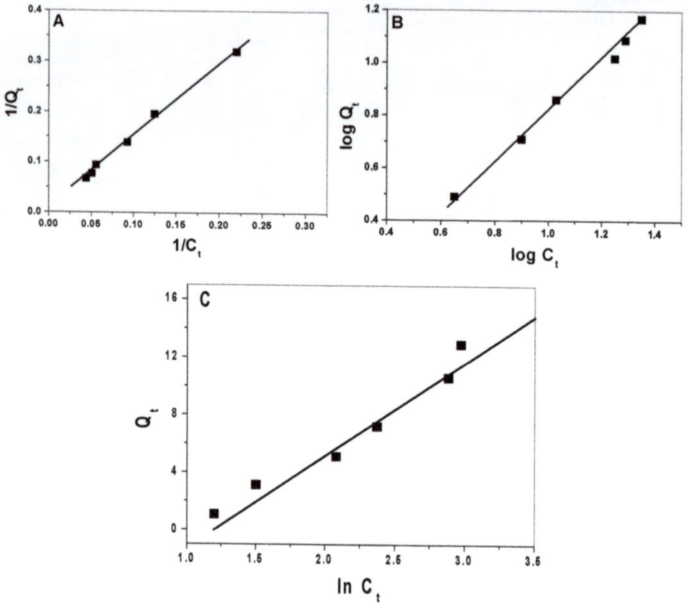

Figure 8. (A) Langmuir, (B) Freundlich, and (C) Temkin isotherms for mercury removal using $CuFe_2O_4$ nano-particles (NPs).

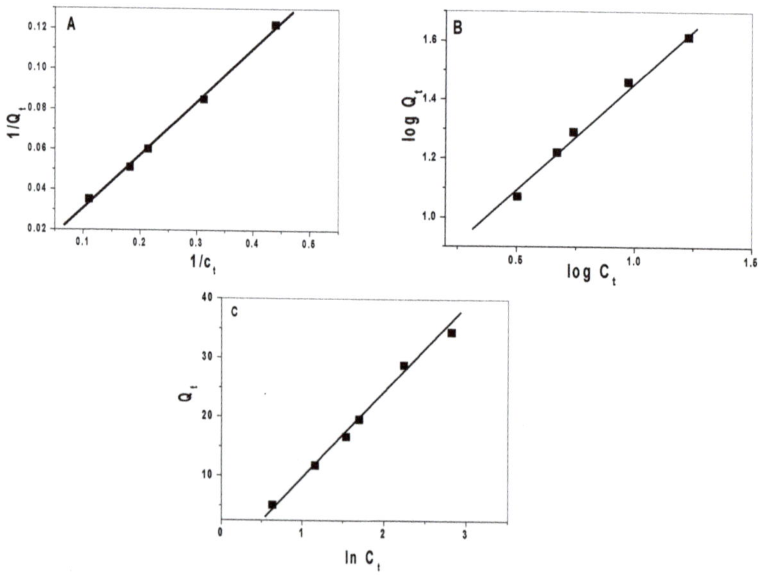

Figure 9. (A) Langmuir, (B) Freundlich, and (C) Temkin isotherms for mercury removal using $CuFe_2O_4$/PANi nano-composite.

Table 3. Isotherm constants for the adsorption of mercury onto $CuFe_2O_4$ and $CuFe_2O_4$/PANI nano-composites.

Model	Parameters	$CuFe_2O_4$	$CuFe_2O_4$/PANI	Unit
Langmuir	X_m	12.5	157.1	mg/g
	B	0.561	0.153	L/mg
	R^2	0.998	0.999	
Freundlich	N	1.06	1.34	mg/g
	K_f	2.75	5.24	mg/g
	R^2	0.997	0.995	
Temkin	K_T	0.34	0.744	L/mg
	b_T	0.371	0.162	K J/mol
	R^2	0.980	0.997	

2.4. Competitive Adsorption of Different Heavy Metals

The adsorption of some metal ions such as Hg^+, Hg^{2+}, Fe^{2+}, Cu^{2+}, Cr^{6+}, Pb^{2+}, and Ag^+ was investigated. $CuFe_2O_4$NPs revealed an affinity order: $Fe^{2+} > Hg^{2+} > Hg^+ \sim Cr^{6+} > Ag^+ >> Pb^{2+}$. No remarkable adsorption for Cu^{2+} ions using $CuFe_2O_4$NPs. For $CuFe_2O_4$/PANI nano-composite, the affinity order was: $Hg^{2+} > Hg^+ > Fe^{2+} \sim Cr^{6+} > Cu^{2+} > Pb^{2+}$. No remarkable adsorption for Ag^+ ions using $CuFe_2O_4$/PANI nano-composite. The removal percentage of the studied ions using $CuFe_2O_4$NPs and $CuFe_2O_4$/PANI nano-composite is shown in Figure 10. From the mentioned results, $CuFe_2O_4$/PANI nano-composite revealed an enhanced removal power towards inorganic mercury than $CuFe_2O_4$NPs only.

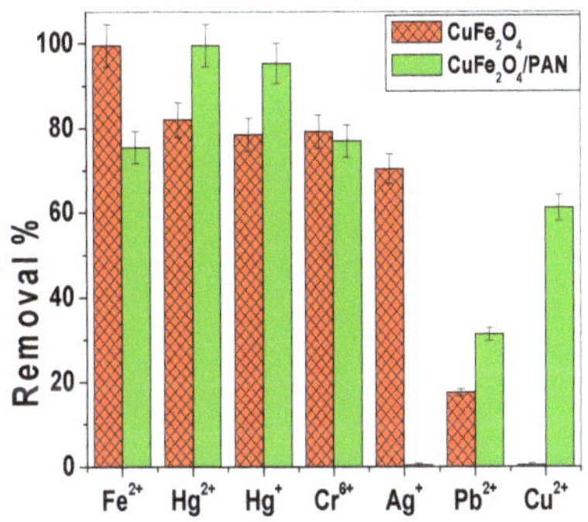

Figure 10. Removal of different metal ions with $CuFe_2O_4$ and $CuFe_2O_4$/PANI nano-composite.

2.5. Regeneration

The adsorbent material was regenerated after each adsorption cycle of mercury by washing with 0.1 M acetic acid. After five cycles of regeneration, the efficiency of $CuFe_2O_4$ NPs for the removal of Hg^+ ions remains 82.0% however, there was a decrease in the removal efficiency of the $CuFe_2O_4$/PANI sorbent reached to 85.3% as shown in Figure 11.

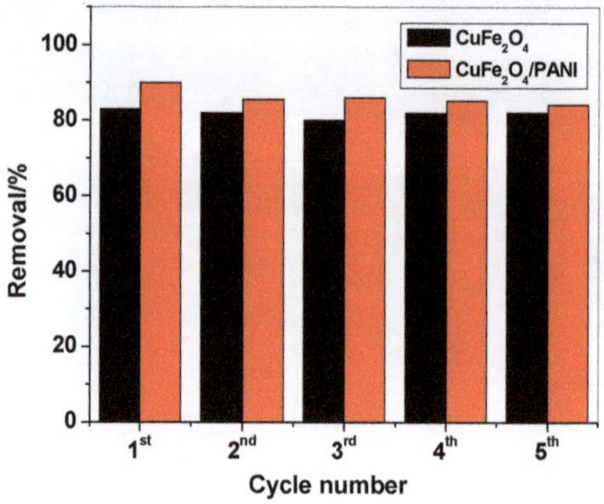

Figure 11. Mercury removal by $CuFe_2O_4$ and $CuFe_2O_4$/PANI nano-composites after regeneration.

2.6. Comparison with Other Sorbents for Mercury Removal

Water pollution becomes a critical issue around the world, and heavy metals contribute to major pollution in water. The application of nano-materials for the removal of mercuric ions from water has attracted significant attention. Table 4 summarizes some of reported sorbents used for mercury removal included the present work. The prepared copper ferrite loaded by polyaniline provides higher maximum adsorption capacity [12,17,18,20,26], short contact time [12,20,23,27], andhigher removal percentage [12,18–21,24–27].

Table 4. Some materials used for the removal of mercuric ion.

Adsorbent Type	Maximum Adsorption Capacity mg/g	Contact Time	Removal %	Ref.
Poly(vinylalcohol)/poly(vinylimidazole) complexing membrane	120	125 min	99.4	[12]
Dithiocarbamate-incorporated mono size polystyrene	33.2	30 min	NR	[17]
Magnetic iron oxide nanoparticles modified with 2-mercaptobenzothiazole	0.59	4 min	98.6	[18]
Thiolated multi-walled carbon nanotubes	204.64	40 min	98	[19]
Amidoamine functionalized multi-walled carbon nanotubes (MWCNT-AA)	101.35	180 min	80	[20]
Mercaptopropyl-coated cobalt ferrite ($CoFe_2O_4$) magnetic nanoparticles	NR	30 min	97	[21]
Poly(aniline-co-5-sulfo-2-anisidine) nanoparticles	2063	48 h	99.8	[23]
Gold Nanoparticle–Aluminum Oxide	676	30 min	>97	[24]
Mercaptoamine-functionalised silica-coated magnetic nanoparticles (MAF-SCMNPs)	355	120 min	NR	[25]
Polyaniline Nanotubes	0.8239	60 min	90	[26]
Iron oxide nanoparticles	NR	24 h	87	[27]
$CuFe_2O_4$	12.5	120 min	82	This work
$CuFe_2O_4$/PAN	157.1	60 min	99.5	

2.7. Mechanism of Adsorption

The adsorption mechanism of Hg^{2+} ions using $CuFe_2O_4$ and $CuFe_2O_4$/PANI composite is shown in Figure 12. The adsorption mechanism can be explained in two ways. Physical adsorption can be occurred on the surface of PANI layer or in the porosity of the adsorbent or chemical adsorption through the interaction between the PANI base layers with mercuric ions. In addition, at the working pH value, oxygen-containing groups (e.g., –OH) in $CuFe_2O_4$ can be ionized to $-O^-$, forming negative charges on the $CuFe_2O_4$ surface and enhance the favorable adsorption of Hg^{2+} ions.

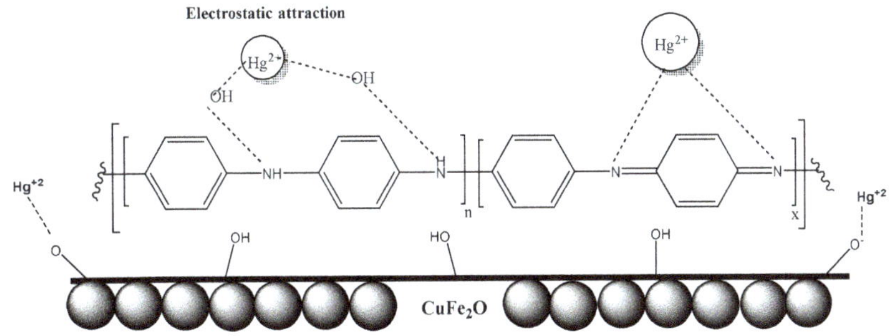

Figure 12. Schematic of Hg^{2+} adsorption mechanism.

3. Materials and Methods

3.1. Materials

For the experimental purpose, all chemicals used were of analytical reagent grade, 98–99%. Metal nitrates, chloride, and sulfate were of the highest purity and supplied by Sigma-Aldrich (St. Louis, MO, USA). Polyethylene glycol 6000 (PEG), potassium hydroxide, and ammonium peroxydisulfate were purchased from Fluka (Ronkonoma, NY, USA). Aniline was purchased from Central Drug House Ltd. (New Delhi, India) and distilled prior to use. All the chemicals were used as received without any further purification.

3.2. Apparatus

High-resolution transmission electron microscopy (HRTEM) images were taken by JEOL-JEM-2100 electron microscope instrument (Osaka, Japan). The prepared adsorbents were characterized by X-ray diffraction (XRD) which were carried out by BRUKER D2 PHASER 2nd generation X-ray diffractometer (Berline, Germany) using CuKα,β radiation (λ = 0.154 nm) in the angular region of 2θ = 4–80°. Operation conditions were 40KV, 40 mA and scanning speed of 8°/min. The Brunauer–Emmett–Teller (BET) surface area measurements were carried out by N_2 adsorption–desorption at 77 K using Nova 3200 s (Florida, FL, USA) unite instrument, in the relative pressure (p/p^0) at 0.25104. Fourier Transform Infrared (ATR-FTIR) was used to obtain the spectra in a spectral range of 4000–500 cm^{-1}. Inductively coupled argon plasma (ICAP 6500 Duo, Thermo Scientific, Abingdon, UK) as used for mercury ion evaluation.

3.3. Preparation of $CuFe_2O_4$Nano-Particles

The synthesis of the nano-particles was done by using the co-precipitation technique [45]. Briefly 11.7 mmol $CuSO_4$ and 14.98 mmol $FeCl_3$ were dissolved in 200 mL 1 wt.% PEG solution. The solution was kept under stirring for about one hour to insure the equilibrium between all the components. To the above mixture, 4M KOH was added drop-wise with vigorous stirring until reaching a pH 9. The mixture was kept under magnetic stirring for another two hours then aged overnight. The precipitate was filtered, washed with distilled water until it was free from Cl^- and SO_4^{2-} ions and dried at 70 °C for two h. The precipitated was then calcined at 600 °C in air for 3 h and then ground using agate motor to obtain a fine powder.

3.4. Preparation of $CuFe_2O_4$/PANI Nano-Composite

The polyaniline copper ferrite nano-composite was prepared using chemical polymerization method by dispersing 2 g of the previously prepared $CuFe_2O_4$ nano-particles in 200 mL of 2M HCl and stirred vigorously at room temperature for 10 min. A 4.5 mL aliquot of distilled aniline monomer was added under continuous stirring for 30 min. To the above suspension, 20 mL of 19.7 mmol $(NH_4)_2S_2O_8$ solution was added drop-wisely, as a polymerization initiator. An immediate color change of the solution to blue green was observed. The suspension was stirred to complete the polymerization process for about 1 h. The copper ferrite doped PANI was separated on a filter paper, rinsed with distilled water, and finally dried at 100°C in an electrical oven. The produced powder has a green color which represents emeraldine salt of polyaniline.

3.5. Removal of Mercury from Waste Water

A range of mercury (II) ion (10–120 µg/mL) was prepared. For the adsorption studies different amounts of either $CuFe_2O_4$ or $CuFe_2O_4$/PANI nano-composite (ranged from 0.05 to 0.3 g) were added to 30 mL of the prepared solution at room temperature and pH 7. These solutions were stirred for a contact time varied from 15 min to 2 h. After adsorption, the solutions were filtered and the adsorbent material was separated. The concentration of Hg^{2+} ion was evaluated before and after the removal of mercury by inductively coupled argon plasma.

The removal percentage of mercury was calculated using the following equation:

$$Removal\% = ((C_0 - C_t))/C_0 \times 100 \tag{5}$$

where, C_0 and C_t are the mercury concentration in µg/mL at initial and after time t, respectively.

4. Conclusions

$CuFe_2O_4$/PANI nano-composite was successfully prepared and its adsorption properties towards Hg^{2+} ions removal were checked. An X-ray diffractometer, TEM, and BET were used to characterize the prepared nano-composites. The crystallite size of the synthesized $CuFe_2O_4$ and $CuFe_2O_4$/PANI nano-composite was 10.2 and 23.4 nm, respectively. Under the optimum conditions, $CuFe_2O_4$/PANI offer higher removal efficiency than $CuFe_2O_4$ for Hg^+/ Hg^{+2} ions which were 95.3 and 99.5%, respectively. Both adsorbents followed the second order model and Langmuir model with adsorption capacity of 12.5 and 157.1 mg/g for $CuFe_2O_4$ and $CuFe_2O_4$/PANI composite, respectively. After five cycles of regeneration, the efficiency of $CuFe_2O_4$NPs for the removal of Hg^+ ions remains 82.0% however, there was a decrease in the removal efficiency of the $CuFe_2O_4$/PANI sorbent reached to 85.3%with lower efficiency and good performance when used again after five cycles. These materials were successfully applied for the removal of Hg^{2+} ions with a high efficiency over other studied heavy metals.

Supplementary Materials: The following are available online, Figure S1: FTIR spectra of (A) $CuFe_2O_4$ and (B) $CuFe_2O_4$/PANI nanocomposites; Figure S2: Thermal-gravimetric analysis (TGA) of $CuFe_2O_4$/PANI.

Author Contributions: The listed authors contributed to this work as described in the following: H.A.E.-N., A.A.H., S.S.M.H., and A.H.K. gave the concepts of the work, interpretation of the results, the experimental part and prepared the manuscript; A.H.K., S.S.M.H., and A.E.-G.E.A. cooperated in the preparation of the manuscript and A.H.K. and S.S.M.H. performed the revision before submission. A.E.-G.E.A., M.A.A.-O., A.Y.A.S. revealed the financial support for the work. All authors have read and agreed to the published version of the manuscript.

Funding: This research was funded by King Saud University, Project No. RSP-2020/66.

Acknowledgments: Authors are grateful to King Saud University for funding the work through Researchers Supporting Project (Project No. RSP-2020/66).

Conflicts of Interest: The authors declare no conflict of interest.

References

1. Mahmud, H.N.M.E.; Huq, A.K.O.; Yahya, R.B. The removal of heavy metal ions from wastewater/aqueous solution using polypyrrole-based adsorbents: A review. *RSC Adv.* **2016**, *6*, 14778–14791. [CrossRef]
2. Gautam, R.K.; Sharma, S.K.; Mahiya, S.; Chattopadhyaya, M.C. Contamination of heavy metals in aquatic media: Transport, toxicity and technologies for remediation. In *Heavy Metals in Water: Presence, Removal and Safety*; RSC Publishing: Cambridge, UK, 2014.
3. Wang, J.; Feng, X.; Anderson, C.W.N.; Xing, Y.; Shang, L. Remediation of mercury contaminated sites—A review. *J. Hazard. Mater.* **2012**, *221–222*, 1–18. [CrossRef] [PubMed]

4. Basha, S.; Murthy, Z.V.P.; Jha, B. Sorption of Hg(II) onto *Carica papaya*: Experimental studies and design of batch sorber. *Chem. Eng. J.* **2009**, *147*, 226–234. [CrossRef]
5. Fu, X.; Feng, X.; Sommar, J.; Wang, S. A review of studies on atmospheric mercury in China. *Sci. Total Environ.* **2012**, *421–422*, 73–81. [CrossRef] [PubMed]
6. Agency for Toxic Substances and Disease Registry (ATSDR). *CERCLA Priority List of Hazardous Substances*; US Department of Health and Human Services, Public Health Service: Atlanta, GA, USA, 2001.
7. Caner, N.; Sarı, A.; Tuzen, M. Adsorption characteristics of mercury(II) ions from aqueous solution onto chitosan-coated diatomite. *Ind. Eng. Chem. Res.* **2015**, *54*, 7524–7533. [CrossRef]
8. Saleh, T.A. Isotherm, kinetic, and thermodynamic studies on Hg(II) adsorption from aqueous solution by silica-multiwall carbon nanotubes. *Environ. Sci. Pollut. Res.* **2015**, *22*, 16721–16731. [CrossRef]
9. Danmaliki, G.I.; Saleh, T.A.; Shamsuddeen, A.A. Response surface methodology optimization of adsorptive desulfurization on nickel/activated carbon. *Chem. Eng. J.* **2017**, *313*, 993–1003. [CrossRef]
10. Danmaliki, G.I.; Saleh, T.A. Effects of bimetallic Ce/Fe nanoparticles on the desulfurization of thiophenes using activated carbon. *Chem. Eng. J.* **2017**, *307*, 914–927. [CrossRef]
11. Ekinci, E.; Budinova, T.; Yardim, F.; Petrov, N.; Razvigorova, M.; Minkova, V. Removal of mercury ion from aqueous solution by activated carbons obtained from biomass and coals. *Fuel Proce. Technol.* **2002**, *77–78*, 437–443. [CrossRef]
12. Bessboussea, H.; Rhlalou, T.; Verchère, J.F.; Lebrun, L. Mercury removal from wastewater using a poly(vinylalcohol)/poly(vinylimidazole) complexing membrane. *Chem. Eng. J.* **2010**, *164*, 37–48. [CrossRef]
13. Chiarle, S.; Ratto, M.; Rovatti, M. Mercury Removal from Water by Ion Exchange Resins Adsorption. *Water Res.* **2000**, *34*, 2971–2978. [CrossRef]
14. Oehmen, A.; Vergel, D.; Fradinho, J.; Reis, M.A.M.; Crespo, J.G.; Velizarov, S. Mercury removal from water streams through the ion exchange membrane bioreactor concept. *J. Hazar. Mat.* **2014**, *264*, 65–70. [CrossRef] [PubMed]
15. Li, Y.; Li, W.; Liu, Q.; Meng, H.; Lu, Y.; Li, C. Alkynyl carbon materials as novel and efficient sorbents for the adsorption of mercury(II) from wastewater. *Environ. Sci.* **2018**, *68*, 169–176. [CrossRef] [PubMed]
16. Hadi, P.; To, M.; Hui, C.; Lin, C.S.K.; McKay, G. Aqueous Mercury Adsorption by Activated Carbons. *Water Res.* **2015**, *15*, 37–55. [CrossRef] [PubMed]
17. Denizli, A.; Kesenci, K.; Arica, Y.; Piskin, E. Dithiocarbamate-incorporated monosize polystyrene microspheres for selective removal of mercury ions. *Reactive Funct. Poly.* **2000**, *44*, 235–243. [CrossRef]
18. Mahmoud, M.E.; Hassan, S.S.M.; Kamel, A.H.; Elserw, M.I.A. Development of microwave-assisted functionalized nanosilicas for instantaneous removal of heavy metals. *Powder Technol.* **2018**, *326*, 454–466. [CrossRef]
19. Hadavifar, M.; Bahramifar, N.; Younesi, H.; Rastakhiz, M.; Li, Q.; Yu, J.; Eftekhari, E. Removal of mercury(II) and cadmium(II) ions from synthetic wastewater by a newly synthesized amino and thiolated multi-walled carbon nanotubes. *Taiwan Inst. Chem. Eng.* **2016**, *67*, 397–405. [CrossRef]
20. Deb, A.S.; Dwivedi, V.; Dasgupta, K.; Ali, S.M.; Shenoy, K.T. Novel Amidoamine Functionalized Multi-Walled Carbon Nanotubes for removal of Mercury(II) Ions from Wastewater: Combined Experimental and Density Functional Theoretical. *Chem. Eng. J.* **2017**, *313*, 899–911. [CrossRef]
21. Viltužnik, B.; Lobnik, A.; Košak, A. The removal of Hg(II) ions from aqueous solutions by using thiol-functionalized cobalt ferrite magnetic nanoparticles. *Sol Gel Sci. Technol.* **2015**, *74*, 199–207. [CrossRef]
22. Vélez, E.; Campillo, G.E.; Morales, G.; Hincapié, C.; Osorio, J.; Arnache, O.; Uribe, J.I.; Jaramillo, F. Mercury removal in wastewater by iron oxide nanoparticles. *J. Phys. Conf. Ser.* **2016**, *687*, 012050. [CrossRef]
23. Li, X.; Feng, H.; Huang, M. Strong Adsorbability of Mercury Ions on Aniline/Sulfoanisidine Copolymer Nanosorbents. *Chem. A Eur. J.* **2009**, *15*, 4573–4581. [CrossRef]
24. Lo, S.; Chen, P.; Huang, C.; Chang, H. Gold Nanoparticle–Aluminum Oxide Adsorbent for Efficient Removal of Mercury Species from Natural Waters. *Environ. Sci. Technol.* **2012**, *46*, 2724–2730. [CrossRef] [PubMed]
25. Baoa, S.; Lia, K.; Ninga, P.; Peng, J.; Jina, X.; Tang, L. Highly effective removal of mercury and lead ions from wastewater by mercaptoamine-functionalised silica-coated magnetic nano-adsorbents: Behaviours and mechanisms. *Appl. Surf. Sci.* **2017**, *393*, 457–466. [CrossRef]
26. Morsi, R.E.; Elsabee, M.Z. Polyaniline Nanotubes: Mercury and Competative Heavy Metals Uptake. *Am. J. Polym. Sci.* **2015**, *5*, 10–17.

27. Parham, H.; Zargar, B.; Shiralipour, R. Hazardous Materials, Fast and efficient removal of mercury from water samples using magnetic iron oxide nanoparticles modified with 2-mercaptobenzothiazole. *J. Hazard Mat.* **2012**, *205–206*, 94–100. [CrossRef] [PubMed]
28. Homaeigohar, S. The Nanosized Dye Adsorbents for Water Treatment. *Nanomaterials* **2020**, *10*, 295. [CrossRef]
29. Homaeigohar, S.; Botcha, N.K.; Zarie, E.S.; Elbahri, M. Ups and downs of water photodecolorization by nanocomposite polymer nanofibers. *Nanomaterials* **2019**, *9*, 250. [CrossRef]
30. Homaeigohar, S.; Elbahri, M. An amphiphilic, graphitic buckypaper capturing enzyme biomolecules from water. *Water* **2019**, *11*, 2. [CrossRef]
31. Mahmoodi, N.M. Photocatalyticozonation of dyes using copper ferrite nanoparticle prepared by co-precipitation method. *Desalination* **2011**, *279*, 332–337. [CrossRef]
32. Sun, J.; Xu, R.; Zhang, Y.; Ma, M.; Gu, N. Magnetic nanoparticles separation based on nanostructures. *J. Magn. Magn. Mat.* **2007**, *312*, 354–358. [CrossRef]
33. Ngomsik, A.F.; Bee, A.; Draye, M.; Cote, G.; Cabuil, V. Magnetic nano- and microparticles for metal removal and environmental applications: A review. *Comptes Rendus Chimie.* **2005**, *8*, 963–970. [CrossRef]
34. Song, Y.; Lu, M.; Huang, B.; Wang, D.; Wang, G.; Zhou, L. Decoration of defective MoS_2 nanosheets with Fe_3O_4nano particles as superior magnetic adsorbent for highly selective and efficient mercury ions (Hg^{2+}) removal. *J. Alloys Comp.* **2018**, *737*, 113–121. [CrossRef]
35. Ngwenya, S.; Guyo, U.; Zinyama, N.P.; Chigondo, F.; Nyamunda, B.C.; Muchanyereyi, N. Response surface methodology for optimization of Cd (II) adsorption from wastewaters by fabricated tartaric acid-maize tassel magnetic hybrid sorbent. *Bioint. Res. in Appl. Chem.* **2019**, *9*, 3996–4005.
36. Sarma, M.K.; Abdul Quadir, M.G.; Bhaduri, R.; Kaushik, S.; Goswami, P. Composite polymer coated magnetic nanoparticles based anode enhances dye degradation and power production in microbial fuel cells. *Biosens.Bioelectr.* **2018**, *119*, 94–102. [CrossRef] [PubMed]
37. Gonzalez-Casamachin, D.A.; De la Rosa, J.R.; Lucio–Ortiz, C.J.; Sandoval-Rangel, L.; García, C.D. Partial oxidation of 5-hydroxymethylfurfural to 2,5-furandicarboxylic acid using O_2 and a photocatalyst of a composite of ZnO/PPy under visible-light: Electrochemical characterization and kinetic analysis. *Chem. Eng. J.* **2020**, *393*, 124699. [CrossRef]
38. Kim, J.; Wainaina, J.; Na, J.S. Synthesis of amphiphilic silica/polymer composite nanoparticles as water-dispersible nano-absorbent for hydrophobic pollutants. *J. Indust. Eng. Chem.* **2011**, *17*, 681–690. [CrossRef]
39. Wan, M.X.; Zhou, W.X. Studies on magnetic properties of film of polyaniline. *Acta Phys. Sin.* **1992**, *41*, 347–352.
40. Wu, K.H.; Chao, C.M.; Liu, C.H.; Chang, T.C. Characterization and corrosion resistance of organically modified silicate–NiZn ferrite/polyaniline hybrid coatings on aluminum alloys. *Corrosion Sci.* **2007**, *49*, 3001–3014. [CrossRef]
41. Pouget, J.P.; Jozefowicz, M.E.; Epstein, A.J.; Tang, X.; Macdiarmid, A.G. X-ray structure of polyaniline. *Macromolecules* **1991**, *24*, 779–789. [CrossRef]
42. Zhou, G.; Guo, K. *Diffraction of Crystal and Pseudo Crystal*; Beijing Universities Public House: Beijing, China, 1999.
43. Gregg, S.J.; Sing, K.S.W. *Adsorption, Surface Area and Porosity*, 2nd ed.; Academic Press: London, UK, 1982.
44. Dada, A.O.; Olalekan, A.P.; Olatunya, A.M.; Dada, O. Langmuir, Freundlich, Temkin and Dubinin-Radushkevich Isotherms Studies of Equilibrium Sorption of Zn^{2+} unto Phosphoric Acid Modified Rice Husk. *IOSR J. Appl. Chem. (IOSR-JAC)* **2012**, *3*, 38–45.
45. Balaji, M.; Raja, M.M.; Asokan, K.; Kanjilal, D.; Rajasekaran, T.R.; Padiyan, D.P. Effect of thermal spike energy created in CuFe2O4 by 150 MeV Ni11+ swift heavy ion irradiation. *Nucl. Instrum. Methods Phys. Res. Section B Beam Interact. Mater. Atoms* **2011**, *269*, 1088–1093. [CrossRef]

Sample Availability: Samples of the compounds are available from the authors.

molecules

MDPI

Article

Porous Activated Carbon from Lignocellulosic Agricultural Waste for the Removal of Acetampirid Pesticide from Aqueous Solutions

Somaia G. Mohammad [1], Sahar M. Ahmed [2], Abd El-Galil E Amr [3,4,*] and Ayman H. Kamel [5,*]

1 Agriculture Research Center (ARC), Central Agricultural Pesticides Laboratory, Pesticide Residues and Environmental Pollution Department, Dokki, Giza 12618, Egypt; sommohammad2015@gmail.com

2 Petrochemical Department, Egyptian Petroleum Research Institute, Ahmed El-Zomor St., Nasr City, Cairo 11727, Egypt; saharahmed92@hotmail.com

3 Pharmaceutical Chemistry Department, Drug Exploration & Development Chair (DEDC), College of Pharmacy, King Saud University, Riyadh 11451, Saudi Arabia

4 Applied Organic Chemistry Department, National Research Center, Dokki, Giza 12622, Egypt

5 Chemistry Department, Faculty of Science, Ain Shams University, Abbasia, Cairo 11566, Egypt

* Correspondence: aamr@ksu.edu.sa (A.E.-G.E.A.); ahkamel76@sci.asu.edu.eg (A.H.K.); Tel.: +966-565-148-750 (A.E.-G.E.A.); +201-000-361-328 (A.H.K.)

Academic Editor: Chiara Bisio

Received: 25 April 2020; Accepted: 13 May 2020; Published: 17 May 2020

check for updates

Abstract: A facile eco-friendly approach for acetampirid pesticide removal is presented. The method is based on the use of micro- and mesoporous activated carbon (TPAC) as a natural adsorbent. TPAC was synthesized via chemical treatment of tangerine peels with phosphoric acid. The prepared activated carbon was characterized before and after the adsorption process using Fourier- transform infrared (FTIR), X-ray diffraction (XRD), particle size and surface area. The effects of various parameters on the adsorption of acetampirid including adsorbent dose (0.02–0.2 g), pH 2–8, initial adsorbate concentration (10–100 mg/L), contact time (10–300 min) and temperature (25–50 °C) were studied. Batch adsorption features were evaluated using Langmuir and Freundlich isotherms. The adsorption process followed the Langmuir isotherm model with a maximum adsorption capacity of 35.7 mg/g and an equilibration time within 240 min. The adsorption kinetics of acetamiprid was fitted to the pseudo-second-order kinetics model. From the thermodynamics perspective, the adsorption was found to be exothermic and spontaneous in nature. TPAC was successfully regenerated and reused for three consecutive cycles. The results of the presented study show that TPAC may be used as an effective eco-friendly, low cost and highly efficient adsorbent for the removal of acetamiprid pesticides from aqueous solutions.

Keywords: green-removal; tangerine peels activated carbon; agriculture waste; acetamiprid pesticide

1. Introduction

Pesticides are carcinogens in nature, and because of their toxicity, they pollute both land and water environments. Pesticide detection in groundwater has increased interest in finding suitable techniques used for their removal from aqueous solution to minimize their concentrations to the permissible level [1].

Acetamiprid (Figure 1) belongs to a new neonicotinoid class of systemic broad-spectrum insecticides. Due to the relatively low and chronic mammalian toxicity, and no long-term cumulative toxicity, it has been used to take the place of organophosphorus and other conventional insecticides for controlling the insects such as Hemiptera, Thysanoptera and Lepidoptera in agricultural products [2].

It is also used to protect crops, especially vegetables, fruits from different insect pests, such as aphids and thrips.

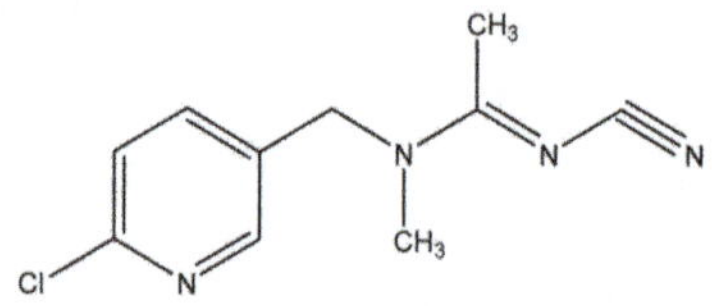

Figure 1. Chemical structure of acetamiprid.

Acetamiprid is widely used due to its high water-solubility, low toxicity for humans and high activity against insects [3]. It is susceptible to be found in water according to the United States Environmental Protection Agency (EPA) due to its high solubility [4]. It contaminates various matrices (water, soil, plants and aquatic species) [5]. In addition, it revealed potential health risk to human beings who are exposed to the primary route of food and water polluted by acetamiprid. However, scientific literature regarding the removal of acetamiprid by means of non-conventional treatment technologies is still incomplete [6].

There are many different methods used for the removal of pesticides from aqueous solutions such as oxidation, photocatalysis, electrochemical and adsorption [7]. The biosorption process is a physicochemical and alternative phenomenon and has many benefits such as abundant in nature, cheap, simplicity, efficient and recycling [6,7]. The mechanism of adsorption involved adsorption via ion exchange, chelation and complexation. It depends on different factors such as the chemical nature of the pollutants, type of the biosorbents, temperature, pH and ionic strength [8].

Different methods were reported for acetamiprid removal from aqueous solutions such as advanced oxidation processes (AOPs) Fenton-based treatments [9,10], heterogeneous photocatalysis [11,12], static sorption based on different clays [13] and other related technologies like the innovative low temperature plasma [14] have been published in the last few years. All of these methods demonstrated their potential towards the efficient removal of acetamiprid from different water matrices.

Activated carbon is considered as an effective adsorbent for removing organic contaminants from aqueous media because of its large specific surface area, excellent stability, environment-friendly, favorable physical/chemical surface characteristics, easy preparation, management and high removal efficiency [15,16]. Activated carbon has been considered to be a promising and effective sorbent for removing of toxic metals [17] and pesticides such as oxamyl [18,19], butachlor [20], diazinon [21] and carbofuran [22] from aqueous solution. However, limited works have been found for the removal of acetamiprid insecticide by agricultural waste [23].

The adsorption efficiency of carbon materials can be enhanced through chemical activation using phosphoric acid. The use of this acid can increase the carbon content to (35%–50%) in the adsorbent. In addition, it is easy to be recovered and has less environmental impacts from its use. The role of the phosphoric acid in converting the precursor to the activated carbon is promoting the decomposition of the precursor and forming the cross-linked structure at low temperatures that make pores more open and then enlarging the surface area [17].

In the present study, porous activated carbon from tangerine peel waste was introduced as an effective biosorbent for acetamiprid pesticide removal from aqueous media. Phosphoric acid was used for the chemical activation of these agriculture wastes. The prepared activated carbon was characterized by FTIR, surface area, XRD and particle size. Different parameters were used in this study through batch adsorption such as different mass, pH, initial concentration of acetamiprid pesticide and contact time. The removal capacity was studied by fitting the adsorption data for isotherm, kinetics and thermodynamics.

2. Results and Discussion

2.1. Characterization of TPAC

2.1.1. FTIR Analysis

The FT-IR spectrum of acetamiprid is shown in Figure 2a. Stretching vibrations of both –CN and C=N groups were observed at 2175 and 1567 cm^{-1}, respectively. Sharp stretching vibrations of C–H bond were observed at 2600–3000 cm^{-1}. Olefinic =CH- groups in the pyridine group were observed at 1495 cm^{-1} and 1463 cm^{-1}, respectively [24]. Stretching C–Cl bond is located at 1099 cm^{-1}. Typical stretching vibrations for N–H and C–N bonds are located at 3300–3487 cm^{-1} and 1341 cm^{-1}, respectively. The FTIR spectra of the TPAC before and after adsorption of acetamiprid were shown in Figure 2b,c. The data showed a broad band at 3423 cm^{-1} which is assigned to stretching vibration of O–H bond. The assigned C–H peak of acetamiprid appeared as a broad weak peak at 3028 cm^{-1}. The C=N and olefinic =CH- groups in acetamiprid appeared at 1567 and 1495 cm^{-1} are now located in Figure 2c at 1535 and 1506 cm^{-1}, respectively. The stretching C–Cl bond which is located at 1099 cm^{-1} is now appeared as weak peak at 1078 cm^{-1}. The band at 1691 cm^{-1} is due to the amide group. Stretching vibration of C=O bond in the carboxylic group is observed at 1742 cm^{-1}. The difference in spectra of TPAC before and after adsorption of acetamiprid is attributed to the interaction between the hydroxyl, amine and carboxylic groups of TPAC and the acetamiprid pesticide.

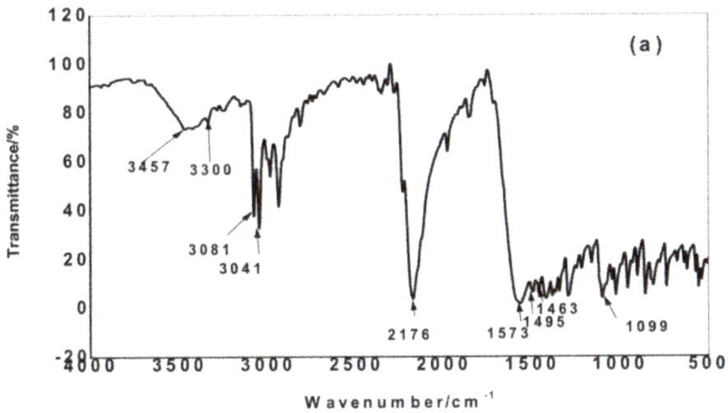

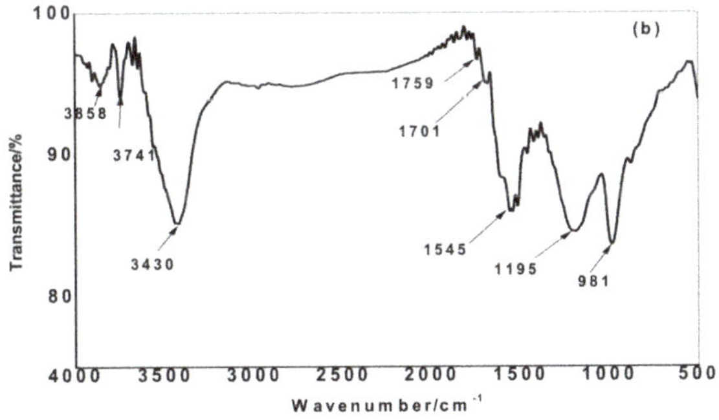

Figure 2. *Cont.*

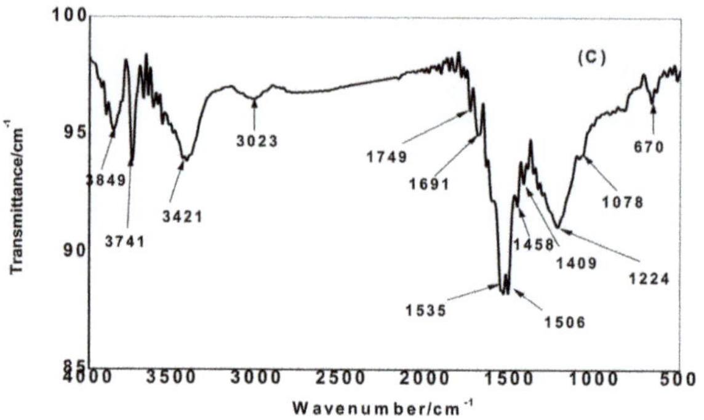

Figure 2. Fourier- transform infrared (FT-IR) of micro and mesoporous activated carbon (TPAC) (**a**) acetamiprid (**b**) before, (**c**) after the adsorption of acetamiprid.

2.1.2. Specific Surface Area and Particle Size

The N_2 adsorption/desorption isotherms and pore size distribution curves were evaluated to characterize the behavior of the prepared TPAC before and after adsorption of acetamiprid (Figure 3). The obtained adsorption isotherm of TPAC sample can be deemed as a combination of type I and IV isotherm curves. As shown in Figure 3a, the Type I isotherm is concave to the P/P^0 axis and the amount adsorbed approaches a limiting value. This limiting uptake is governed by the accessible micropore volume rather than by the internal surface area. A steep uptake at very low P/P^0 is due to enhanced adsorbent–adsorptive interactions in narrow micropores (micropores of molecular dimensions), resulting in micropore filling at very low P/P^0. Type IV isotherms are given by mesoporous adsorbents. The adsorption behavior in mesopores is determined by the adsorbent–adsorptive interactions and also by the interactions between the molecules in the condensed state. In this case, the initial monolayer–multilayer adsorption on the mesopore walls [25]. According to *t*-plot method, the textural features of TPAC before and after acetamiprid adsorption were presented in Table 1. From the date obtained, the structure of TPAC was a combination of meso- and micropores with a larger proportion of mesoporous [26]. It is observed that the total surface area and the total pore volume of TPAC after the adsorption of the pesticide decreased from 687.8 to 296.4 m^2/g and from 0.64 to 0.38 cc/g, respectively. This indicates that the free pores are congested by the pesticide molecules and indicates the existence of both micro- and meso-porous in the TPAC structure.

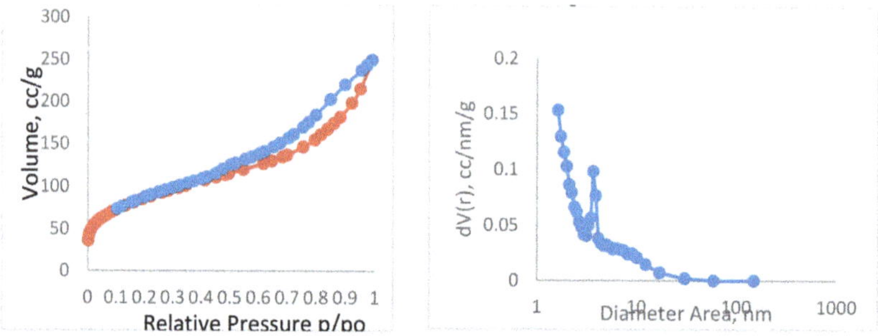

(a)

Figure 3. *Cont.*

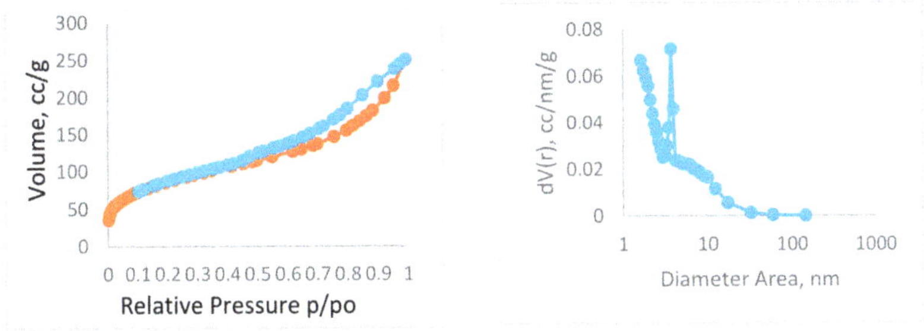

(b)

Figure 3. (**a**) Nitrogen adsorption–desorption isotherms and the corresponding pore size distribution curve of the prepared TPAC; (**b**) nitrogen adsorption–desorption isotherms and the corresponding pore size distribution curve of the prepared TPAC after adsorption of acetamiprid.

Table 1. Textural features and elemental analysis of TPAC before (A) and after (B) acetamiprid adsorption.

Samples	Mesopore Size, nm	Micropore Size, nm	Mesopore Volume, cc/g	Micropore Volume, cc/g	Total Pore Volume, cc/g	Mesopore Area, m^2/g	Micropore Area, m^2/g	Total Surface Area, m^2/g
TPAC (A)	3.6	1.61	0.47	0.17	0.64	385.9	301.9	687.8
TPAC (B)	3.7	1.6	0.35	0.03	0.38	196.4	99.9	296.4

Dynamic light scattering (DLS) is used to determine the particle size distribution of the prepared TPAC as shown in Figure 4. The data showed that about 62.4% of the particles have sizes between 80 and 205 nm with a mean size of 142.5 nm. It is also showed that about 25% of the particles have sizes lower than or equal to 100 nm; 4.2% of the particles have sizes between 295 and 1055 nm, and less than 0.5% have sizes greater than 1000 nm.

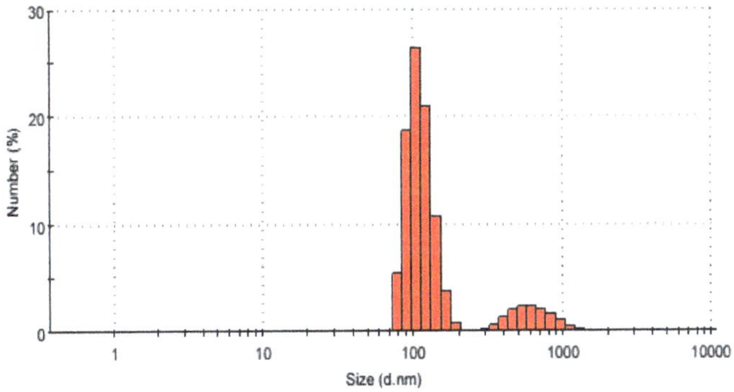

Figure 4. The Particle size of TPAC.

2.1.3. XRD Analysis

The crystal structure and the phase purity of the prepared TPAC were examined by XRD analysis. The XRD pattern for TPAC is shown in Figure 5. It exhibited one broad diffraction peak at 2θ = 25.18 (d002) and a small peak at 45.85 (d100) which are attributed to the presence of graphite crystallite in TPAC. The diffraction patterns agree well with the standard JCPDS for the graphite phase with the hexagonal structure (JCPDS no. 7440-44-0).

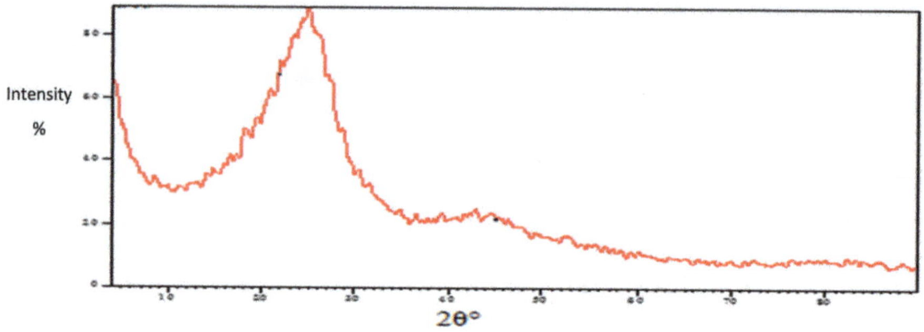

Figure 5. X-ray diffraction (XRD) of TPAC.

2.2. *Effect of Different Factors on the Adsorption Process*

The effect of different parameters on the removal of acetamiprid pesticide by TPAC such as adsorbent dose, pH and initial concentrations of acetamiprid and contact time was investigated. Adsorbent dosage is an important parameter in the adsorption process because it determines the sorbent–sorbate equilibrium of the system and the capacity of the adsorbent for a given concentration [27]. The effect of adsorbent dose on the removal of acetamiprid pesticide by TPAC was studied at different adsorbent doses varying from 0.02–0.2 g/100 mL as shown in Figure 6a. As the adsorbent dose of TPAC increases, the percentage removal of acetamiprid increased from 27.8% to 92.5%. This can be attributed to the increase in the number of active sites [28]. At higher adsorbent dosage, there is incomplete usage of the adsorption sites due to the formation of aggregates. In contrast to the percentage removal, the adsorption capacity of acetamiprid by TPAC was declined in the range of 34.7–11.5 mg/g with increasing adsorbent dose. It was noted that after an adsorbent dosage of 0.1 g, the adsorption efficiency did not show remarkable increase that indicates the saturation of the adsorption sites. An adsorbent dose of 0.1 g seems to be optimum for the removal of acetamiprid experiments by TPAC.

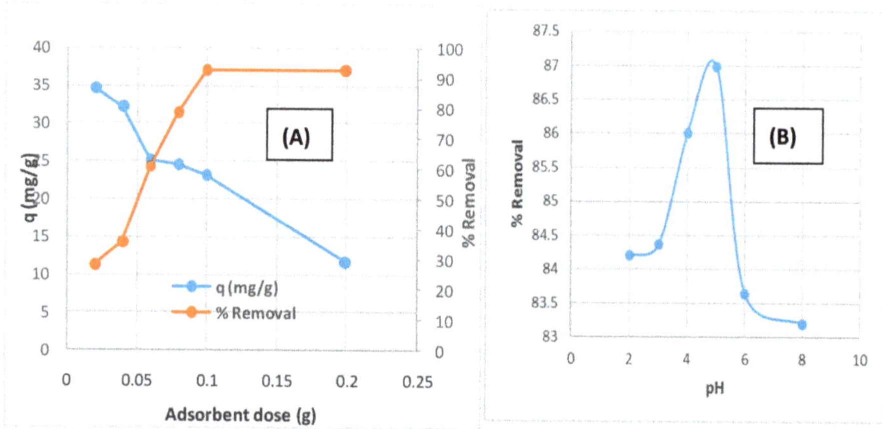

Figure 6. *Cont.*

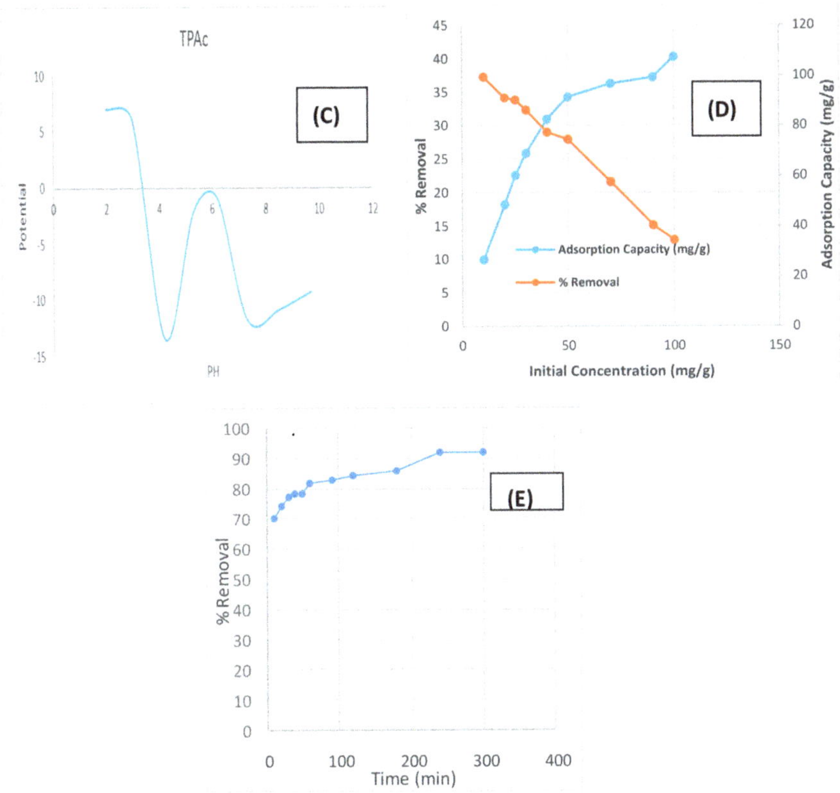

Figure 6. Effect of (**A**) adsorbent dose (initial concentration 25 mg/L, stirring 200 rpm and contact time 5 h); (**B**) pH (initial concentration 25 ppm, mass 0.1 g and contact time 5 h); (**C**) zeta potential of TPAC; (**D**) initial concentration of acetamiprid (mass 0.1 g and contact time 5 h); and (**E**) contact time (mass 0.1 g, stirring 200 rpm and initial concentration 25 ppm) on the removal of acetamiprid by TPAC.

The effect of pH on the removal of acetamiprid pesticide by TPAC was studied at the pH range from 2–8 at initial concentration of 25 μg/mL and adsorbent dose 0.1 g. The effect of pH on the removal of acetamiprid onto TPAC from aqueous solution at different pH values (2–8) and zeta potential of TPAC are shown in Figure 6b,c. The removal percentage of acetamiprid reaches its maximum value (86.99%) at pH 5.0. At pH > 5, the percentage removal decreases and reaches 83.2% at pH 8. These results indicated that pH 5 is suitable for the maximum removal of acetamiprid than the higher pH values. Figure 6c illustrates that PZC for TPAC was found to be 6.0 and its surface carries a negative charge at pH 5. Hence, pH 5 was chosen for all subsequent removal steps of positively charged acetamiprid molecules. The maximum sorption is attributed to the strong van der Waals attraction between the surface groups and pesticide molecules [29].

The effect of the initial concentration of acetamiprid was carried out by increasing acetamiprid concentration from 10–100 μg/mL. As shown in Figure 6D, the adsorption capacity increased from 9.94 to 36.31 mg/g and the removal efficiency decreased from 99.37% to 40.275% upon changing the initial concentration of the pesticide from 10 to 100 mg/L. These results showed that the adsorption of acetamiprid pesticide by TPAC depends on the initial concentration. Similar trends have been reported by the previous studies using activated carbon from pumpkin seed hulls for the removal of 2,4-dichlorophenoxyacetic acid [30,31].

Contact time is an important parameter for the removal of acetamiprid by TPAC. As shown in Figure 6E, the adsorption process of acetamiprid by TPAC was rapid in the first 10 min with percentage removal of 70.1% and reaches its maximum value (91%) after 240 min. The adsorption of acetamiprid attained equilibrium at 240 min due to the fully occupied of the available adsorption sites on the surface of the adsorbent. Further increase in the contact time did not affect the adsorption.

2.3. Adsorption Kinetic Modeling

The kinetics of acetamiprid adsorption by TPAC was studied by two common models, pseudo-first-order [32] and pseudo-second-order models [33].

The pseudo-first-order is given by Equation (1).

$$Log\ (q_e - q_t) = log\ q_e - (k_1/2.303)\ t \quad (1)$$

The pseudo-second-order is given by Equation (2):

$$t/q_t = 1/k_2 q_{e2} + (1/q_e)t \quad (2)$$

where q_e and q_t are the amount of acetamiprid adsorbed by TPAC at equilibrium and at time t, respectively. The adsorption rate constants of pseudo-first and pseudo-second-order were represented as k_1 and k_2, respectively. The kinetic parameters are given in Table 2. As can be seen from Figure 7, the pseudo-second-order kinetic model adsorption capacity calculated is 23.364 mg/g. The low value of R^2 for the pseudo-first-order kinetic model could be an indication that the rate of the adsorption of acetamiprid by TPAC depends on both concentration and time [34]. The results showed that (R^2) values of the pseudo-second-order kinetic model were closer to unity, indicating that the adsorption of acetamiprid by TPAC followed the pseudo-second-order kinetic model. Similar trend was observed for the adsorption of bentazon on activated carbon prepared from *Lowosoniainerm* wood [35].

Table 2. Kinetic parameters and correlation coefficients for the removal of acetamiprid by TPAC.

Kinetic Model	Parameter	Values
Pseudo-first-order	k_1(1/min)	0.01 ± 0.02
	q_e (mg/g)	7.3 ± 0.01
	R^2	0.884
Pseudo-second-order	k_2 (g/mg.min)	0.005 ± 0.001
	q_e (mg/g)	23.3 ± 0.717
	R^2	0.998

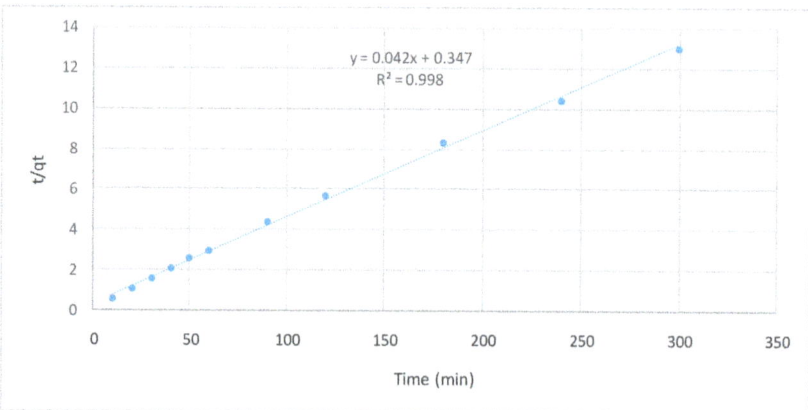

Figure 7. Pseudo-second-order kinetic model for adsorption of acetamiprid by TPAC.

2.4. Adsorption Isotherms Modeling

The adsorption isotherms are important to describe the interaction between adsorbate and adsorbent. In this study, different isotherm models were used to study the adsorption of acetamiprid pesticide onto TPAC. The most commonly applied isotherms such as Langmuir [36] and Freundlich [37] isotherm models were tested. The linear form of Langmuir isotherm is given by Equation (3).

$$1/Q_t = (1/X_m\ b)\ C_t + 1/X_m \tag{3}$$

where Q_t is the adsorption capacity at equilibrium (mg/g); C_t is the equilibrium concentration of the acetamiprid solution (mg/L); t (min) is the contact time; X_m (mg/g) is the maximum monolayer adsorption capacity and b (L/mg) is the adsorption equilibrium constant.

The Langmuir isotherm model assumes that monolayer adsorption occurs on homogeneous surfaces containing a finite number of adsorption sites and no lateral interactions between the adsorbed molecules exist [38]. The dimensionless constant (R_L) is a very important parameter from Langmuir isotherm which suggests that the removal of acetamiprid onto TPAC is favorable or not and can be determined by the following equation:

$$R_L = 1/(1 + bC^0) \tag{4}$$

where C^0 is the initial acetamiprid level (mg/L) and b is the Langmuir isotherm constant. RL value describes the adsorption mechanisms, which is unfavorable ($R_L > 1$), linear ($R_L = 1$), favorable ($0 < R_L > 1$) or irreversible ($R_L = 0$). The value of R_L in the present investigation was found to be 0.09 indicating that the adsorption of acetamiprid onto TPAC is favorable.

The Freundlich isotherm model assumes that multilayer adsorption occurs on heterogeneous surfaces of the adsorbent [39]. In addition, assuming the stronger binding sites were occupied first, and the strength of the binding decreased with the increasing of the occupation of sites. The linear form of Freundlich isotherm is given by:

$$\log q_e = \log K_f + 1/\mathrm{n} \log C_e \tag{5}$$

where K_f is an indicator of the adsorption capacity of TPAC (i.e., the greater the value of K_f the greater adsorption capacity). The value of $1/n$ is the heterogeneity factor of site energies and reflected the intensities of adsorption [40]. These suggested that the ability for adsorption of acetamiprid pesticide by TPAC was strong. In this study, different isotherms have been used to explain the adsorption mechanism. The isotherm parameters are presented in Table 3. The different isotherm plots for the adsorption of acetamiprid onto TPAC are shown in Figure 8A,B. The best fit of these isotherm models was selected according to the value of the correlation coefficients (R^2). It was found that the Langmuir isotherm model is the best fit for the adsorption of acetamiprid, which indicated that the acetamiprid pesticide is adsorbed by TPAC as a monolayer with the homogeneous surface. The maximum adsorption capacity was found to be 35.71 mg/g.

Table 3. Parameters of isotherm models for the removal of acetamiprid by TPAC.

Isotherm Model	Parameter	Values
Freundlich	K_f	4.8 ± 0.03
	R^2	0.919
	$1/n$	0.02 ± 0.001
Langmuir	B (L/mg)	1.01 ± 0.02
	q_m (mg/g)	35.7 ± 0.05
	R^2	0.997

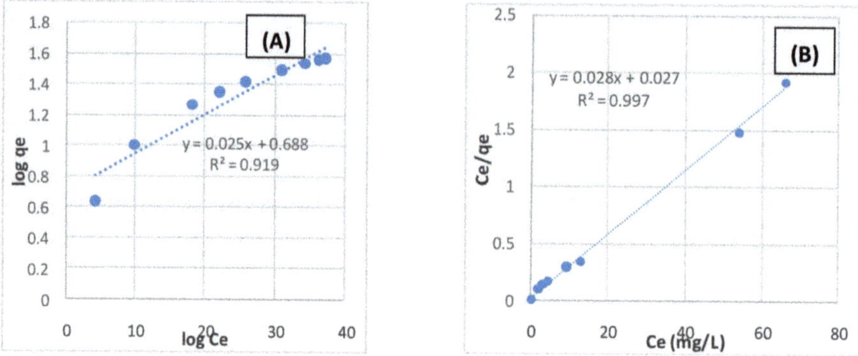

Figure 8. Adsorption isotherm models (**A**) Langmuir and (**B**) Freundlich.

The energy of adsorption was less than 8 KJ/mol indicating that the adsorption of acetamiprid onto TPAC is physical adsorption (Van der Waals forces). Similar trends have been reported by [18], they studied the adsorption of oxamyl pesticide onto Egyptian apricot stones.

2.5. Thermodynamics

Thermodynamic is essential to find out whether the process is spontaneous or not. The thermodynamic parameters such as enthalpy changes ($\Delta H°$), entropy changes ($\Delta S°$), (K_D) binding affinity and Gibbs free energy changes ($\Delta G°$) were used to evaluate the influence of temperature on adsorption process and determine the nature and spontaneity of the adsorption of acetamiprid pesticide onto TPAC were calculated by the following equations:

$$\Delta G° = -RT\ ln\ K_D \tag{6}$$

$$ln\ K_D = -\Delta H°/RT + \Delta G°/R \tag{7}$$

The values of thermodynamic parameters are in Table 4. The negative value of $\Delta H°$ confirmed the exothermic nature of the adsorption process. The value of $\Delta H°$ was found to be in the range between 8 and 40 KJ/mol which suggests the physico-sorption process [41]. The negative values of $\Delta G°$ were −71.7, −72.8 and −74.5 KJ/mol, respectively at all temperatures and indicate that the adsorption of acetamiprid by TPAC is spontaneous.

Table 4. Thermodynamic parameters for the adsorption of acetamiprid pesticide onto TPAC.

Temperature (K)	$\Delta G°$ (KJ/mol)	$\Delta H°$ (KJ/mol)	$\Delta S°$ (KJmol^{-1} K^{-1})
298	−71.7		
308	−72.8	−37.8	−113.5
323	−74.5		

2.6. Comparison of Acetamiprid Adsorption Capacity with Other Adsorbents

The adsorption capacity of acetamiprid pesticide on TPAC was compared with different adsorbents reported in the previous studies. According to Table 5, the adsorption capacity in the present work of acetamiprid pesticide on tangerine peels-derived activated carbon was higher in comparison with some studies [13] while lower in comparison with other studies [23]. The reason for this variation is due to the adsorbent dose and the initial concentration used. The adsorption capacities varied from different adsorbents according to different parameters such as, source of the activated carbon, surface characterization, pH, particle size and functional groups.

Table 5. Comparison of acetamiprid removal by different adsorbents.

Adsorbent Type	Maximum Adsorption Capacity (mg/g)	Initial Concentration (mg/L)	Contact Time (min)	Optimum pH	Isotherm Fitted	Adsorbent Dose (g)	% Removal	Ref.
Bentonite	9.1	100	30	7.0	Langmuir	1	-	[13]
Bentonite and clay	7.8	100	30	7.0	Langmuir	1	-	[13]
Kaolin	7.7	100	30	7.0	Langmuir	1	-	[13]
Orange peels activated carbon	151.5	300	120	5.6	Freundlich	1	99.4	[23]
Almond shells activated carbon	370.3	300	120	5.6	Freundlich	0.5	99.4	[23]
Tangerine peels activated carbon	35.7	25	240	5.6	Langmuir	0.1	92.0	This Study

2.7. Reusability of the Activated Carbon

Reusability or recycling is very important benefits of the adsorbent for the removal of different contaminants due to the lower cost. Figure 9. shows the performance of TPAC after four cycles of adsorption–desorption. The percentage removal of acetamiprid by TPAC was the same during each cycle. We measured 80.05% during the first cycle, 80.05% during the second cycle; 80.05% during the third cycle and 80.05% during the fourth cycle. These results indicated that the TPAC has been reused for four cycles at a commercial scale level in order to lower the cost of the adsorbing materials and reduce the need for new adsorbents.

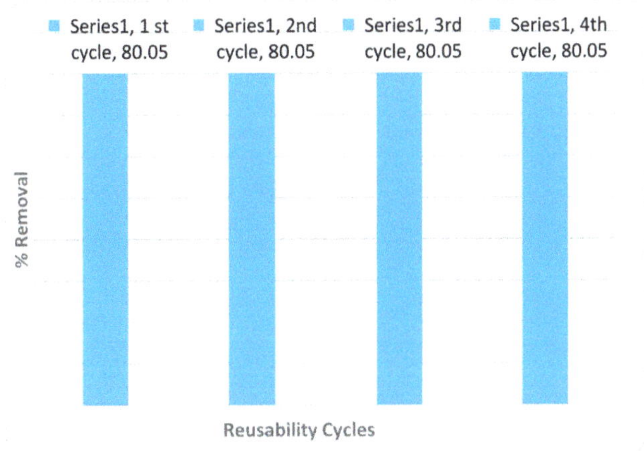

Figure 9. Reusability of TPAC.

3. Materials and Methods

3.1. Apparatus

The concentrations of acetamiprid in solution before and after the adsorption process was quantized using a reversed-phase high performance liquid chromatography (HPLC) (Agilent 1200, WA, USA) coupled with a variable wavelength diode array detector (DAD), C18 column (5 μm particles, 250 mm × 4.6 mm i.d.) and an auto-sampler with an electric sample valve. The mobile-phase solvent was 70% Milli-Q ultrapure water and 30% acetonitrile at a constant flow rate of 1.0 mL/min. The wavelength of measuring acetamiprid was 246 nm. The surface functional groups of TPAC before and after adsorption of acetamiprid pesticide were determined by Fourier Transform infrared spectrometer (PerkinElmer 1720). The textural characteristics of TPAC including surface area, pore volume and pore diameter were determined by BELSORP-mini II instrument. The X-ray diffraction of the prepared activated carbon was performed with a XPERT PRO (Almelo, The Netherlands) diffractometer in the range of 2θ (4–80°). Zeta plus Zeta potential analyzer (Nanozs, Malvern Company, Malvern, UK) determined the particle size of the prepared activated carbon. Prior to the analysis, all samples were degassed at around 200 °C under vacuum overnight to clean their surface.

3.2. Chemicals

All chemicals used in this work were of analytical reagent grade. Acetamiprid (97% purity), acetonitrile (HPLC grade) and orthophosphoric acid (85%) were obtained from Sigma-Aldrich (St. Lois, MO, USA). Deionized water was obtained from a Millipore Milli-Q Academic Ultra-Pure Water Purification System.

3.3. Preparation of Activated Carbon

All tangerine peel samples were collected from the local fruit markets in Egypt. The peels were washed with distilled water and dried in an oven at 80 °C for 48 h. Chemical activation was carried out by impregnation of the tangerine peels with 1:1-(*w*/*w*) H_3PO_4 for 24 h then dried at 100 °C for further 24 h. Pyrolysis was carried out in a muffle furnace at 500 °C in the absence of air for 2 h. The produced activated carbon after pyrolysis was washed with deionized water and dried in an oven at 80 °C for 24 h. The yield of the produced activated carbon was 48.4% and calculated by:

$$Yield,b\% = (\text{emphDry weight of the activated carbon/Dry weight of the precursor}) \times 100 \quad (8)$$

3.4. Acetamiprid Uptake Study

A 100 mg of TPAC adsorbent was inserted into 100-mL aliquots of acetamiprid solutions (10–100 μg/mL). The solutions were stirred for 240 min at room temperature and filtered. Acetamiprid concentrations, before and after the treatment, were measured using HPLC measurements. The removal percentage of acetamiprid was calculated using the following equation:

$$Removal\% = [(C_0 - C_t)/C_0] \times 100 \quad (9)$$

where: C_0 and C_t are the acetamiprid concentration in μg/mL at initial and after time t, respectively.

3.5. Batch Adsorption Experiments

Aliquots (100 mL) of 25-μg/mL acetamiprid solution were treated with different adsorbent doses (0.02–0.2 gm), under different pH values (2–8) for varying contact times (10–300 min). The acetamiprid solutions were stirred, filtered and the concentration was measured using HPLC measurements.

3.6. Adsorption Isotherms and Kinetics

Adsorption equilibrium was performed using the different initial concentrations of acetamiprid (10–100 μg/mL) in presence of 0.1 g of TPAC under constant stirring at 200 rpm. The equilibrium was analyzed using different isotherm models such as Langmuir, Freundlich and Dubinin–Radushkevich (D-R) adsorption isotherm models. The adsorption capacity of acetamiprid by TPAC was calculated from the following equation:

$$q_e = \frac{(C_0 - C_e)\,V}{m} \quad (10)$$

The adsorption kinetics data were evaluated with 0.1 g of TPAC at initial concentration of acetamiprid of 25 μg/mL and the samples were taken after time varied from 10 to 300 min.

3.7. Adsorption Thermodynamics

Adsorption of acetamiprid was investigated at different temperatures in the range 25–50 °C using initial concentration 25 μg/mL and 0.1 g of TPAC. Prior to the analysis, the samples were centrifuged at 4000 rpm for 5 minutes and filtered using 0.45-μm Millipore filter. Then the concentration of acetamiprid was determined by HPLC method.

3.8. Reusability

The reusability TPAC was tested by performing five alternating adsorption–desorption cycles. Each adsorption cycle was performed as mentioned previously then the adsorbent was filtered, washed and dried. The collected adsorbent was then dispersed into 20 mL of 0.5-M HCl and shaken at 180 rpm for 10 minutes to regenerate the adsorbent. Prior to the next cycle, the regenerated adsorbent was washed repeatedly with ultrapure water until the effluent was neutral (pH 7). The regenerated TPAC was used for sequential adsorption–desorption cycles and the acetamiprid adsorption efficiency was obtained.

4. Conclusions

Tangerine peels-derived activated carbon (TPAC) was synthesized and characterized by FTIR, surface area, XRD and particle size analysis. TPAC was applied to the removal of acetamiprid pesticide from aqueous solution. The adsorption of acetamiprid was influenced by adsorbent dose, pH, initial concentration, contact time and temperature. The adsorption isotherm was well fitted by Langmuir with a maximum adsorption capacity of 35.71 mg/g. The adsorption process was found to be rapid in the first 20 min and attained equilibrium at 240 min. The pseudo-second-order kinetic model describes the adsorption of acetamiprid by TPAC better than the pseudo-first-order kinetic model. The thermodynamics indicates that the adsorption of acetamiprid onto TPAC was exothermic and spontaneous in nature. Furthermore, the potential of regenerating of the tangerine peels activated carbon was investigated and successfully regenerated and reused for three cycles. It is concluded that the tangerine peels activated carbon could be a good activated carbon for the removal of acetamiprid pesticide from aqueous media.

Author Contributions: The listed authors contributed to this work as described in the following: S.G.M., S.M.A. and A.H.K., gave the concepts of the work, interoperated the results, the experimental part and prepared the manuscript, A.H.K., S.G.M., S.M.A. and A.E.-G.E.A. cooperated in the preparation of the manuscript and A.H.K. and A.E.-G.E.A. performed the revision before submission. A.E.-G.E.A. revealed the financial support for the work. All authors read and approved the final manuscript.

Funding: King Saud University, Researchers Supporting Project (Project No. RSP-2019/66).

Acknowledgments: Authors are grateful to King Saud University for funding the work through Researchers Supporting Project (Project No. RSP-2019/66).

Conflicts of Interest: The authors declare no conflict of interest.

References

1. EPA. *U.S. Office of Pesticide Programs, Environmental Fate and Efect Division. Revised EFED Risk Assessment of Carbaryl in Support of the Registration Eligibility Decision*; EPA: Washington, DC, USA, 2003.
2. Seccia, S.; Fidente, P.; Barbini, D.A.; Morrica, P. Multiresidue determination of nicotinoid insecticide residues in drinking water by liquid chromatography with electrospray ionization mass spectrometry. *Anal. Chim. Acta* **2005**, *553*, 21–26. [CrossRef]
3. Morrissey, C.A.; Mineau, P.; Devries, J.H.; Sanchez-Bayo, F.; Liess, M.; Cavallaro, M.C.; Liber, K. Neonicotinoid contamination of global surface waters and associated risk to aquatic invertebrates: A review. *Environ. Int.* **2015**, *74*, 291–303. [CrossRef] [PubMed]
4. Pesticide Fact Sheet: Acetamiprid, EPA (United States Environmental Protection Agency) 2002. Available online: https://www3.epa.gov/pesticides/%20chem_search/reg_actions/%20registration/%20fs_PC-099050_15-Mar-02.pdf (accessed on 25 April 2020).
5. Hoyle, S.C.A. *Neonicotinoids in California's Surface Waters: A Preliminary Review of Potential Risk to Aquatic Invertebrates*; The Xerces Society for Invertebrate Conservation: Portland, OR, USA, 2016; pp. 1–17.
6. Barbosa, M.O.; Moreira, N.F.F.; Ribeiro, A.R.; Pereira, M.F.R.; Silva, A.M.T. Occurrence and removal of organicmicropollutants: an overview of the watch list of EU Decision 2015/495. *Water Res.* **2016**, *94*, 257–279. [CrossRef] [PubMed]

7. Liu, G.; Li, L.; Huang, X.; Zheng, S.; Xu, X.; Liu, Z.; Zhang, Y.; Wang, J.; Lin, H.; Xua, D. Adsorption and removal of organophosphorus pesticides from environmental water and soil samples by using magnetic multi-walled carbon nanotubes@organic framework ZIF-8. *J. Mater. Sci.* **2018**, *53*, 10772–10783. [CrossRef]
8. Saeed, M.; Nadeem, R.; Yousaf, M. Removal of industrial pollutant (Reactive Orange 122 dye) using environment-friendly sorbent Trapa bispinosa's peel and fruit. *Int. J. Environ. Sci. Technol.* **2014**, *12*, 1223–1234. [CrossRef]
9. Carra, I.; Sánchez Pérez, J.A.; Malato, S.; Autin, O.; Jefferson, B.; Jarvis, P. Application of high intensity UVC-LED for the removal of acetamipridwith the photo-Fenton process. *Chem. Eng. J.* **2015**, *264*, 690–696. [CrossRef]
10. Mitsika, E.E.; Christophoridis, C.E.; Fytianos, K. Fenton and Fenton-like oxidation of pesticide acetamiprid in water samples: Kinetic study of the degradation and optimization using response surface methodology. *Chemosphere* **2013**, *93*, 1818–1825. [CrossRef]
11. Fenoll, J.; Garrido, I.; Hellín, P.; Flores, P.; Navarro, S. Photodegradation of neonicotinoid insecticides in water by semiconductor oxides. *Environ. Sci. Pollut. Res.* **2015**, *22*, 15055–15066. [CrossRef]
12. Guzsvány, V.; Rajic, L.; Jović, B.; Orčić, D.; Csanádi, J.; Lazić, S.; Abramović, B.F. Spectroscopic monitoring of photocatalytic degradation of the insecticide acetamiprid and its degradation product 6-chloronicotinic acid on TiO2 catalyst. *J. Environ. Sci. Heal. Part. A* **2012**, *47*, 1919–1929. [CrossRef]
13. Choumane, F.Z.; Benguella, B. Removal of acetamiprid from aqueous solutions with low-cost sorbents. *DESALINATION Water Treat.* **2014**, 1–12. [CrossRef]
14. Li, S.; Ma, X.; Jiang, Y.; Cao, X. Acetamiprid removal in wastewater by the low-temperature plasma using dielectric barrier discharge. *Ecotoxicol. Environ. Saf.* **2014**, *106*, 146–153. [CrossRef] [PubMed]
15. Daneshvar, N. Investigation of adsorption kinetics and isotherms of imidacloprid as a pollutant from aqueous solution by adsorption onto industrial granular activated carbon. *J. Food Agric. Environ.* **2007**, *5*, 425–429.
16. Zhang, J.; Zhou, A.-N.; Shao, L.; He, P. The use of biochar-amended composting to improve the humification and degradation of sewage sludge. *Bioresour. Technol.* **2014**, *168*, 252–258. [CrossRef] [PubMed]
17. Li, H.; Gao, P.; Cui, J.; Zhang, F.; Wang, F.; Cheng, J. Preparation and Cr(VI) removal performance of corncob activated carbon. *Environ. Sci. Pollut. Res.* **2018**, *25*, 20743–20755. [CrossRef] [PubMed]
18. Ahmed, S. Egyptian Apricot Stone (*Prunus armeniaca*) as a Low Cost and Eco-friendly Biosorbent for Oxamyl Removal from Aqueous Solutions. *Am. J. Exp. Agric.* **2014**, *4*, 302–321. [CrossRef]
19. Mohammad, S.; Ahmed, S.M.; Badawi, A.F.M. A comparative adsorption study with different agricultural waste adsorbents for removal of oxamyl pesticide. *DESALINATION Water Treat.* **2014**, *55*, 1–12. [CrossRef]
20. Haq, A.U.; Saeed, M.; Usman, M.; Muneer, M.; Adeel, S.; Abbas, S.; Iqbal, A. Removal of butachlor from aqueous solution using cantaloupe seed shell powder: kinetic, equilibrium and thermodynamic studies. *Int. J. Environ. Sci. Technol.* **2018**, *16*, 6029–6042. [CrossRef]
21. Dehghani, M.H.; Kamalian, S.; Shayeghi, M.; Yousefi, M.; Heidarinejad, Z.; Agarwal, S.; Gupta, V.K. High-performance removal of diazinon pesticide from water using multi-walled carbon nanotubes. *Microchem. J.* **2019**, *145*, 486–491. [CrossRef]
22. Vithanage, M.; Mayakaduwa, S.; Herath, I.; Ok, Y.S.; Mohan, D. Kinetics, thermodynamics and mechanistic studies of carbofuran removal using biochars from tea waste and rice husks. *Chemosphere* **2016**, *150*, 781–789. [CrossRef]
23. Mohammad, S.; Ahmed, S.M. Adsorptive removal of acetamiprid pesticide from aqueous solution using environmentally friendly natural and agricultural wastes. *DESALINATION Water Treat.* **2019**, *145*, 280–290. [CrossRef]
24. Qiu, K.; Song, X.; Tang, G.; Wu, L.; Min, S. Determination of Fipronil in Acetamiprid Formulation by Attenuated Total Reflectance-Mid-Infrared Spectroscopy Combined with Partial Least Squares Regression. *Anal. Lett.* **2013**, *46*, 2388–2399. [CrossRef]
25. Thommes, M.; Kaneko, K.; Neimark, A.V.; Olivier, J.P.; Rodriguez-Reinoso, F.; Rouquerol, J.; Sing, K.S. Physisorption of gases, with special reference to the evaluation of surface area and pore size distribution (IUPAC Technical Report). *Pure Appl. Chem.* **2015**, *87*, 1051–1069. [CrossRef]
26. Xiuli, H.; Haixia, J.; Yong, Z.; Weifeng, H.; Yangfan, Z.; Ping, G.; Rui, D.; Enhui, L. A high performance nitrogen-doped porous activated carbon for supercapacitor derived from pueraria. *J. Alloys Comp.* **2018**, *744*, 544–551.

27. Kılıç, M.Ö.; Apaydın-Varol, E.; Pütün, A.E. Adsorptive removal of phenol from aqueous solutions on activated carbon prepared from tobacco residues: Equilibrium, kinetics and thermodynamics. *J. Hazard. Mater.* **2011**, *189*, 397–403. [CrossRef] [PubMed]
28. Abdelhafez, A.A.; Li, J. Removal of Pb(II) from aqueous solution by using biochars derived from sugar cane bagasse and orange peel. *J. Taiwan Inst. Chem. Eng.* **2016**, *61*, 367–375. [CrossRef]
29. Sahithya, K.; Das, D.; Das, N. Efective removal of dichlorvos from aqueous solution using biopolymer modifed MMT–CuO composites: equilibrium, kinetic and thermodynamic studies. *J. Mol. Liq.* **2015**, *211*, 821–830. [CrossRef]
30. Njoku, V.; Foo, K.Y.; Hameed, B. Microwave-assisted preparation of pumpkin seed hull activated carbon and its application for the adsorptive removal of 2,4-dichlorophenoxyacetic acid. *Chem. Eng. J.* **2013**, *215*, 383–388. [CrossRef]
31. Doczekalska, B.; Kuśmierek, K.; Swiatkowski, A.; Bartkowiak, M. Adsorption of 2,4-dichlorophenoxyacetic acid and 4-chloro-2-metylphenoxyacetic acid onto activated carbons derived from various lignocellulosic materials. *J. Environ. Sci. Heal. Part. B* **2018**, *53*, 290–297. [CrossRef]
32. Długosz, O.; Banach, M. Kinetic, isotherm and thermodynamic investigations of the adsorption of Ag^+ and Cu^{2+} on vermiculite. *J. Mol. Liq.* **2018**, *258*, 295–309. [CrossRef]
33. Simonin, J.-P. On the comparison of pseudo-first order and pseudo-second order rate laws in the modeling of adsorption kinetics. *Chem. Eng. J.* **2016**, *300*, 254–263. [CrossRef]
34. Kumar, M.; Tamilarasan, R. Kinetics, equilibrium data and modeling studies for the sorption of chromium by Prosopis juliflora bark carbon. *Arab. J. Chem.* **2017**, *10*, S1567–S1577. [CrossRef]
35. Omri, A.; Wali, A.; Benzina, M. Adsorption of bentazon on activated carbon prepared from Lawsonia inermis wood: Equilibrium, kinetic and thermodynamic studies. *Arab. J. Chem.* **2016**, *9*, S1729–S1739. [CrossRef]
36. Ghaedi, M.; Hajati, S.; Karimi, F.; Barazesh, B.; Ghezelbash, G. Equilibrium, kinetic and isotherm of some metal ion biosorption. *J. Ind. Eng. Chem.* **2013**, *19*, 987–992. [CrossRef]
37. Freundlich, H.M.F. Over the adsorption in solution. *J. Phys. Chem.* **1906**, *57*, 385–470.
38. Tan, X.; Liu, Y.; Zeng, G.; Wang, X.; Hu, X.; Gu, Y.; Yang, Z. Application of biochar for the removal of pollutants from aqueous solutions. *Chemosphere* **2015**, *125*, 70–85. [CrossRef]
39. Li, F.; Shen, K.; Long, X.; Wen, J.; Xie, X.; Zeng, X.; Liang, Y.; Wei, Y.; Lin, Z.; Huang, W.; et al. Preparation and Characterization of Biochars from Eichornia crassipes for Cadmium Removal in Aqueous Solutions. *PLoS ONE* **2016**, *11*, e0148132. [CrossRef]
40. Fan, S.; Wang, Y.; Wang, Z.; Tang, J.; Tang, J.; Li, X. Removal of methylene blue from aqueous solution by sewage sludge-derived biochar: Adsorption kinetics, equilibrium, thermodynamics and mechanism. *J. Environ. Chem. Eng.* **2017**, *5*, 601–611. [CrossRef]
41. Lima, E.C.A.M.; Machado, F.M. Chapter 3: Kinetic and equilibrium models of adsorption. In *Carbon Nanomaterials as Adsorbents for Environmental and Biological Applications*; Bergmann, C.P., Machado, F.M., Eds.; Springer International Publishing: Berlin/Heidelberg, Germany, 2015; pp. 33–69.

Sample Availability: Samples of the compounds are not available from the authors.

molecules

Article

Application of Pineapple Leaves as Adsorbents for Removal of Rose Bengal from Wastewater: Process Optimization Operating Face-Centered Central Composite Design (FCCCD)

Siham S. Hassan, Ahmed S. El-Shafie, Nourhan Zaher and Marwa El-Azazy *

Department of Chemistry and Earth Sciences, College of Arts and Sciences, Qatar University, Doha 2713, Qatar; s.hersi@qu.edu.qa (S.S.H.); aelshafie@qu.edu.qa (A.S.E.-S.); nm1601246@qu.edu.qa (N.Z.)

* Correspondence: marwasaid@qu.edu.qa

Academic Editors: Chiara Bisio and Monica Pica

Received: 4 July 2020; Accepted: 5 August 2020; Published: 18 August 2020

Abstract: Adsorptive removal of rose bengal (RB) from contaminated water samples was approached using pineapple leaves (PAL). Three adsorbents were utilized for that purpose; raw pineapple leaves (RPAL) and the thermally activated bio-waste leaves at 250 and 500 °C. Two measures were executed to evaluate the functionality of exploited biomasses; percentage removal (%R) and adsorption capacity (q_e). Face-centered central composite design (FCCCD) was conducted to experiment the influence of variables on the %R. Dose of PAL as adsorbent (AD), concentration of RB (DC), pH and contact time (CT), were the inspected factors. Existence of functional groups and formation of activated carbon was instigated employing Fourier-transform infrared (FT-IR) and Raman spectroscopies. Scanning electron microscopy (SEM) and energy-dispersive X-ray spectroscopy (EDX) analyses were used to explore surface features. Thermal behavior of adsorbents was studied using thermogravimetric analysis (TGA). The surface area and other surface structural properties were established using the Brunauer Emmett-Teller (BET) analysis. An amount of 92.53% of RB could be removed with an adsorption capacity of 58.8 mg/g using a combination of pH 5.00 ± 0.20, RPAL dose of 0.05 mg/50 mL, and 10-ppm RB for 180 min. Equilibrium studies divulge a favorable adsorption that follows the Freundlich isotherm. Pseudo-second-order model explains the observed adsorption kinetics.

Keywords: green adsorbents; pineapple leaves; rose bengal (RB) dye; face-centered central composite design (FCCCD), percentage removal (%R); adsorption capacity (q_e)

1. Introduction

Water is perceived as the most important renewable source of life, where surface and ground water play major roles in agriculture, livestock production, hydropower generation, etc. The rate of growth of the world population is increasing day after day. This escalating growth is logically associated with several environmental concerns. Water pollution is one of the most serious apprehensions that living creatures have ever faced, if not the most challenging at all. The quality of water is particularly significant for human health. As per the World Health Organization (WHO) reports, poor water quality is responsible for 2.2 million deaths annually. Moreover, more than 2/3 of infant deaths stem from waterborne diseases [1–4].

Numerous contaminants contribute to water pollution. Among these pollutants, heavy metals, anions (sulfates, phosphates, fluoride, etc.), dyes, pesticides, fertilizers, and pharmaceuticals are the most common [4–10]. Dyes, the topic of the current investigation, are widely applied in various industries, e.g., paper, cosmetics, paint and textiles production, food processing, etc. Discharge of the industrial effluents into water bodies causes not only a direct mutilation of water physicochemical features (such as color, pH, salinity, organic carbon content, etc.), but also instigates detrimental effects on the ecosystem and consequently the human health. This effect is exacerbated by the diverse chemical structure of these dyes and their resistance to biodegradation [11–13].

Rose bengal (RB), a basic xanthene dye, also known as 'C.I. 45440 and C.I. Acid Red 94', is chemically recognized as disodium-4,5,6,7-tetrachloro-3′,6′-dihydroxy-2′,4′,5′,7′-tetraiodo-3*H*-spiro [isobenzofuran-1,9′-xanthen]-3-one (molar mass: 1017.64 g/mol). Sodium salt of RB is commonly used in diagnosing eye damage via staining the corneal and conjunctival cells. Other applications of RB include treatment of certain cancers (melanoma and breast cancers), skin conditions such as psoriasis and also as antibacterial. Moreover, RB is extensively used in fabric and photochemical manufacturing. Nonetheless, RB has shown serious impacts on human health, especially when it gets in contact with skin and eyes causing discomfort, irritation, redness and blistering [14–18]. Removal of RB from wastewater has been done implementing various methods such as photo-degradation, nanofiltration and adsorption [19–22].

Adsorption is one of the most promising strategies for wastewater treatment. On one hand, adsorption is convenient, easy to maneuver, and can be conducted using readily available materials. Conversely, dyes in specific are premeditated to be chemically stable with long–standing photolytical properties. Most of the used strategies for removal of dyes require pre- and post-treatment steps. In addition, majority of these methods are either impractical (request a tedious experimental setup) or expensive with reduced removing capabilities [6–11,18,23,24]. Moreover, some of these techniques might not be efficient at low pollutant concentrations. Adsorption is therefore a reasonable choice. Developing the model adsorbent and how the adsorption process is conducted are the keywords in managing the adsorption process. Agricultural as well as industrial wastes represent a real burden on the ecosystem if not appropriately recycled and reprocessed. Sources of agricultural wastes are variable. Yet, by-products of the agricultural processing such as peels, pits, shells, leaves, etc. represent important naturally occurring resources that are copiously available and should be thoroughly thought of for the production of value-added materials [24].

Pineapples (PA, *Ananas comosus*, Family: Bromeliaceae) is a perennial herbaceous plant. PA fruit is mostly planted in coastline and tropical areas. In India, for example, PA fruits are grown on around 2,250,000 acres of land. The first bud of the leaves looks attractive. Later on, leaves become stiff; sword shaped and spirally assembled around the fruit [25,26]. Leaves represent the waste biomass of PA fruits and are commonly used as a source of natural fibers. Amount of waste produced from PA (leaf waste) is worrying, where approximately 20,000–25,000 tons per acre are left out after the harvesting process [27]. Leaf fibers consist of mainly holocellulose and lignin with minor amounts of ash [28]. Raw and activated PAL have shown a promising removal potential for different kinds of contaminants. Table 1 shows an evaluation for the performance of PA through different studies with different adsorbates [29–35].

As previously mentioned, having an ideal adsorption process could be managed by not only developing the model adsorbent, but also by engineering the adsorption process and more specifically the influencing variables. Different parameters are known to affect the interaction between the adsorbent and the adsorbate such as adsorbent dosage (AD), concentration of the adsorbate (DC), contact time (CT), pH, surface area, as well as the nature of the adsorbent and the pollutant. The conventional strategy for investigating the influence of these variables on the adsorption capability of an adsorbent is to scrutinize the effect of a single variable per time (univariate analysis). This stratagem and in addition for being time and effort consuming, involves several experimentations, an issue that jeopardizes the method greenness. Moreover, this univariate-based strategy does not yield the adequate amount of data that enable the researcher to draw the full picture for the adsorption process. Nevertheless, and since the objective is to build a green bioremediation strategy, coupling of the adsorption process to factorial designs would overcome these concerns [36].

Table 1. Evaluation of the performance of pineapple leaf (PAL) processed in current work compared with other studies used PLP as adsorbent for removal different adsorbates.

Absorbent	Modification Method	Analytical Approach Used	Surface Area (m^2/g)	Adsorbate	Adsorption Capacity (mg/g)	%Removal	References
Raw pineapple leaves (RPAL)	Please check the experimental part of this paper.	FCCCD	4.79	Rose Bengal	58.8	92.53%	Current work
Pineapple leaf powder (PLP)	Leaves were washed, dried at 80 °C for 24 h, at pressure 70 kPa, pulverized, and grinded to fine PLP to be used.	Single variate analysis	ND *	Cu (II)	9.28	90%	[29]
NaOH-treated pineapple waste	Leaves were washed several times, dried in oven at 105 °C for 24 h, grinded and screened by 60 mesh sieves to use.	Single variate analysis (Batch experiments)	ND *	Pb (II) and Cd (II)	ND *	<95%	[30]
Surface modified pineapple crown leaves (PCL)	Leaves were washed, dried at 70 °C for 48 h, pretreated by isopropyl alcohol and NaOH to produce (P)PCL, modified by acetic acid and hydrogen peroxide to produce (M)PCL.	Single variate analysis	(P)PCL: 32.90	Cr (VI) and Cr (III)	Cr(VI) on (M)PCL: 3.91 Cr(VI) on (P)PCL: 2.69 Cr(III) on (M)PCL: 2.54 Cr(III) on (P)PCL: 1.82	ND *	[31]
Pineapple leaf powder (PLP)	Leaves were washed several times, dried in oven at 80 °C for 48 h, grinded into powder for further use.	Single variate analysis	5.24	Methylene Blue	ND *	<95%	[32]
Pineapple crown activated carbon and $ZnCl_2$	Leave were washed with distilled water, dried at 110 °C, chopped into small pieces, and mixed at ratio 1:1 with zinc chloride.	Single variate analysis (Batch experiments)	914.7	Methylene Blue	288.34	* ND	[33]
Pineapple leaf powder	Leaves were washed several times, dried in oven at 105 °C for 24 h, grinded and sieved to fine powder to use.	Single variate analysis	ND *	Remazol Brilliant Blue R	9.58	<90%	[34]
Pineapple leaf powder	Leaves were washed with distilled water, dried in oven at 105 °C, crushed, and sieved.	Single variate analysis (Batch experiments)	ND *	Methylene Blue	78.13	* ND	[35]

* ND: Not Determined.

Offering irresistible advantages including saving of time, efforts, and resources, a response surface methodological approach (RSM)–face-centered central composite design (FCCCD) will be utilized in the current approach to optimize the investigated responses. Factorial levels for four independent variables will be adjusted with the target being set to maximize the removal of the studied contaminant (RB) using PAL (raw and thermally treated at 250 and 500 °C, labelled as TTPAL250 and TTPAL500, respectively) as adsorbents. The amount of RB dye adsorbed will be analyzed using spectrophotometry. TGA, FT-IR, SEM, Raman, EDX, CHN, and BET analyses will be used to characterize the prepared adsorbents. To further study the nature of the adsorbents and adsorption process, both kinetic and equilibrium studies will be performed.

2. Results and Discussion

2.1. Selection of the Best Performing Adsorbent

Performance of the three prepared adsorbents was measured in terms of %R and the adsorption capacity (q_e) and using Equations (1) and (2), respectively. Table 2 shows a comparison between the three prepared adsorbents under the same conditions. As per the results revealed in Table 2, RPAL had the highest %R and q_e and therefore was further used in the subsequent studies:

$$(\%R) = \frac{C_0 - C_e}{C_0} \times 100\%, \tag{1}$$

$$(q_e) = \frac{C_0 - C_e}{W} V \tag{2}$$

where C_0 (mg L^{-1}) denotes the initial concentration of RB solution, C_e is the concentration of the RB solution at equilibrium, V stands to the volume of the solution (L), and W is the weight of the adsorbent used (g).

Table 2. Performance of PAL—based adsorbents in terms of %R and q_e. Testing adsorption performance was conducted using a variable blend of pH = 7.00 ± 0.20, DC = 50 ppm, AD = 50 mg/15 mL, CT = 30 min. The responses shown were calculated using Equations (1) and (2).

Adsorbent Type	Percentage Removal (%R)	Adsorption Capacity (q_e, mg/g)
RPAL	42.96	6.44
TTPAL250	18.99	2.85
TTPAL500	18.53	2.78

2.2. Response Surface Methodology (RSM): FCCCD

As previously mentioned, the purpose of the current approach is to investigate and optimize the adsorption capability of PAL to RB dye from artificially contaminated water samples. The novelty of the current approach stems from using a multivariate platform that surmounts all the previous cons of the univariate approach. FCCCD, as mentioned, was the design of choice, where the impact of four variables on a single response was assessed and optimized. Central composite designs (CCD) usually contain built-in points from preceding full/fractional designs. In the current case, a full factorial design was the preceding design and the value of alpha (α) or the distance between the axial points and the center was equal to one, denoting a FCCCD [36]. The measured response (%R) was calculated using the formula shown in Equation (1). Conducted experimental runs (as executed by the design setup) as well as the working factorial limits accompanied by the observed and predicted responses are shown in Table 3.

Table 3. Independent factors and their levels together with the observed and predicted dependent variable and the FCCCD matrix.

Factors	Low Level	Medium Level	High Level
pH	5	8	11
Adsorbent Dose (AD, B, g/50 mL)	0.01	0.03	0.05
Dye Concentration (DC, C, ppm)	10	20	30
Contact Time (CT, D, min)	5	92.5	180

Experimental Runs, Observed and Predicted Responses

Expt No	Blk *	pH	AD	DC	CT	%R Obs. **	%R Pred. **	RE ***
01	1	5(−)	0.03(0)	20(0)	92.5(0)	38.19	36.20	0.05
02	1	8(0)	0.01(−)	20(0)	92.5(0)	1.43	2.83	0.49
03	1	8(0)	0.05(+)	20(0)	92.5(0)	47.06	35.19	0.34
04	1	8(0)	0.03(0)	20(0)	92.5(0)	33.00	25.36	0.30
05	1	8(0)	0.03(0)	20(0)	5(−)	38.19	40.15	0.05
06	1	8(0)	0.03(0)	10(−)	92.5(0)	34.72	30.87	0.12
07	1	11(+)	0.03(0)	20(0)	92.5(0)	23.41	15.59	0.50
08	1	8(0)	0.03(0)	30(+)	92.5(0)	28.15	20.14	0.40
09	1	8(0)	0.03(0)	20(0)	180(+)	62.73	50.91	0.23
10	1	8(0)	0.03(0)	20(0)	92.5(0)	19.32	25.36	0.24
11	1	5(−)	0.01(−)	10(−)	5(−)	25.46	24.28	0.05
12	1	5(−)	0.01(−)	30(+)	5(−)	7.84	12.17	0.36
13	1	11(+)	0.05(+)	10(−)	5(−)	21.14	26.06	0.19
14	1	8(0)	0.03(0)	20(0)	92.5(0)	19.24	25.36	0.24
15	1	5(−)	0.01(−)	30(+)	180(+)	10.40	10.18	0.02
16	2	5(−)	0.05(+)	10(−)	5(−)	86.64	92.53	0.06
17	2	11(+)	0.01(−)	10(−)	180(+)	26.86	30.36	0.11
18	2	5(−)	0.05(+)	10(−)	180(+)	88.40	92.60	0.04
19	2	5(−)	0.01(−)	10(−)	180(+)	27.07	34.02	0.20
20	2	11(+)	0.01(−)	10(−)	5(−)	1.88	3.88	0.51
21	2	5(−)	0.05(+)	30(+)	5(−)	71.46	74.71	0.04
22	2	8(0)	0.03(0)	20(0)	92.5(0)	23.71	25.36	0.06
23	2	11(+)	0.01(−)	30(+)	180(+)	17.04	22.04	0.23
24	2	8(0)	0.03(0)	20(0)	92.5(0)	20.07	25.36	0.21
25	2	11(+)	0.01(−)	30(+)	5(−)	5.14	6.29	0.18
26	2	5(−)	0.05(+)	30(+)	180(+)	50.70	59.65	0.15
27	2	8(0)	0.03(0)	20(0)	92.5(0)	23.41	25.36	0.08
28	2	11(+)	0.05(+)	10(−)	180(+)	44.26	49.17	0.10
29	2	11(+)	0.05(+)	30(+)	180(+)	31.96	39.67	0.19
30	2	11(+)	0.05(+)	30(+)	5(−)	25.45	29.75	0.14

* Blk: Block; ** Obs: observed readings; ** Pred.: predicted readings; *** RE = Relative error = / (*Measured value – Actual value*)/Actual value /.

2.3. Investigation of Statistically Significant Variables

In order to investigate the statistical significance of tested variables, Pareto chart of standardized effects, normal and half-normal probability plots alongside with analysis of variance (ANOVA) were implemented. Pareto chart (Figure 1) shows that AD (B) is the most statistically effective factor, followed by the effect of pH (A). It can be also observed that the CT is not that much effective compared to the other factors, however the squared interaction (CT × CT) was the third most influencing variable. The interaction of the CT × AD was the least effective factor on the %R of RB dye. Similar conclusions were obtained using the analysis of variance test (ANOVA) at 95.0 confidence interval (95.0 CI). ANOVA results are shown in Table 4. F-value is shown for every model term and is sufficiently large in case of statistically significant variables. As shown in the table as well, variables with a significance level (p-value) less than 0.05 are statistically significant, and the opposite is true. Table 4 also shows that lack-of-fit has a p-value of 0.633 (statistically not significant) inferring goodness-of-fit.

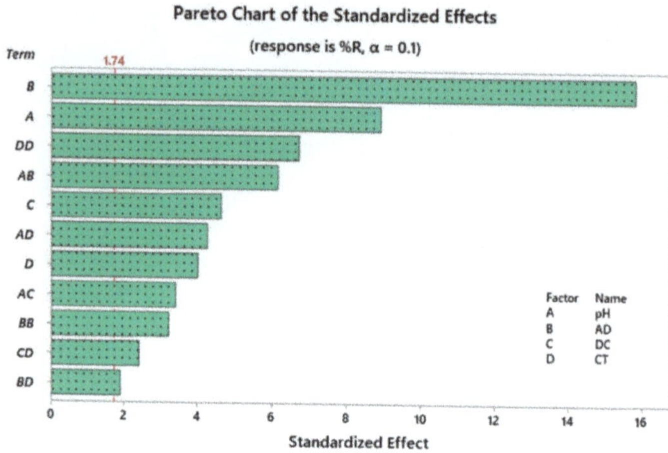

Figure 1. Pareto chart of standardized effects following response transformation.

Table 4. Analysis of variance (ANOVA) for the transformed response.

	DF *	Adj SS	Adj MS	F-Value	*p*-Value
Model	12	1329.53	110.794	40.79	0
Blocks	1	27.58	27.583	10.15	0.005
Linear	4	993.74	248.436	91.45	0
pH	1	215.12	215.12	79.19	0
AD	1	677.08	677.082	249.24	0
DC	1	58.02	58.024	21.36	0
CT	1	43.52	43.518	16.02	0.001
2-Way Interactions	5	209.07	41.813	15.39	0
pH × AD	1	102.61	102.609	37.77	0
pH × DC	1	31.67	31.665	11.66	0.003
pH × CT	1	49.25	49.254	18.13	0.001
AD × CT	1	9.81	9.814	3.61	0.074
DC × CT	1	15.72	15.724	5.79	0.028
Squared Interactions	2	123.74	61.872	22.78	0
AD × AD	1	28.23	28.234	10.39	0.005
CT × CT	1	121.83	121.825	44.85	0
Error	17	46.18	2.717		
Lack-of-Fit	13	33.92	2.609	0.85	0.633
Pure Error	4	12.27	3.066		
Total	29	1375.71			

* DF is degrees of freedom SS is sum of squares and MS is mean of squares.

It is noteworthy to mention that response surface regression was performed versus blocks, pH, AD, DC as well as the CT employing Box-Cox transformation [37] where the transformation factor, $\lambda = 0.75$ and backward elimination of terms (α to remove = 0.1) was used, Equation (3):

$$\text{(Transformed response) } Y' = (Y_{\lambda} - 1)/\lambda \text{ (transformation factor)} \quad (3)$$

The outcome of the response surface regression is the following mathematical paradigm shown in Equation (4):

$$\begin{aligned} \%R^{0.75} = {} & 12.46 - 1.442\,\text{pH} + 1134\,\text{AD} - 0.450\,\text{DC} - 0.1495\,\text{CT} - 7465\,\text{AD} \times \text{AD} \\ & + 0.000810\,\text{CT} \times \text{CT} - 42.21\,\text{pH} \times \text{AD} + 0.0468\,\text{pH} \times \text{DC} \\ & + 0.00668\,\text{pH} \times \text{CT} - 0.448\,\text{AD} \times \text{CT} - 0.001133\,\text{DC} \times \text{CT}, \end{aligned} \quad (4)$$

Equation (4) shows that increasing the pH value would reduce the %R. Conversely, increasing the dose of RPAL would enhance the removal of RB. Model summary shows that the value of R^2 was relatively high (R^2 = 96.95%) and close to the value of R^2—adjusted (R^2 (adj) = 94.27%), indicating the linearity of the proposed model. The value of R^2—predicted was also high (R^2(pred) = 89.99%), implying that the proposed model is significantly capable of detecting new observations. This finding could be further confirmed by referring to Table 3 where both experimental and predicted values are revealed together with the difference between the experimental and actual values, relative to the actual values expressed as the relative error (RE). The shown error is relatively small reflecting a close match between observed and predicted responses.

2.4. Contour Plots of %R and Surface Optimization

Figure 2 illustrates the two–dimensional (2D) plots for the measured fitted response surface. Each of the shown panels reveals the effect of two factors on %R. As shown in the attached legend, the dark red color implies a lower %R, while dark grey color means higher %R. Having the upper left panel as an example (AD × pH), having an AD of 0.042–0.048 g/50 mL and a pH level of 5.00–5.20, the %R is in the range of 50%–60%. Similar conclusions can be obtained from the rest of panels for each factorial combination.

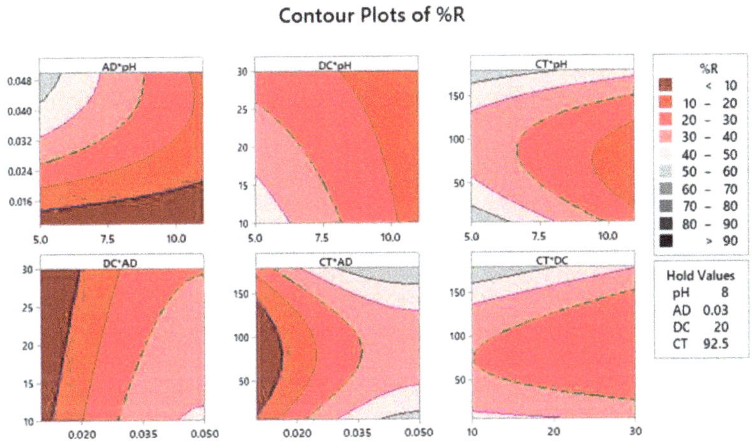

Figure 2. Response contour plots for the %R. Dark grey regions represent regions where maximum %R could be obtained using the factorial combination in each panel.

A typical strategy to deal with how a mixture of factorial settings satisfies the destinations that they were setup for is the use of "optimization plot"—a tool offered by Minitab to optimize the measured response. As shown in Figure 3, the objective was set to attain a 100% removal of RB, and the variable settings were fluctuated to achieve the objective. As shown, a blend of the tested variables at the level denoted as 'Cur' would produce a response value of 92.53%. The desirability value (*d*) was high enough, entailing the favorability of the mentioned blend. Figure 3 also shows that increasing the dose of RPAL enhances the adsorption process. This can be attributed to the increase in the number of adsorption sites available for the uptake of RB. The figure also shows that increasing the pH would decrease the %R, and similarly the DC. Impact of CT and as shown in the figure has a varying effect, where increasing the CT from 5 to 92.5 min. has resulted in reduced %R, while with increasing the time from 92.5 to 180 min., removal was improved. These findings are similar to the conclusions obtained from the mathematical paradigm described in Equation (4). Yet, explanation of these findings will be considered in lights of surface chemistry and nature of the dye throughout the next subsections.

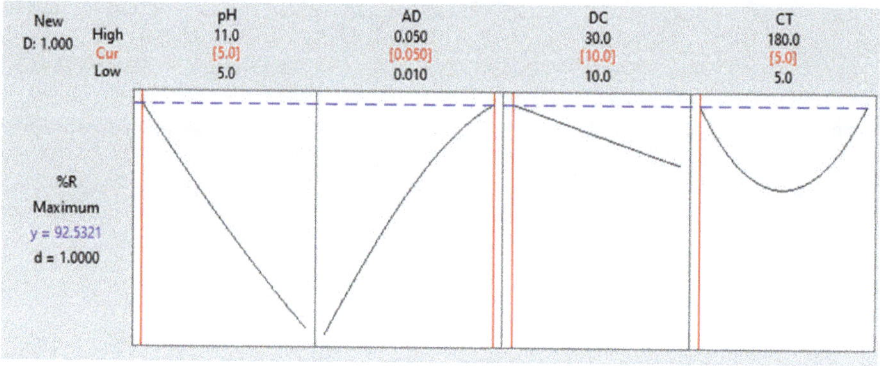

Figure 3. Optimization plot. A factorial combination of pH = 5.00 ± 0.20, AD = 0.050 g/50 mL, DC = 10 ppm and CT of 5 min. would achieve %R = 92.53%.

2.5. *Adsorbent Characterization*

2.5.1. Thermogravimetric Analysis (TGA)

Thermogravimetric analysis of RPAL was done under N_2 with a heating rate of 10 °C/min. The data represented in Figure 4 shows that the weight loss for the RPAL sample occurs over three steps as follows:

(1) First step: loss of adsorbed water molecules at a temperature range of 25–100 °C and represents 6.08% of the sample,

(2) In this step, >49% of the sample is decomposed between (200–500 °C) including the loss of crystalline water at ~200 °C and part of the organic matter, as represented by a major peak at 305 °C, which could be ascribed to the decomposition of the organic material in RPAL,

(3) The last step at ~525 °C where 14.46% of the RPAL sample was lost at this stage and it could be related to the carbonization of PAL.

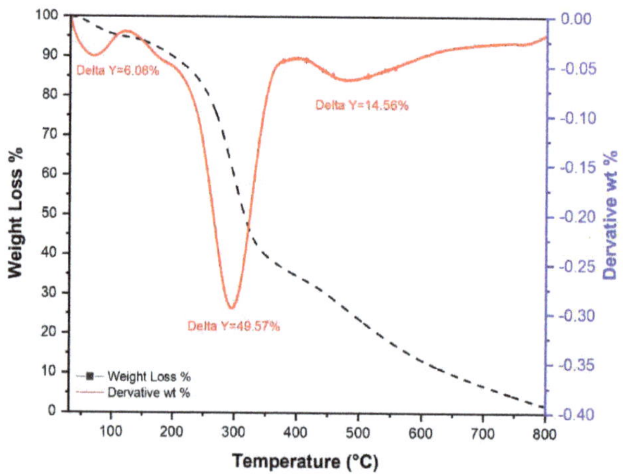

Figure 4. TGA graph of air—dried pineapple leaves (RPAL).

As shown from the TGA findings, thermal treatment of the RPAL might have resulted in the evaporation of the small molecules such as H_2O, CO, and CO_2. Absence of these functionalities in the thermally treated samples would explain their diminished adsorption capabilities compared to the RPAL sample.

2.5.2. Fourier Transform Infrared Spectroscopic Analysis (FT-IR)

FT-IR spectra of RPAL and TTPAL250 are given in Figure 5. As previously indicated, PAL are mainly composed of lignocellulosic material [28,29,32]. The obtained spectra show the existence of almost the same peaks in the two samples but with lower intensity in the thermally treated one due to the decomposition of lignocellulosic material, a finding that explains the subordinate adsorption capability of the later compared to the former [33]. The obtained data show a broad absorption band centered at 3325 cm^{-1} for the RPAL and 3318.4 cm^{-1} for TTPAL250. This peak could be assigned to the hydrogen–bonded—OH vibration of the cellulosic structure of the RPAL. In addition, it could be attributed to N-H group which is confirmed later by the presence of a high concentration of nitrogen in both raw and thermally treated samples in the CHN analysis. The spectra also show the presence of the absorption band at 2913–2920 cm^{-1} in both samples, which could be ascribed to the C-H stretching of aliphatic—CH groups. The absorption bands at 1595–1585.8 cm^{-1} confirm the presence of bending N-H of amines. The two bands at 1365 and 1375 cm^{-1} can be assigned to bending—OH. The absorption band at 1034.3 cm^{-1} for the RPAL and 1033.5 cm^{-1} for TTPAL250 can be ascribed to the presence of C–O stretching. The FT-IR results confirm the presence of surface functional groups that should have played an important role in the adsorption of RB onto PAL.

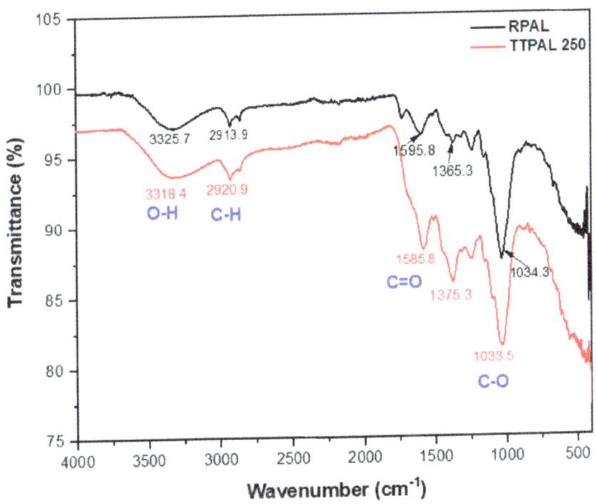

Figure 5. FT-IR spectra of RPAL and TTPAL250.

By combining the FT-IR data together with the FCCCD analysis findings, it can be recognized that the pH has a substantial influence on RB sorption process. Measurements were made at three pH values 5.00, 8.00, and 11.00. These values were carefully selected, where RB had the same absorption maxima in the three solutions. Moreover, the color of RB disappeared at pH less than 4.00. As per the design analysis, biosorption of RB onto RPAL was maximum at the acidic side (pH = 5.00 ± 0.20) and further elevation in the pH has resulted in a diminutive removal, Figure 3. RB and as previously reported, is an anionic dye with a pKa value of 4.50 [38,39]. Therefore, at pH > pKa, RB will start to be ionized (deprotonated, negatively charged). On the other hand, the surface of RPAL at the acidic side and as per the FT-IR analysis might have some positively charged functionalities. The existence of negatively charged RB on the positively charged RPAL surface would encourage electrostatic interaction. Conversely, at pH = 11.00 surface of RPAL will be negatively charged, therefore, less interaction between RB and RPAL surface. Similar results for better sorption of RB in acidic media has been previously reported using different adsorbents such as Fe (III)–montmorillonite [16], chitosan–TiO_2 nanocomposite [17], and bottom ash [40].

2.5.3. Raman Analysis

Raman spectra of raw and thermally treated pineapples are shown in Figure 6. The obtained spectra show the absence of any peaks in the range between 1000 to 2000 cm^{-1} in the raw sample. This could be explicated taking in consideration that carbon in the raw sample exists in the form of organic matter. Contrariwise, the Raman spectra of the burnt samples (TTPAL250 and TTPAL500) show two peaks which could be ascribed to the D–and G–bands at approximately 1351 cm^{-1} (D–band) and 1585 cm^{-1} (G–band). It is imperative to mention that these two bands are characteristic peaks for carbon materials. In addition, the resulted D–, and G–bands pattern is close to the bands present in graphene oxide [41]. Besides, the D–band reflects the carbon lattice properties including defects and sizes, but the G–band shows the stretching of C-C in sp^2 system [42]. Furthermore, the ration between intensity of D–band to G–band was calculated (I_D/I_G) and compared for the two thermally treated samples. Interestingly, the I_D/I_G for TTPAL250 was 0.90 compared to 1.07 for TTPAL500. This finding confirms the fact that the number of defects has increased by increasing the burning temperature. Yet, it can be also observed that the burning process (carbonization) might have resulted in the elimination of some essential functional groups, which in turn might have an important role in the diminished removal efficiency of the TTPAL250 and TTPAL500 compared to RPAL sample.

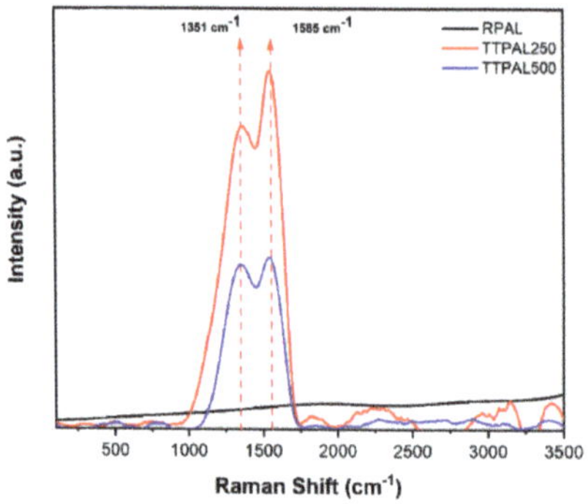

Figure 6. Raman spectra of the raw pineapple leaves (RPAL) and the thermally treated samples (TTPAL250 and TTPAL500).

2.5.4. Scanning Electron Microscopy Analysis (SEM)

The surface structure of the raw and the thermally treated PAL was explored using the scanning electron microscope (SEM). The SEM micrographs presented in Figure 7 showed that the RPAL (Figure 7A) has plain surface without any pores and the same was also observed following the burning process at 250 °C (Figure 7B). On the other hand, the surface has completely changed after burning at 500 °C. Figure 7C shows the presence of high porous surface compared to the raw material, confirming the formation of carbonaceous material with advanced pore structure and the loss of organic matter after burning at 500 °C. These findings are in a good match with the obtained data by FT-IR and TGA analyses. Furthermore, EDX analysis shows the effect of the burning process on the concentration of carbon and oxygen. Results show that carbon content has increased from 75.79% in the RPAL to 82.90% in the burnt sample (Figure 7D,E). In addition, the oxygen content has decreased from 22.91% in the RPAL to 10.27% in RPAL500. This decrease might be attributed to the loss of water oxygen during the burning process, an issue that might have a negative impact on the removal efficiency of the thermally treated samples and as was confirmed by the FT-IR and Raman analyses.

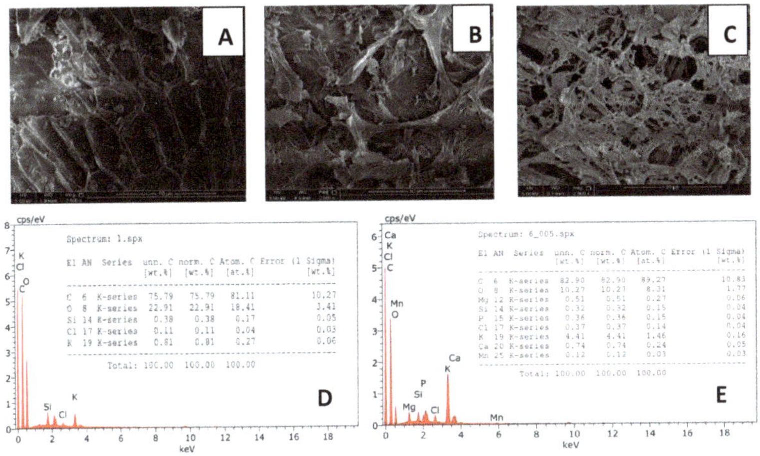

Figure 7. The upper panel is the SEM micrographs of RPAL (**A**), TTPAL250 (**B**), and TTPAL500 (**C**). The lower panel is the EDX analysis of RPAL (**D**), and TTPAL500 (**E**).

2.5.5. Carbon, Hydrogen, and Nitrogen Analysis (CHN)

Data shown in Table 5 represent a comparison between three samples RPAL, TTPAL250, and TTPAL500 in terms of the percentage Carbon, Hydrogen, and Nitrogen. The collected data show that the %C and %N has increased following the thermal treatment in contrast to the %H. These findings indicate that the burning process might cause the loss of hydrogen in crystalline and physical water in contrast to the carbon concentration, which has increased because of the conversion of the biomass into carbon during the burning process.

Table 5. CHN Elemental analysis of the prepared adsorbents.

Adsorbent	%C	%H	%N
RPAL	39.555	4.991	2.447
TTPAL250	52.140	4.942	3.117
TTPAL500	50.353	2.529	2.438

2.5.6. Brunauer–Emmett–Teller (BET) Surface Area Analysis

Table 6 shows the measured BET surface area and the total pore volume of the three adsorbents using N_2 adsorption–desorption measurements. The obtained data show that the surface area of RPAL is 4.59 m^2/g and this area has increased (almost doubled) following thermal treatment to 9.81 m^2/g for TTPAL500 with no much difference between TTPAL250 and TTPAL500. On the other hand, the total pore volume has increased from 0.016 to 0.041 cm^3/g for RPAL and TTPAL500, respectively. This increase in the pore volume is confirmed by the SEM micrographs. Conversely, the pore radius has decreased in the thermally treated samples compared to the raw one. These findings together with the FT-IR and Raman, and FCCCD analyses might explain the superiority of RPAL as adsorbent compared to the TTPAL250 and TTPAL500 samples, and confirm that the adsorption process is controlled by the chemical structure of the adsorbent surface, which in turn is affected by the adsorption conditions. Figure 8 displays that the three adsorbents show a type III adsorption isotherm with H3—hysteresis loop, indicating the unrestricted multilayer formation and that lateral interactions between the adsorbate molecules are stronger than the interactions between adsorbent and the adsorbate. The H3—hysteresis indicates the aggregation of plate–like particles to form slit–like pores in loose assemblies. Furthermore, it also shows the presence of two types of pores including mesopores (2–50 nm diameter) and macropores

(>50 nm diameter, according to the IUPAC classification), in alignment with the analysis of SEM micrographs, Figure 7 [43].

Table 6. Brunauer–Emmett–Teller (BET) analysis of RAPL and thermally treated samples.

Parameters	RPAL	TTPAL250	TTPAL500
Langmuir SA (m^2/g)	4.59	8.43	9.81
Total pore volume (cm^3/g)	0.016081	0.02674	0.040636
Average pore radius (°A)	105.5	81.4	96

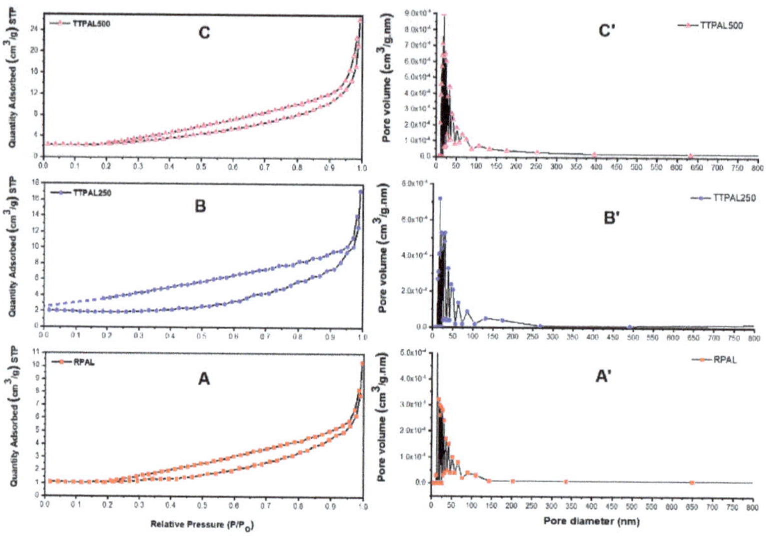

Figure 8. BET analysis of (**A**) RPAL, (**B**) TTPAL250, and (**C**) TTPAL500.

2.6. *Equilibrium and Kinetics Studies of the Adsorption of RB onto PAL*

The data displayed in Table 2 prove that RPAL has higher adsorption efficiency compared to the thermally treated samples, hence, the equilibrium isotherms and kinetics studies were carried out using the RPAL sample. Important information, such as the maximum quantity adsorbed, the type of interaction (chemi—or physisorption) between the adsorbate and the adsorbent surface, are by and large obtained using adsorption isotherms. Kinetics studies, on the other hand, are used to find the different factors affecting the adsorption process including adsorption rate, type of the layer formed on the surface of the adsorbent (mono or multilayer), and the type of the adsorption mechanisms. The data given below will show the kinetics and adsorption isotherms of the adsorption of RB dye onto the RPAL sample.

2.6.1. Equilibrium Isotherms

The biosorption of RB dye onto the RPAL was studied using four isotherms: (1) Langmuir, (2) Freundlich, (3) Temkin, and (4) Dubinin–Radushkevich (DR) paradigms [44–47]. Single-layer homogeneous adsorption on the surface of the adsorbent was explained by Langmuir isotherm as shown in Figure 9A and Table 7. The Langmuir equation is shown below:

$$q_e = \frac{q_m\, K_L\, C_e}{1 - K_L\, C_e} \quad (5)$$

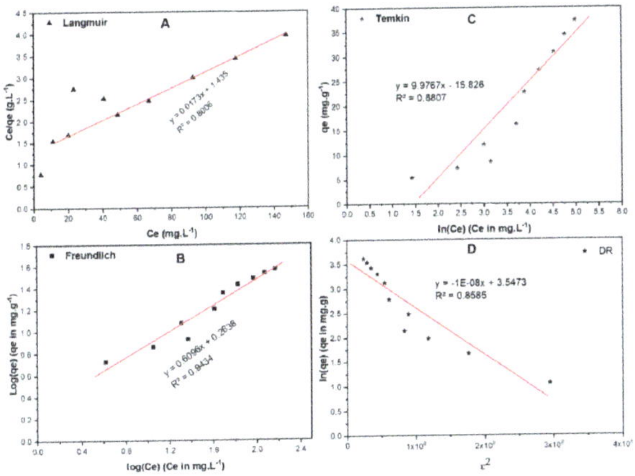

Figure 9. Adsorption isotherms of RB on RPAL including (**A**) Langmuir, (**B**) Freundlich, (**C**) Temkin, and (**D**) Dubinin–Radushkevich (DR).

Table 7. General and linearized equation of Langmuir, Freundlich, Temkin and Dubinin–Radushkevich isotherms, beside their parameters for the adsorption of RB on ADPP.

Isotherm	Equations (Generalized/Linearized Forms)	Parameters	Value
Langmuir	$q_e = \frac{q_m\ K_L\ C_e}{1-K_L\ C_e}$	q_m (mg/g)	58.80
		K_L (L·mole^{-1})	0.012
	$\frac{C_e}{q_e} = \frac{1}{q_m\ K_L} + \frac{C_e}{q_m}$	R^2	0.801
Freundlich	$q_e = K_F C_e^{\frac{1}{n}}$	$\frac{1}{n}$	0.609
		K_F (mole/g) (L/mole)$^{1/n}$	1.835
	$\log(q_e) = \log(K_F) + \left(\frac{1}{n}\right)\log(C_e)$	R^2	0.943
Temkin	$q_e = \frac{RT}{b_T}\ln(A_T\ C_e)$	b_T (J/mole)	248.4
		A_T (L/mole)	0.205
	$q_e = \frac{RT}{b_T}\ln(A_T) + \frac{RT}{b_T}\ln(C_e)$	R^2	0.881
DR	$\ln(q_e) = \ln(q_m) - \beta\epsilon^2$	β	1×10^{-8}
	$\epsilon = RT(1 + \frac{1}{C_e})$	E (kJ/mole)	7.07
	$E = 1/\ \sqrt{2\beta}$	q_s (mg·g)	34.72
		R^2	0.858

In Equation (5), q_m and K_L stand for the maximum adsorption capacity and the Langmuir equilibrium coefficient, respectively. Langmuir equation can be expressed using the following formula:

$$R_L = \frac{1}{1 - K_L\ C_0} \tag{6}$$

where R_L and C_0 represent the separation factor and the initial concentration (mg/L) respectively. The R_L value reflects the feasibility of the sorption process. Therefore, if R_L is higher than 1, the adsorption process is counted as unfavorable and if R_L is equal to 1, the adsorption isotherm is linear. In cases where the R_L value is in the range between 0 and 1, then the adsorption process is favorable, and it occurs spontaneously, while if R_L is equal to 0, the adsorption is expressed as irreversible process [47]. Based on the obtained data for the current work, the R_L value was found to be less than 1 and higher than 0, indicating that the biosorption of RB onto RPAL was spontaneous and the monolayer maximum adsorption capacity (q_{max}) = 58.80 mg/g.

The heterogeneous adsorption is usually portrayed using the Freundlich isotherm described by the following equation:

$$q_e = K_F C_e^{\frac{1}{n}} \tag{7}$$

where C_e is the equilibrium concentration of RB (mg L^{-1}); q_e is the amount of RB adsorbed/unit mass (mg·g^{-1}), while K_F (mole·g^{-1}) (L·$mole^{-1}$)$^{1/n}$ and $1/n$, are the Freundlich coefficients. This model presumes neither homogenous adsorption nor restricted level of biosorption. According to the obtained data shown in Figure 9B and Table 7, the highest R^2 value (0.943)–among the four studied models—was obtained using the Freundlich isotherm, implying that this model holds for the RB—RPAL system. Freundlich coefficient $1/n = 0.609$ and $n = 1.642$, signifying that the biosorption of RB is favorable where the value of $1/n < 1$. This isotherm also designates that the adsorption might not be monolayer and that adsorption sites with higher affinity might be inhabited first. This finding also explains why the removal efficiency (%R) has decreased with increasing [RB].

The adsorbate–adsorbent interaction was also studied using the Temkin isotherm as shown in Figure 9C and Table 7. Temkin isotherm, however, cannot be used to explain the adsorption of RB onto RPAL since the R^2 value = 0.881. The DR isotherm, (Figure 9D and Table 7) was used to detect the type of adsorption on a heterogeneous surface [47]. Based on the reported information on the relation between the free energy value and the adsorption mechanism, where if the free energy value is <8.0 kJ/mol, the adsorption process is physisorption while if the free energy is >8.0 kJ/mol then the adsorption process will be chemisorption. According to the data revealed in Table 7, the free energy for adsorption of RB onto RPAL is physisorption where the amount of free energy equals 7.07 kJ/mole. Yet, this type of isotherm might not be applicable in the current investigation where data did not show an excellent goodness–of–fit with $R^2 = 0.858$. These findings together with the characterization outcomes show that and though free energy implies physisorption, occurrence of chemisorption cannot be ruled out [48].

2.6.2. Biosorption Kinetics

In this study, four models were tested; pseudo–first order (PFO), pseudo–second order (PSO), Elovich and Weber–Morris (W–M) to explain the kinetics of the adsorption process of RB onto RPAL. The data shown in Figure 10A,B represent the plots of [ln(q_e–q_t) vs. time] and [time/q_t vs. time] for the two tested kinetic models; PFO and PSO, respectively. Other parameters together with their values are listed in Table 8. By comparing the linearity and the calculated adsorption capacity at equilibrium for these two models, it can be detected that the PSO model is more applicable in explaining the adsorption of RB onto RPAL [49–51]. Therefore, the reaction of RB with RPAL can be expressed as:

$$RB + RPAL \xrightarrow{k} \{RB - RPAL\} \tag{8}$$

Table 8. The kinetics study results corresponding to Figure 10.

Model	Parameter	Value
Pseudo–first order (PFO) $\ln(q_e - q_t) = \ln(q_e) - k_1 t$	K_1 (min^{-1})	0.493
	q_e (mg/g)	12.61
	R^2	0.863
Pseudo–second order (PSO) $\frac{1}{q_e} = \frac{1}{k_2 q_e^2} + \frac{1}{q_e}t$ *where K_2 is rate constant (g·mg^{-1}·min^{-1})*	K_2 (g·mg^{-1}·min^{-1})	0.019
	q_e (mg/g)	25.91
	R^2	0.965
Elovich model $q_t = \beta \ln(\alpha\beta) + \beta \ln(t)$ *where q_t is adsorbed quantity at time t, α is initial sorption concentration rate (mg·g^{-1}·min^{-1}), and β is desorption constant (g/mg).*	A	3.79×10^{12}
	B	1.817
	R^2	0.953
Weber–Morris intra–particle diffusion model $q_t = K_I t^{0.5} + C$ *where K_I is intra–particle diffusion rate constant (mg·g^{-1}·$min^{-0.5}$), and C is boundary thickness effect.*	K_I	1.262
	C	53.66
	R^2	0.888

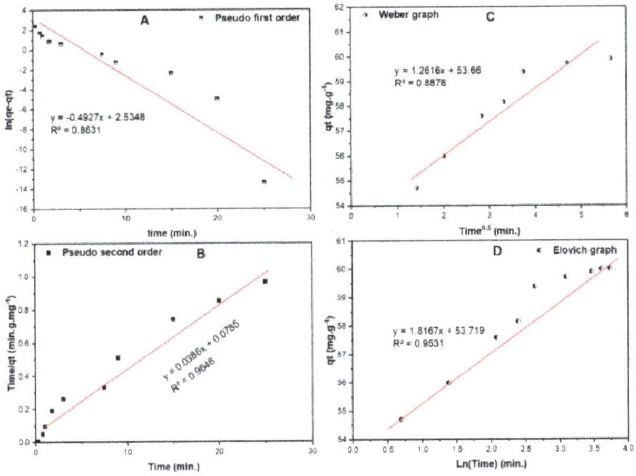

Figure 10. Kinetic models for the adsorption of RB on RPAL including (**A**). Pseudo first order, (**B**). Pseudo second order, (**C**). Elovich and (**D**). intra–particle diffusion (Weber–Morris) curves.

Therefore, the rate of the reaction can be expressed as: k[RB][RPAL], implying that the adsorption rate depends mainly on both RB and RPAL concentrations. Weber–Morris intra–particle diffusion model, Figure 10C, indicates that the diffusion rate is very fast with the value of K_1 = 1.262. The mechanism of adsorption process using this model involves the formation of a layer of RB around the particles of RPAL, which will prevent any penetration of more RB and form a boundary layer (53.66 mg/g). This value is close to the q_{max} obtained from the Langmuir isotherm. Finally, the Elovich model, Figure 10D, shows a low R^2 value (0.953) compared to PSO model. This model shows that the initial adsorption rate ($\alpha = 3.79 \times 10^{12}$ mg·g^{-1}·min^{-1}) is higher than the desorption rate (β = 1.817 g·mg^{-1}). Therefore, the adsorption of RB onto RPAL involves a second–order uptake rate vs. the existing surface sites.

3. Materials and Methods

3.1. Materials and Reagents

The chemicals used were of the analytical grade and were used as acquired with no additional purification. Sodium hydroxide, sodium tetraborate–10–hydrate and hydrochloric acid were purchased from Sigma–Aldrich (Eschenstrasse, Taufkirchen, Germany). Rose bengal (RB) was a product of BDH Laboratory Supplies (Poole, UK). Values of pH were adjusted as previously mentioned [10]. Pure water was used for diluting the RB dye solutions to 1000 ppm. Pineapple leaves (PAL) were used after drying as will be described in their method of preparation.

3.2. Instrumentation and Software

A Jenway pH meter was used for the preparation of different pH dye solution. An ST8 Benchtop Centrifuge (Thermo Scientific, Waltham, MA, USA) was used for separating the components of each sample mixture. The absorbance was measured using an UV–Vis spectrophotometer (Agilent DAD, Agilent, Santa Clara, CA, USA). The surface morphology of the prepared pineapple leaves was identified using a scanning electron microscope (SEM– Quanta 200, Thermo Scientific, Waltham, MA, USA) and energy–dispersive X–ray spectroscopy (EDX, Thermo Scientific, Waltham, MA, USA). Fourier transform infrared radiation (FT-IR, Bruker Alpha, MA, USA) was used to determine the functional groups on the surface of pineapple leaf. The Raman spectrum was recorded in the range from 50–3500 cm^{-1} using a Raman microscope (DXR Raman Microscope, Thermo Scientific, Waltham, MA, USA), with a laser beam at 532 nm as excitation source. Furthermore, a thermal gravimetric analyzer (TGA400, PerkinElmer, Waltham, MA, USA was utilized to inspect the thermal stability of the pineapple leaf. Finally, Minitab®19 software (Minitab Inc., Chicago, IL, USA) was used to construct the face–centered central composite design (FCCCD).

3.3. *Face—Centered Central Composite Design (FCCCD)*

The design of experiment chosen to conduct the current study is FCCCD. The percentage removal (%R) as a single response was optimized as a function for four independent variables, pH, DC, AD, and CT (Table 3). The design matrix involved conducting 30 basic runs in one replicate over two blocks with α = one. Design points involved 16 cube points, eight axial points, and total of six center points. The full design matrix as shown in Table 3.

3.4. *Preparation of RB*

Ultra–pure water was artificially contaminated with RB dye to have a stock solution of 1000 ppm. Serial dilutions of the RB solution were prepared by adjusting the desired pH value using the previously prepared pH adjusting solutions. Three calibration curves were prepared, therefore, at three pH values, Table 3, and measured at 548 nm.

3.5. *Adsorbent Preparation*

3.5.1. Air–Dried Raw Pineapple Leaves (RPAL)

Pineapples were purchased from a local market in Doha–Qatar. The leaves of the pineapple were separated from the bottom of the pineapple fruit using a metal blade. The crown base was detached, then the pineapple leaves were cut into small pieces approximately 1 × 1 cm. These pieces were rinsed with tap water followed by distilled water to remove any impurities or pollutants present on their surface. The cut leaves were then dried and exposed to the sunlight directly for three consecutive days until they are completely dry. Dry leaves were allotted as three portions. The first portion was further dried in air and labeled as raw pineapple leaves (RPAL).

3.5.2. Thermal Treatment of Pineapple Leaves

Portions 2 and 3 were activated in the oven at 250 °C and 500 °C for 1 h, and labeled as, thermally treated pineapple leaves; (TTPAL250), and (TTPAL500), respectively. The three portions and after the previous treatment were chopped well with electrical grinder until it becomes fine powder.

3.6. *Evaluation of the Adsorption Perfomance of the Prepared Adsorbents*

Two batches of 15 mL centrifuge tubes were prepared. The first set was the sample and the second set was for the blanks. In each tube for both sets, 30–150 mg of RPAL was added. The pH value of the RB solutions was adjusted to the desired figure (Table 3). Next, the two sets of samples and blanks were centrifuged at 4200 rpm for the time specified in Table 3 to facilitate obtaining the supernatant. UV–Vis spectrophotometer was used to measure the absorbance of the supernatant.

4. Conclusions

The present work has emphasized that economic PAL adsorbents could be efficiently used for the adsorption of rose bengal (RB) from wastewater. Three types of adsorbents were developed for that purpose, raw (RPAL) and thermally treated PAL at 250 °C and 500 °C. Results showed that RPAL is more efficient for the removal of RB. A smart and ecofriendly platform has been proposed to engineer the removal process. In this context, a response surface methodological approach (face–centered central composite design, FCCCD) was used to optimize the variables influencing the adsorption process. The response (%R) was measured as a function of four factors (pH, AD, DC, and CT). As per the response surface regression model, increasing the dose of RPAL improves the adsorption of the dye, in contrast to pH and DC. FT-IR and Raman spectra were used to examine the prepared adsorbents. FT-IR data showed the presence of–OH, N–H, C–H, and C–O function groups in RPAL as well as in the thermally treated sample but with a lower intensity. Raman spectra showed the formation of carbonaceous material after the burning process as confirmed by the presence of D– and G–bands. The equilibrium studies revealed that the biosorption of RB on RPAL could be represented by the Freundlich isotherm. The maximum monolayer adsorption capacity was 58.80 mg/g as determined by the Langmuir isotherm. Furthermore, the adsorption of RB onto RPAL is physisorption with free energy equals 7.07 kJ/mol as calculated by the Dubinin–Radushkevich (DR) isotherm. However, and considering the SEM and BET analyses together with the FT-IR findings, occurrence of chemisorption cannot be ruled out. The kinetic studies showed that the adsorption process was a second–order reaction and adsorption rate depends mainly on both RB and RPAL concentrations.

Author Contributions: Conceptualization, M.E.-A. and S.S.H.; methodology, A.S.E.-S. and N.Z.; software, M.E.-A.; validation, M.E.-A., A.S.E.-S. and S.S.H.; formal analysis, M.E.-A., S.S.H., A.S.E.-S.; investigation, M.E.-A., S.S.H., A.S.E.-S., and N.Z.; resources, M.E.-A.; data curation, M.E.-A., N.Z., A.S.E.-S. and S.S.H.; writing—original draft preparation, M.E.-A. and N.Z.; writing—review and editing, M.E.-A., S.S.H., A.S.E.-S. and N.Z.; visualization, M.E.-A., and S.S.H.; supervision, M.E.-A.; project administration, M.E.-A. All authors have read and agreed to the published version of the manuscript.

Funding: This research received no external funding.

Acknowledgments: The project members would like to extend their special thanks to the Central Lab Unit (CLU) at Qatar University.

Conflicts of Interest: The authors declare no conflict of interest.

References

1. Hinrichsen, D.; Tacio, H.D. Finding the source: The Coming Freshwater Crisis is Already Here. Available online: http://www.wilsoncenter.org/topics/pubs/popwawa2 (accessed on 26 April 2020).
2. *Global Water Supply and Sanitation Assessment 2000 Report*; World Health Organization: Geneva, Switzerland, 2000. Available online: http://www.who.int/water_sanitation_health/monitoring/jmp2000 (accessed on 4 May 2020).
3. Ullah, R.; Malik, R.; Qadir, A. Assessment of groundwater contamination in an industrial city, Sialkot, Pakistan. *Afr. J. Environ. Sci. Technol.* **2009**, *3*, 429–446. [CrossRef]
4. Owa, F.D. Water pollution: Sources, effects, control and management. *Mediterr. J. Soc. Sci.* **2013**, *4*, 65–68. [CrossRef]
5. El-Azazy, M.; Kalla, R.N.; Issa, A.A.; Al-Sulaiti, M.; El-Shafie, A.S.; Shomar, B.; Al-Saad, K. Pomegranate peels as versatile adsorbents for water purification: Application of Box–Behnken design as a methodological optimization approach. *Environ. Prog. Sustain. Energy* **2019**, *38*, 13223. [CrossRef]
6. Al-Saad, K.; El-Azazy, M.; Issa, A.A.; Al-Yafei, A.; El-Shafie, A.S.; Al-Sulaiti, M.; Shomar, B. Recycling of date pits into a green adsorbent for removal of heavy metals: A fractional factorial design-based approach. *Front. Chem.* **2019**, *7*, 552. [CrossRef]
7. El-Azazy, M.; El-Shafie, A.S.; Issa, A.A.; Al-Sulaiti, M.; Al-Yafie, J.; Shomar, B.; Al-Saad, K. Potato peels as an adsorbent for heavy metals from aqueous solutions: Eco-structuring of a green adsorbent operating Plackett-Burman Design. *J. Chem.* **2019**, *2019*, 4926240. [CrossRef]
8. El-Azazy, M.; Dimassi, S.; El-Shafie, A.S.; Issa, A.A. Bio-Waste Aloe vera Leaves as an efficient adsorbent for Titan Yellow from Wastewater: Structuring of a novel adsorbent using Plackett-Burman factorial design. *J. Appl. Sci.* **2019**, *9*, 4856. [CrossRef]
9. El-Azazy, M.; El-Shafie, A.S.; Ashraf, A.; Issa, A.A. Eco-structured biosorptive removal of basic fuchsin using pistachio nutshells: A definitive screening design—Based approach. *J. Appl. Sci.* **2019**, *9*, 4855. [CrossRef]
10. Avci, A.; Inci, I.; Baylan, N. Adsorption of ciprofloxacin hydrochloride on multiwall carbon nanotube. *J. Mol. Struct.* **2020**, *1206*, 127711. [CrossRef]
11. Hao, O.J.; Kim, H.; Chiang, P.-C. Decolorization of wastewater. *Crit. Rev. Environ. Sci. Technol.* **2000**, *30*, 449–505. [CrossRef]
12. Benkhaya, S.; Harfi, S.; El Harfi, A. Classifications, properties and applications of textile dyes: A review. *Appl. J. Environ. Eng. Sci.* **2017**, *3*, 311–320. [CrossRef]

13. Shabir, G.; Saeed, A.; Channar, P.A. A review on the recent trends in synthetic strategies and applications of xanthene dyes. *Mini Rev. Org. Chem.* **2018**, *14*. [CrossRef]
14. Tabery, H.M. Toxic effect of rose bengal dye on the living human corneal epithelium. *Acta Ophthalmol. Scand.* **1998**, *76*, 142–145. [CrossRef] [PubMed]
15. Chilvers, K.F.; Reed, R.H.; Perry, J.D. Phototoxicity of rose bengal in mycological media-implications for laboratory practice. *Lett. Appl. Microbiol.* **2008**, *28*, 103–107. [CrossRef] [PubMed]
16. Vinuth, M.; Bhojya, N.; Vinoda, B.M.; Pradeepa, S.M.; Kumar, A.; Chandra, S. Rapid removal of hazardous rose bengal dye using Fe(III)—Montmorillonite as an effective adsorbent in aqueous solution. *J. Environ. Anal. Toxicol.* **2016**, *6*, 2. [CrossRef]
17. Ahmed, M.; Abdelbar, N.M.; Mohamed, A.A. Molecular imprinted chitosan-TiO_2 nanocomposite for the selective removal of Rose Bengal from wastewater. *Int. J. Biol. Macromol.* **2018**, *107*, 1046–1053. [CrossRef] [PubMed]
18. Dutta, A.K.; Maji, S.K.; Adhikary, B. γ-Fe_2O_3 nanoparticles: An easily recoverable effective photo-catalyst for the degradation of rose Bengal and methylene blue dyes in the waste-water. *Mater. Res. Bull.* **2014**, *49*, 28–34. [CrossRef]
19. Elfeky, A.S.; Salem, S.S.; Elzaref, A.S.; Owda, M.E.; Eladawy, H.A.; Saeed, A.M.; Awad, M.A.; Abou-Zeid, R.E.; Fouda, A. Multifunctional cellulose nanocrystal /metal oxide hybrid, photodegradation, antibacterial and larvicidal activities. *Carbohydr. Polym.* **2020**, *230*, 115711. [CrossRef]
20. Rasool, M.A.; Goethem, V.C.; Vankelecom, I.F.J. Green preparation process using methyl lactate for cellulose-acetate-based nanofiltration membranes. *Sep. Purif. Technol.* **2019**, *232*, 115903. [CrossRef]
21. Karimi, A.; Khataee, A.; Safarpour, M.; Vatanpour, V. Development of mixed matrix ZIF-8/polyvinylidene fluoride membrane with improved performance in solvent resistant nanofiltration. *Sep. Purif. Technol.* **2019**, *237*, 116358. [CrossRef]
22. Shen, G.; Pan, L.; Zhang, R.; Sun, S.; Hou, F.; Zhang, X.; Zou, J. Low-spin-state hematite with superior adsorption of anionic contaminations for water purification. *Adv. Mater.* **2020**, *32*, 1905988. [CrossRef]
23. Sivamani, S.; Leena, G.B. Removal of dyes from wastewater using adsorption—A review. *Int. J. Biosci. Technol.* **2009**, *2*, 47–51. [CrossRef]
24. Jiahao, M.; Yang, Q.; Zhang, N.; Zhang, W.; Zheng, Y.; Zhang, Z. A review on agro-industrial waste (AIW) derived adsorbents for water and wastewater treatment. *J. Environ. Manag.* **2018**, *227*, 395–405. [CrossRef]
25. Asim, M.; Abdan, K.; Jawaid, M.; Nasir, M.; Dashtizadeh, Z.; Ishak, M.R.; Hoque, M.E. A review on pineapple leaves fibre and its composites. *Int. J. Polym. Sci.* **2015**, *2015*, 4–5. [CrossRef]
26. Bartholomew, D.P.; Paull, R.E.; Rohrbach, K.G. *The Pineapple: Botany, Production, and Uses*, 1st ed.; CAB International: New York, NY, USA, 2003; pp. 3–5.
27. Nanthaya, K.; Taweechai, A. A new approach to "Greening" plastic composites using pineapple leaf waste for performance and cost effectiveness. *Mater. Des.* **2014**, *55*, 292–299.
28. Mishra, S.; Misra, M.; Tripathy, S.S.; Nayak, S.K.; Mohanty, A.K. Potentiality of pineapple leaf fibre as reinforcement in PALF-polyester composite: Surface modification and mechanical performance. *J. Reinf. Plast. Comp.* **2001**, *20*, 321–334. [CrossRef]
29. Weng, C.H.; Wu, Y.C. Potential low-cost biosorbent for copper removal: Pineapple leaf powder. *J. Environ. Eng.* **2012**, *138*, 286–292. [CrossRef]
30. Mopoung, R.; Kengkhetkit, N. Lead and cadmium removal efficiency from aqueous solution by NaOH treated pineapple waste. *Int. J. Appl. Chem.* **2016**, *12*, 23–35.
31. Gogoi, S.; Chakraborty, S.; Saikia, M.D. Surface modified pineapple crown leaf for adsorption of Cr(VI) and Cr(III) ions from aqueous solution. *J. Environ. Chem. Eng.* **2018**, *6*, 2492–2501. [CrossRef]
32. Weng, C.H.; Lin, Y.T.; Tzeng, T.W. Removal of methylene blue from aqueous solution by adsorption onto pineapple leaf powder. *J. Hazard. Mater.* **2009**, *170*, 417–424. [CrossRef]
33. Mahamad, M.N.; Zaini, M.A.A.; Zakaria, Z.A. Preparation and characterization of activated carbon from pineapple waste biomass for dye removal. *Int. Biodeterior. Biodegrad.* **2015**, *102*, 274–280. [CrossRef]
34. Rahmat, N.A.; Ali, A.A.; Salmiati, H.N.; Muhamad, M.S.; Kristanti, R.A.; Hadibarata, T. Removal of Remazol Brilliant Blue R from aqueous solution by adsorption using pineapple leaf powder and lime peel powder. *Water Air Soil Pollut.* **2016**, *227*, 1–11. [CrossRef]

35. Lutpi, N.A.; Yin, T.H.; Shian, W.Y. Removal of methylene blue using pineapple peel powder as adsorbent. In Proceedings of the 3rd CUTSE International Conference Miri, Sarawak, Malaysia, 8–9 November 2011; pp. 352–356.
36. Elazazy, M.S. Factorial design and machine learning strategies: Impacts on pharmaceutical analysis. In *Spectroscopic Analyses*; Zafar, F., Ed.; IntechOpen: London, UK, 2017. Available online: https://www.intechopen.com/books/spectroscopic-analyses-developments-and-applications/factorial-design-and-machine-learning-strategies-impacts-on-pharmaceutical-analysis (accessed on 1 May 2020). [CrossRef]
37. Ho, L.H.; Feng, S.Y.; Yen, T.M. Using modified IPA to improve service quality of standard hotel in Taiwan, an analysis of transformations. *J. Serv. Sci. Manag.* **2014**, *7*, 211–252.
38. Narayanaswamy, R.; Wolfbeis, O.S. *Optical Sensors: Industrial Environmental and Diagnostic Applications*; Springer: Berlin/Heidelberg, Germany, 2004. [CrossRef]
39. Sabnis, R.W. *Handbook of Biological Dyes and Stains: Synthesis and Industrial Applications*, 1st ed.; John Wiley & Sons: Hoboken, NJ, USA, 2010. [CrossRef]
40. Gupta, V.K.; Mittal, A.; Jhare, D.; Mittal, J. Batch and bulk removal of hazardous colouring agent rose bengal by adsorption over bottom Ash. *RSC Adv.* **2012**, *2*, 8381–8389. [CrossRef]
41. Stankovich, S.; Dikin, D.A.; Piner, R.D. Synthesis of graphene-based nanosheets via chemical reduction of exfoliated graphite oxide. *Carbon* **2007**, *45*, 1558–1565. [CrossRef]
42. Childres, I.; Jauregui, L.A.; Park, W.; Cao, H.; Chen, Y.P. Raman spectroscopy of graphene and related materials. In *New Developments in Photon and Materials Research*, 1st ed.; Jang, J.I., Ed.; Nova Science Publishers, Inc.: Oxford, UK, 2013; pp. 403–418.
43. Thommes, M.; Kaneko, K.; Neimark, A.V. Physisorption of gases, with special reference to the evaluation of surface area and pore size distribution (IUPAC Technical Report). *Pure Appl. Chem.* **2015**, *87*, 1052–1069. [CrossRef]
44. Langmuir, I. Adsorption of gases on glass, mica and platinum. *J. Am. Chem. Soc.* **1918**, *40*, 1361–1403. [CrossRef]
45. Guo, X.; Wang, J. Comparison of linearization methods for modeling the Langmuir adsorption isotherm. *J. Mol. Liq.* **2019**, *296*, 111850. [CrossRef]
46. Araújo, C.S.T.; Almeida, I.L.S.; Rezende, H.C.; Marcionilio, S.M.L.O.; Léon, J.J.L.; de Matos, T.N. Elucidation of mechanism involved in adsorption of Pb(II) onto lobeira fruit (*Solanum lycocarpum*) using Langmuir, Freundlich and Temkin isotherms. *Microchem. J.* **2018**, *137*, 348–354. [CrossRef]
47. Moussavi, G.; Barikbin, B. Biosorption of chromium (VI) from industrial wastewater onto pistachio hull waste biomass. *Chem. Eng. J.* **2010**, *162*, 893–900. [CrossRef]
48. Ali, R.M.; Hamad, H.A.; Hussein, M.M.; Malash, G.F. Potential of using green adsorbent of heavy metal removal from aqueous solutions: Adsorption kinetics, isotherm, thermodynamic, mechanism and economic analysis. *Ecol. Eng.* **2016**, *91*, 317–332. [CrossRef]
49. Sarma, G.K.; Khan, A.; El-Toni, A.M.; Rashid, M.H. Shape-tunable CuO-Nd$(OH)_3$ nanocomposites with excellent adsorption capacity in organic dye removal and regeneration of spent adsorbent to reduce secondary waste. *J. Hazard. Mater.* **2019**, *380*, 120838. [CrossRef] [PubMed]
50. Sun, B.; Yuan, Y.; Li, H.; Li, X.; Zhang, C.; Guo, F.; Liu, X.; Wang, K.; Zhao, X. Waste-cellulose-derived porous carbon adsorbents for methyl orange removal. *Chem. Eng. J.* **2019**, *371*, 55–63. [CrossRef]
51. Hubbe, M.A.; Azizian, S.; Douven, S. Implications of apparent pseudo-second-order adsorption kinetics onto cellulosic materials: A review. *BioResources* **2019**, *14*, 7582–7626.

molecules

MDPI

Article

Evidence for Phytoremediation and Phytoexcretion of NTO from Industrial Wastewater by Vetiver Grass

Abhishek RoyChowdhury [1], Pallabi Mukherjee [2], Saumik Panja [2], Rupali Datta [3], Christos Christodoulatos [4] and Dibyendu Sarkar [2,*]

[1] Environmental Science and Natural Resources Program, School of Science, Navajo Technical University, Crownpoint, NM 87313, USA; aroychowdhury@navajotech.edu
[2] Department of Civil, Environmental and Ocean Engineering, Stevens Institute of Technology, Hoboken, NJ 07030, USA; pallabi.mkrje@gmail.com (P.M.); spanja1@stevens.edu (S.P.)
[3] Department of Biological Sciences, Michigan Technological University, Houghton, MI 49931, USA; rupdatta@mtu.edu
[4] Center for Environmental Systems, Stevens Institute of Technology, Hoboken, NJ 07030, USA; christod@stevens.edu
* Correspondence: dsarkar@stevens.edu; Tel.: +1-201-2168028

Abstract: The use of insensitive munitions such as 3-nitro-1,2,4-triazol-5-one (NTO) is rapidly increasing and is expected to replace conventional munitions in the near future. Various NTO treatment technologies are being developed for the treatment of wastewater from industrial munition facilities. This is the first study to explore the potential phytoremediation of industrial NTO-wastewater using vetiver grass (*Chrysopogon zizanioides* L.). Here, we present evidence that vetiver can effectively remove NTO from wastewater, and also translocated NTO from root to shoot. NTO was phytotoxic and resulted in a loss of plant biomass and chlorophyll. The metabolomic analysis showed significant differences between treated and control samples, with the upregulation of specific pathways such as glycerophosphate metabolism and amino acid metabolism, providing a glimpse into the stress alleviation strategy of vetiver. One of the mechanisms of NTO stress reduction was the excretion of solid crystals. Scanning electron microscopy (SEM), electrospray ionization mass spectrometry (ESI-MS), and Fourier-transform infrared spectroscopy (FTIR) analysis confirmed the presence of NTO crystals in the plant exudates. Further characterization of the exudates is in progress to ascertain the purity of these crystals, and if vetiver could be used for phytomining NTO from industrial wastewater.

Keywords: insensitive munitions; 3-nitro-1,2,4-triazol-5-one (NTO); industrial wastewater; vetiver grass; phytoremediation; phytoextraction

check for updates

Citation: RoyChowdhury, A.; Mukherjee, P.; Panja, S.; Datta, R.; Christodoulatos, C.; Sarkar, D. Evidence for Phytoremediation and Phytoexcretion of NTO from Industrial Wastewater by Vetiver Grass. *Molecules* **2021**, *26*, 74. https://dx.doi.org/10.3390/molecules26010074

Academic Editors: Chiara Bisio and Monica Pica

Received: 26 November 2020
Accepted: 22 December 2020
Published: 26 December 2020

1. Introduction

Conventional explosives such as 2,4,6-trinitrotoluene (TNT) and 1,3,5-trinitro-1,3,5-triazine (RDX) have been used in weapons for decades. Since the 1990s, however, the focus has shifted to developing formulations of insensitive munitions (IMs), which are safer to handle but remain as effective as conventional explosives [1–3]. 3-nitro-1,2,4-triazol-5-one (NTO) is one of the main constituents of IM formulations [4]. It is less sensitive to heat and shock and is safer to handle and transport. Detailed chemical properties of NTO are presented in Supplementary Information (Table S1). NTO is more water-soluble than conventional explosives such as RDX and TNT. The solubility of NTO in water increases from 9.97 to 1989.67 g/L when the temperature increases from 11 to 33 °C [5]. Due to its high solubility, wastewater generated in munition plants containing NTO requires physical, chemical, and/or biological treatment according to regulatory standards before being released into the environment.

As wastewaters produced in industrial munition facilities have the potential to contain residues of explosive compounds and their transformation products, they are subjected to extensive industrial waste treatment processes under regulatory discharge guidelines [6–9]. These processes can be expensive and inefficient. Aerobic and anaerobic biological treatment processes have been explored for the remediation of NTO in wastewater. Under anaerobic conditions, NTO is biotransformed into ATO (3-amino-1,2,4-triazol-5-one), which requires further treatment based on regulatory standards. In a sequential anaerobic-aerobic biodegradation study, while NTO biotransformed into ATO under anaerobic conditions, ATO later mineralized under aerobic conditions [10,11]. Sorbents such as granular activated carbon (GAC) are ineffective, as NTO carries an electrostatic charge in aqueous solutions and sorbs very poorly to GAC [1,12]. Other processes such as reverse osmosis (RO) and electrochemical degradation for NTO removal either produce concentrated waste streams or additional regulated byproducts [1]. A Fe/Cu bimetal system was used to remove NTO from an aquatic medium, and a pH and a solid-to-liquid ratio-based removal of NTO was reported [4,13]. The phototransformation of NTO in an aqueous medium was also tested [14]. The fate of NTO during biological wastewater treatment was also studied, and the ability of wastewater sludges to promote the biotransformation of NTO to ATO was documented [15].

It is important to develop more effective treatment technologies for wastewater streams containing NTO, since the processes tested so far are expensive, ineffective, or produce harmful byproducts. The objective of this study was to evaluate the potential use of vetiver grass (*Chrysopogon zizanioides* L.) to remove NTO from wastewater. Vetiver is high biomass, fast-growing, perennial grass. It has an extensive root system that can penetrate deeply (3–4 m). Vetiver's ability to remove various environmental contaminants including various metals and antibiotics is well studied [16–20]. Vetiver was also shown to be effective in the remediation of various explosive compounds from the environment [21,22]. Studies showed that vetiver grass has the potential to remove TNT, RDX, HMX (1,3,5,7-tetranitro-1,3,5,7-tetrazocane), DNAN (2,4-dinitroanisole), and NQ (nitroguanidine) from hydroponic media [23–25]. Vetiver's ability to uptake TNT from the soil in the presence of urea has also been reported [26,27]. Studies showed that plant–microbe interaction plays a significant role in vetiver's ability to remove TNT from soil [28]. While vetiver has been recognized as an effective candidate for the phytoremediation of several explosive compounds, its potential for removing NTO from water or soil has not yet been tested.

2. Results and Discussion

2.1. Uptake of NTO by Vetiver Grass

The untreated NTO-wastewater used in this study was alkaline, with a pH of 10.2 ± 0.1 ($\pm$standard deviation, SD). NTO concentration in the wastewater was measured as $23{,}161 \pm 135$ mg/L ($\pm$ SD). Nitrate, nitrite, and ammonium-nitrogen concentrations in the wastewater were 1680 ± 185 ($\pm$ SD), 1.07 ± 0.05 ($\pm$ SD), and 19.8 ± 2.3 mg/L ($\pm$ SD), respectively. The wastewater also contained 1.5 ± 0.3 mg/L ($\pm$ SD) of Na and 44 ± 3 mg/L ($\pm$SD) of Ca.

Two different treatments were used for this study: (1) vetiver grown in NTO-wastewater (NV), and (2) NTO-wastewater without plants (negative control, NC). NV and NC were set up in triplicate, resulting in six containers in total. The duration of the entire experiment was 100 days. Vetiver batches were replaced every 20 days in NV treatment resulting in five successive batch studies. Figure 1 and Table S2 present the trend of average NTO reduction in both NV and NC treatments. At the end of the 100-day study, the average NTO concentration was reduced by $83.68 \pm 0.43\%$ ($\pm$ SD) of its initial concentration in NV treatments whereas only $5.0 \pm 1.3\%$ ($\pm$ SD). NTO reduction was estimated in NC treatments. For NV treatments, the NTO concentration reduced steadily from the first to the fourth successive batches and eventually stayed steady from days 80–100 for the 5th batch of vetiver. Varying amounts of NTO were removed in the NV treatments by successive batches of vetiver. While the first batch removed $22.73 \pm 1.07\%$ ($\pm$ SD) of the initial NTO, later batches removed $51.97 \pm 0.56\%$ ($\pm$ SD), $68.9 \pm 0.9\%$ ($\pm$ SD),

83.22 ± 1.06% (± SD), and 83.68 ± 0.43% (± SD). NTO reduction was measured at the end of the second, third, fourth, and fifth batches, respectively. No significant NTO reduction in NC treatments indicated that NTO reduction was caused by the vetiver plants. Microbiological transformation of NTO to ATO to urea, CO2, and N2 has been reported earlier [2,10,11]. For this study, no transformation products of NTO were detected in the NTO-wastewater in NV treatments, indicating that NTO had been taken up by vetiver over time. Previous studies showed that plants such as big bluestem (*Andropogon gerardii*), Indiangrass (*Sorghastrum nutans*), and switchgrass (*Panicum virgatum*) uptake DNAN and NQ as nitrogen sources and store them in their roots and shoots [29]. RDX bioaccumulates in the edible parts of plants such as lettuce, tomatoes, and corn [30–38]. We previously reported that vetiver is capable of taking up TNT from soil and water and degrading it within its tissue [22,26]. A decrease in NTO concentration in NV treatments indicate NTO uptake by vetiver. However, since the NTO-wastewater also contained high levels of nitrate, it is not clear if the vetiver used NTO as a nitrogen source.

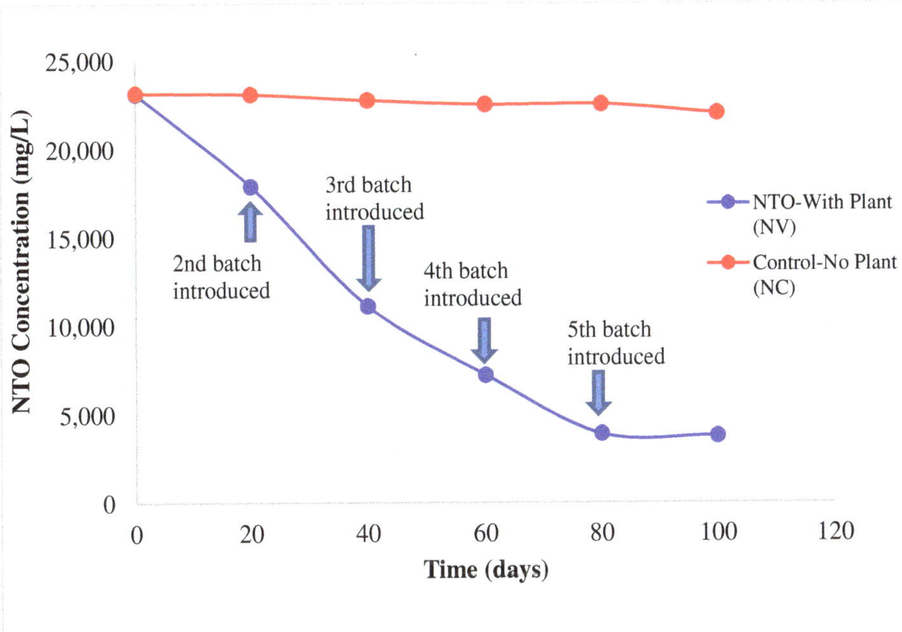

Figure 1. Change in 3-nitro-1,2,4-triazol-5-one (NTO) concentration in NTO-wastewater during the 100 d experiment. New batches of vetiver plants were introduced in NTO with plant (NV) treatments every 20 days. In control experiments (NC), containers of NTO were maintained without vetiver plants. At the end of the 100-day study, the average NTO concentration was reduced by 83.68 ± 0.43% (±SD) of its initial concentration in NV treatments, whereas only 5.0 ± 1.3% (±SD) NTO reduction was estimated in NC treatments.

The presence of NTO in vetiver roots and shoots was detected in plants collected from NV treatments. It was estimated that on average, vetiver shoots and roots contained 830.9 ± 95.1 (±SD) and 747.8 ± 73.5 mg/kg (±SD) NTO, respectively. The calculated translocation factor (TF) for NTO in vetiver was 1.11, which indicated that vetiver translocated NTO from root to shoot. The presence of NTO inside vetiver roots and shoots indicated that NTO had been taken up by vetiver, and a translocation factor above 1 demonstrates that vetiver translocated NTO from its roots to its shoots. Many studies have previously reported the translocation of explosives such as DNAN, NQ, TNT, and RDX by plants to their above-ground biomass [22,23,27,29,31]. This is the first study, to our knowledge, to report the translocation of NTO in a plant.

Figure 2 presents the change in nitrate concentration in NTO-wastewater over time for NV and NC treatments. For NV treatments, 37.26 ± 4.52% (±SD) reduction in nitrate in the NTO-wastewater in comparison to its initial value was observed during this study. Nitrogen is an essential nutrient for plant growth, and the reduction in nitrate in NV treatments is due to its uptake by vetiver plants. In contrast, for the NC treatments, nitrate concentration increased by 14.88 ± 3.34% (±SD) on average from its initial concentration during the study. Nitrate efflux from plants is common and occurs in both stressed and non-stressed plants. Efflux increased in plants that are subjected to mechanical or transplantation stress or changes in the pH of the media [32].

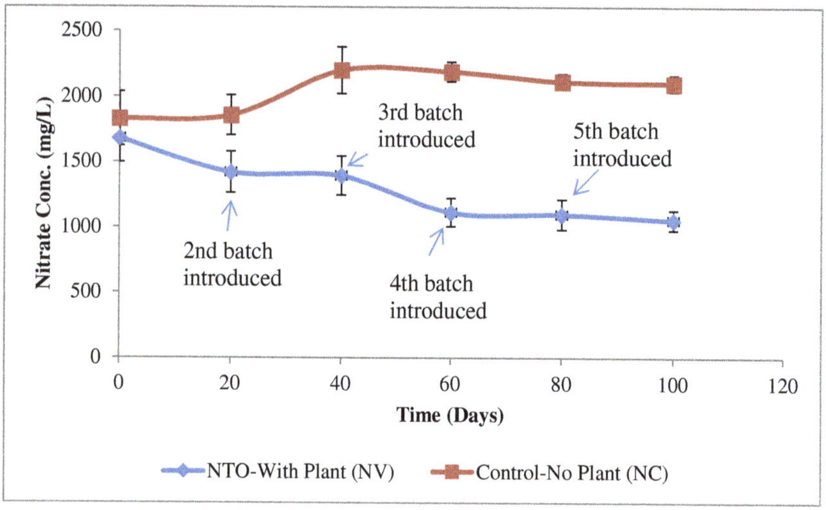

Figure 2. Change in nitrate concentration in NTO-wastewater during the 100 d experiment. New batches of vetiver plants were introduced in NTO with plant (NV) treatments every 20 days. In control experiments (NC), containers of NTO were maintained without vetiver plants. For NV treatments, 37.26 ± 4.52% (±SD) reduction in nitrate in the NTO-wastewater in comparison to its initial value was observed during this study. In contrast, for the NC treatments, nitrate concentration increased by 14.88 ± 3.34% (±SD) on average from its initial concentration during the study.

2.2. NTO Phytotoxicity Analysis

At the beginning of each successive batch study, vetiver plants were individually weighed before introducing them to the experimental treatment, NV (vetiver grown in NTO-wastewater). Each successive batch of plants introduced was of approximately the same size and weight (21.1 ± 0.7 g (±SD)) as the previous batch. Figure 3A presents the change in plant biomass over time. At the end of each successive batch, on average, vetiver lost 16.07 ± 6.83% (±SD) of its initial biomass in NV treatments. It was observed that in the first three successive batches, plant biomass loss was higher (22.01 ± 0.20% (±SD), 18.2 ± 0.6% (±SD), and 22.5 ± 0.5% (±SD), respectively). However, for the fourth and fifth batches, the loss in biomass was much lower (8.59 ± 0.30% (±SD), and 9.03 ± 0.40% (±SD) respectively). NTO concentration was reduced by 69–83% of its initial value when the last two batches of plants were introduced, which might be within the tolerance range of NTO for vetiver. From the loss in vetiver biomass in the NV samples, it is clear that NTO is toxic to vetiver plants. Our results are similar to other studies on the effect of explosive compounds on plants. Studies showed that the biomass of *L. sativa* was significantly reduced by TNT at a concentration higher than 32 mg/kg [33]. Plant biomass was reduced by 40% and 70% at a TNT concentration of 100 and 1000 mg/kg in comparison to the control, respectively. It was also reported that the growth of *Morella cerifera* was impacted significantly at 30 and 100 mg/L TNT and RDX concentrations [34]. Other studies also reported significant growth inhibition in smooth bromegrass (*Bromus* sp.), switchgrass (*Panicum virgatum*), big bluestem (*Andropogon geraldii*), and blue grama (*Bouteloua gracilis*), due to TNT toxicity [34–37].

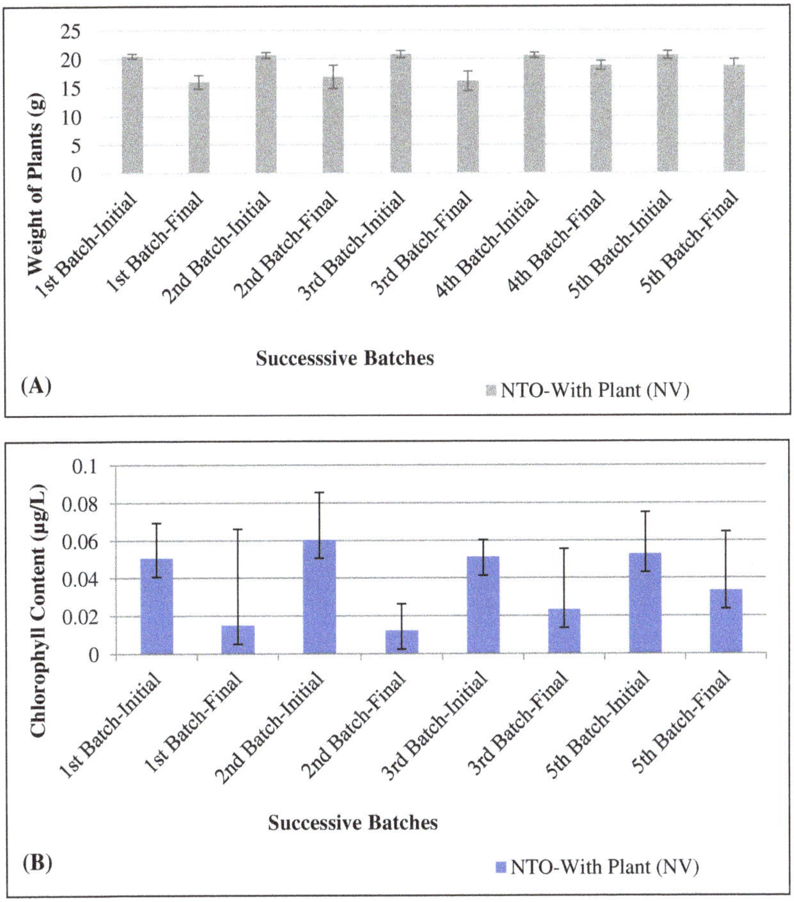

Figure 3. Phytotoxicity of NTO on vetiver grown in NTO-wastewater. (**A**) Plant biomass, (**B**) chlorophyll content. Successive batches of vetiver plants were introduced every 20 days. At the end of each successive batch, on average, vetiver lost 16.07 ± 6.83% (±SD) of its initial biomass in NV treatments. A significant loss in chlorophyll content was observed in vetiver in NV treatments for all batches. An average chlorophyll reduction of 60.18 ± 18.79% (±SD) was noted for the four successive batches.

Figure 3B presents the change in chlorophyll content of vetiver during the successive batches for NV treatment (vetiver grown in NTO-wastewater). A significant loss in chlorophyll content was observed in vetiver in NV treatments for all batches. Signs of chlorosis were visible in all the NV treatments. An average chlorophyll reduction of 60.18 ± 18.79% (±SD) was noted for the four successive batches (data for the fourth batch were not analyzed). These results show that NTO is toxic to vetiver and impacts its chlorophyll content. Other studies also reported significant chlorophyll loss in various plants due to TNT and RDX toxicity [33,38].

The metabolic profiles of vetiver shoot and root tissues exposed to NTO-wastewater (NV treatments) were compared to control tissues of healthy vetiver grown in hydroponic plant growth media (Figure 4). Control tissues showed statistically significant differences in response when compared to treated shoot and root tissue in the PLS-DA model (Figure 4A,B). The major pathways showing upregulation in shoot include (1) glycerophospholipid (GLP) metabolism, (2) galactose metabolism, (3) linoleic acid metabolism, and (4) sphingolipid metabolism (Figure 4A). In the root, the major pathways affected include (1) pyrimidine and purine metabolism, (2) amino acid metabolism (cysteine and methionine) (3) glycerophospholipid metabolism, and (4) linoleic acid metabolism. Figure 4A,B show the significance of the major upregulated metabolites ranked using the variable importance in projection (VIP) score (>1) from the PLS-DA model. The overall metabolic response resembles the osmotic stress response generated by metal or salt stress in plants. Enhancement of galactose and amino acid metabolism could serve to provide osmoprotectants. An increase in lipid peroxidation and membrane damage is indicated by the presence of high levels of phospholipids and linoleic acid. Glycerophospholipids are generated as a result of osmotic stress caused by salt or dehydration [39–41]. They act as signaling molecules that trigger several downstream effects that help plants respond to stress. Large increases in the levels of various amino acids have been reported to combat salt and metal stress in vetiver [32,33].

2.3. *Plant Exudates Analysis*

During the five successive batch studies, all NTO-wastewater-treated plants exuded an unknown solid material at the junction of their root and shoot (Figure S1). The amounts of plant exudates varied for the individual batches. The amounts of exudates showed a decreasing trend from the first to the fifth batches. While the highest amount was exuded in the second batch, very little exudation was seen in the fourth and fifth batches. This result indicates that the exudation correlated with the level of NTO in the wastewater. Halophytes excrete salts as well as metals from their salt glands or trichomes on leaves when exposed to high salt or metal-containing media [34,41–43]. It was reported that as much as 30%–50% of toxic compounds the plants take up are excreted as a detoxification mechanism to protect sensitive photosynthetic tissue from damage [34,40]. Plant exudates were collected and analyzed by SEM, electrospray ionization mass spectrometry (ESI-MS), and FTIR, and the results were compared with pure NTO to decipher any similarities in structure and composition between them.

2.4. *Optical Microscopy and Scanning Electron Microscope (SEM) Analysis*

Before performing the scanning electron microscope (SEM) imaging, plant exudates were initially inspected under an optical microscope. An AmScope digital microscope imaging camera was used to capture pictures under the optical microscope. Figure S2 presents the optical microscope image of plant exudates. Clear crystalline structures can be seen in this picture. NTO is known to form an agglomeration of rod-like large crystals once exposed to air [44], and our findings are in agreement with the earlier report. Figure 5 represents the SEM image of pure solid NTO. Distinctive block-like structures were visible when NTO particles were examined. To our knowledge, no earlier study has reported an SEM image of pure NTO; hence, it was not possible to compare our result with any other study. Figure 5 also presents the SEM image of plant exudates (dried and ground). It is clear from the picture that the plant exudates were a mixture of many different substances. No specific distinctive structural feature was found under the SEM to identify the composition of this material. The presence of block-like structures was seen under 1000× magnification, which showed that NTO is a part of the exudates.

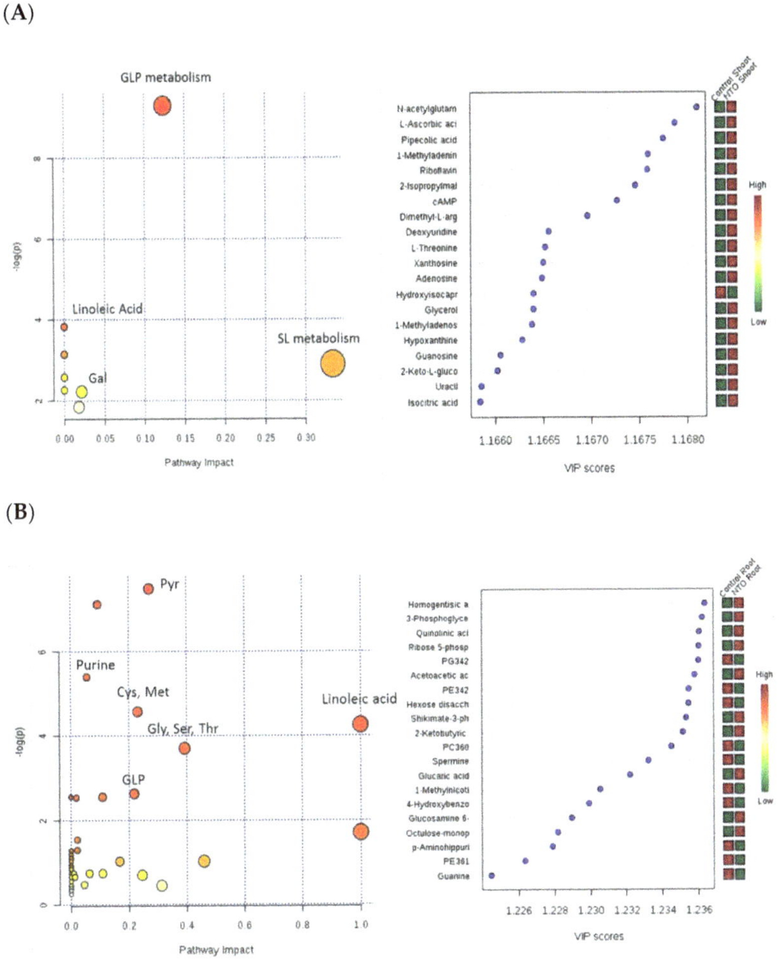

Figure 4. Metabolic profile of vetiver (**A**) shoot and (**B**) root grown in NTO-wastewater treatments (NV) compared to control plants grown in a nutrient medium. GLP metabolism—glycerophospholipid metabolism, Gal—galactose metabolism; Linoleic Acid—linoleic acid metabolism; SL metabolism—sphingolipid metabolism.

2.5. Electrospray Ionization Mass Spectrometry (ESI-MS) Analysis

The ESI-MS analysis of NTO-containing wastewater and exudates was done both in negative and positive mode. The mass to charge ratio (m/z) was calculated to identify the peaks obtained from the samples. The literature showed that in the negative mode, NTO appears at an m/z ratio of 129 Da [2]. Figure 6A presents the ESI-MS results of both NTO-wastewater and plant exudates (dissolved in DI water) in the negative mode. The NTO peak was recorded at an m/z ratio of 129 Da in both NTO-wastewater and plant exudates under negative mode. This result shows that exudates contain NTO particles. As ESI-MS is a qualitative tool, no measurement could be done to quantify the NTO. Several peaks were found in both NTO-wastewater and plant exudate samples. As our NTO-wastewater was an industrial sample, the presence of many other impurities was recorded by ESI-MS spectra. Most of the peaks were found to be adducts of sodium salts (sodium nitrate and sodium carbonate). Figure 6B shows the ESI-MS results of NTO-wastewater and plant exudates in the positive mode. In the positive mode, peaks at m/z ratio of 23 Da and 39 Da position represent sodium and potassium, respectively, which were present in both NTO-wastewater and plant exudates. Our results show that NTO-wastewater contained 1.5 mg/L of total sodium (Na) throughout the study. So, it was clear that sodium, present in the plant exudates, came from the wastewater media.

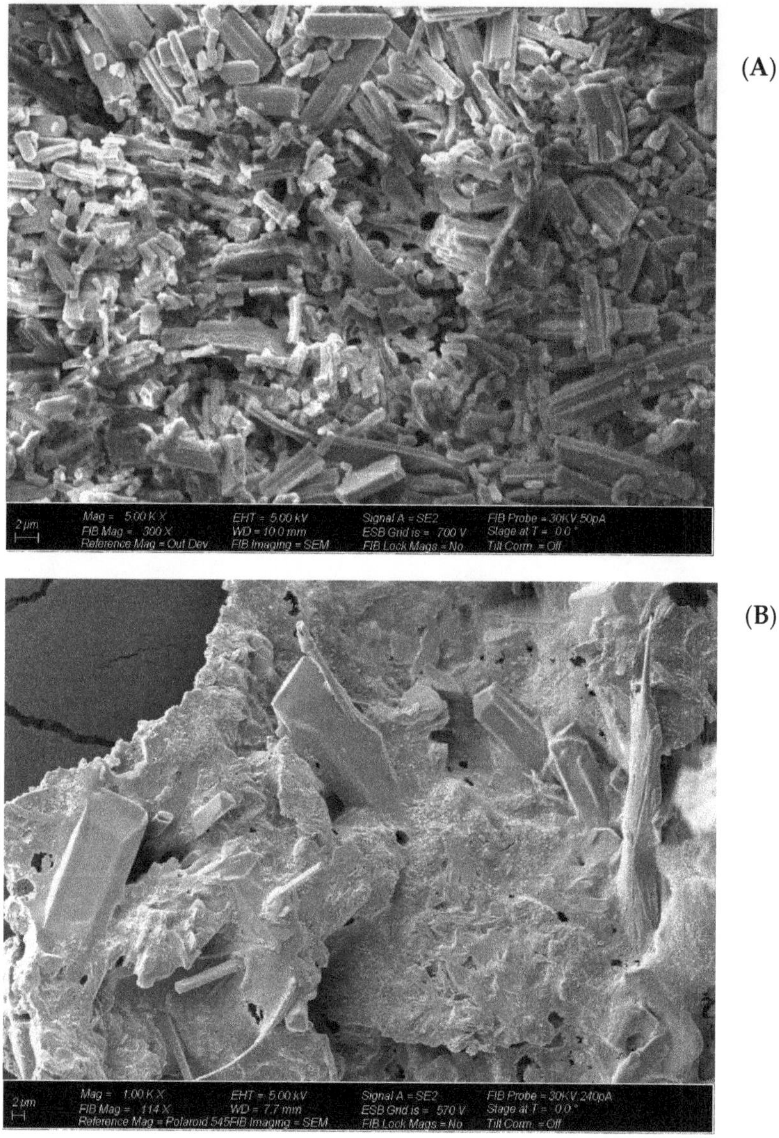

Figure 5. Scanning electron microscope (SEM) image of (**A**) pure NTO solid (5000×) and (**B**) plant exudates (1000×). Distinctive block-like structures were visible when pure NTO particles were examined. The presence of block-like structures was seen in plant exudates, which showed that NTO is a part of the exudates.

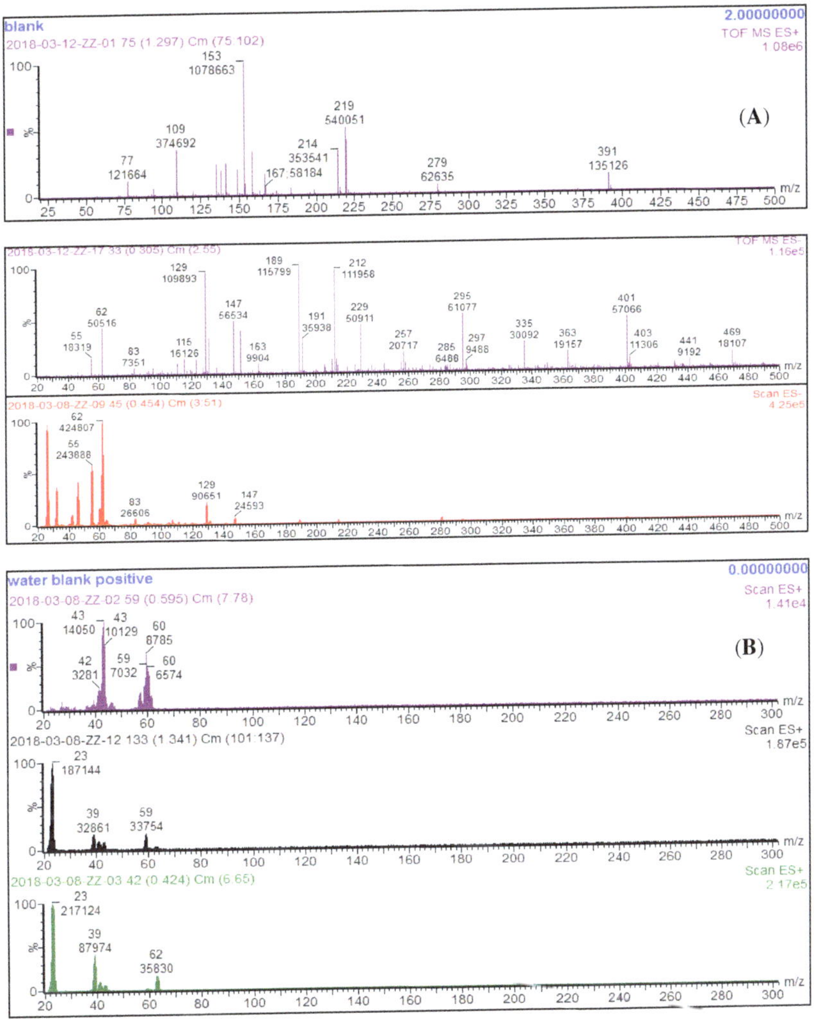

Figure 6. Electrospray ionization mass spectrometry (ESI-MS) analysis of NTO-wastewater and plant exudates at (**A**) negative mode and (**B**) positive mode. Both figures contain a blank (top), NTO-wastewater (middle), and plant exudates (bottom). NTO peak was recorded at an *m/z* ratio of 129 Da in both NTO-wastewater and plant exudates under negative mode.

2.6. Fourier-Transform Infrared Spectroscopy (FTIR) Analysis

Figure 7A shows the FTIR spectra of pure NTO solids. Based on the chemical structure of NTO, three distinct peaks can be expected from an NTO molecule: 1800–1600 cm^{-1} for C=O, 1550–1500 cm^{-1}, and 1372–1290 cm^{-1} for N-O bonds. These three distinct peaks were observed in the pure NTO solids (Figure 7). Figure 7B shows a comparison of FTIR spectra between pure NTO solids and plant exudates. The figure shows these three peaks in plant exudates with a slight shift in position. Several studies have shown that the FTIR peak shift can occur for various reasons, including specific molecular interactions, such as hydrogen bonding, presence of water molecule in the chemical structure, and dipole–dipole interactions [45,46]. Our analysis showed the presence of Na and Ca ions in the plant exudates. The interaction of these co-existing ions with the original NTO molecules could have attributed to the observed peak shift in the FTIR spectra. In addition, vetiver was grown in NTO wastewater and the interaction between NTO molecules and other ions occurred in the hydroponic media, which might have resulted in the introduction of the water molecule(s) in the structure, which might have resulted in the peak shift. FTIR analysis also confirmed similarities in chemical structure between pure NTO and plant exudates, which establishes the presence of NTO in plant exudates.

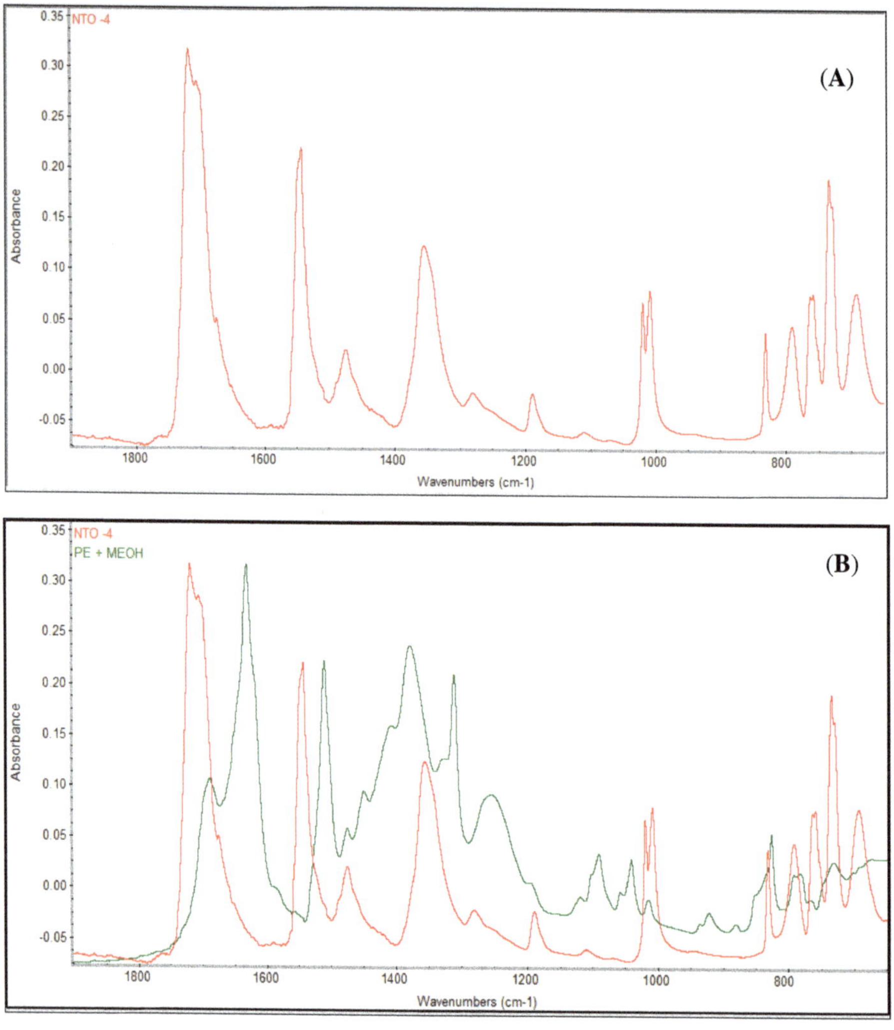

Figure 7. FTIR spectra of (**A**) pure NTO solid, and (**B**) both solid NTO and plant exudates. Peaks at 1800–1600 cm^{-1} (for C=O bonds), 1550–1500 cm^{-1}, and 1372–1290 cm^{-1} (for N-O bonds) are distinct for NTO. All three distinct peaks were also visible for plant exudates.

Salt-tolerant plant species have been reported to detoxify metals in their tissue by phytoexcreting toxic metals through salt glands or trichomes on their leaves [42,43]. Understanding this process would help in 'mining' the exuded metals, which would be an added benefit for phytoremediation applications. We report for the first time the extrusion of a munition compound. Further studies are required to find out if the vetiver system could be used to recover and reuse NTO discarded in the waste stream of industrial munition facilities.

3. Materials and Methods

3.1. Wastewater Characterization

NTO-wastewater and pure NTO solids were obtained from an industrial munition facility in the US. The detailed characterization of NTO-wastewater was performed before the study. The pH of the wastewater was measured using an PC 700 pH meter, Oakton, Vernon Hills, IL, USA. HACH test kits, HACH Company, Loveland, CO, USA were used to measure the total nitrogen (TN), ammonia-nitrogen (N-NH_4^+), and total phosphorus (TP) concentrations of the wastewater sample. Nitrate (NO_3^-) and nitrite (NO_2^-) concentrations of the wastewater were measured using Dionex ion chromatography (IC) with IonPac AS16 (4 mm × 250 mm, Dionex, Thermo Fisher Scientific, Sunnyvale, CA, USA), equipped with a guard column IonPac AG16 (4 mm×50 mm, Dionex, Thermo Fisher Scientific, Sunnyvale, CA, USA). The total organic carbon (TOC) concentration of the wastewater sampled was measured using a UV-Persulfate TOC Analyzer Phoenix 8000 (Teledyne Tekmar, Mason, OH, USA). In addition, NTO-containing wastewater sample was analyzed for Na, Ca, K, and Mg using an inductively coupled plasma optical emission spectrometry (ICP-OES, 5100 SVDV, Agilent Technologies, Santa Clara, CA, USA). NTO concentration in wastewater samples was measured using a high-performance liquid chromatography (HPLC, Agilent Technologies, Santa Clara, CA, USA, Infinity Series 1260, equipped with a ProStar 410 Auto-sampler and a DAD detector and coupled with a porous graphite column Hypercarb 7 ram, 100 × 4.6 mm). The flow rate of the mobile phase was at 1 mL/min with an isocratic mixture of water: acetonitrile + 0.1% trifluoroacetic acid of 70:30 (*v/v*). The sample injection volume was 35 µL. The analytical wavelengths were 215 nm. Under these conditions, NTO elutes at 4.2 min. A calibration range from 1 to 50 mg/L was used for the analysis and the wastewater samples were diluted as required. Dilution factors were considered while calculating the final amount. A known quality check (QC) standard was inserted after every 10 samples to validate the efficiency of the analytical procedure. All analyses were done in triplicate.

3.2. Experimental Setup and Analyses

Vetiver grass (*Chrysopogon zizanioides* L.) was purchased from Agriflora Tropical, Puerto Rico, USA. Plants were initially potted in garden soil and grown there for 30 days. The plants were then placed in a hydroponic system in half-strength Hoagland's solution for 14 days for acclimatization. After 14 days, the plants were removed from the Hoagland's solution, dried completely using paper towels, weighed, and used for the experiment. The experiment was conducted in 1 L plastic bottles with a working volume of 500 mL. Two different treatments were used: vetiver in NTO-wastewater (NV) (triplicates), and NTO-wastewater without any vetiver plant (negative control, NC) (triplicates). A 4% plant to solution ratio was maintained for each treatment. All vetiver plants were trimmed from their shoots and roots in such a way that all of them were of approximately the same size and weight. No plant growth nutrients were provided for NV treatments. The 100-day-long experiment was conducted in five successive batch studies of 20 days each. After every 20 days, the old batches of vetiver plants were replaced by new batches of plants. Wastewater samples including all replicates were collected at the same time from each of the bottles periodically. Before collecting the samples, each bottle was mixed thoroughly by swirling so that a homogenous solution can be obtained inside the bottles. Samples were collected by submerging the pipette in the liquid part of the bottles. Samples were analyzed for their NTO concentration, and nitrate concentration. For each measurement (NTO and nitrate concentration), samples were analyzed in triplicates, and analyzed concentrations were compared in Microsoft Excel (version 2007) by calculating mean and standard deviation values.

Vetiver plants collected from NV treatments were also tested for NTO translocation inside the vetiver's body. Both vetiver roots and shoots were collected (in triplicates) at the end of each successive batch and were analyzed separately to determine NTO concentration in them. A 0.5 g sample (root and shoot) was initially ground to a powder with liquid nitrogen. The powdered tissue was transferred to a tube and 5 mL of acetonitrile was added to each sample. The tubes were kept on a tube rotator for 24 h. Subsequently, the samples were filtered using a 0.45 μm syringe filter and were analyzed for NTO using HPLC. NTO translocation inside the vetiver was measured by calculating the translocation factor (TF) following the standard protocol [29,47]. The analyzed data were compared in Microsoft Excel (version 2007) and mean and standard deviation values were calculated.

3.3. Phytotoxicity Analysis

At the end of every successive batch, phytotoxicity analysis was performed on the plant samples (collected from NTO treatment, NV) by conducting a total chlorophyll study and a plant biomass study. Total chlorophyll (as a combination of chlorophyll a and b) extraction from the vetiver samples was performed following standard protocols [30,48], and the absorbance was measured at 645 and 663 nm using a Cytation 3 microplate reader, Biotek Instruments, Winooski, VT, USA. The weight of each plant was recorded before and after each successive batch study. Before weighing the plants, the roots were dried thoroughly with paper towels.

3.4. Plant Metabolomics Study

Vetiver samples were frozen in liquid nitrogen and were stored at −80°C until the metabolomics studies were performed. Metabolites were extracted according to a standard protocol [40] with a few modifications adopted by the earlier published literature [48,49]. Ampicillin (0.5 mg/mL) was added as an internal standard before extraction. Methanol: acetonitrile (50:50) with 0.125% formic acid was used as an extraction buffer. LC-MS/MS analysis was performed on the extracted samples using an ABSciex Qtrap 5500 mass spectrophotometer (Sciex, Framingham, MA, USA) equipped with a Turbo V electrospray ionization (ESI) source, a Shimadzu LC-20A system, and a PAL CTC autosampler following a standard protocol [33,49]. A total of 325 metabolites were targeted in multiple reaction monitoring (MRM) mode. Two injections, one for negative mode (ESI−) and one for positive mode (ESI+), were performed. The dwell time was set at 5 ms. Purified standards were used to optimize the compound-specific MS/MS parameters. Peaks were manually reviewed, and the peak area of each metabolite was intergraded through Multiquant v3.0 (Sciex). All data processing was done following standard protocol 3450. MetaboAnalyst 2.0 (http://www.metaboanalyst.ca) was used for all statistical analyses. Partial least-squares discriminant analysis (PLS-DA) was chosen for multivariate analysis. A VIP score >1.5 was considered as significant.

3.5. Plant Exudates Analysis

In all the successive batches, it was observed that vetivers grown in NTO-wastewater treatments exuded substances from their shoots, which eventually deposited at the junction of the plant's roots and shoots (Figure S1). During the first successive batch study, the plants exuded the material starting from the fifth day of the experiment. For different batches, the amount of total exuded material varied. Plants exuded the highest amount of material in the second batch of the study. The rate of exuded material subsequently decreased, and a very small amount was collected during the fourth and fifth batches of the study.

At the end of every successive batch study, plant exudates were carefully scraped off the plant surface and properly stored. The weight of the collected solids from every treatment was measured and noted. Special attention was given during the collection process so that no plant shoot part was scraped off with the exudates.

Plant exudates were completely dried and used for further analysis. Initially, plant exudates were inspected under an optical microscope. An AmScope digital microscope imaging camera (AmScope, Irvine, CA, USA) was used to capture pictures under the optical microscope. As many researchers reported that electron microscopy is a good tool to check the purity and morphology of the energetic compounds, the microscopy scans were performed on both exuded solids and pure NTO solids using a field-emission scanning electron microscope Auriga 40 (ZEISS) (SEM, LEO DSM 982, LEO Electron Microscopy, Thornwood, NY, USA).

Electrospray ionization mass spectrometry (ESI-MS) analysis was also performed on both plant exudates (collected from the first two batches of successive batch studies) and NTO-wastewater (obtained from the industrial facility) using a Micromass Quattro Ultima mass spectrometer (Waters Micromass, Manchester, UK) equipped with an electrospray ion source. Many researchers reported that ESI-MS is a reliable qualitative tool to reflect solution-phase structures [50,51]. The ESI-MS comparison of both plant exudates and NTO-containing wastewater was performed to find similarities in chemical structures between these two samples. All analyses were done in triplicates.

Fourier-transform infrared spectroscopy (FTIR) of both plant exudates and pure NTO solids was performed using a Nicolet iS50 FT-IR (Thermo Scientific, Waltham, MA, USA). Both solids were ground to prepare finer particles using a mortar pestle. As plant exudates were moist initially, they were dried in an oven (60 °C for 2 h) before use for FTIR analysis. All analyses were done in triplicates.

4. Conclusions

We evaluated the potential of using vetiver grass to remove NTO from wastewater collected from a munition manufacturing facility. In addition to a high concentration of NTO, the wastewater also contained a high concentration of nitrate. The wastewater was treated with five successive batches of vetiver hydroponically, and the batches were replaced every 20 days. The average NTO concentration decreased by 84% in 100 days. In control tanks without vetiver, the reduction was about 5% during the same period. NTO was detected in root and shoot tissues of vetiver, and high translocation from root to shoot was observed. The vetiver plants showed toxicity symptoms such as a reduction in biomass and a decline in chlorophyll when exposed to NTO. Metabolomic studies indicated an increase in lipid peroxidation, membrane damage, and osmotic stress in vetiver exposed to NTO. During the batch studies, NTO-treated plants produced an exudate at the junction of root and shoot. The amounts of exudates showed a decreasing trend from the first to the fifth batches. While the highest amount was exuded in the second batch, very little exudation was seen in the fourth and fifth batches, as NTO levels declined. SEM, ESI-MS, and FTIR spectroscopic analysis confirmed the presence of NTO crystals in the plant exudates, indicating vetiver exudation of NTO as a mechanism to relieve stress in vetiver. Further studies are needed to understand whether any plant or microorganism-mediated biotransformation or degradation of NTO occurs in vetiver. Further studies are also needed to test the feasibility of this technology in large-scale applications under controlled greenhouse environments. If proven feasible in scaled-up settings, existing NTO wastewater holding tanks can be retrofitted with floating treatment platforms of vetiver. At regular intervals, vetiver biomass can be removed and incinerated under controlled conditions. A significant reduction in the total amount of energetics waste is possible by applying this technology at an expense that is much lower than conventional hazardous waste treatment technologies.

Supplementary Materials: The following are available online, Table S1: Relevant chemical properties of NTO. Table S2: Change in NTO concentration in NTO-wastewater during the 100-d experiment in vetiver grown in NTO-wastewater. Figure S1: Plant exudates deposited at the junction of vetiver root and shoot. Figure S2: Plant exudates under optical microscope. Photos show the presence of crystalline structures that correspond to NTO.

Author Contributions: Conceptualization and research design, D.S. and R.D.; investigation—batch analysis, A.R., P.M., S.P., investigation—SEM, ESI-MS and FTIR analysis, P.M., S.P. A.R; investigation—metabolomics, R.D.; data interpretation, A.R., D.S., and R.D.; writing—original draft, A.R.; writing—review and editing, D.S., R.D., and C.C.; supervision and project management, funding acquisition, D.S., C.C. All authors have read and agreed to the published version of the manuscript.

Funding: This research was funded by the Consortium for Energy, Environment, and Demilitarization (CEED) contract number SINIT-15-0013. The APC was funded by MDPI.

Data Availability Statement: The data presented in this study are available on request from the corresponding author.

Acknowledgments: The authors would like to thank Tsan-Liang Su, and Andrew Mai for their analytical help; Zhaoyu Zheng, and Athula Attygalle and the Center for Mass Spectrometry (Stevens Institute of Technology) for their help with ESI-MS analysis; and Tseng-Ming Chou for his help with SEM analysis (the Laboratory for Multiscale Imaging, Stevens Institute of Technology).

Conflicts of Interest: The authors declare no conflict of interest.

References

1. Felt, D.; Johnson, J.; Larson, S.; Hubbard, B.; Henry, K.; Nestler, C.; Ballard, J. *Evaluation of Treatment Technologies for Wastewater from Insensitive Munitions Production. Phase 1: Technology Down-Selection*; ERDC/EL TR-13-20; Technical Report; US Army, Armament Research, Development and Engineering Center (ARDEC): Picatinny Arsenal, NJ, USA, 2013.
2. Richard, T.; Weidhass, J. Biodegradation of IMX-101 explosive formulation constituents: 2,4-Dinitroanisole (DNAN), 3-nitro-1,2,4-triazol-5-one (NTO), and nitroguanidine. *J. Hazard. Mater.* **2014**, *280*, 372–379. [CrossRef] [PubMed]
3. Jangid, S.K.; Sarangapani, R.; Solanki, V.J.; Singh, M.K.; Pandit, G.; Vijayalakshmi, R.; Sinha, R.K. Evaluation studies on partial replacement of RDX by spherical NTO in HTPB-based insensitive sheet explosive formulation. *J. Energetic Mater.* **2019**, *37*, 320–330. [CrossRef]
4. Kitcher, E.; Braida, W.; Koutsospyros, A.; Pavlov, J.; Su, T.-L. Characteristics and products of the reductive degradation of 3-nitro-1,2,4-triazol-5-one (NTO) and 2,4-dinitroanisole (DNAN) in a Fe-Cu bimetal system. *Environ. Sci. Pollut. Res.* **2016**, *24*, 2744–2753. [CrossRef] [PubMed]
5. Kim, K.-J.; Kim, M.J.; Lee, J.-M.; Kim, S.-H.; Kim, H.-S.; Park, B.-S. Solubility, Density, and Metastable Zone Width of the 3-Nitro-1,2,4-triazol-5-one + Water System. *J. Chem. Eng. Data* **1998**, *43*, 65–68. [CrossRef]
6. Abraham, J.; Lin, Y.; Roychowdhury, A.; Christodoulatos, C.; Conway, M.; Smolinski, B.; Braida, W. Algae toxicological assessment and valorization of energetic-laden wastewater streams using Scenedesmus obliquus. *J. Clean. Prod.* **2018**, *202*, 838–845. [CrossRef]
7. RoyChowdhury, A.; Abraham, J.; Abimbola, T.; Lin, Y.; Christodoulatos, C.; Lawal, A.; Arienti, P.; Smolinski, B.; Braida, W. From waste to energy: Optimizing growth of microalgae Scenedesmus obliquus in untreated ener-getic-laden wastewater streams from an ammunition facility for bioenergy production. *Protect. Restor. Environ.* **2018**, *14*, 1085–1094.
8. Roychowdhury, A.; Abraham, J.; Abimbola, T.; Lin, Y.; Christodoulatos, C.; Lawal, A.; Koutsospyros, A.; Braida, W. Assessing Oil Content of Microalgae Grown in Industrial Energetic-Laden Wastewater. *Environ. Process.* **2019**, *6*, 969–983. [CrossRef]
9. Lin, Y.; Abraham, J.; Roychowdhury, A.; Su, T.-L.; Braida, W.; Christodoulatos, C. Ecotoxicological response of Scenedesmus obliquus to pure energetic compounds and metal ions found in wastewater streams from munitions manufacturing. *Algal Res.* **2020**, *48*, 101927. [CrossRef]
10. Madeira, C.L.; Speet, S.A.; Nieto, C.A.; Abrell, L.; Chorover, J.; Sierra-Alvarez, R.; Field, J.A. Sequential anaerobic-aerobic biodegradation of emerging insensitive munitions compound 3-nitro-1,2,4-triazol-5-one (NTO). *Chemosphere* **2017**, *167*, 478–484. [CrossRef]
11. Madeira, C.L.; Jog, K.V.; Vanover, E.T.; Brooks, M.D.; Taylor, D.K.; Sierra-Alvarez, R.; Waidner, L.A.; Spain, J.C.; Krzmarzick, M.J.; Field, J.A. Microbial Enrichment Culture Responsible for the Complete Oxidative Biodegradation of 3-Amino-1,2,4-triazol-5-one (ATO), the Reduced Daughter Product of the Insensitive Munitions Compound 3-Nitro-1,2,4-triazol-5-one (NTO). *Environ. Sci. Technol.* **2019**, *53*, 12648–12656. [CrossRef]
12. Fawcett-Hirst, W.; Temple, T.; Ladyman, M.K.; Coulon, F. Adsorption behaviour of 1,3,5-trinitroperhydro-1,3,5-triazine, 2,4-dinitroanisole and 3-nitro-1,2,4-triazol-5-one on commercial activated carbons. *Chemosphere* **2020**, *255*, 126848. [CrossRef]
13. Koutsospyros, A.; Pavlov, J.; Fawcett, J.; Strickland, D.; Smolinski, B.; Braida, W. Degradation of high energetic and insensitive munitions compounds by Fe/Cu bimetal reduction. *J. Hazard. Mater.* **2012**, *219*, 75–81. [CrossRef]
14. Becher, J.B.; Beal, S.A.; Taylor, S.; Dontsova, K.; Wilcox, D.E. Photo-transformation of aqueous nitroguanidine and 3-nitro-1,2,4-triazol-5-one: Emerging munitions compounds. *Chemosphere* **2019**, *228*, 418–426. [CrossRef]
15. Jog, K.V.; Sierra-Alvarez, R.; Field, J.A. Rapid biotransformation of the insensitive munitions compound, 3-nitro-1,2,4-triazol-5-one (NTO), by wastewater sludge. *World J. Microbiol. Biotechnol.* **2020**, *36*, 67. [CrossRef] [PubMed]
16. Andra, S.S.; Datta, R.; Sarkar, D.; Makris, K.C.; Mullens, C.P.; Sahi, S.V.; Bach, S.B. Induction of lead-binding phytochelatins in vetiver grass [*Vetiveria zizanioides* (L.)]. *J. Environ. Qual.* **2009**, *38*, 868–877. [CrossRef] [PubMed]
17. Roychowdhury, A.; Sarkar, D.; Datta, R. Remediation of Acid Mine Drainage-Impacted Water. *Curr. Pollut. Rep.* **2015**, *1*, 131–141. [CrossRef]

18. Roychowdhury, A.; Datta, R.; Sarkar, D. Heavy Metal Pollution and Remediation. In *Green Chemistry*; Elsevier: Amsterdam, The Netherlands, 2018; pp. 359–373.
19. Roychowdhury, A.; Sarkar, D.; Datta, R. A combined chemical and phytoremediation method for reclamation of acid mine drainage–impacted soils. *Environ. Sci. Pollut. Res.* **2019**, *26*, 14414–14425. [CrossRef] [PubMed]
20. Sengupta, A.; Sarkar, D.; Das, P.; Panja, S.; Parikh, C.; Ramanathan, D.; Bagley, S.; Datta, R. Tetracycline uptake and metabolism by vetiver grass (*Chrysopogon zizanioides* L. Nash). *Environ. Sci. Pollut. Res.* **2016**, *23*, 24880–24889. [CrossRef]
21. Panz, K.; Miksch, K. Phytoremediation of explosives (TNT, RDX, HMX) by wild-type and transgenic plants. *J. Environ. Manag.* **2012**, *113*, 85–92. [CrossRef]
22. Kiiskila, J.D.; Das, P.; Sarkar, D.; Datta, R. Phytoremediation of Explosive-Contaminated Soils. *Curr. Pollut. Rep.* **2015**, *1*, 23–34. [CrossRef]
23. Makris, K.C.; Shakya, K.M.; Datta, R.; Sarkar, D.; Pachanoor, D. High uptake of 2,4,6-trinitrotoluene by vetiver grass-Potential for phytoremediation? *Environ. Pollut.* **2007**, *146*, 1–4. [CrossRef] [PubMed]
24. Doskey, C.M. Uptake and Fate of Hexahydro-1,3,5-Trinitro-1,3,5-Triazine by *Chrysopogon zizanioides*. Master's Thesis, Michigan Technological University, Houghton, MI, USA, 2012.
25. Panja, S.; Sarkar, D.; Datta, R. Vetiver grass (*Crysopogon zizanioides*) is capable of removing insensitive high explosives from munition industry wastewater. *Chemosphere* **2018**, *209*, 920–927. [CrossRef] [PubMed]
26. Das, P.; Datta, R.; Makris, K.C.; Sarkar, D. Vetiver grass is capable of removing TNT from soil in the presence of urea. *Environ. Pollut.* **2010**, *158*, 1980–1983. [CrossRef] [PubMed]
27. Das, P.; Sarkar, D.; Makris, K.C.; Datta, R. Urea-facilitated uptake and nitroreductase-mediated transformation of 2,4,6-trinitrotoluene in soil using vetiver grass. *J. Environ. Chem. Eng.* **2015**, *3*, 445–452. [CrossRef]
28. Rylott, E.L.; Lorenz, A.; Bruce, N.C. Biodegradation and biotransformation of explosives. *Curr. Opin. Biotechnol.* **2011**, *22*, 434–440. [CrossRef]
29. Richard, T.; Weidhaas, J. Dissolution, sorption, and phytoremediation of IMX-101 explosive formulation constituents: 2,4-dinitroanisole (DNAN), 3-nitro-1,2,4-triazol-5-one (NTO), and nitroguanidine. *J. Hazard. Mater.* **2014**, *280*, 561–569. [CrossRef]
30. Price, R.; Pennington, J.C.; Neumann, D.; Hayes, C.L. *Plant Uptake of Explosives from Contaminated Soil and Irrigation Water at the Former Nebraska Ordnance Plant, Mead, Nebraska*; Technical Report; EL-97-11; US Army Corps of Engineers, Waterways Experiment Station: Washington, DC, USA, 1997.
31. Makris, K.C.; Sarkar, D.; Datta, R. Coupling indigenous biostimulation and phytoremediation for the restoration of 2,4,6-trinitrotoluene-contaminated sites. *J. Environ. Monit.* **2010**, *12*, 399–403. [CrossRef]
32. Segonzac, C.; Boyer, J.-C.; Ipotesi, E.; Szponarski, W.; Tillard, P.; Touraine, B.; Sommerer, N.; Rossignol, M.; Gibrat, R. Nitrate Efflux at the Root Plasma Membrane: Identification of an Arabidopsis Excretion Transporter. *Plant Cell* **2007**, *19*, 3760–3777. [CrossRef]
33. Ali, N.A.; Dewez, D.; Robidoux, P.Y.; Popovic, R. Photosynthetic parameters as indicators of trinitrotoluene (TNT) inhibitory effect: Change in chlorophyll a fluorescence induction upon exposure of Lactuca sativa to TNT. *Ecotoxicology* **2006**, *15*, 437–441. [CrossRef]
34. Peterson, M.; Horst, G.; Shea, P.; Comfort, S. Germination and seedling development of switchgrass and smooth bromegrass exposed to 2,4,6-trinitrotoluene. *Environ. Pollut.* **1998**, *99*, 53–59. [CrossRef]
35. Krishnan, G.; Horst, G.L.; Darnell, S.; Powers, W.L. Growth and development of smooth bromegrass and tall fescue in TNT-contaminated soil. *Environ. Pollut.* **2000**, *107*, 109–116. [CrossRef]
36. Best, E.P.H.; Smith, T.; Hagen, F.L.; Dawson, J.; Torrey, A.J. *Candidate Herbaceous Plants for Phytoremediation of Energetics on Ranges*; ERDC Report TR-07-11; Engineer Research and Development Center: Vicksburg, MS, USA, 2007.
37. Via, S.M.; Zinnert, J. Impacts of explosive compounds on vegetation: A need for community scale investigations. *Environ. Pollut.* **2016**, *208*, 495–505. [CrossRef] [PubMed]
38. Vila, M.; Lorber-Pascal, S.; Laurent, F. Fate of RDX and TNT in agronomic plants. *Environ. Pollut.* **2007**, *148*, 148–154. [CrossRef] [PubMed]
39. Hou, Q.; Ufer, G.; Bartels, D. Lipid signalling in plant responses to abiotic stress. *Plant, Cell Environ.* **2016**, *39*, 1029–1048. [CrossRef] [PubMed]
40. De Vos, R.C.H.; Moco, S.; Lommen, A.; Keurentjes, J.J.B.; Bino, R.J.; Hall, R.D. Untargeted large-scale plant metabolomics using liquid chromatography coupled to mass spectrometry. *Nat. Protoc.* **2007**, *2*, 778–791. [CrossRef]
41. Zhou, Q.; Yu, B.J. Accumulation of Inorganic and Organic Osmolytes and Their Role in Osmotic Adjustment in NaCl-Stressed Vetiver Grass Seedlings. *Russ. J. Plant Physiol.* **2009**, *56*, 678–685. [CrossRef]
42. Lefévre, I.; Marchal, G.; Meerts, P.; Corréal, E.; Lutts, S. Chloride salinity reduces cadmium accumulation by the Mediterranean halophyte species *Atriplex halimus* L. *Environ. Exp. Bot.* **2009**, *65*, 142–152. [CrossRef]
43. Kadukova, J.; Manousaki, E.; Kalogerakis, N. Pb and Cd Accumulation and Phyto-Excretion by Salt Cedar (Tamarix Smyrnensis Bunge). *Int. J. Phytoremediation* **2008**, *10*, 31–46. [CrossRef]
44. Yang, G.; Nie, F.; Li, J.; Guo, Q.; Qiao, Z. Preparation and Characterization of Nano-NTO Explosive. *J. Energetic Mater.* **2007**, *25*, 35–47. [CrossRef]
45. Ryu, S.R.; Noda, I.; Jung, Y.M. Positional Fluctuation of IR Absorption Peaks: Frequency Shift of a Single Band or Relative Intensity Changes of Overlapped Bands? American Laboratory Technical Article. 2011. Available online: www.americanlaboratory.com/913-Technical-Articles/1244-Positional-Fluctuation-of-IR-Absorption-Peaks-Frequency-Shift-of-a-Single-Band-or-Relative-Intensity-Changes-of-Overlapped-Bands (accessed on 3 November 2020).

46. Kannan, P.P.; Karthick, N.K.; Mahendraprabu, A.; Shanmugam, R.; Elangovan, A.; Arivazhagan, G. Red/blue shifting hydrogen bonds in acetonitrile-dimethyl sulphoxide solutions: FTIR and theoretical studies. *J. Mol. Struct.* **2017**, *1139*, 196–201. [CrossRef]
47. Marchiol, L.; Assolari, S.; Sacco, P.; Zerbi, G. Phytoextraction of heavy metals by canola (*Brassica napus*) and radish (*Raphanus sativus*) grown on multicontaminated soil. *Environ. Pollut.* **2004**, *132*, 21–27. [CrossRef] [PubMed]
48. Li, K.; Wang, X.; Pidatala, V.; Chang, C.-P.; Cao, X. Novel Quantitative Metabolomic Approach for the Study of Stress Responses of Plant Root Metabolism. *J. Proteome Res.* **2014**, *13*, 5879–5887. [CrossRef] [PubMed]
49. Pidatala, V.; Li, K.; Sarkar, D.; Ramakrishna, W.; Datta, R. Identification of Biochemical Pathways Associated with Lead Tolerance and Detoxification in *Chrysopogon zizanioides* L. Nash (Vetiver) by Metabolic Profiling. *Environ. Sci. Technol.* **2016**, *50*, 2530–2537. [CrossRef] [PubMed]
50. Hibbs, J.A.; Jariwala, F.B.; Weisbecker, C.S.; Attygalle, A.B. Gas-Phase Fragmentations of Anions Derived from N-Phenyl Benzenesulfonamides. *J. Am. Soc. Mass Spectrom.* **2013**, *24*, 1280–1287. [CrossRef] [PubMed]
51. Xia, H.; Attygalle, A.B. Effect of Electrospray Ionization Source Conditions on the Tautomer Distribution of Deprotonated p-Hydroxybenzoic Acid in the Gas Phase. *Anal. Chem.* **2016**, *88*, 6035–6043. [CrossRef] [PubMed]

molecules

Article

Stepwise Ethanol-Water Fractionation of Enzymatic Hydrolysis Lignin to Improve Its Performance as a Cationic Dye Adsorbent

Wenjie Sui [1,*,†], Tairan Pang [2,†], Guanhua Wang [2,3,*], Cuiyun Liu [2], Ashak Mahmud Parvez [4], Chuanling Si [2] and Chao Li [3,5]

1 State Key Laboratory of Food Nutrition and Safety, College of Food Science and Engineering, Tianjin University of Science & Technology, Tianjin 300457, China
2 Tianjin Key Laboratory of Pulp and Paper, College of Light Industry Science and Engineering, Tianjin University of Science and Technology, Tianjin 300457, China; pangtr1995@163.com (T.P.); zclcy1214@163.com (C.L.); sichli@tust.edu.cn (C.S.)
3 Hunan BISEN Environmental & Energy Co., Ltd., Changsha 410100, China; chaoli_tu@tsinghua.edu.cn
4 Department of Mechanical Engineering, University of New Brunswick, Fredericton, NB E3B 5A3, Canada; Ashak.Parvez@unb.ca
5 School of Environment, Tsinghua University, Beijing 100084, China
* Correspondence: wjsui@tust.edu.cn (W.S.); ghwang@tust.edu.cn (G.W.); Tel.: +86-02260601313 (G.W.)
† These authors contributed equally to the work.

Academic Editors: Chiara Bisio and Monica Pica
Received: 11 May 2020; Accepted: 2 June 2020; Published: 3 June 2020

Abstract: In this work, lignin fractionation is proposed as an effective approach to reduce the heterogeneity of lignin and improve the adsorption and recycle performances of lignin as a cationic dye adsorbent. By stepwise dissolution of enzymatic hydrolysis lignin in 95% and 80% ethanol solutions, three lignin subdivisions (95% ethanol-soluble subdivision, 80% ethanol-soluble subdivision, and 80% ethanol-insoluble subdivision) were obtained. The three lignin subdivisions were characterized by gel permeation chromatography (GPC), FTIR, 2D-NMR and scanning electron microscopy (SEM), and their adsorption capacities for methylene blue were compared. The results showed that the 80% ethanol-insoluble subdivision exhibited the highest adsorption capacity and its value (396.85 mg/g) was over 0.4 times higher than that of the unfractionated lignin (281.54 mg/g). The increased adsorption capacity was caused by the enhancement of both specific surface area and negative Zeta potential. The maximum monolayer adsorption capacity of 80% ethanol-insoluble subdivision by adsorption kinetics and isotherm studies was found to be 431.1 mg/g, which was much higher than most of reported lignin-based adsorbents. Moreover, the 80% ethanol-insoluble subdivision had much higher regeneration yield (over 90% after 5 recycles) compared with the other two subdivisions. Consequently, the proposed fractionation method is proved to be a novel and efficient non-chemical modification approach that significantly improves adsorption capacity and recyclability of lignin.

Keywords: enzymatic hydrolysis lignin; sequential dissolution fractionation; methylene blue adsorption capacity

1. Introduction

Nowadays, synthetic organic dyes are widely applied in various industries, such as printing, paper, textile, electroplating, pulp mill, food and cosmetic productions [1]. During the dyeing process, 10–15% of dye is discharged in the effluent [2,3], which not only creates an aesthetic nuisance to the environment but also increases the overall loading of chemical oxygen demand (COD) in the receiving water. In order to remove dyes from industrial effluents, numerous technologies have been developed, such as adsorption [4,5], coagulation [6], chemical oxidation [7], photocatalytic degradation [8] and biological treatment [2]. Among them, adsorption has been preferred owing to its low operation cost and high efficiency [4,9]. Various kinds of adsorbents have been reported for dyes removing from water effluent, such as activated carbon [10,11], bentonite [12], and synthesized polymers [13]. Currently, there are increasing interests in utilizing biomass derived materials as a dye adsorbent due to their benign properties, such as acceptable specific strength, low cost, no health risk and sustainability [9,14].

Lignin, as the dominant aromatic component of lignocellulosic biomass, has attracted considerable attention in the pollutant adsorption field since it contains a number of functional groups (e.g., phenolic hydroxyl and carboxyl) that serve as potential active sites for adsorption [15–19]. Owing to the negative charge of these functional groups in water environment, unmodified lignin possesses preferable adsorption ability for cationic dyes [16,20–22]. Zhang et al. investigated the methylene blue adsorption by organosolv lignin from rice straw and found that the adsorption capacity was 40.02 mg/g at 20 °C, which was comparable with other lignocellulose adsorbents [23]. Menkiti et al. utilized alkaline lignin extracted from elephant grass to remove aqueous crystal violet dye and their results indicated that the crystal violet uptake capacity of lignin was about 25.0 mg/g at 30 °C [24]. Feng et al. analyzed the methylene blue removal behaviors of deacetylated lignin from acetic acid pulping of eucalyptus and suggested that the modification and isolation processes of lignin could improve the adsorption ability from 18.2 to 63.3 mg/g owing to the enhanced contents of phenolic hydroxyl and carboxylic groups [22].

However, lignin is a typically heterogeneous biopolymer and the heterogeneity of lignin leads to its inhomogeneous properties [25], which may also involve the adsorption capacities. Although many efforts have been dedicated to the study of different sources/types of lignin as potential absorbents for removal of cationic dyes from wastewater, until now, no attempts to explore the effect of heterogeneity of one particular lignin sample on its dye adsorption capacity has been reported. It is known that lignin fractionation is a simple but efficient process to decrease the lignin inhomogeneity by subdividing the heterogeneous lignin into different fractions (subdivisions), which show obviously reduced heterogeneity in both molecular weight and chemical structure [26]. Thus, the effect of heterogeneity on lignin adsorption performance can be then investigated by comparing the adsorption performances of the subdivisions obtained from lignin fractionation.

In this work, enzymatic hydrolysis lignin (EHL), a main byproduct from cellulosic ethanol production, was fractionated into three subdivisions by stepwise dissolution in 95% and 80% ethanol solvents. The effect of lignin heterogeneity on its dye adsorption capacity was investigated by a comparison of the methylene blue (MB, a typical cationic dye pollutant) adsorption capacities of the three obtained lignin subdivisions. Interestingly, it was found that the three lignin subdivisions exhibited remarkably different adsorption capacities under the optimized conditions (e.g., pH, adsorbent dosage, temperature and time) and the 80% ethanol-insoluble subdivision had much higher adsorption capacity than the other two subdivisions. Moreover, the specific surface area and the Zeta potential of the three lignin subdivisions were measured to analyze the possible formation mechanism of different adsorption capacities among the three subdivisions. With respect to the recyclability study, the MB loaded lignin adsorbents were regenerated using a common ethanol washing process and afterwards reused for MB removal from a fresh solution.

2. Results and Discussion

2.1. Lignin Fractionation by Stepwise Dissolution

By stepwise dissolution in 95% and 80% ethanol solvents, the enzymatic hydrolysis lignin (EHL) was fractionated into three subdivisions (Figure 1a), namely, 95% ethanol soluble subdivision (S1), 80% ethanol soluble subdivision (S2), and 80% ethanol insoluble subdivision (S3). The yields of the three lignin subdivisions were 22.87%, 27.803% and 45.34%, respectively (Table 1). As shown in Figure 1a, the color of the lignin subdivisions was varied, which turned from light brown (S1), to dark brown (S2), and finally to black (S3, Figure 1a). The molecular weight of the three subdivisions increased from S1 to S3 (Figure 1b) and all the three subdivisions exhibited reduced polydispersity compared with the original EHL (Table 1). These results suggest that the fractionation method using sequential dissolution in 95% and 80% ethanol realizes the separation of different molecular weight lignin subdivisions from the heterogeneous EHL. The fractionation rationale lies in the different solubility of lignin subdivisions with different molecular weights in the selective ethanol/water solvents [25].

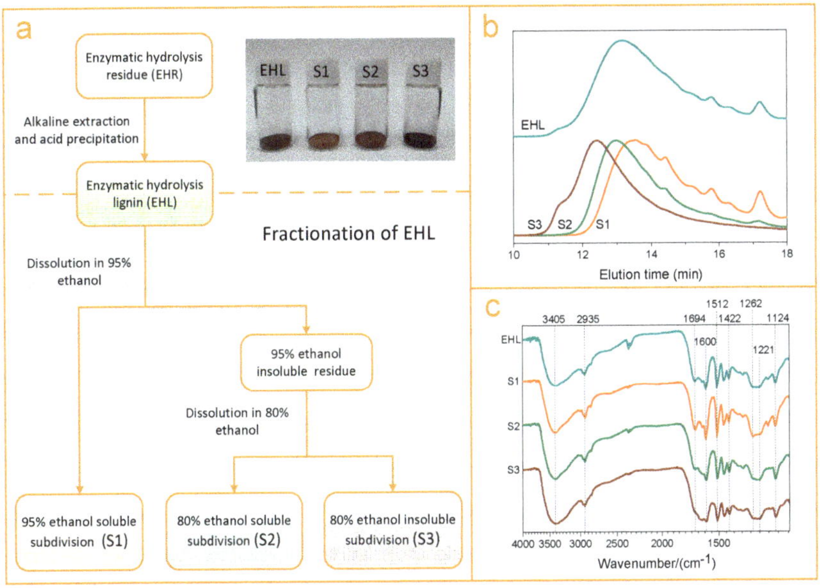

Figure 1. (**a**) Procedure for stepwise ethanol-water fractionation of enzymatic hydrolysis lignin (EHL), inset: pictures of S1, S2, S3, and EHL; (**b**) Molecular weight distributions of S1, S2, S3, and EHL; (**c**) FTIR spectra of S1, S2, S3, and EHL.

Table 1. Yield, molecular weight distribution, specific surface area and Zeta potential of lignin subdivisions, as well as those of enzymatic hydrolysis lignin (EHL).

	S1	S2	S3	EHL (S1 + S2 + S3)
Yield (%)	22.87	27.80	45.34	-
M_w	4050	6270	12230	7280
M_n	2238	3800	7194	3569
Polydispersity	1.81	1.65	1.70	2.04
Specific surface area	2.44	4.06	4.84	2.89
Zeta potential [1]	−19.31 ± 1.63	−28.69 ± 2.29	−35.24 ± 3.08	−24.71 ± 1.77

[1] Detected at pH 6.0.

The FTIR spectra of EHL and its three subdivisions basically conformed to the typical absorption bands of lignin (Figure 1c). The broad band at 3405 cm^{-1} was assigned to the stretching of hydroxyl and the signal at 2935 cm^{-1} was attributed to C-H stretching vibration from methyl and methane [25,27]. Three sharp peaks observed at 1600, 1512 and 1422 cm^{-1} were caused by the aromatic skeleton vibrations, indicating the primary structure of lignin [28]. The bands at 1221 and 1124 cm^{-1} reflected syringyl structures in lignin, while the band at 1262 cm^{-1} was associated with guaiacyl units. One obvious difference among the three subdivisions was that the S1 had much higher absorption intensity at 1694 cm^{-1} compared to those noticed in the other two subdivisions. The signal at 1694 cm^{-1} was assigned to the unconjugated β-ketone in the side chain of lignin unit, which was principally formed by the breakage of the β-*O*-4 bond during the steam explosion process [28]. Thus, the results indicate that the depolymerized lignin produced during the steam explosion process is enriched in S1, which agree well with the results from molecular weight analysis.

2D-NMR analysis was also conducted to acquire a more detailed understanding of the chemical structure of the three lignin subdivisions (Figure 2). In the side chain region (δC/δH 50–90/2.6–5.8), three typical interunit linkages, including β-*O*-4′ structure (A), pinoresinol structure (B), and dibenzodioxocin structure (C), were identified [29,30]. It was found that the signal intensities for aryl ester linkage (β-*O*-4′ structure, A) were stronger for S2 and S3 while the C-C bonding (structure B and C) signals were more noticeable in S1. This phenomena could be explained by the fact that it is relatively easier to break the aryl ester linkage during the steam explosion pretreatment [31–33] and the extraction process using NaOH aqueous solution [34]. Thus, the degraded lignin with low-molecular-weight lignin (S1) exhibited a lower amount of β-*O*-4′ aryl ester linkage. The aromatic region in the 2D-HSQC spectra exhibited obvious guaiacyl (G), syringyl (S), and *p*-hydroxyphenyl (H) rings, suggesting that the lignin used in this work was a GSH-type lignin. Additionally, the phenolic acid signals were much more evident in S1 (Figure 2). This result supports the previously reported findings that the phenolic acids are enriched in the low-molecular-weight subdivision since they show good solubility in 95% ethanol solution [25].

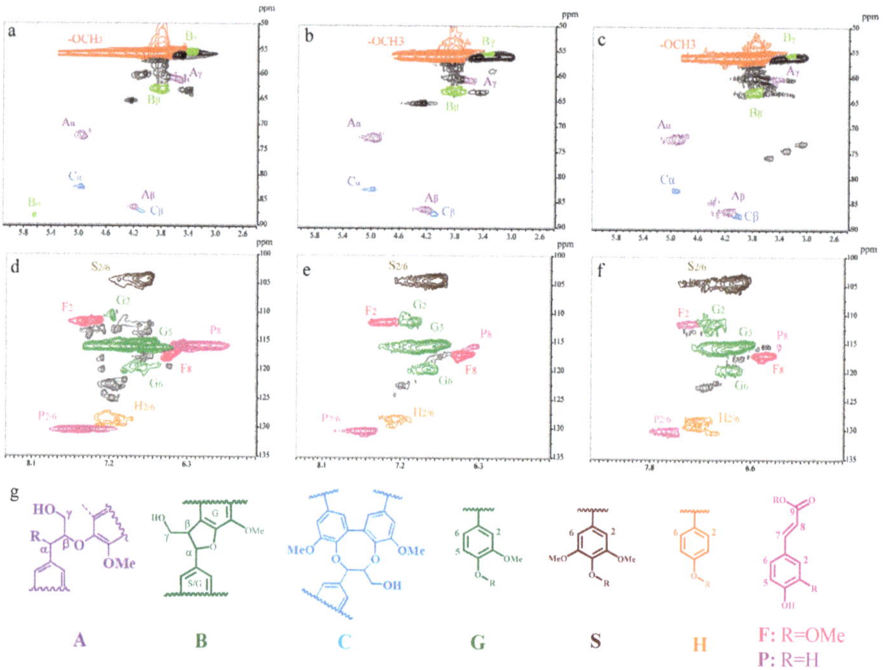

Figure 2. Side-chain (**a–c**) and aromatic regions (**d–f**) of 2D-NMR (HSQC) spectra of three lignin subdivisions (S1: **a** and **d**; S2: **b** and **e**; S3: **c** and **f**). Main substructures identified: (A) β-*O*-4′ aryl ether linkage; (B) phenyl-coumaran structure formed by β-5′ and α-*O*-4′ linkages; (C) dibenzodioxocin structure; (G) guaiacyl unit; (S) syringyl unit; (H) *p*-hydroxyphenyl unit; (F) ferulic acid; (P) *p*-coumaric acid.

The surface morphology of the three lignin subdivisions as well as the parent EHL is presented in Figure 3. It was observed that the particles of EHL, S1, and S2 were large in size with smooth surface and sharp edges. Nair et al. observed the microcosmic surface of alkaline lignin using SEM and they also found that the alkali lignin particles were larger in size with sharp edges and lesser surface roughness [35]. However, S3 exhibited visibly decreased particle size and enhanced surface roughness. This result can be explained by the preparation processes of the lignin samples. EHL, S1 and S2 were prepared by precipitation from lignin solution using pH adjusting or solvent evaporation (Sections 3.2 and 3.3). This precipitation processes resulted in the aggregation of small lignin particles and the formation of large lignin particles with a smooth surface. However, the S3 was obtained from EHL after sequential dissolution of EHL in 95% and 80% ethanol solutions. The dissolution of low molecular weight lignin (S1 and S2) resulted in the formation of a porous structure on the lignin particle and even the fragmenting of lignin particles, which increased the surface roughness and decreased the particle size (Figure 3e).

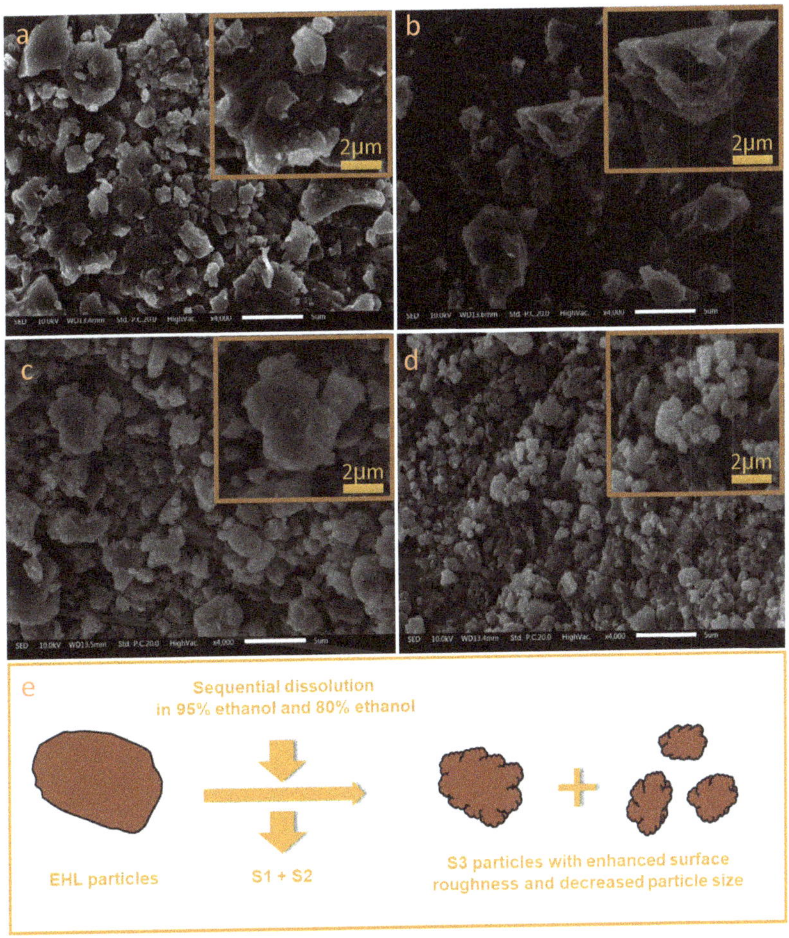

Figure 3. Scanning electron microscopy (SEM) images of EHL (**a**), S1 (**b**), S2 (**c**), and S3 (**d**). The scale bar for the SEM images is 5 μm. Schematic showing morphological changes from EHL particles to the 80% ethanol insoluble subdivision (S3) particles (**e**).

Due to the increase of the surface roughness and the decrease of particle size, the specific surface area of S3 (4.84 m^2/g) was the highest among the four lignin samples, which increased 67.57% compared with the parent EHL (2.89 m^2/g, Table 1). The specific surface area of EHL was slightly higher than those of alkali lignin [35] and kraft lignin [36], probably due to the reduced particle size after grinding. The increased specific surface area of S3 contributes to the high adsorption capacity of MB molecules compared with the other two subdivisions. Besides, the Zeta potential of lignin particles, which is critical for the adsorption based on the electrostatic attraction, was also determined (Table 1). All lignin particles exhibited negative charges on their surface owing to the deprotonation of acidic groups (e.g., phenolic OH and COOH) in lignin molecule. Among the three lignin subdivisions, S3 had the highest negative charge (−35.24 mv) on its surface followed by S2 (−28.69 mv) while S1 had the lowest negative charge (−19.31 mv). These results can be explained by the larger specific surface area of S3, which results in more active sites (deprotonated groups) and higher negative charges [37,38]. The large negative charge of S3 increases the electrostatic attraction of cationic MB on the particle surface and the MB adsorption capacity.

2.2. Adsorption Capacities of Lignin Subdivisions

Batch experiments were performed in order to assess the performance of the three lignin subdivisions as absorbents for MB removal from simulated wastewater. Figure 4a shows the effect of initial MB concentration on the adsorption ability of the three lignin samples as well as the parent EHL. At low MB concentrations (100 mg/L), all lignin samples could adsorb MB completely, resulting in the same adsorption capacity (about 100 mg/g) and almost 100% removal efficiency. When the MB concentration was increased to 200 mg/L, S1 showed the lowest MB adsorption capacity (117.39 mg/g) among the three lignin subdivisions. Besides, the MB adsorption capacities of S2 and S3 were almost the same since both of them removed all MB in the solution. Once the MB concentration further increased to 300 mg/L, only S3 was able to realize the complete removal of MB from the aqueous solution while the adsorption capacity of S2 was 240.13 mg/g. These results indicate that the three lignin subdivisions show different MB adsorption capacities: S3 has the highest adsorption capacity, followed by S2, while S1 has the lowest adsorption capacity. This result is possibly caused by the high specific surface area and high Zeta potential of S3 particles, as presented in Section 2.1. The further increase of MB concentration up to 400 mg/L and the adsorption capacities of all the three lignin samples increased due to the enhanced mass transfer driving force [15]. At this concentration (400 mg/L), the MB removal efficiency of S3 was 81.46%, suggesting the achievement of saturation adsorption. Thus, MB concentration of 400 mg/L was chosen for further adsorption experiments in this study in order to avoid the insufficient adsorption of MB.

Generally, the solution pH is critical to the charge distribution of the adsorbent and the adsorbate, which determines the electrostatic or molecular interactions between the absorbent and the adsorbate and accordingly, the adsorption capacity [35,39]. The effect of solution pH on the MB adsorption capacity of the lignin samples is presented in Figure 4b. It was found that the solution pH had a considerable impact on MB adsorption of the lignin samples. The MB adsorption capacity of lignin samples increased distinctly with the increasing of the pH value from 4.0 to 8.0, while decreased with the further rise of the pH to 9.0. These results can be explained by the fact that with the increasing of the solution pH from 4.0 to 8.0, more phenolic OH and COOH groups are deprotonated [23]. Thus, the negative charge on the surface of lignin particles increases and, therefore, the MB adsorption amount enhances. However, the further increase of pH to 9.0 may result in the partial dissolution of lignin, which reduces the adsorption capacity of the lignin sample.

It was also found from Figure 4b that S3 exhibited the highest adsorption capacity and MB removal efficiency among the three subdivisions at all pH values. The effect of adsorption temperature and contacting time on the MB adsorption capacity of EHL and its three subdivisions were also investigated (Figure 4c,d). Elevating the temperature (30 °C to 50 °C) and prolonging the adsorption time (0.5 to 4 h) were helpful to increase MB adsorption capacity of lignin samples. As expected, under all conditions in Figure 4c,d, S3 exhibited the highest adsorption capacity and MB removal efficiency among the three investigated adsorbents. After optimization of the adsorption conditions (lignin concentration 1000 mg/L, MB concentration 400 mg/L, pH 8.0, temperature 50 °C and time 120 min), the maximum adsorption capacity and removal efficiency of S3 were found to be 396.85 mg/g and 99.21%, respectively. At this condition, the adsorption capacity of S3 increased about 41% compared with the parent EHL (adsorption capacity 281.54 mg/g), which means that the simple fractionation process of EHL is an efficient route to obtain a lignin subdivision with a significantly enhanced MB adsorption capacity.

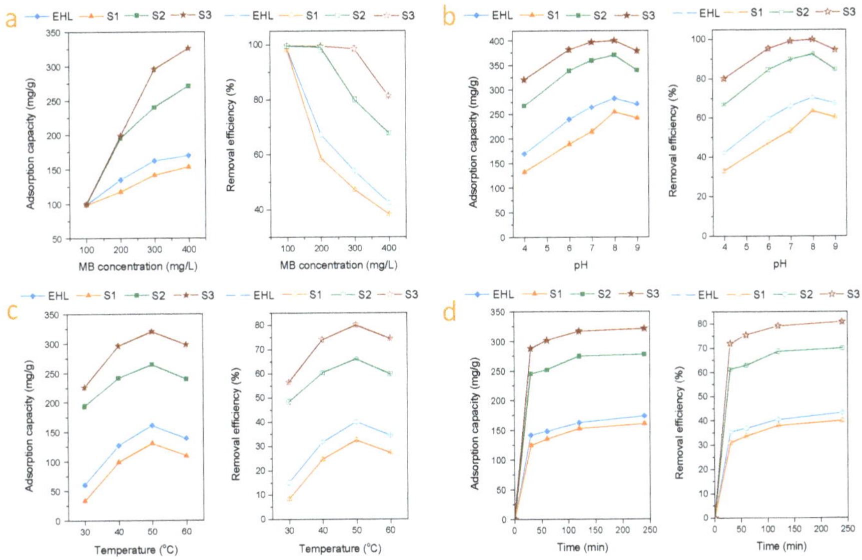

Figure 4. Methylene blue (MB) adsorption on EHL and its three subdivisions: (**a**) effect of MB concentration on adsorption capacity and removal efficiency (lignin concentration 1000 mg/L, temperature 50 °C, time 120 min, and pH value 4.0); (**b**) effect of pH (lignin concentration 1000 mg/L, MB concentration 400 mg/mL, temperature 50 °C and time 120 min); (**c**) effect of temperature (lignin concentration 1000 mg/L, MB concentration 400 mg/L, time 120 min, and pH value 4.0); (**d**) effect of time (lignin concentration 1000 mg/L, MB concentration 400 mg/L, pH value 4.0, and temperature 50 °C).

The surface morphologies of S3 particles and MB adsorbed S3 particles were compared (shown in Figure S1). It was found that the MB-loaded S3 particles exhibited bigger particle size compared with the original S3 particles. The morphological change of S3 particles after adsorbing MB is possibly due to the aggregation of S3 particles. During the adsorption, the cationic MB interacts with the negatively charged S3 particles and neutralizes the charge on the particle surface, which results in the aggregation of lignin particles. The EDS chemical analyses of S3 before and after MB adsorption were also analyzed and shown in Figure S1. The trace Na and Cl detected from the EDS of S3 possibly originated from the lignin isolation process that used NaOH solution to extract lignin from EHR and precipitates lignin by the addition of HCl. The EDS spectra of MB-adsorbed S3 showed the existence of N and S elements, which clearly demonstrates the adsorption of MB by the lignin particles. In order to confirm the adsorption of MB by other lignin samples, their EDS spectra were also recorded (Figure S2) and the element compositions based on the EDS spectra after normalization were tabulated as seen in Table S1. The highest contents of N and S elements on the surface of S3 particles further confirmed the best adsorption performance of S3.

2.3. *Adsorption Kinetics and Isotherms Studies*

The adsorption kinetics of MB on S3 at different temperatures are presented in Figure 5a. The adsorption amount enhanced sharply at the first 10 min, then increased slowly and later tended to stabilize after 120 min. The kinetic coefficients of pseudo-first-order and pseudo-second-order models are listed in Table 2. It was observed that the calculated equilibrium adsorption capacities (Qe) from pseudo-second-order kinetics were closer to the experimental values at different temperatures. Moreover, R^2 values obtained from the pseudo-second-order model were higher compared with those from the pseudo-first-order model. These results suggested that the MB adsorption on S3 took place according to the pseudo-second-order model, which agreed with the previous studies [22,23]. The overall rate of adsorption process is controlled by the chemical adsorption involving charge interaction between MB and S3 [23].

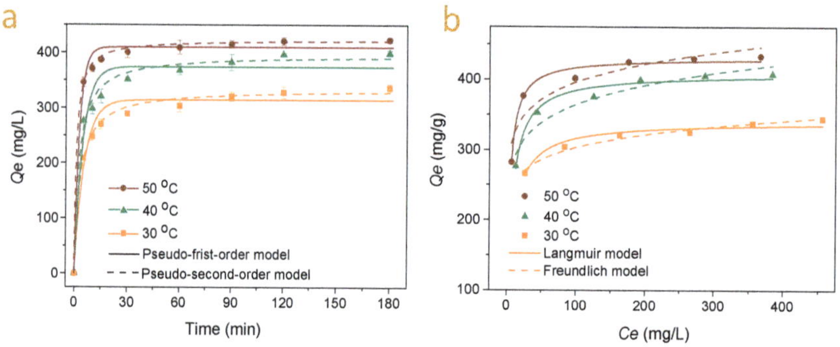

Figure 5. (**a**) Adsorption kinetics of MB on S3 (MB: 400 mg/L, 120 mL; adsorbent dose: 0.1 g; pH: 8) and the non-linear fitting curves using pseudo-first-order and pseudo-second-order models. (**b**) Adsorption isotherms of MB on S3 (adsorbent dose: 1 mg/mL, 10 mL; pH: 8; contact time 120 min) and the non-linear fitting curves using Freundlich and Langmuir models.

Table 2. Kinetic and isotherm parameters for adsorption of methylene blue on 80% ethanol insoluble subdivision (S3).

			Pseudo-First-Order Kinetics			Pseudo-Second-Order Kinetics		
	T (°C)	Q_e (mg/g)	K_1 (min^{-1})	Q_1 (mg/g)	R^2	K_2*10^{-4} (g·mg^{-1}·min^{-1})	Q_2 (mg/g)	R^2
Kinetic parameters	30	337.7	0.18	312.5	0.966	9.3	331.1	0.994
	40	395.1	0.21	372.0	0.952	9.7	392.5	0.989
	50	418.9	0.35	406.9	0.987	20.2	420.1	0.999
	T (°C)		Freundlich Constant			Langmuir Constant		
			n	K_F	R^2	b (L/mg)	Q_m (mg/g)	R^2
Isotherm parameters	30		11.8	205.4	0.972	0.13	341.0	0.931
	40		9.5	225.6	0.916	0.16	408.5	0.969
	50		11.0	260.8	0.858	0.29	431.1	0.979

The MB adsorption isotherms on S3 at various temperatures are shown in Figure 5b. The equilibrium adsorption amount Q_e enhanced with the increasing of MB equilibrium concentration C_e in solution. The curves of the adsorption isotherms fitted by Freundlich and Langmuir models are also presented in Figure 5b and the fitting parameters of the Freundlich and Langmuir equations are listed in Table 2. By comparing the correlation coefficients (R^2), it was found that the adsorption was better described by the Langmuir model compared to the Freundlich model. Besides, the Langmuir constant values of Q_m matched very well to the experimental values (Table 2). Therefore, these results suggested that the adsorption of MB on S3 was a monolayer adsorption. Zhang et al. obtained similar results when they used organosolv lignin from rice straw to remove MB from aqueous solutions [23]. The maximum monolayer adsorption capacity of S3 was found to be 431.1 mg/g, which was much higher than those of most bio-absorbents from lignin and lignin-contained biomass (Table 3). Moreover, it is notable that the MB adsorption capacity of S3 is comparable to those of biomass derived activated carbons (Table 3). Thus, the result indicates that the S3 obtained through the simple fractionation approach has a similar MB adsorption capacity as the biomass derived activated carbons prepared using a thermochemical modification process [40,41].

Table 3. The maximum monolayer adsorption (Q_m) of methylene blue onto various adsorbents.

Adsorbents	Adsorption Capacity (mg/g)	References
80% insoluble subdivision of EHL (corn stalk)	431.1	This work
Deacetylated acetic acid lignin (eucalyptus)	63.3	[22]
Organosolv lignin (rice straw)	40.0	[23]
Formic lignin (sugar cane bagasse)	34.2	[42]
Fe_3O_4@lignosulfonate/phenolic microsphere	292.6	[43]
Straw based adsorbents	274.7	[44]
Swede rape straw	246.4	[45]
Bamboo-based activated carbon	454.2	[46]
cork waste-based activated carbon	350.0	[40]
Coconut husk-based activated carbon	434.8	[41]

2.4. Recycling Studies

To evaluate the adsorbent recyclability, the used lignin adsorbents including the EHL and its three subdivisions, were regenerated by washing with ethanol and then reused to adsorb MB. The recoveries of the four lignin adsorbents are presented in Figure 6a. The S1 recovery after ethanol washing was only 5.36%. This was because S1 is the soluble fraction of EHL in 95% ethanol, and after ethanol washing, most of S1 dissolves. Similarly, S2 also exhibited a low recovery since it was soluble in 80% ethanol. Compared with S1 and S2, S3 had much higher recovery (over 90% after 5 recycles) due to its insolubility in both 95% and 80% ethanol solutions. The recovered S3 was then reused for up to six operation cycles and the corresponding MB adsorption capacities are shown in Figure 6b. Although the adsorption capacity of regenerated S3 was slightly decreased, it still exhibited over 75% of the original adsorption capacity after 5 cycles of regeneration. These results indicate that S3 exhibits satisfactory recycling performance via ethanol washing regeneration, showing excellent potential in practical applications.

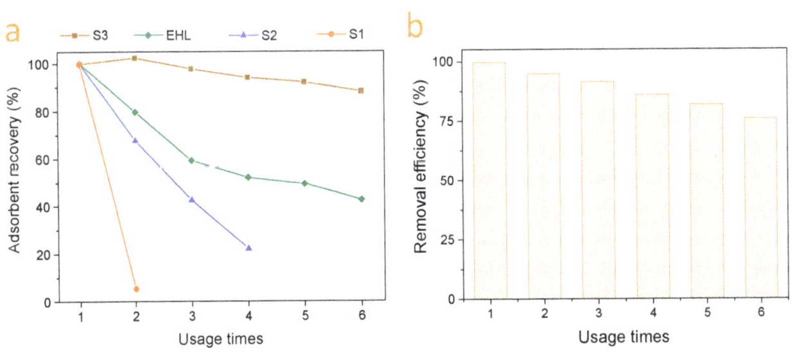

Figure 6. (**a**) Recovery of lignin adsorbents (S1, S2, S3, and EHL) under different recycle times; (**b**) adsorption capacities of S3 under different recycle times.

3. Materials and Methods

3.1. Material

Enzymatic hydrolysis residue (EHR) was kindly supplied by Songyuan Laihe Chemical Co. Ltd. (Jilin, China). The enzymatic hydrolysis (cellulase loading 30 U/g, temperature 50 °C and time 48 h) was carried out using steam exploded corn stalk (steam pressure 1.5 MPa and treatment time 5 min) as the raw material [28]. All laboratory reagents used in this work were analytically pure.

3.2. Isolation of Enzymatic Hydrolysis Lignin from Enzymatic Hydrolysis Residue

The isolation of EHL from EHR was based on alkaline extraction followed by acid precipitation [28]. Specifically, 20.0 g of EHR was added to 400 mL of 1% NaOH (*w*/*v*) solution and then the mixture was heated at 80 °C for 2 h with mechanical stirring (120 rpm). After alkaline extraction, the liquid and solid fractions were separated by a centrifugation at 5000 rpm for 10 min. The lignin extracted liquor (supernatant) was obtained and then adjusted to pH 2.0 using diluted HCl (1 mol/L) with vigorous agitation at ambient temperature. The precipitated EHL was then obtained by centrifugation (6000 rpm, 5 min) and washed three times with 400 mL distilled water. Afterward, the cleaned EHL was dried at 55 °C for 12 h using a vacuum oven (VOS-30A, STIK, Shanghai, China) and ground by a mortar to pass through a 120-mesh sieve.

3.3. Fractionation of Enzymatic Hydrolysis Lignin

The EHL was subdivided by stepwise dissolution in 95% and 80% ethanol solvents to obtain three lignin subdivisions according to the previous study [25]. Briefly, 5.0 g of EHL was added into 250 mL of 95% ethanol solution (ethanol/ethanol+water, *v*/*v*) and the mixture was stirred mechanically at 200 rpm for 10 min at ambient temperature. Afterward, the mixture was centrifuged at 6000 rpm for 10 min to obtain the supernatant (S1) and the insoluble lignin. The insoluble lignin was subsequently added in 250 mL of 80% ethanol solution (ethanol/ethanol+water, *v*/*v*). After stirring at 200 rpm for 10 min, the mixture was centrifuged at 5000 rpm for 5 min to obtain the supernatant (S2) and the insoluble lignin subdivision (S3). The S1 and S2, dissolved in 95% and 80% ethanol solutions, respectively, were recovered by vacuum rotary evaporation. All three subdivisions were further dried in a vacuum oven at 55 °C for 12 h and ground by a mortar to pass through a 120-mesh sieve.

3.4. Characterization of EHL and Lignin Subdivisions

Molecular weight distribution was measured by gel permeation chromatography (GPC) using a hydrophilic gel column (TSK G3000PWxl column, Tosoh Co., Tokyo, Japan).) [25]. The lignin samples were dissolved into 1% NaOH solution and diluted in tris-acetate buffer (20 mmol/L, pH 7.4). The column was operated at 25 °C and eluted with tris-acetate buffer (20 mmol/L, pH 7.4) at a flow rate of 0.5 mL/min. FTIR spectra were acquired using the KBr technique through a FTIR spectrophotometer (FTIR-650, Gangdong Sci. & Tech. Tianjin, China) [28]. The samples were firstly mixed with KBr by a ratio of 1:100 (sample to KBr, *w*/*w*), and the mixture was then ground in an agate mortar and compressed to obtain the KBr disc for FTIR determination. The wavenumber region was between 4000 and 400 cm^{-1} with a resolution of 4 cm^{-1} and 20 scans were recorded. Zeta potential and particle size distribution of lignin samples were analyzed by dynamic light scattering measurement (Malvern Instruments Ltd., Malvern, UK). Here, lignin solutions were prepared with an identical concentration and adjusted to the same pH value for Zeta potential and particle size determination. The specific surface area of lignin samples was measured through N_2 adsorption using a specific surface area analysis instrument (Quantachrome Autosorb-IQ, Boynton, FL, USA) at a liquid nitrogen temperature [43]. The morphology of lignin adsorbents was observed using Scanning electron microscopy (SEM) measurements (SHIMADZU SSX-550, Kyoto, Japan), combined with Energy-dispersive X-ray spectroscope (EDS) for the determination of chemical compositions of lignin adsorbents before and after the adsorption. Before the analysis, a small amount of lignin samples was evenly spread on the conductive adhesive and a thin gold coating was deposited onto the uniformly.

3.5. Batch experiments for Methylene Blue Adsorption

MB Adsorption experiments from aqueous solutions by the lignin samples (S1, S2, S3, and EHL) were carried out in triplicate using 50 mL Erlenmeyer flasks containing 10 mL of MB solution with different initial concentrations (C_0, 100–400 mg/L). The pH of the MB solution ranging from 2.0 to 9.0 was adjusted by adding 0.1 M NaOH and 0.1 M HCl solutions. Lignin sample with a fixed amount was added into the MB solution and the mixture was shaken in a thermostatic shaking incubator (120 rpm) at predetermined temperatures. After adsorption, the mixture was centrifuged at 5000 rpm for 5 min to collect the lignin adsorbent from the MB aqueous solution. The residual MB in the solution was measured using a UV-visible Spectrophotometer (Shimadzu UV-2500, 664 nm, Kyoto, Japan) [43]. The adsorption capacity (Q_t) and percentage of MB removal (E) were obtained according to Equations (1) and (2), respectively [47]:

$$Q_t(\text{mg/g}) = \frac{(C_0 - C_t)V}{m} \quad (1)$$

$$E(\%) = \frac{(C_0 - C_t)}{C_0} \times 100 \quad (2)$$

where C_0 is the initial MB concentration (mg/L) and C_t is the MB concentration in mg/L at time t (min), m is the weight of lignin adsorbent in g, and V is the volume of the MB solution in L.

3.6. Adsorption Kinetics and Isotherms

The kinetics of the MB adsorption process by the 80% ethanol insoluble subdivision (S3) was studied by contacting 120 mL 400 mg/L MB solution (pH 8.0) with 100 mg lignin at 30, 40 and 50 °C. The pseudo-first and second-order models were employed to simulate the adsorption kinetics and the equations were represented as the following [48]:

Pseudo-first-order models:

$$Q_t = Q_e(1 - \exp^{-k1t}) \quad (3)$$

Pseudo-second-order models:

$$Q_t = \frac{k_2 Q_e{}^2 t}{1 + k_2 Q_e t} \quad (4)$$

where Q_t (mg/g) is the amount of MB adsorbed at time t (min), k_1 and k_2 are the rate constants of pseudo-first-order and pseudo-second-order models, respectively.

The isotherms of MB adsorption on lignin were investigated by contacting 10 mL MB solution (pH 8.0, 300–800 mg/L) with 10 mg lignin at 30, 40 and 50 °C for 120 min. Freundlich and Langmuir isotherm models were used to simulate the adsorption data [49]:

Freundlich model:

$$Q_e = k_F C_e{}^{1/n} \quad (5)$$

Langmuir model:

$$Q_e = \frac{Q_m k_L C_e}{1 + k_L C_e} \quad (6)$$

where Q_e is the equilibrium MB adsorption amount of lignin (mg/g), C_e is the equilibrium MB concentration (mg/L), and Q_m (mg/g) is the monolayer maximum adsorption ability. In the Freundlich model, n and k_F represent the intensity of adsorption and the multilayer adsorption capacity (mg/g), while k_L in Langmuir model is the Langmuir constant (L/mg).

3.7. Lignin Recycling

Adsorbent recyclability was also studied by regeneration of MB loaded adsorbent in ethanol solution [50,51]. Typically, 0.01 g lignin was added into a 50 mL Erlenmeyer flask with 10 mL of 400 mg/mL MB solution. The pH of the MB solution was 8.0. The mixture was shaken in a thermostatic shaking incubator (120 rpm) at 50 °C for 180 min. Once the adsorption equilibrium was achieved, the used lignin was collected by centrifugation (5000 rpm, 5 min). The MB desorption from the lignin adsorbents was conducted by ultrasonically washing with 10 mL ethanol for 4 times. Finally, the lignin adsorbent was washed with distilled water and ground to pass through a 120-mesh sieve after vacuum drying. The regenerated lignin was used for the next adsorption run following the identical experimental conditions. The adsorption-desorption process was repeated at least 6 times using the same lignin adsorbent.

4. Conclusions

Three lignin subdivisions with decreased heterogeneity of molecular weight and structure were obtained by stepwise dissolution of EHL in 95% and 80% ethanol solutions. The 80% ethanol-insoluble subdivision (S3) showed the highest adsorption capacity (396.85 mg/g). The dissolution of S1 and S2 resulted in the decrease of S3 particle size and the formation of a rough surface, which led to the increase of specific surface area (4.84 m^2/g) and the negative Zeta potential (−35.24 mv), and, therefore, the adsorption capacity. The MB adsorption on S3 followed pseudo-second-order kinetics and based on the adsorption isotherms fitted by Langmuir model, the maximum monolayer adsorption capacity of MB on S3 was 431.1 mg/g, which was much higher than those of most reported lignin-based adsorbents. Moreover, S3 demonstrated much better recyclability owing to its high regeneration yield (over 90% after 5 recycles) by ethanol desorption. Overall, the adsorption capacity and recyclability of EHL was significantly improved using the simple stepwise ethanol-water fractionation process.

Supplementary Materials: The following are available online, Figure S1: SEM (a and c) and EDS analyses (b and d) of S3 (a and b) and MB adsorbed S3 (c and d), Figure S2: EDS analyses of lignin before and after methylene blue adsorption, Table S1: Element compositions of lignin and MB adsorbed lignin based on EDS spectra after normalization.

Author Contributions: Conceptualization, W.S. and G.W.; methodology, T.P. and C.L. (Cuiyun Liu); software, T.P.; formal analysis, T.P. and C.L. (Cuiyun Liu); investigation, T.P. and C.L. (Cuiyun Liu); data curation, W.S. and T.P.; writing—original draft preparation, W.S. and T.P.; writing—review and editing, G.W., A.M.P., C.S. and C.L. (Chao Li); supervision, S.W. and G.W.; funding acquisition, G.W. All authors have read and agreed to the published version of the manuscript.

Funding: This research was funded by Major Science and Technology Innovation Project of Shandong Province, grant number 2018CXGC022, Natural Science Foundation of China, grant number 21808171 and 31700515, Research and Development Plan of Key Areas in Hunan Province, grant number 2019NK2032, and the Hunan (Hu-Xiang) Youth Talent Program, grant number 2018RS3114.

Acknowledgments: This work was supported by the Major Science and Technology Innovation Project of Shandong Province (2018CXGC0220), Natural Science Foundation of China (21808171 and 31700515), Research and Development Plan of Key Areas in Hunan Province (2019NK2032), and the Hunan (Hu-Xiang) Youth Talent Program (2018RS3114).

Conflicts of Interest: The authors declare no conflict of interest. The funders had no role in the design of the study; in the collection, analyses, or interpretation of data; in the writing of the manuscript, or in the decision to publish the results.

References

1. Forgacs, E.; Cserháti, T.; Oros, G. Removal of synthetic dyes from wastewaters: A review. *Environ. Int.* **2004**, *30*, 953–971. [CrossRef] [PubMed]
2. Jin, X.-C.; Liu, G.-Q.; Xu, Z.-H.; Tao, W.-Y. Decolorization of a dye industry effluent by Aspergillus fumigatus XC6. *Appl. Microbiol. Biotechnol.* **2007**, *74*, 239–243. [CrossRef] [PubMed]
3. Brown, D.; Hitz, H.R.; Schäfer, L. The assessment of the possible inhibitory effect of dyestuffs on aerobic waste-water bacteria experience with a screening test. *Chemosphere* **1981**, *10*, 245. [CrossRef]
4. Yagub, M.T.; Sen, T.K.; Afroze, S.; Ang, H.M. Dye and its removal from aqueous solution by adsorption: A review. *Adv. Colloid Interface Sci.* **2014**, *209*, 172–184. [CrossRef]
5. Hu, L.; Guang, C.; Liu, Y.; Su, Z.; Gong, S.; Yao, Y.; Wang, Y. Adsorption behavior of dyes from an aqueous solution onto composite magnetic lignin adsorbent. *Chemosphere* **2020**, *246*, 125757. [CrossRef]

6. Verma, A.K.; Dash, R.R.; Bhunia, P. A review on chemical coagulation/flocculation technologies for removal of colour from textile wastewaters. *J. Environ. Manage.* **2012**, *93*, 154–168. [CrossRef]
7. Al-Kdasi, A.; Idris, A.; Saed, K.; Guan, C.T. Treatment of textile wastewater by advanced oxidation processes-a review. *Global nest: The Int. J.* **2004**, *6*, 222–230.
8. Akpan, U.G.; Hameed, B.H. Parameters affecting the photocatalytic degradation of dyes using TiO2-based photocatalysts: A review. *J. Hazard. Mater.* **2009**, *170*, 520–529. [CrossRef]
9. Salleh, M.A.M.; Mahmoud, D.K.; Karim, W.A.W.A.; Idris, A. Cationic and anionic dye adsorption by agricultural solid wastes: A comprehensive review. *Desalination* **2011**, *280*, 1–13. [CrossRef]
10. You, X.; Li, E.; Liu, J.; Li, S. Using Natural Biomacromolecules for Adsorptive and Enzymatic Removal of Aniline Blue from Water. *Molecules* **2018**, *23*, 1606. [CrossRef]
11. Chen, F.; Hu, X.; Tu, X.; Chen, L.; Liu, X.; Tan, L.; Mao, Y.; Shi, J.; Teng, X.; He, S.; et al. High-Yield Production of Lignin-Derived Functional Carbon Nanosheet for Dye Adsorption. *Polymers* **2020**, *12*, 797. [CrossRef]
12. Tahir, S.; Rauf, N. Removal of a cationic dye from aqueous solutions by adsorption onto bentonite clay. *Chemosphere* **2006**, *63*, 1842–1848. [CrossRef] [PubMed]
13. Abbasian, M.; Jaymand, M.; Niroomand, P.; Farnoudian-Habibi, A.; Karaj-Abad, S.G. Grafting of aniline derivatives onto chitosan and their applications for removal of reactive dyes from industrial effluents. *Int. J. Biol. Macromol.* **2017**, *95*, 393–403. [CrossRef] [PubMed]
14. Srinivasan, A.; Viraraghavan, T. Decolorization of dye wastewaters by biosorbents: A review. *J. Environ. Manage.* **2010**, *91*, 1915–1929. [CrossRef] [PubMed]
15. Li, Z.; Kong, Y.; Ge, Y. Synthesis of porous lignin xanthate resin for Pb^{2+} removal from aqueous solution. *Chem. Eng. J.* **2015**, *270*, 229–234. [CrossRef]
16. Supanchaiyamat, N.; Jetsrisuparb, K.; Knijnenburg, J.T.N.; Tsang, D.C.W.; Hunt, A.J. Lignin materials for adsorption: Current trend, perspectives and opportunities. *Bioresour. Technol.* **2019**, *272*, 570–581. [CrossRef] [PubMed]
17. González-López, M.E.; Robledo-Ortíz, J.R.; Rodrigue, D.; Pérez-Fonseca, A.A. Highly porous lignin composites for dye removal in batch and continuous-flow systems. *Mater. Lett.* **2020**, *263*. [CrossRef]
18. Wang, G.; Qi, S.; Xia, Y.; Parvez, A.M.; Si, C.; Ni, Y. Mild One-Pot Lignocellulose Fractionation Based on Acid-Catalyzed Biphasic Water/Phenol System to Enhance Components' Processability. *ACS Sustain. Chem. Eng.* **2020**, *8*, 2772–2782. [CrossRef]
19. Carrott, P.J.M.; Carrott, M.R. Lignin–from natural adsorbent to activated carbon: A review. *Bioresour. Technol.* **2007**, *98*, 2301–2312. [CrossRef]
20. Sipponen, M.H.; Pihlajaniemi, V.; Littunen, K.; Pastinen, O.; Laakso, S. Determination of surface-accessible acidic hydroxyls and surface area of lignin by cationic dye adsorption. *Bioresour. Technol.* **2014**, *169*, 80–87. [CrossRef]
21. Feng, Q.; Cheng, H.; Li, J.; Wang, P.; Xie, Y. Adsorption behavior of basic dye from aqueous solution onto alkali extracted lignin. *BioResources* **2014**, *9*, 3602–3612. [CrossRef]
22. Feng, Q.; Cheng, H.; Chen, F.; Zhou, X.; Wang, P.; Xie, Y. Investigation of cationic dye adsorption from water onto acetic acid lignin. *J. Wood Chem. Technol.* **2016**, *36*, 173–181. [CrossRef]
23. Zhang, S.; Wang, Z.; Zhang, Y.; Pan, H.; Tao, L. Adsorption of methylene blue on organosolv lignin from rice straw. *Procedia Environ. Sci.* **2016**, *31*, 3–11. [CrossRef]
24. Menkiti, M.; Aniagor, C.; Agu, C.; Ugonabo, V. Effective adsorption of crystal violet dye from an aqueous solution using lignin-rich isolate from elephant grass. *Water Conserv. Sci. Eng.* **2018**, *3*, 33–46. [CrossRef]
25. Wang, G.; Chen, H. Fractionation and characterization of lignin from steam-exploded corn stalk by sequential dissolution in ethanol–water solvent. *Sep. Purif. Technol.* **2013**, *120*, 402–409. [CrossRef]
26. Jiang, P.; Li, Q.; Gao, C.; Lu, J.; Cheng, Y.; Zhai, S.; An, Q.; Wang, H. Fractionation of alkali lignin by organic solvents for biodegradable microsphere through self-assembly. *Bioresour. Technol.* **2019**, *289*, 121640. [CrossRef]
27. Prado, R.; Erdocia, X.; Labidi, J. Lignin extraction and purification with ionic liquids. *J. Chem. Technol. Biotechnol.* **2013**, *88*, 1248–1257. [CrossRef]
28. Liu, C.; Si, C.; Wang, G.; Jia, H.; Ma, L. A novel and efficient process for lignin fractionation in biomass-derived glycerol-ethanol solvent system. *Ind. Crop. Prod.* **2018**, *111*, 201–211. [CrossRef]

29. Yuan, T.-Q.; Sun, S.-N.; Xu, F.; Sun, R.-C. Characterization of lignin structures and lignin–carbohydrate complex (LCC) linkages by quantitative 13C and 2D HSQC NMR spectroscopy. *J. Agric. Food Chem.* **2011**, *59*, 10604–10614. [CrossRef]
30. An, L.; Si, C.; Wang, G.; Sui, W.; Tao, Z. Enhancing the solubility and antioxidant activity of high-molecular-weight lignin by moderate depolymerization via in situ ethanol/acid catalysis. *Ind. Crop. Prod.* **2019**, *128*, 177–185. [CrossRef]
31. Maniet, G.; Schmetz, Q.; Jacquet, N.; Temmerman, M.; Gofflot, S.; Richel, A. Effect of steam explosion treatment on chemical composition and characteristic of organosolv fescue lignin. *Ind. Crop. Prod.* **2017**, *99*, 79–85. [CrossRef]
32. Wang, G.; Chen, H. Enhanced lignin extraction process from steam exploded corn stalk. *Sep. Purif. Technol.* **2016**, *157*, 93–101. [CrossRef]
33. Wen, J.L.; Sun, S.L.; Xue, B.L.; Sun, R.C. Recent Advances in Characterization of Lignin Polymer by Solution-State Nuclear Magnetic Resonance (NMR) Methodology. *Materials* **2013**, *6*, 359–391. [CrossRef] [PubMed]
34. Min, D.; Jameel, H.; Chang, H.; Lucia, L.; Wang, Z.; Jin, Y. The structural changes of lignin and lignin–carbohydrate complexes in corn stover induced by mild sodium hydroxide treatment. *RSC Adv.* **2014**, *4*, 10845–10850. [CrossRef]
35. Nair, V.; Panigrahy, A.; Vinu, R. Development of novel chitosan–lignin composites for adsorption of dyes and metal ions from wastewater. *Chem. Eng. J.* **2014**, *254*, 491–502. [CrossRef]
36. Sciban, M.; Klasnja, M. Study of the adsorption of copper (II) ions from water onto wood sawdust, pulp and lignin. *Adsorpt. Sci. Technol.* **2004**, *22*, 195–206. [CrossRef]
37. Amer, F.; Mahmoud, A.A.; Sabet, V. Zeta Potential and Surface Area of Calcium Carbonate as Related to Phosphate Sorption1. *Soil Sci. Soc. Am. J.* **1985**, *49*, 1137–1142. [CrossRef]
38. Singh, S.; Barick, K.C.; Bahadur, D. Surface engineered magnetic nanoparticles for removal of toxic metal ions and bacterial pathogens. *J. Hazard. Mater.* **2011**, *192*, 1539–1547. [CrossRef]
39. Bayat, M.; Javanbakht, V.; Esmaili, J. Synthesis of zeolite/nickel ferrite/sodium alginate bionanocomposite via a co-precipitation technique for efficient removal of water-soluble methylene blue dye. *Int. J. Biol. Macromol.* **2018**, *116*, 607–619. [CrossRef] [PubMed]
40. Novais, R.M.; Caetano, A.P.F.; Seabra, M.P.; Labrincha, J.A.; Pullar, R.C. Extremely fast and efficient methylene blue adsorption using eco-friendly cork and paper waste-based activated carbon adsorbents. *J. Clean. Prod.* **2018**, *197*, 1137–1147. [CrossRef]
41. Tan, I.; Ahmad, A.; Hameed, B. Adsorption of basic dye on high-surface-area activated carbon prepared from coconut husk: Equilibrium, kinetic and thermodynamic studies. *J. Hazard. Mater.* **2008**, *154*, 337–346. [CrossRef]
42. Consolin Filho, N.; Venancio, E.; Barriquello, M.; Hechenleitner, A.; Pineda, E. Methylene blue adsorption onto modified lignin from sugar cane bagasse. *Eclética Química* **2007**, *32*, 63–70. [CrossRef]
43. Wang, G.; Liu, Q.; Chang, M.; Jang, J.; Sui, W.; Si, C.; Ni, Y. Novel Fe_3O_4@ lignosulfonate/phenolic core-shell microspheres for highly efficient removal of cationic dyes from aqueous solution. *Ind. Crop. Prod.* **2019**, *127*, 110–118. [CrossRef]
44. Zhang, W.; Yan, H.; Li, H.; Jiang, Z.; Dong, L.; Kan, X.; Yang, H.; Li, A.; Cheng, R. Removal of dyes from aqueous solutions by straw based adsorbents: Batch and column studies. *Chem. Eng. J.* **2011**, *168*, 1120–1127. [CrossRef]
45. Feng, Y.; Zhou, H.; Liu, G.; Qiao, J.; Wang, J.; Lu, H.; Yang, L.; Wu, Y. Methylene blue adsorption onto swede rape straw (*Brassica napus L.*) modified by tartaric acid: Equilibrium, kinetic and adsorption mechanisms. *Bioresour. Technol.* **2012**, *125*, 138–144. [CrossRef] [PubMed]
46. Hameed, B.; Din, A.M.; Ahmad, A. Adsorption of methylene blue onto bamboo-based activated carbon: Kinetics and equilibrium studies. *J. Hazard. Mater.* **2007**, *141*, 819–825. [CrossRef] [PubMed]
47. Hsu, D.; Lu, C.; Pang, T.; Wang, Y.; Wang, G. Adsorption of Ammonium Nitrogen from Aqueous Solution on Chemically Activated Biochar Prepared from Sorghum Distillers Grain. *Appl. Sci.* **2019**, *9*, 5249. [CrossRef]
48. Li, J.; Li, H.; Yuan, Z.; Fang, J.; Chang, L.; Zhang, H.; Li, C. Role of sulfonation in lignin-based material for adsorption removal of cationic dyes. *Int. J. Biol. Macromol.* **2019**, *135*, 1171–1181. [CrossRef]
49. Yang, Y.; Wei, X.; Sun, P.; Wan, J. Preparation, Characterization and Adsorption Performance of a Novel Anionic Starch Microsphere. *Molecules* **2010**, *15*, 2872–2885. [CrossRef]

50. Ai, L.; Jiang, J. Removal of methylene blue from aqueous solution with self-assembled cylindrical graphene–carbon nanotube hybrid. *Chem. Eng. J.* **2012**, *192*, 156–163. [CrossRef]
51. Zhao, M.; Tang, Z.; Liu, P. Removal of methylene blue from aqueous solution with silica nano-sheets derived from vermiculite. *J. Hazard. Mater.* **2008**, *158*, 43–51. [CrossRef] [PubMed]

Sample Availability: Not available

molecules

MDPI

Article

An Analysis of Exhaust Emission of the Internal Combustion Engine Treated by the Non-Thermal Plasma

Ming-Hsien Hsueh [1,*], Chia-Nan Wang [1], Meng-Chang Hsieh [2], Chao-Jung Lai [3], Shi-Hao Wang [1], Chia-Hsin Hsieh [1], Tsung-Liang Wu [1] and Jo-Hung Yu [1]

1 Department of Industrial Engineering and Management, National Kaohsiung University of Science and Technology, Kaohsiung 807, Taiwan; cn.wang@nkust.edu.tw (C.-N.W.); shwang@nkust.edu.tw (S.-H.W.); charlie820906@gmail.com (C.-H.H.); wut@nkust.edu.tw (T.-L.W.); henry@nkust.edu.tw (J.-H.Y.)
2 Institute of Undersea Technology, National Sun Yat-Sen University, Kaohsiung 804, Taiwan; pedro@g-mail.nsysu.edu.tw
3 Department of Fashion Design and Management, Tainan University of Technology, Tainan 71002, Taiwan; t30129@mail.tut.edu.tw
* Correspondence: mhhsueh@nkust.edu.tw

Academic Editors: Chiara Bisio and Monica Pica
Received: 18 November 2020; Accepted: 18 December 2020; Published: 21 December 2020

check for updates

Abstract: Industries' air pollution causes serious challenges to modern society, among them exhaust gases from internal combustion engines, which are currently one of the main sources. This study proposes a non-thermal plasma (NTP) system for placement in the exhaust system of internal combustion engines to reduce the toxic contaminants (HC, CO, and NO_x) of exhaust gases. This NTP system generates a high-voltage discharge that not only responds to the ion chemical reaction to eliminate NO_x and CO, but that also generates a combustion reaction at the local high temperature of plasma to reduce HC. The NTP system was designed on both the front and rear of the exhaust pipe to analyze the difference of different exhaust flow rates under the specified frequency. The results indicate that the NTP system can greatly reduce toxic contaminants. The NTP reactor placed in the front of exhaust pipe gave HC and CO removal efficiency of about 34.5% and 16.0%, respectively, while the NTP reactor placed in the rear of exhaust pipe gave NO_x removal efficiency of about 41.3%. In addition, the voltage and material directly affect the exhaust gases obviously. In conclusion, the proposed NTP system installed in the exhaust system can significantly reduce air pollutants. These results suggest that applying NTP to the combustion engine should be a useful tool to simultaneously reduce both emissions of NO_x and CO.

Keywords: non-thermal plasma (NTP); exhaust emission; internal combustion engine; ion chemical reaction

1. Introduction

The last few years have witnessed rapid development in industrial technology, with an increasing demand for fuel-based energy. The combustion of fuel is the primary source of energy for vehicle engines, aircraft engines, and industrial machinery. Fuel combustion produces a combination of chemical energy, thermal energy, and exhaust gases. Portions of combustion gases harmful to the environment such as unburned hydrocarbon, nitrogen oxides, carbon monoxide, and particulate matter are released into the environment. The exhaust gases emitted into the atmosphere cause air pollution due to the incomplete combustion of carbonaceous fuel. Inhaling gases are harmful to humans and cause health issues such as lung cancer, asthma, cardiovascular diseases [1–3]. Inhaling the exhaust gases damages the airway and may impair their function, especially the lungs [4]. Therefore, with mounting environmental concerns, the purification of exhaust gases from the internal combustion engine is becoming increasingly urgent to address. Balki et al. (2014) investigated the exhaust emission of an SI engine using gasoline, ethanol, and methanol at different engine speeds and observed that the use of alcohol instead of gasoline as the fuel in a low power engine caused a decline mainly in NO_x, hydrocarbons (HC) and CO emissions. In addition, when methanol and ethanol were used in the combustion engine, the emissions of NO_x, CO, and HC were reduced by 49% and 47.6%, 22.6% and 21.25%, and 21.6% and 19.13%, respectively [5]. Hsueh et al. (2016) studied a fuel temperature control device using a thermoelectric module (TEC) chip to measure engine performance and exhaust emission at various fuel temperatures and air/fuel (A/F) ratios. The emission ratios of HC and CO decreased but that of NO_x increased as the fuel temperature increased [6]. Doğan et al. (2017) investigated ethanol-gasoline blends as fuel for a four-cylinder and four-stroke spark-ignition engine and found that ethanol added to gasoline in the combustion engine fuel caused a reduction in CO and NO_x emission ratios [7].

In recent years, the development of non-thermal plasma (NTP) has attracted high interest and is considered as the green strategy for exhaust pollutant remediation. It can be used to remove various air pollutants such as SO_2, NO_x, HC, CO, and VOC [8,9]. In NTP, the electron temperature is higher than the gas temperature, which is close to room temperature. High energy electrons collide with surrounding gas molecules to produce reactive species such as dissociated molecules, ions, free radicals, and secondary electrons [10]. The reactive species actively combine with the environmental pollutants to form a decomposition product [11]. Various NTP systems such as the spark, dielectric barrier discharges, gliding arc, corona, microwave, and glow have been investigated for the direct conversion of exhaust gases [12–16]. The use of a plasma reactor with a catalyst to treat exhaust gases from gasoline engines was investigated by Dan et al. (2005), who observed that particulate matter removal efficiency ranged approximately from 25 to 57% [17]. Kim et al. (2017) investigated the effects of NTP on a lean premixed model gas turbine combustor of NO_x and CO by changing the mixing nozzle exit velocity and the equivalence ratio. The result showed a reduction in NO_x and CO emissions [18]. Adnan et al. (2017) tried to increase the rate of NTP decomposition by increasing the flow rates of the exhaust gases and could reduce the concentrations of CO, CO_2, HC, and NO_x by more than 95% [19].

In this study, the variation of toxic contaminants in the exhaust system of the internal gasoline engine was investigated by using the NTP system. The NTP reactor was placed in the front or rear of the exhaust pipe to treat exhaust gas and the variation of emission was analyzed. The exhaust gas variation of the engine was observed by controlling the position, voltage, and types of NTP reactors with different engine speeds.

2. Experimental Apparatus and Techniques

2.1. Experimental Setup

The schematic of the experimental setup is shown in Figure 1, which mainly includes three sections, the production of exhaust gas by the motorcycle engine, the device for exhaust emissions reduction by the NTP system, and the exhaust gas analyzer of the detection system. The engine used for this experimental test was the model GP-125 manufactured by KYMCO Co. Ltd. (Taiwan), and there was no catalyst inside the exhaust pipe. The specifications of the engine in the power and torque measurement test are mentioned in Table 1. The experimental control conditions of the engine are mentioned in Table 2. The engine speeds were set in the range of 4000–6000 rpm, a frequently used range for practical motorcycle engines. The detailed experimental setup of the NTP system is shown in Figure 2, which is consisted mainly of three parts, the NTP reactor, the voltage converter, and the NTP control device. The NTP reactors used for this experiment test were model CR8EGP and CR8EIX manufactured by NGK Spark Plug Co., Ltd. (Aichi, Japan). The specifications of NTP reactors are mentioned in Table 3. NTP reactor was placed in the model A, in the front of the exhaust pipe or model

B, in the rear of the exhaust pipe, as described in Figures 2 and 3. The voltage converter used for this experiment test was model 3051A-LGL3–900 (KYMCO Co. Ltd., Taiwan). The NTP control device used for this experiment test was model DP-30032 (HILA International Inc, Taiwan). The specifications of the NTP control device are mentioned in Table 4. Pulsed voltages with a frequency of 10 Hz and adjustable amplitudes from 2 to 5 V were applied to the electrodes to produce a spark. The exhaust gas analyzer used for this experiment test was model EF-306EN (Exford, Taiwan), which measures the range of each exhaust gas, as mentioned in Table 5. The exhaust gas was measured at the end of the exhaust pipe and the data were recorded for 5 min under normal conditions to calculate an average. The A/F ratio was maintained at 13.7 to fix the stable fuel supply condition and observe the change in engine power performance and emission.

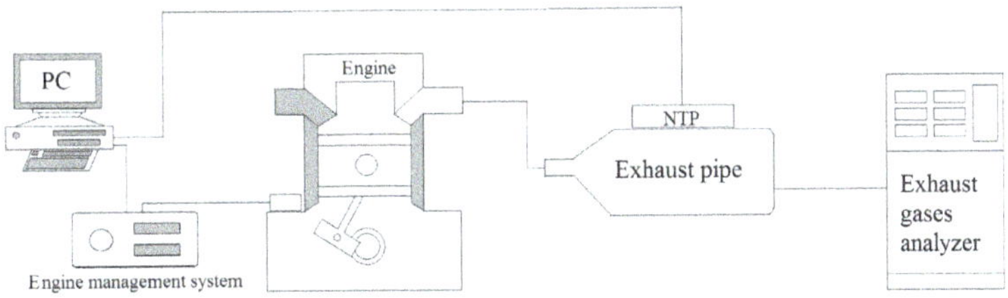

Figure 1. Schematic of this experiment and setup location.

Table 1. The engine specification of motorcycle.

Parameter	Value
Model	GP125
Stoke	four-stoke
Engine type	Single cylinder
Displacement (c.c.)	124.6
Bore × Stoke (mm)	52.4 × 57.8
Compression ratio	9.9:1
Fuel	Unleaded gasoline
Max power (kw/rpm)	7.8/7500
Max torque (N-m/rpm)	10.3/5500
Wheelbase (mm)	1220
Air/fuel ratio (A/F)	13.7

Table 2. The control conditions of engine.

Parameter	Value
Engine speed (rpm)	4000, 5000, 6000
Fuel octane rating	95
Engine temperature (°C)	100–110
Intake air temperature (°C)	22–26

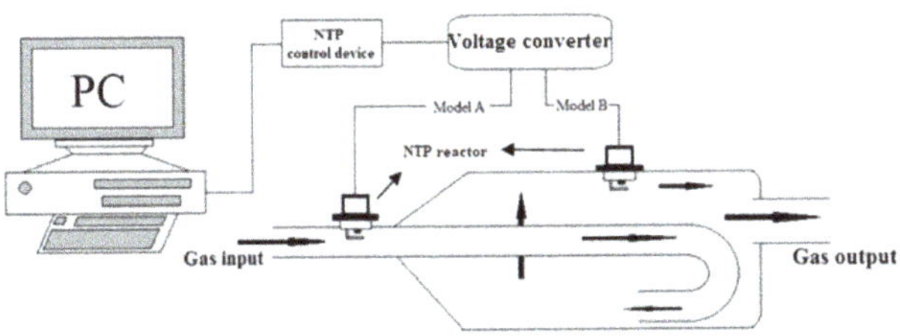

Figure 2. The setup of non-thermal plasma (NTP) system.

Table 3. The types of NTP reactor.

Parameter	Value	
	Type One	**Type Two**
Model	NGK-CR8EGP	NGK-CR8EIX
Material	Platinum	Iridium alloy
Discharging Gap (before used) (mm)	0.7	0.8
Discharging Gap (after used) (mm)	0.7	0.7
Center electrode diameter (mm)	0.8	0.8
Outer electrode diameter (mm)	2.20	1.85

Figure 3. Photographic view of the device set up on (**a**) model A and (**b**) model B of the experimental motorcycle.

Table 4. The specification of NTP control device.

Parameter	Value
Model	DP-30032
Main output voltage (V)	0–30
Main output current (A)	0–3
Fixed output voltage (V)	2.5/3.3/5
Fixed output current (A)	3
Resolution (mV/mA)	100/10
Precision	±(1% reading + 2 digits)
Power supply (V, Hz)	AC110/220 ± 10% selectable, 50/60
Dimensions (mm)	250W × 150H × 310D

Table 5. The range of each exhaust gas.

Parameter	Value	
	Measuring Range	**Tolerance**
HC (ppm)	0–2000	±12
CO (%)	0–10	±0.06
CO_2 (%)	0–20	±0.5
NO_x (ppm)	0–5000	±25

2.2. Experimental Procedure

The measurement equipment was adjusted and corrected before the experiment began. The preparation and measurement procedures were as follows:

1. The engine condition of the motorcycle, such as engine oil, air filter, and fuel was checked.
2. The exhaust gas analyzer was calibrated and warmed up.
3. The engine was initiated for 30 min for the temperature to reach 100 °C.
4. The NTP system was checked and the type of NTP reactor was selected.
5. Before each experiment, a new NTP reactor was replaced.

In this study, the engine speed was 4000, 5000, and 6000 rpm, and the engine combustion produced exhaust gas when the engine started. The exhaust gas was discharged to the exhaust gas analyzer after passing through the NTP reactor. The data were directly recorded with the help of Microsoft Excel. The sparking frequency of the NTP control device was 10 Hz. The experimental procedure is presented in Figure 4. In these experiments, the different rpm of the engine were measured and compared to determine the optimal type and location of the NTP system. Based on the results of this experiment, the optimal condition of the NTP system was obtained and later applied to an actual vehicle.

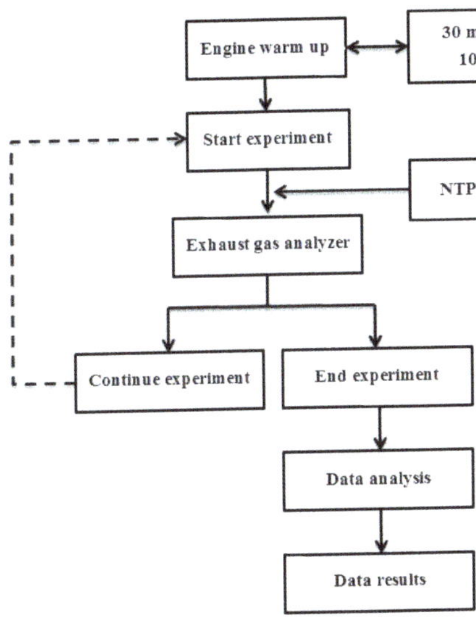

Figure 4. The procedure of experimental.

The dissociation process of exhaust gas was defined as:
The removal efficiency of pollutants (RE) for HC, CO, and NO_x:

$$RE\ (\%) = \frac{[Gas]_{off} - [Gas]_{on}}{[Gas]_{off}} \times 100\% \tag{1}$$

The conversion efficiency of pollutants (CE) for CO_2:

$$CE\ (\%) = \frac{[Gas]_{on} - [Gas]_{off}}{[Gas]_{off}} \times 100\% \tag{2}$$

where $[Gas]_{off}$ is the exhaust emission value without the NTP system; $[Gas]_{on}$ is the exhaust emission value with the NTP system.

2.3. NTP Reactor

The spark is generated by the electrode gap of the NTP reactor. (Tables 3 and 4). The combustible gas is ignited in the exhaust gas. The exhaust gas composition of engine combustion includes HC, CO, CO_2, NO_x, SO_2, and PM2.5, etc. But the instrument used in this article can only measure HC, CO, CO_2, and NO_x. Therefore, the major reactions in the exhaust gas reduction by the NTP reactor are listed below [20–23].

The short high current spark with excited atomic radicals, excited molecules, and ions are generated, and the temperature during the spark phase can be as high as 3000 K. Therefore, we considered the Zeldovich thermal mechanism (see the equations below) of NO_x generation.

$$N_2 + O \rightarrow NO + N \tag{3}$$

$$N + O_2 \rightarrow NO + O \tag{4}$$

Oxygen molecules are broken down into oxygen radicals by the free electrons. The reactions of these oxygen molecules are listed below [20–23].

$$O_2 + e^- \rightarrow O + O \tag{5}$$

$$O + O \rightarrow O_2 \tag{6}$$

$$O_3 + e^- \rightarrow O_2 + O \tag{7}$$

$$O + O_2 \rightarrow O_3 \tag{8}$$

$$O + O_3 \rightarrow O_2 + O_2 \tag{9}$$

Under a higher energy electron collision, N_2 transforms into N radicals and the water molecules partially dissociate to form active species, such as OH and H radicals. The results of H_2O and N_2 molecules at higher energy electron collision reactions are listed below [21,23].

$$N_2 + e^- \rightarrow N + N \tag{10}$$

$$N + N \rightarrow N_2 \tag{11}$$

$$H_2O + e^- \rightarrow OH + H \tag{12}$$

$$H_2O + O \rightarrow OH + OH \tag{13}$$

$$OH + O \rightarrow O_2 + H \tag{14}$$

$$OH + O_3 \rightarrow HO_2 + O_2 \tag{15}$$

$$OH + HO_2 \rightarrow O_2 + H_2O \tag{16}$$

The mechanism for the destruction of CO_2 in a non-thermal plasma involves the electron-induced dissociation of CO_2 to give CO and O followed by recombination of the oxygen atoms to give the final products, CO and O_2. CO can also recombine with atomic oxygen to reform CO_2. The reactions are listed below [19,20].

$$CO_2 + e^- \rightarrow CO + O \tag{17}$$

$$CO_2 + e^- \rightarrow C + O_2 \tag{18}$$

$$CO + O \rightarrow CO_2 \tag{19}$$

$$CO + e^- \rightarrow C + O \tag{20}$$

$$CO + H_2O \rightarrow CO_2 + H_2 \tag{21}$$

$$C + O_3 \rightarrow CO + O_2 \tag{22}$$

The process of NO_x removal mainly occurs through the oxidation pathway. Once N and O radicals are formed, the most important reactions of the oxidation of NO to NO_2 are listed below [21–23].

$$N + O \rightarrow NO \tag{23}$$

$$N + O_3 \rightarrow NO + O_2 \tag{24}$$

$$NO + e^- \rightarrow N + O \tag{25}$$

$$NO_2 + e^- \rightarrow N + O + O \tag{26}$$

$$NO + O \rightarrow NO_2 \tag{27}$$

$$NO + O_3 \rightarrow NO_2 + O_2 \tag{28}$$

$$NO + N \rightarrow N_2 + O \tag{29}$$

$$NO + H_2O \rightarrow NO_2 + 2H \tag{30}$$

$$NO_2 + O_3 \rightarrow NO_3 + O_2 \tag{31}$$

$$NO_2 + HO_2 \rightarrow HNO_3 + O \tag{32}$$

$$NO_2 + OH \rightarrow HNO_3 \tag{33}$$

3. Results and Discussion

3.1. HC

Figures 5–8 present the HC emission value and the removal efficiency using different materials (Table 3) and at various locations (Figures 2 and 3) of the NTP reactor. The HC emissions are known to be a result of the poor combustion of the fuel [24]. We found that the control voltage of 2 to 5 V in the NTP system could reduce the HC emissions at different engine speeds. The minimum HC removal efficiency was 1.2% at the control voltage 2 V for platinum (type one) at 6000 rpm and for reactor in the front of the exhaust system (model A). The control voltage 1 V exhibited little change in the HC removal efficiency, possibly because the voltage and the spark were too small. Therefore, it does not show in the figure. The control voltage 0 V indicates the original exhaust gas without the NTP system. Our results show that HC removal efficiency increased with increasing voltage because of the increase in shape and density of the spark with increasing voltage. With the increase in the contact area of the large spark and high density with the HC exhaust gas, the re-burning probability of the HC exhaust gas increased. Particularly, the shape and density of the spark approached the critical value at the control voltage 4 V, and the spark variation was small from the control voltage of 5 to 4 V.

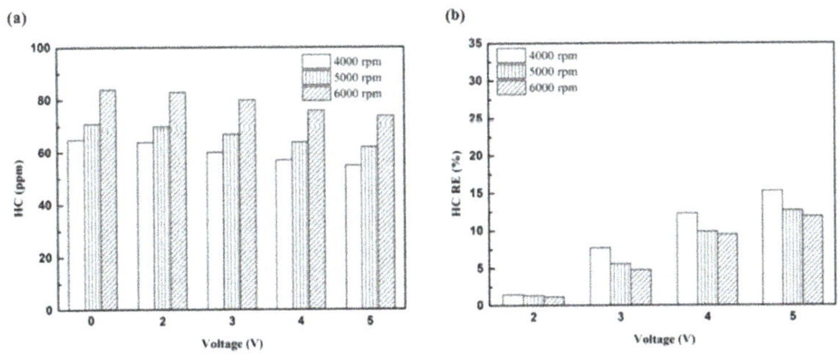

Figure 5. The HC (**a**) emission values and (**b**) removal efficiency (RE) using platinum in the front of the exhaust pipe.

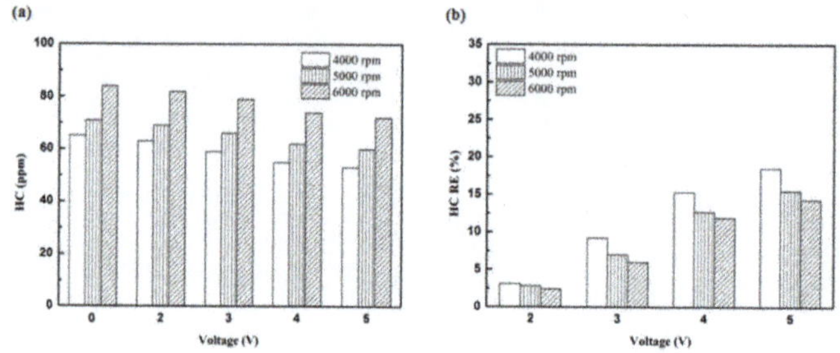

Figure 6. The HC (**a**) emission values and (**b**) RE using iridium alloy in the front of the exhaust pipe.

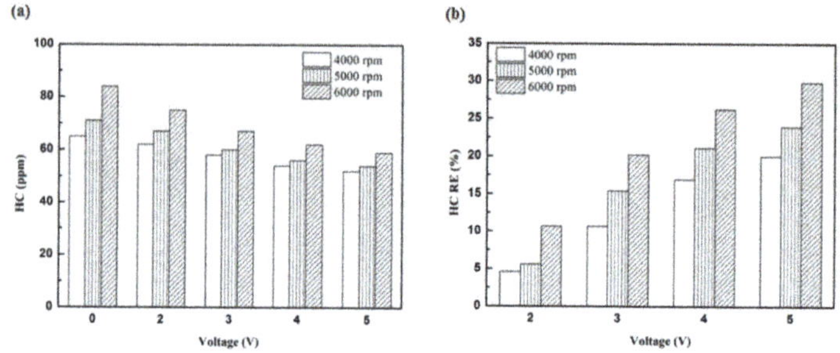

Figure 7. The HC (**a**) emission values and (**b**) RE using platinum in the rear of the exhaust pipe.

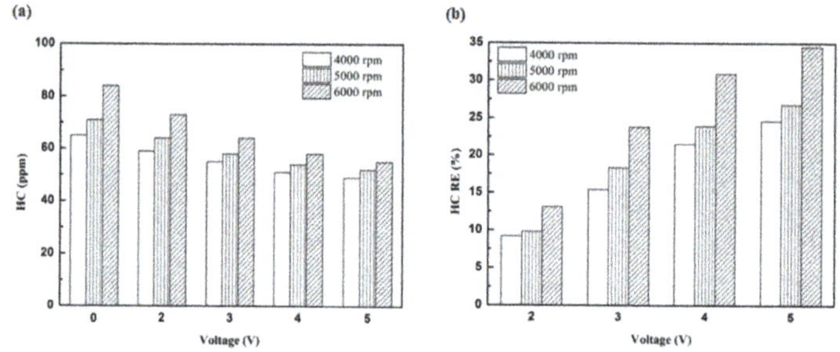

Figure 8. The HC (**a**) emission values and (**b**) RE using iridium alloy in the rear of the exhaust pipe.

Figures 9–11 show the HC emission value and removal efficiency at control voltage 5 V using different materials and at various locations of the NTP reactor. Table 6 shows the maximum and minimum values of HC removal efficiency at control voltage 5 V using different materials and at various locations of the NTP reactor. The maximum HC removal efficiency was 34.5% for iridium alloy (type two) at 6000 rpm and for reactor in the rear of the exhaust system (model B). The minimum HC removal efficiency was 11.9% for platinum (type one) at 6000 rpm and for the reactor in the front of the exhaust system (model A). These results indicate that HC exhaust gas concentration increases with the increase in engine speed. This is because, with the increase in the engine speed, the gasoline fuel gets no time to burn in the engine. The NTP reactor materials of type one and type two reduce the HC exhaust gas concentration and type two exhibited a better effect than the type one. This is because the low resistance and small electrode area of type two material cause the formation of a larger spark with a higher internal density. With the increase in the contact area of large spark and high density with the HC exhaust gas, the re-burning probability of HC exhaust gas increased. The NTP reactor location in models A and B reduced the HC exhaust gas concentration. In model A, the HC removal efficiency increased with decreasing engine speed because of the slow flow rate of exhaust gas with decreasing engine speed. The re-burning probability of HC exhaust gas increased with the increase in contact time between the spark and HC exhaust gas. In model B, the HC removal efficiency increased with an increase in engine speed. This is because the flow rate of exhaust gas was almost the same in model B. The re-burning probability of HC exhaust gas increased with an increase in HC gas flow per unit area. The reduction in HC removal efficiency of model B was higher than that of model A. This is because of the larger cross-sectional area in model B, which leads to the complete release of spark and slow gas flow rate, increasing the re-burning probability of HC exhaust gas. Therefore, HC exhaust gas emissions get reduced.

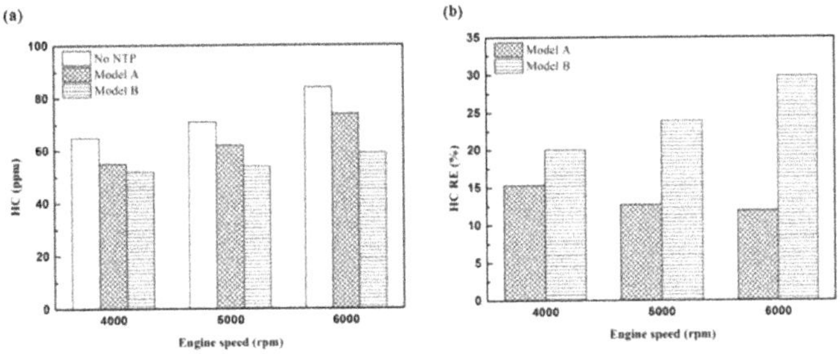

Figure 9. The HC (**a**) emission values and (**b**) RE of control voltage 5 V using platinum.

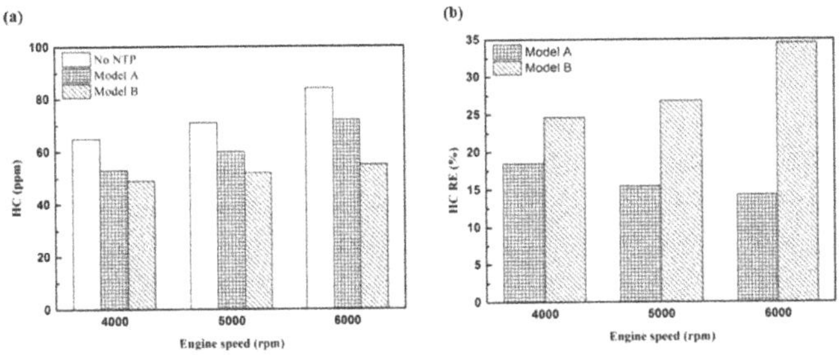

Figure 10. The HC (**a**) emission values and (**b**) RE of control voltage 5 V using iridium alloy.

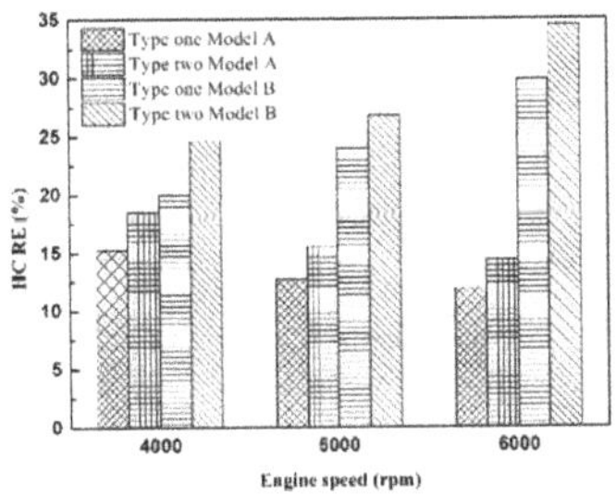

Figure 11. The HC RE at control voltage 5 V.

Table 6. The maximum and minimum values of HC removal efficiency at control voltage 5 V.

Parameter	HC RE Value	
	Maximum	Minimum
Platinum and in the front of the exhaust pipe (Type one and model A)	15.3%	11.9%
Iridium alloy and in the front of the exhaust pipe (Type two and model A)	18.5%	14.3%
Platinum and in the rear of the exhaust pipe (Type one and model B)	29.8%	20%
Iridium alloy and in the rear of the exhaust pipe (Type two and model B)	34.5%	24.6%

3.2. CO

Figures 12–15 present the CO emission value and removal efficiency of the different types and locations of NTP reactor. The control voltage of 2 to 5 V in the NTP system could reduce the CO emissions at different engine speeds. The minimum CO removal efficiency was 1.2% at the control voltage 2 V for Platinum (type one) and for in the front of the exhaust system (model A) at 6000 rpm. The results show that the removal efficiency of CO increases with increasing voltage. This is because with increasing voltage, the shape of the spark gets larger, leading to an increase in the production of free radicals. Abundant reactive radicals and higher combustion reaction temperature caused the increase in the dissociation probability of CO exhaust gas.

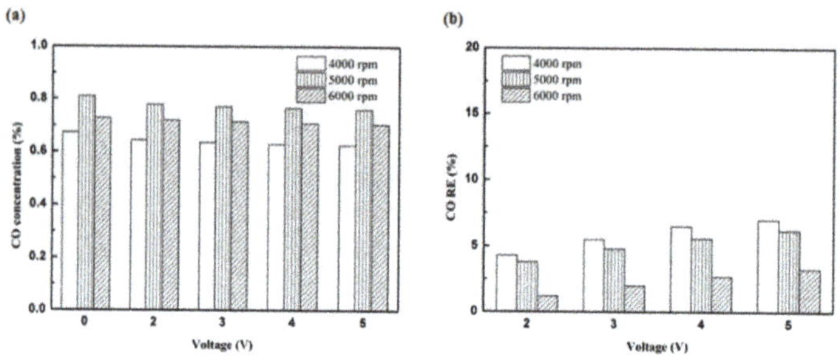

Figure 12. The CO (**a**) emission values and (**b**) RE using platinum in the front of the exhaust pipe.

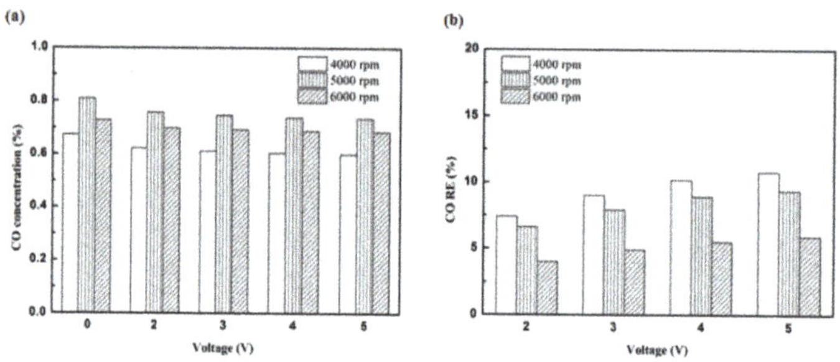

Figure 13. The CO (**a**) emission values and (**b**) RE using iridium alloy in the front of the exhaust pipe.

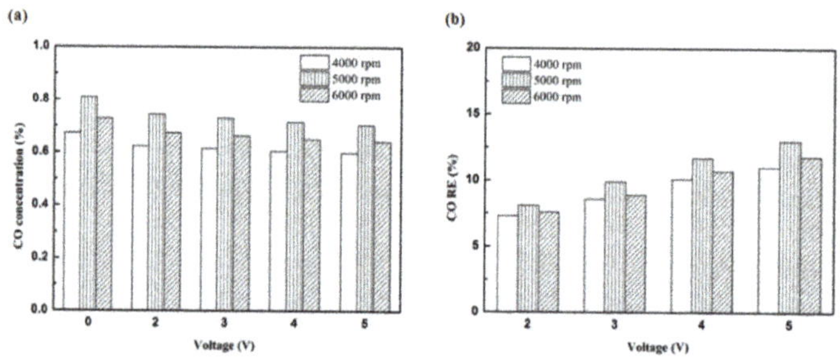

Figure 14. The CO (**a**) emission values and (**b**) RE using platinum in the rear of the exhaust pipe.

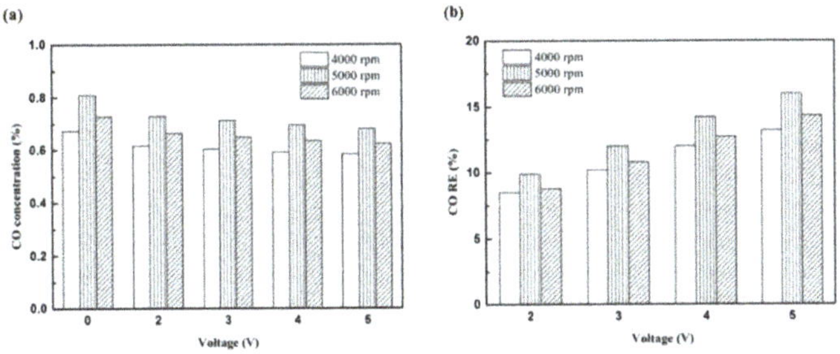

Figure 15. The CO (**a**) emission values and (**b**) RE using iridium alloy in the rear of the exhaust pipe.

Figures 16–18 present the CO emission value and removal efficiency of control voltage 5 V using different materials and at various locations of the NTP reactor. Table 7 shows the maximum and minimum values of CO removal efficiency at control voltage 5 V using different materials and at various locations of the NTP reactor. The maximum CO removal efficiency was 16.0% for iridium alloy (type two) at 5000 rpm and for reactor in the rear of the exhaust system (model B). The minimum CO removal efficiency was 3.3% for platinum (type one) at 6000 rpm and for reactor in the front of the exhaust system (model A). The NTP reactor material of types one and type two reduced the CO exhaust gas concentration. The effect of type two NTP reactor material was found to be better than that of type one because, low resistance forms larger spark and high internal density. The production of free radicals increases by increasing the spark size. The larger amount of radicals and higher combustion reaction temperature caused an increase in the dissociation probability of CO exhaust gas. The dissociation of CO is broadly divided into two parts. In the first part, the dissociation of CO gives carbon and oxygen atoms. Carbon atoms get deposited on the surface of the NTP reactor, as shown in Figure 19. The second part involves the recombination of CO and oxygen atoms to form the CO_2. The location of the NTP reactor in models A and B reduce the CO exhaust gas concentration. In model A, the CO removal efficiency increased with a decrease in the engine speed, which caused the flow rate of exhaust gas to be slow. The dissociation probability of CO exhaust gas increased with the increase in contact time between the spark and CO exhaust gas. The CO removal efficiency was the highest in model B at 5000 rpm because the flow rate of exhaust gas was almost the same in model B. The dissociation probability of CO exhaust gas increased with an increase in CO gas flow per unit area. The reduction in CO removal efficiency of model B is higher than that of model A. This is because in model B, the dissociation of free radical and CO exhaust gas has a larger space, and the flow rate of exhaust gas was slow. This reduced the CO exhaust gas emissions. However, the amount of change in the CO exhaust gas measurement is small, there may be random errors in this experiment which may be caused by ambient temperature, humidity and dust. The random errors of CO exhaust values is about 0.1.

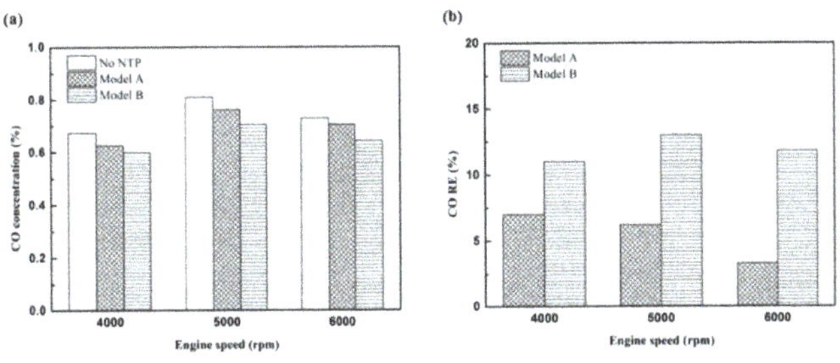

Figure 16. The CO (**a**) emission values and (**b**) RE of control voltage 5 V using platinum.

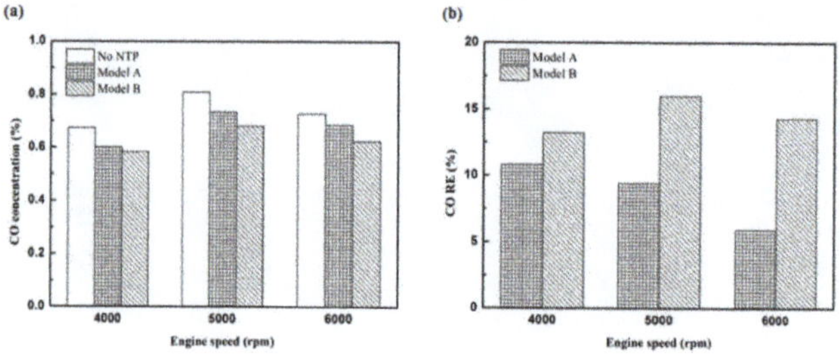

Figure 17. The CO (**a**) emission values and (**b**) RE of control voltage 5 V using iridium alloy.

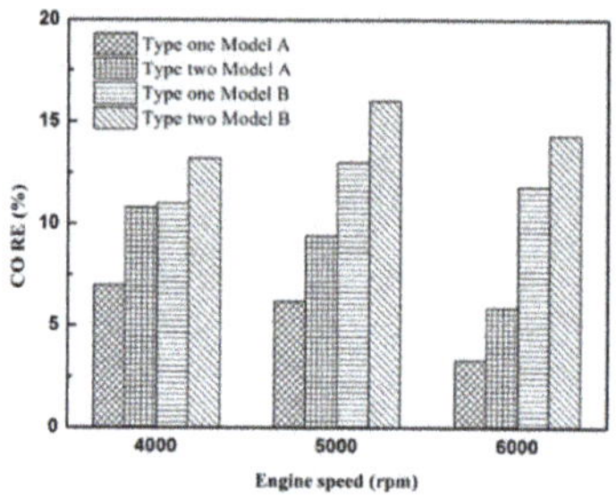

Figure 18. The CO RE at control voltage 5 V.

Table 7. The maximum and minimum values of CO removal efficiency at control voltage 5 V.

Parameter	CO RE Value	
	Maximum	Minimum
Platinum and in the front of the exhaust pipe (Type one and model A)	7%	3.3%
Iridium alloy and in the front of the exhaust pipe (Type two and model A)	10.8%	5.9%
Platinum and in the rear of the exhaust pipe (Type one and model B)	13%	10.1%
Iridium alloy and in the rear of the exhaust pipe (Type two and model B)	16%	13.2%

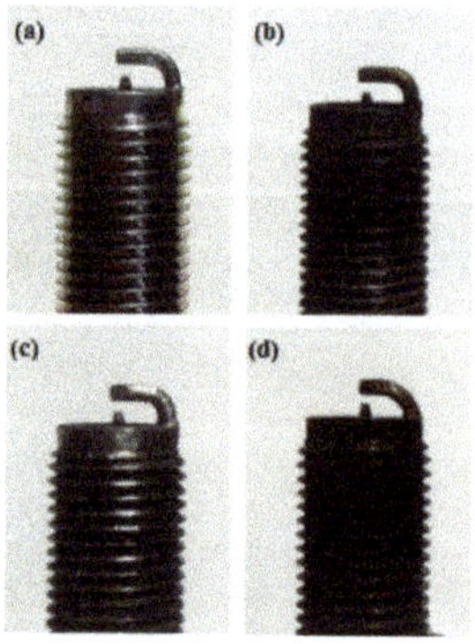

Figure 19. A photographic view of NTP reactor (**a**) platinum and (**c**) iridium alloy before this experiment and (**b**) platinum and (**d**) iridium alloy after this experiment.

3.3. CO_2

Figures 20–23 present the CO_2 emission value and conversion efficiency using the different materials and at various locations of the NTP reactor. The control voltage of 2 to 5 V in the NTP system could increase the CO_2 emissions at different engine speeds. The maximum CO_2 conversion efficiency was 0.9% at the control voltage 2 V for platinum (type one) at 4000 rpm and for reactor in the front of the exhaust system (model A). The results show that CO_2 conversion efficiency increases with decreasing voltage. This is because decreasing voltage leads to a smaller shape of spark and this reduction causes a decline in the production of free radicals. Ultimately, this reduces the dissociation probability of CO_2 exhaust gas.

Figures 24–26 show the CO_2 emission value and conversion efficiency at control voltage 5 V using different materials and at various locations of the NTP reactor. Table 8 shows the maximum and minimum values of CO_2 conversion efficiency at control voltage 5 V using different materials and at various locations of the NTP reactor. The maximum CO_2 conversion efficiency was 0.6% for platinum (type one) at 4000 rpm and for reactor in the front of the exhaust system (model A). The minimum CO_2 conversion efficiency was −0.3% for iridium alloy (type two) at 4000 rpm and for reactor in the rear of the exhaust system (model B). Most of the NTP reactor materials of type one and type two increase CO_2 exhaust gas concentration, while the type one exhibits a better effect than the type two. The reduced production of free radicals causes the dissociation probability of CO_2 exhaust gas to decrease. CO_2 is also obtained from the combination of CO and oxygen atoms. The dissociation of CO_2 is divided into two parts. The first part involves the dissociation of CO_2 to give CO and oxygen atoms. In the second part, the dissociation of CO_2 gives carbon and oxygen atoms. Carbon atoms get deposited on the surface of the NTP reactor (Figure 19). Most of the NTP reactor locations of model A and model B increase the CO_2 exhaust gas concentration. This is because the amount of CO_2 dissociation is less than the amount of recombination of CO and oxygen atoms. This result is similar to that of CO exhaust gas emissions. However, the amount of change in the CO_2 exhaust gas measurement was small, thus, there may be errors in measurement.

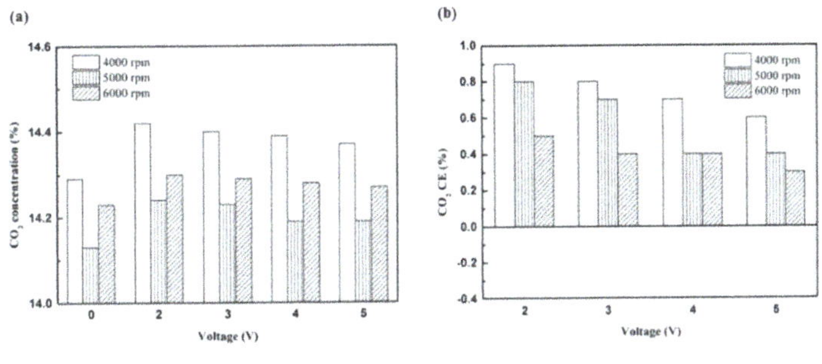

Figure 20. The CO_2 (**a**) emission values and (**b**) CE at platinum in the front of the exhaust pipe.

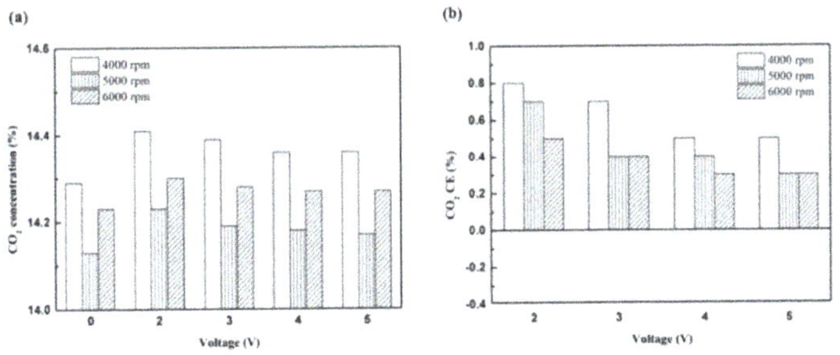

Figure 21. The CO_2 (**a**) emission values and (**b**) CE using iridium alloy in the front of the exhaust pipe.

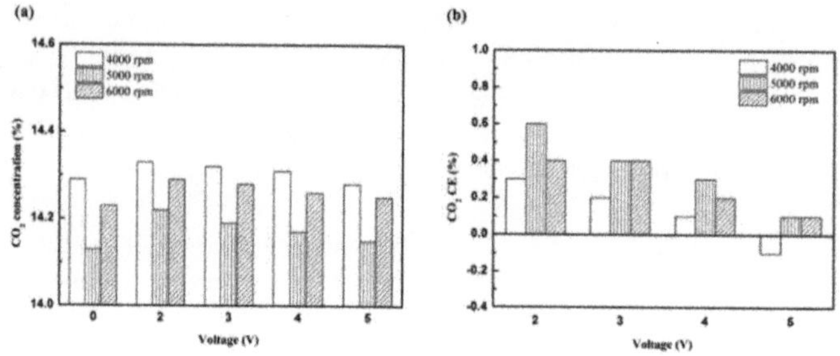

Figure 22. The CO_2 (**a**) emission values and (**b**) CE using platinum in the rear of the exhaust pipe.

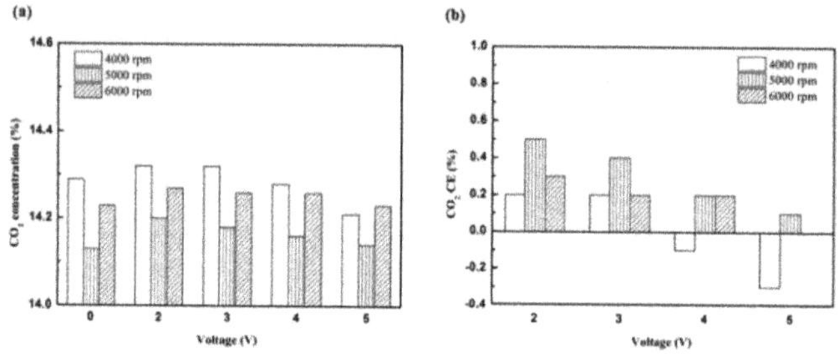

Figure 23. The CO_2 (**a**) emission values and (**b**) CE using iridium alloy in the rear of the exhaust pipe.

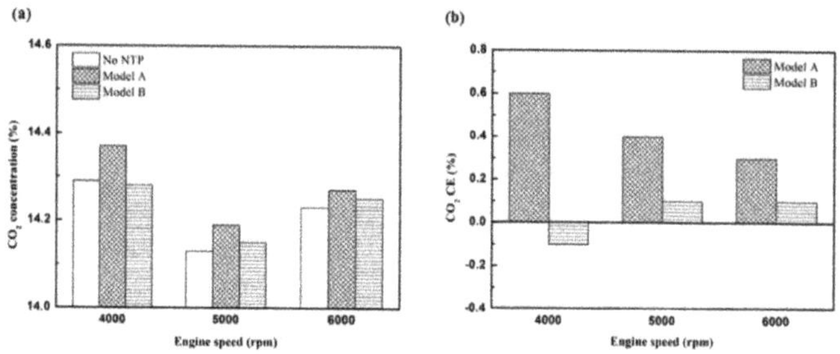

Figure 24. The CO_2 (**a**) emission values and (**b**) CE of control voltage 5 V using platinum.

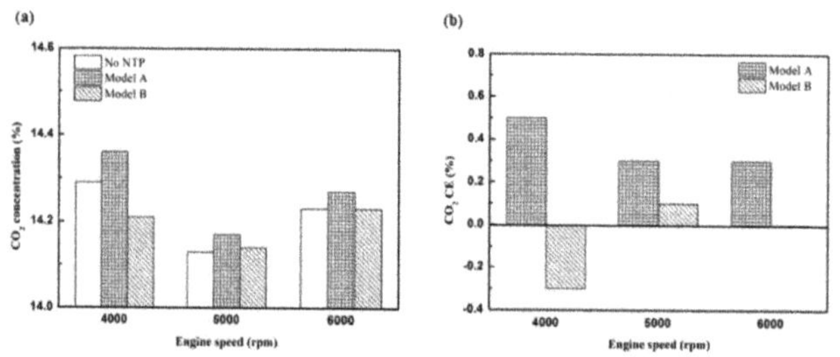

Figure 25. The CO_2 (**a**) emission values and (**b**) CE of control voltage 5 V using iridium alloy.

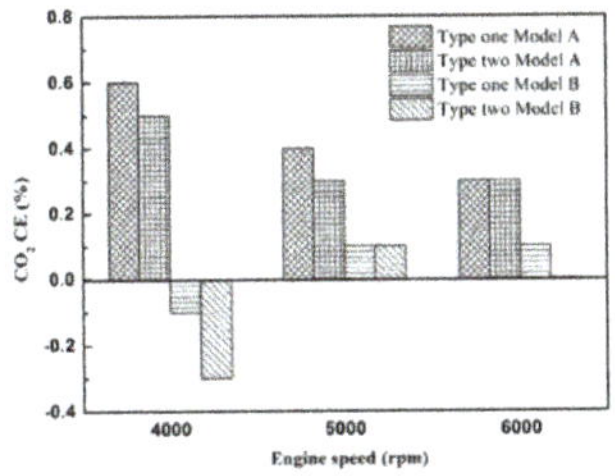

Figure 26. The CO_2 CE at control voltage 5 V.

Table 8. The maximum and minimum values of CO_2 conversion efficiency at control voltage 5 V.

Parameter	CO_2 CE Value	
	Maximum	**Minimum**
Platinum and in the front of the exhaust pipe (Type one and model A)	0.6%	0.3%
Iridium alloy and in the front of the exhaust pipe (Type two and model A)	0.5%	0.3%
Platinum and in the rear of the exhaust pipe (Type one and model B)	0.1%	−0.1%
Iridium alloy and in the rear of the exhaust pipe (Type two and model B)	0%	−0.3%

3.4. NO_x

Figure 27 presents the oil temperature of the engine at different engine speeds. Figures 28–31 present the NO_x emission value and the removal efficiency at different types and locations of the NTP reactor. The principal source of NO_x emissions is the oxidation of atmospheric nitrogen. We found that a control voltage between 2 to 5 V in the NTP system, could reduce NO_x emissions at different engine speeds. The maximum NO_x removal efficiency was 41.3% at the control voltage 4 V for iridium alloy (type two) at 3000 rpm and for reactor in the front of the exhaust system (model A). The minimum NO_x removal efficiency was 4.7% at the control voltage 2 V for platinum (type one) at 6000 rpm and for reactor in the rear of the exhaust system (model B). The results show that NO_x removal efficiency increased with increasing voltage, however, NO_x removal efficiency began decreasing after reaching control voltage 4 V. This is because the change in the spark of control voltage 5 V to the control voltage 4 V was small. The free radicals and exhaust gas degradation appeared to be similar, but the combustion reaction temperature increased. With the increase in combustion reaction temperature, the amount of diatomic nitrogen dissociated from the monatomic nitrogen increased and finally resulted in high NO_x.

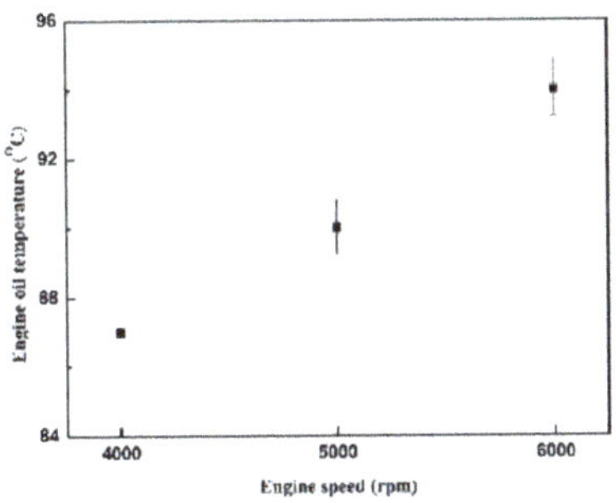

Figure 27. The oil temperature of the engine at different engine speeds.

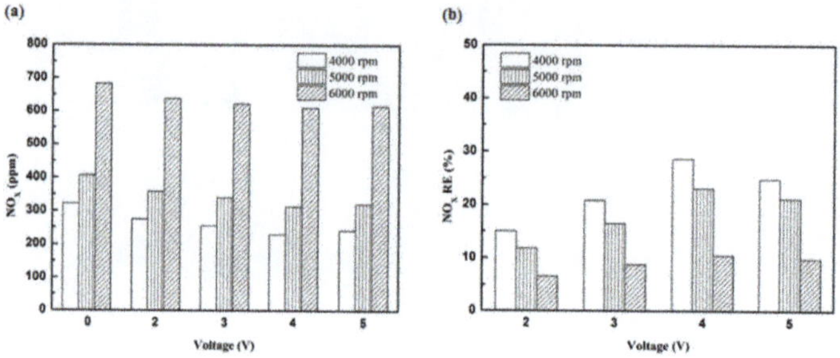

Figure 28. The NO_X (**a**) emission values and (**b**) RE using platinum in the front of the exhaust pipe.

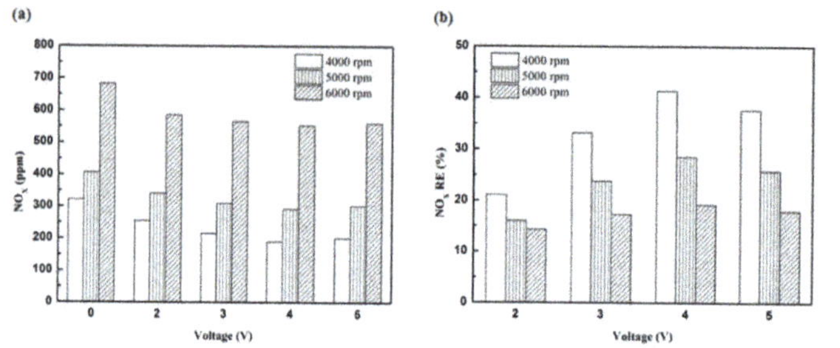

Figure 29. The NO_X (**a**) emission values and (**b**) RE using iridium alloy in the front of the exhaust pipe.

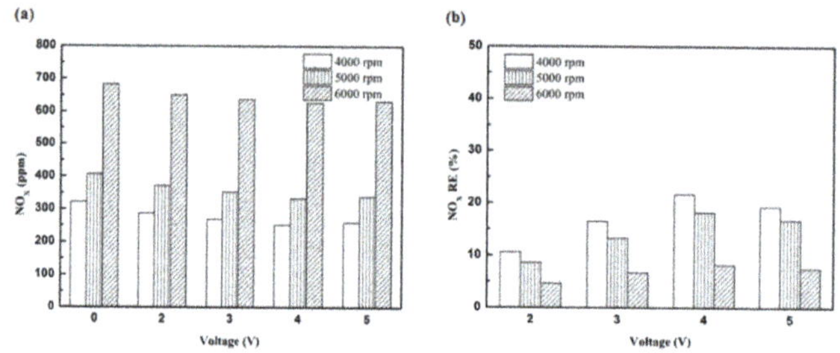

Figure 30. The NO_X (**a**) emission values and (**b**) RE using platinum and in the rear of the exhaust pipe.

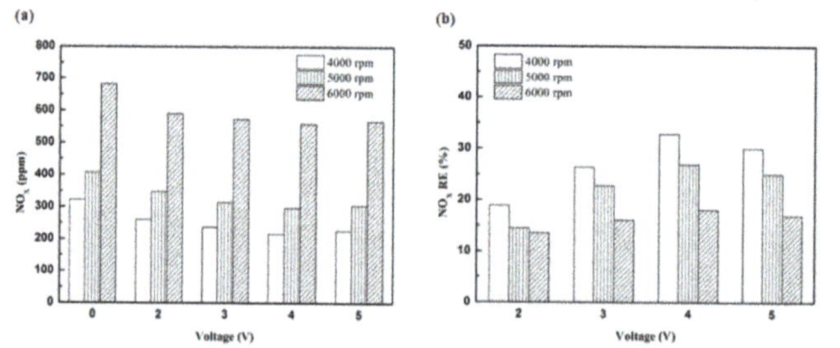

Figure 31. The NO_X (**a**) emission values and (**b**) RE using iridium alloy in the rear of the exhaust pipe.

Figures 32–34 present NO_X emission value and removal efficiency at control voltage 5 V using different materials and at various locations of the NTP reactor. Table 9 shows the maximum and minimum values of NO_X removal efficiency at control voltage 5 V using different materials and at various locations of the NTP reactor. The maximum NO_X removal efficiency was 37.6% for iridium alloy (type two) at 4000 rpm and for reactor in the front of the exhaust system (model A). The minimum

NO_x removal efficiency was 7.5% for platinum (type one) at 6000 rpm and for reactor in the rear of the exhaust system (model B). The results show an increase in NO_x exhaust gas concentration with an increase in engine speed, which then causes the temperature of the combustion chamber to increase. At high temperatures, nitrogen and oxygen atoms easily recombine to form NO_x. The NTP reactor materials of type one and type two reduce the NO_x exhaust gas concentration. The effect of type two material was found to be better than type one. The production of free radicals increased by increasing the spark size and the area of contact between the free radicals and NO_x exhaust gas also increased. This enhanced the probability of dissociation of NO_x exhaust gas to give nitrogen and oxygen atoms. The NTP reactor location in models A and B reduced the NO_x exhaust gas concentration. Moreover, the NO_x removal efficiency increased with decreasing engine speed because of the slower the flow rate of exhaust gas and the lower temperature of exhaust gas. The NO_x removal efficiency of model A was better among the two models because the products of NO_x dissociation can easily recombine to form the NO_x at high temperatures. The combination possibility of nitrogen and oxygen atoms is lower at the faster flow rate of the exhaust gas. Therefore, the NO_x exhaust gas emissions are reduced.

Table 9. The maximum and minimum values of NO_x removal efficiency at control voltage 5 V.

Parameter	NO_x RE Value	
	Maximum	Minimum
Platinum and in the front of the exhaust pipe (Type one and model A)	24.8%	9.8%
Iridium alloy and in the front of the exhaust pipe (Type two and model A)	37.6%	18%
Platinum and in the rear of the exhaust pipe (Type one and model B)	19.3%	7.5%
Iridium alloy and in the rear of the exhaust pipe (Type two and model B)	30.1%	17%

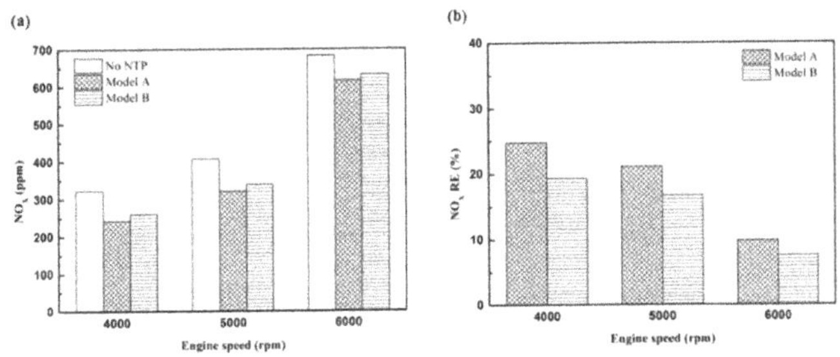

Figure 32. The NO_x (**a**) emission values and (**b**) RE of control voltage 5 V using platinum.

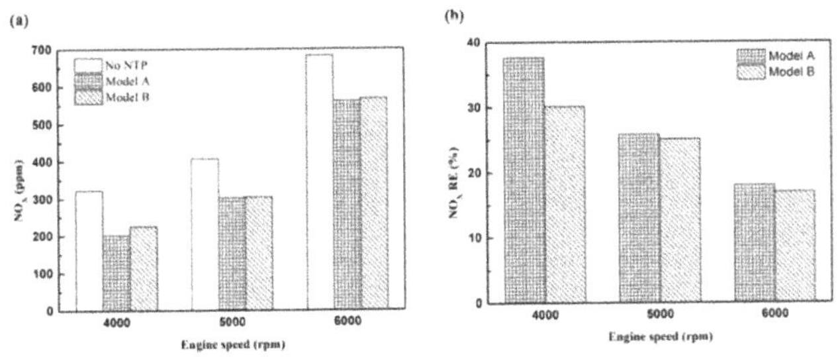

Figure 33. The NO_x (**a**) emission values and (**b**) RE of control voltage 5 V using iridium alloy.

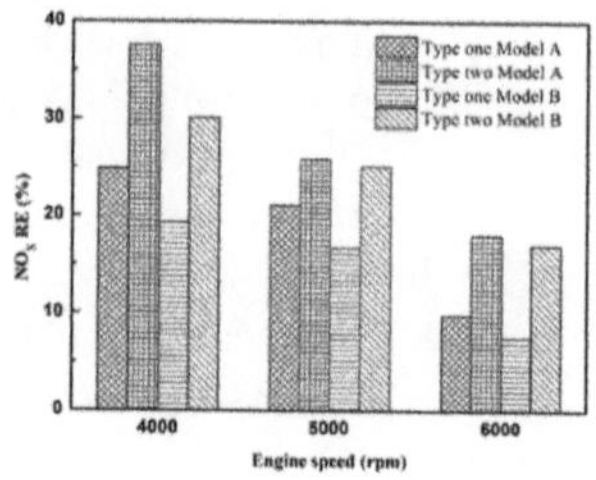

Figure 34. The NO_x RE at control voltage 5 V.

4. Conclusions

In this study, a motorcycle engine was examined using the NTP system to measure exhaust gas emission using different materials, voltage, and locations of the NTP reactor.

Considering the HC exhaust gas emissions, the NTP system re-burns HC exhaust gas to reduce the HC exhaust gas concentration. The HC removal efficiency increases with increasing voltage. The shapes and density of the spark approached critical values at the control voltage 4 V. The HC exhaust gas emissions of model B were lower than model A. The effect of type two NTP reactor material was better than that of type one. The maximum reduction of HC removal efficiency obtained with the NTP system reached 34.5%, while the minimum HC removal efficiency was 1.2%.

Considering the CO exhaust gas emissions, the reaction between free radicals and CO exhaust gas gates mainly dissociated to obtain carbon and oxygen atoms. Carbon gets deposited on the surface of the NTP reactor and affects the life of the NTP reactor. Furthermore, CO and oxygen atoms also recombine to form CO_2. The CO removal efficiency increases with the increase in voltage. The CO removal efficiency of model B was found to be higher than that of model A. The effect of the type two NTP reactor material was better than type one. The maximum reduction of CO removal efficiency obtained with the NTP system reached 16.0%, while the minimum CO removal efficiency was 1.2%.

In terms of the CO_2 exhaust gas emissions, the CO_2 exhaust gas dissociates to form CO and oxygen atoms, and the dissociation of CO_2 obtains carbon atoms and oxygen atoms. Like CO, carbon gets deposited on the surface of the NTP reactor. However, most of the amount of dissociation was found to be less than the amount of HC re-burning and CO reduction. Therefore, the total amount of CO_2 increased and the CO_2 conversion efficiency increased with decreasing voltage. The CO_2 conversion efficiency of model A was higher than that of model B. The effect of type one NTP reactor material was better than that of type two. The maximum CO_2 conversion efficiency obtained with the NTP system reached 0.9%, while the minimum CO_2 conversion efficiency was −0.3%.

Considering the NO_x exhaust gas emissions, the NO_x gets dissociated by the NTP system to reduce the NO_x exhaust gas concentration. The removal efficiency NO_x increased with increasing voltage, but the NO_x exhaust gas removal efficiency began to decrease after reaching control voltage 4 V. As the combustion reaction temperature increased, the amount of diatomic nitrogen dissociated from the monatomic nitrogen increased and finally resulted in high NO_x. The NO_x exhaust gas emissions of model A were lower than that of model B. The effect of type two NTP reactor material was better than that of type one. The maximum reduction of NO_x exhaust gas emissions removal efficiency obtained with the NTP system reached 41.3%, while the minimum NO_x removal efficiency was 4.7%.

Finally, according to the above discussion, the effect of iridium alloy is better than platinum for HC, CO, CO_2, and NO_x emission removal. The effect of the NTP reactor in the rear of the exhaust system is better than in the front of the exhaust system for HC, CO, and CO_2 emission removal. However, the NO_x emission removal for the reactor in front of the exhaust system is better than that in the rear of the exhaust system.

5. Patents

Taiwan Patent M526623.

Author Contributions: M.-H.H. guided the research direction, found the solutions and designed the approach to set up the testing facility; C.-N.W. wrote the initial draft of the paper; M.-C.H. analyzed the experimental results in this study; S.-H.W., C.-H.H., C.-J.L., T.-L.W., and J.-H.Y. conducted the experiments and collected the results. All authors have read and agreed to the published version of the manuscript.

Funding: This paper is supported by the Ministry of Science and Technology (MOST) and the National Kaohsiung University of Science and Technology under MOST-106-2622-E-151-007-CC3.

Acknowledgments: The authors would like their thanks to the Ministry of Science and Technology (MOST) and the National Kaohsiung University of Science and Technology for their financial support during this study.

Conflicts of Interest: The authors declare no conflict of interest.

References

1. Subrahmanyam, C.; Magureanu, M.; Renken, A.; Kiwi-Minsker, L. Catalytic abatement of volatile organic compounds assisted by non-thermal plasma Part 1. A novel dielectric barrier discharge reactor containing catalytic electrode. *Appl. Catal. B Environ.* **2006**, *65*, 150–156. [CrossRef]
2. Durme, J.V.; Dewulf, J.; Leys, C.; Langenhove, H.V. Combining non-thermal plasma with heterogeneous catalysis in waste gas treatment: A review. *Appl. Catal. B Environ.* **2008**, *78*, 324–333. [CrossRef]
3. Saoud, W.A.; Assadi, A.A.; Guiza, M.; Bouzaza, A.; Aboussaoud, W.; Soutrel, I.; Ouederni, A.; Wolbert, D.; Rtimi, S. Abatement of ammonia and butyraldehyde under non-thermal plasma and photocatalysis: Oxidation processes for the removal of mixture pollutants at pilot scale. *Chem. Eng. J.* **2018**, *344*, 165–172. [CrossRef]
4. Moldovan, M.; Palacios, M.A.; Gomez, M.M.; Morrison, G.; Rauch, S.; Mclcod, C.; Ma, R.; Caroli, S.; Alimonti, A.; Petrucci, F.; et al. Environmental risk of particulate and soluble platinum group elements released from gasoline and diesel engine catalytic converters. *Sci. Total Environ.* **2002**, *296*, 199–208. [CrossRef]
5. Balkı, M.K.; Sayın, C.; Canakci, M. The effect of different alcohol fuels on the performance, emissions and combustion characteristic of a gasoline engine. *Fuel* **2014**, *115*, 901–906. [CrossRef]
6. Hsueh, M.H.; Lin, D.F. The performance analysis and design optimization of fuel temperature control for the injection combustion engine. *J. Chin. Inst. Eng.* **2016**, *39*, 777–784. [CrossRef]
7. Doğan, B.; Erol, D.; Yaman, H.; Kodanli, E. The effect of ethanol-gasoline blends on performance and exhaust emissions of a spark ignition engine through exergy analysis. *Appl. Ther. Eng.* **2017**, *120*, 433–443. [CrossRef]
8. Roland, U.; Holzer, F.; Kopinke, F.D. Improved oxidation of air pollutants in a non-thermal plasma. *Catal. Today* **2002**, *73*, 315–323. [CrossRef]
9. Xiao, G.; Xu, W.; Wu, R.; Ni, M.; Du, C.; Gao, X.; Luo, Z.; Cen, K. Non-thermal plasmas for VOCs abatement. *Plasma Chem. Plasma Process.* **2014**, *34*, 1033–1065. [CrossRef]
10. Talebizadeh, P.; Rahimzadeh, H.; Babaie, M.; Anaghizi, S.J.; Ghomi, H.; Ahmadi, G.; Brown, R.; Roy, S. Evaluation of residence time on nitrogen oxides removal in non-thermal plasma reactor. *PLoS ONE* **2015**, *10*, e0140897. [CrossRef]
11. Kim, H.H.; Ogata, A.; Futamura, S. Atmospheric plasma-driven catalysis for the low temperature decomposition of dilute aromatic compounds. *J. Phys. D Appl. Phys.* **2005**, *38*, 1292–1300. [CrossRef]
12. Xu, W.; Li, M.W.; Xu, G.H.; Tian, Y.L. Decomposition of CO_2 using DC corona discharge at atmospheric pressure. *Jpn. J. Appl. Phys.* **2004**, *43*, 8310–8311. [CrossRef]
13. Indarto, A.; Yang, D.R.; Choi, J.W.; Lee, H.; Song, H.K. Gliding arc plasma processing of CO_2 conversion. *J. Hazard. Mater.* **2007**, *146*, 309–315. [CrossRef] [PubMed]
14. Reyes, P.G.; Mendez, E.F.; Osorio-Gonzalez, D.; Castillo, F.; Martinez, H. Optical emission spectroscopy of CO_2 glow discharge at low pressure. *Phys. Status Solidi C* **2008**, *5*, 907–910. [CrossRef]
15. Aerts, R.; Snoeckx, R.; Bogaerts, A. In-situ chemical trapping of oxygen after the splitting of carbon dioxide by plasma. *Plasma Process. Polym.* **2014**, *11*, 985–992. [CrossRef]
16. Uhm, H.S.; Kwak, H.S.; Hong, Y.C. Carbon dioxide elimination and regeneration of resources in a microwave plasma torch. *Environ. Pollut.* **2016**, *211*, 191–197. [CrossRef]
17. Dan, Y.; Dengshan, G.; Gang, Y.; Xianglin, S.; Fan, G. An investigation of the treatment of particulate matter from gasoline engine exhaust using non-thermal plasma. *J. Hazard. Mater.* **2005**, *B127*, 149–155. [CrossRef]

18. Kim, G.T.; Seo, B.H.; Lee, W.J.; Park, J.; Kim, M.K.; Lee, S.M. Effects of applying non-thermal plasma on combustion stability and emissions of NO_x and CO in a model gas turbine combustor. *Fuel* **2017**, *194*, 321–328. [CrossRef]
19. Adnan, Z.; Mir, S.; Habib, M. Exhaust gases depletion using non-thermal plasma (NTP). *Atmos. Pollut. Res.* **2017**, *8*, 338–343. [CrossRef]
20. Mikoviny, T.; Kocan, M.; Matejcik, S.; Mason, N.J.; Skalny, J.D. Experimental study of negative corona discharge in pure carbon dioxide and its mixtures with oxygen. *J. Phys. D Appl. Phys.* **2004**, *37*, 64–73. [CrossRef]
21. Talebizadeh, P.; Babaie, M.; Brown, R.; Rahimzadeh, H.; Ristovski, Z.; Arai, M. The role of non-thermal plasma technique in NO_x treatment: A review. *Renew. Sustain. Energy Rev.* **2014**, *40*, 886–901. [CrossRef]
22. Haddouche, A.; Lemerini, M. Chemical analysis of NO_x removal under different reduced electric fields. *Plasma Sci. Technol.* **2015**, *17*, 589–594. [CrossRef]
23. Jiang, N.; Shang, K.F.; Lu, N.; Li, H.; Li, J.; Wu, Y. High-efficiency removal of NO_x from flue gas by multitooth wheel-cylinder corona discharge plasma facilitated selective catalytic reduction process. *IEEE Trans. Plasma Sci.* **2016**, *44*, 2738–2743. [CrossRef]
24. Mishra, P.C.; Gupta, A.; Kumar, A.; Bose, A. Methanol and petrol blended alternate fuel for future sustainable engine: A performance and emission analysis. *Measurement* **2020**, *155*. [CrossRef]

MDPI
St. Alban-Anlage 66
4052 Basel
Switzerland
Tel. +41 61 683 77 34
Fax +41 61 302 89 18
www.mdpi.com

Molecules Editorial Office
E-mail: molecules@mdpi.com
www.mdpi.com/journal/molecules

www.ingramcontent.com/pod-product-compliance
Lightning Source LLC
Chambersburg PA
CBHW061601220926
51850CB00005B/390
9783036516967